Volume IV

Surveys in Differential Geometry: Integral Systems

Lectures on Geometry and Topology, sponsored by Lehigh University's Journal of Differential Geometry

Edited by

Chuu Lian Terng

Karen Uhlenbeck

International Press

Surveys in Differential Geometry: Integral Systems, Lectures on Geometry and Topology sponsored by the Journal of Differential Geometry, Chuu Lian Terng and Karen Uhlenbeck

International Press Incorporated, Boston
PO Box 38-2872
Cambridge, MA 02238

The paper used in this book is acid-free and falls within the guidelines established to ensure permanence and durability.

Surveys in Differential Geomety

Vol. 1: Lectures given in 1990, edited by S.-T. Yau and H. Blaine Lawson

Vol. 2: Lectures given in 1993, edited by C.C. Hsiung and S.-T. Yau

Vol. 3: Lectures given in 1996, edited by C.C. Hsiung and S.-T. Yau

Vol. 4: Integrable Systems, edited by Chuu Lian Terng and Karen Uhlenbeck

Contents

Introduction

Chuu-Lian Terng
Karen Uhlenbeck

I. The theory of integrable systems has been an active area of mathematics for the past thirty years. Different aspects of the subject have fundamental relations with mechanics and dynamics, applied mathematics, algebraic structures, theoretical physics, analysis and spectral theory and geometry. Most differential geometers have some knowledge and experience with finite dimensional integrable systems as they appear in sympectic geometry (mechanics) or ordinary differential equations, although the reformulation of part of this theory as algebraic geometry is not as commonly known. There are two quite separate methods of extension of these ideas to partial differential equations; one based on algebraic constructions and one based on spectral theory and analysis. These are less familiar still to differential geometers. This volume is a collection of papers intended to introduce and explain some of the more geometric aspects of integrable systems to the community.

This introduction contains a short historical discussion of the early geometers' ideas on "integrable geometric constructions" followed by a brief discussion of the material in the papers. An excellent historical treatment of finite dimensional integrable systems can be found in Moser's article [1980 Mo]. Many expositions on integrable equations (cf. [1991 AC], [1997 Pa]) discuss the history of the Korteveg de Vries (KdV) equation, beginning with the physical observation of Russell [1834 R] and the discovery of the equations by Boussinesq [1871 Bo], Korteveg and de Vries [1895 KdV]. In addition to describing water waves, the KdV equation also arises as a the model for the universal limit of lattice vibrations as the spacing goes to zero. Fermi-Pasta-Ulam's [1955 FPU] surprising numerical experiments on an anharmonic lattice, and the ingenious explanation by Zabuski-Kruskal [1965 ZK] of the results in terms of "solitons" of the KdV equation were quickly followed by a ground-breaking paper of Gardner, Greene, Kruskal and Miura [1967 GGKM] that introduced the method of solving KdV using the inverse scattering transform for the Hill's operator. This represents the approach of applied mathematics to discovering equations, solutions, and their scattering properties. The local differential geometry of the 19th century, on the other hand, foreshadows to some extent modern algebraic constructions. Rather than giving a description of the development of the basic ideas, we describe some of the early geometry and include a bibliography which lists many of the important early papers and a selection from the years between 1975 and 1985 in chronological order. The few recent papers are those we specifically refer to in this introduction.

Many geometric equations are known to have integrable aspects, especially if one takes into account that most experts do not have a good definition of "integrable" as applied to partial differential equations, particularly elliptic examples. In addition to those we mention in our historical discussion, the equations for harmonic maps (sigma models) from surfaces into groups ([1978 ZMi], [1981 Ch], [1989 U]), harmonic tori in symmetric spaces [1993 BFDP], constant mean curvature surfaces in space forms ([1989 PS], [1991 Bo]), isometric immersions of space forms in other space forms ([1919 Ca], [1980 TT]), and the theory of affine spheres ([1910 Tz], [1997 BS]) and affine minimal surfaces ([1980 CT]) are all examples of "elliptic" integrable systems. Likewise, the Yang-Mills equations on C^2 or $E^{2,2}$, their reductions to monopole equations in three variables, and self-dual equations on surfaces are new equations which originated in the physics literature of the late 1970's [1987 Hi]. One atypical development illustrating the great variety in the geometric theory is the reduction of the global $SU(N)$ monopole equation on R^3 to an integrable ordinary differential equation by the physicist Nahm [1984, Do]. The ideas surrounding string theory result in a series of deep and not completely understood connections between representation theory of certain algebras and many of the more classical theories of integrable systems in mathematics. Most recently, supersymmetric quantum field theories produce in a natural way moduli spaces of vacua or ground states which have new geometry generated by the supersymmetry. Since the supersymmetry generalizes the classical symmetries which produce integrals for the Euler-Lagrange equations via Noether's theorem, the connection with integrability is perhaps not surprising. However, this does not explain entirely the use of integrable systems in hyper-Kähler geometry ([1992 Hi]), Seiberg-Witten theory ([1996 DW]), special Kähler geometry ([1996 F], [1997 CRTP], [1997 Fr]) and quantum cohomology ([1991 W], [1994 KM], [1994 RT]).

The 19th century geometers were mainly interested in the local theory of surfaces in R^3, which we might regard as the prehistory of these modern constructions. The sine-Gordon equation arose first through the theory of surfaces of constant Gauss curvature -1, and the reduced 3-wave equation can be found in Darboux's work on triply orthogonal systems of R^3 ([1866 Da]). In 1906, a student of Levi Civitas, da Rios, wrote a master's thesis, in which he modeled the movement of a thin vortex filament in a viscous liquid using the equations of a curve propagating in R^3 along its binormal ([1906 dR]). It was much later that Hasimoto ([1971 Ha]) showed the equivalence of this system with the nonlinear Schrödinger equation. Since the equations were rediscovered somewhat independently of their geometric history, the main contribution of the classical geometers lies in the methods they developed for constructing explicit solutions of these equations. Ribaucour, Bianchi, Lie, and Bäcklund worked out the theory of what are now called Backlund transformations ([1883 Ba]). Lie developed a general theory of transformations known as "Lie transformations" which includes as examples scaling transformations and Lorentz transformations ([1979 L]). Darboux transformations for the Hills operator (the isospectral operator for KdV) were constructed by Darboux ([1882 Da]), but had been found even earlier by Moutard ([1875 M]). Most of the integrable partial differential equations are known to have Bäcklund, Lie and Darboux transformations.

II. The three-wave equation and the sine-Gordon equation arise out of the study of the classical differential geometry of R^3. We take a short trip back to this subject. A triply orthogonal system on R^3 is a local coordinate system $X(x_1, x_2, x_3)$ on R^3 such that the Euclidean metric ds^2 on R^3 has the form

$$ds^2 = a_1(x)^2 dx_1^2 + a_2(x)^2 dx_2^2 + a_3(x)^2 dx_3^2$$

for some smooth functions a_1, a_2, a_3, i.e., $X_{x_i} \cdot X_{x_j} = a_i^2 \delta_{ij}$ for $1 \leq i, j \leq 3$. Then the three one-parameter families of surfaces in R^3, defined by setting the different x_i to be constant, are mutually orthogonal. Already in 1813, Dupin ([1813 Du]) published a proof that the x_i-curves are lines of curvature for these surfaces. The flatness of ds^2 implies that

$$\begin{cases} (\beta_{ij})_{x_j} + (\beta_{ji})_{x_i} + \beta_{ik}\beta_{kj} = 0, \\ (\beta_{ij})_{x_k} = \beta_{ik}\beta_{kj}, \quad i,\ j,\ k \text{ distinct}, \end{cases}$$

where $\beta_{ij} = (a_i)_{x_j}/a_j$ for $i \neq j$. These are the *Equations of Lamé* or *The Lamé System*. Given a solution (β_{ij}) of the Lamé system, finding a triply orthogonal system with these (β_{ij}) means finding the (a_1, a_2, a_3) that satisfy

$$(a_i)_{x_j} = \beta_{ij} a_j, \quad \text{for all } i \neq j.$$

Solutions of this latter system depend on three functions of one variable. Thus, given one triply orthogonal system, we can derive from it infinitely many others, parametrized by three functions of one variable. These transformations of a triply orthogonal system are referred to as Combescure transformations ([1867 Co]).

Triply orthogonal systems are easy to construct, and consequently many solutions to the equations of Lamé are known. For example, let $f(x_1, x_2)$ be a surface in R^3 parametrized by line of curvatures coordinates, and let N be the unit normal field. Then $X(x_1, x_2, x_3) = f(x_1, x_2) + x_3 N(x_1, x_2)$ is an orthogonal coordinate system for a domain in R^3. Moreover, if the two fundamental forms of the surface f are given by

$$I = b_1^2 dx_1^2 + b_2^2 dx_2^2, \quad II = \lambda_1 b_1^2 dx_1^2 + \lambda_2 b_2^2 dx_2^2,$$

then the Euclidean metric of R^3 written in this coordinate is $ds^2 = \sum_{i=1}^3 a_i^2 dx_i^2$, where

$$a_i(x_1, x_2, x_3) = (1 - \lambda_i(x_1, x_2)x_3)b_i(x_1, x_2), \quad \text{for } i = 1, 2, \quad a_3 = 1.$$

In this case, the equations of Lamé for $\beta_{ij} = (a_i)_{x_j}/a_j$ are precisely the Gauss and Codazzi equations of the surface f.

Darboux considered the Lamé system restricted to the case when $\beta_{ij} = \beta_{ji}$. This condition is easily seen to be equivalent to the condition that $ds^2 = \frac{\partial \phi}{\partial x_1} dx_1^2 + \frac{\partial \phi}{\partial x_2} dx_2^2 + \frac{\partial \phi}{\partial x_3} dx_3^2$ for some function ϕ. Darboux called a metric with this property an *Egorof* metric. In this case, The Lamé system is equivalent to the reduced 3-wave equation studied more recently by [1973 ZM]. Local analytic solutions of this system were classified by Cartan (cf. [1945 Ca]) using the methods which he and Kähler developed, now usually refered to as the "Cartan-Kähler theory". In fact, Cartan proved that (i) if c_1, c_2, c_3 are distinct, then the line $t(c_1, c_2, c_3)$ is non-characteristic, and (ii) that if L is a fixed, non-characteristic line, then given any three analytic functions f_1, f_2, f_3 defined on L, there exists a local analytic solution $\beta_{12}, \beta_{13}, \beta_{23}$ of the reduced 3-wave system such that $\beta_{12}, \beta_{13}, \beta_{23}$ are equal to f_1, f_2, f_3 on L. However, the ability to handle the global problem came only with the insights which come from scattering theory, and the solution of the problem with arbitrary smooth and rapidly decaying data is a recent development.

In the first construction we can find of a "Bäcklund transformation", Ribaucour ([1870 Ri]) produced in 1870 an interesting triply orthogonal system. We include a description of this result to give a flavor of the geometry of that time. Note that classical geometers referred to a surface in R^3 of constant Gauss curvature -1 as *pseudospherical*, and we will keep this terminology. Start with a surface M in R^3, and let C_p denote the unit circle in the physical tangent plane T_pM in R^3 centered at p. Ribaucour proved that if there exist a one-parameter family of surfaces that meet these tangent circles orthogonally, then, first of all both M and all the surfaces in the one-parameter family are pseudoshperical. In addition these pseudospherical surfaces are one of the families of surfaces in a triply orthogonal system. To describe this sytem, we let $f(u, v)$ denote the embedding of M in R^3 parametrized by line of curvature coordinates (u, v), with e_1 and e_2 the unit vectors along the lines of curvature in the directions of f_u and f_v respectively. Fix a base point p_0 in M, and find for each angle τ the surface P_τ which is orthogonal to C_{p_0} at the point

$$f(u_0, v_0) + \cos\tau\, e_1(u_0, v_0) + \sin\tau\, e_2(u_0, v_0).$$

Define the function $\theta(u, v, \tau)$ to be the angle between $f(u, v) + e_1(u, v)$ and $P_\tau(u, v) \cap C_{f(u,v)}$. Then

$$X(u, v, \tau) = f(u, v) + \cos\theta(u, v, \tau)e_1(u, v) + \sin\theta(u, v, \tau)e_2(u, v)$$

is an orthogonal coordinate system in R^3, and the surface defined by $\tau = \tau_0$ is P_{τ_0}.

The sine-Gordon equation first arose in the study of pseudospherical surfaces. If M is a surface of negative curvature, then there exist asymptotic coordinates (s, t). The Codazzi equations of a pseudospherical surface implies that the (s, t) can be chosen to be arc length. There is a reduction of the Gauss equation in which the fact that the Gauss curvature is -1 translates into the sine-Gordon equation, $q_{st} = \sin q$, where q is the angle between the s and t-curves. Thus, by The Fundamental Theorem of surfaces, there is a local bijective correspondence between solutions of the sine-Gordon equation and pseudospherical surfaces, up to rigid motion. The surfaces typically have singlarities and self-intersections. Modern computer graphics provide beautiful pictures of such surfaces which one believes the classical geometers would have appreciated very much.

Bäcklund transformations arose from a geometrical construction on pseudospherical surfaces, but to understand them we need to review yet more classical geometry. A *congruence of lines* is a two-parameter family of lines $L(u, v)$ in R^3. We describe the line $L(u, v)$ as the line through $x(u, v)$ and in the direction of $\xi(u, v)$ parameterized by t,

$$L(u, v): \quad x(u, v) + t\xi(u, v).$$

A surface M given by $Y(u, v) = x(u, v) + t(u, v)\xi(u, v)$ for some smooth function t is called a *focal surface* of the congruence of lines if the line $L(u, v)$ is tangent to M at $Y(u, v)$ for all (u, v). Hence $\xi(u, v)$ lies in the tangent plane of M at $Y(u, v)$, which is spanned by $x_u + t_u\xi + t\xi_u$ and $x_v + t_v\xi + t\xi_v$. This implies that t satisfies the following quadratic equation:

$$\det(\xi, x_u + t\xi_u, x_v + t\xi_v) = 0.$$

In general, this quadratic equation has two distinct solutions for t. Hence generically each congruence of lines has two focal surfaces, M and M^*. This results in a diffeomorphism $\ell : M \to M^*$ such that the line joining p and $p^* = \ell(p)$ is tangent to both M and M^*. We will call ℓ a *line congruence*. If you have difficulty understanding this construction, it is helpful to think through the construction of a congruence of lines (a one-parameter family) in R^2 and the single focal curve for this family.

A line congruence $\ell : M \to M^*$ is called *pseudospherical* with constant θ if the distance between p and $p^* = \ell(p)$ is $\sin\theta$ and the angle between the normal of M at p and the normal of M^* at p^* is θ for all $p \in M$. Bäcklund ([1883 Ba]) showed that if ℓ is a pseudospherical line congruence, then both M and M^* are pseudospherical, and ℓ maps asymptotic lines to asymptotic lines. However, the transformations come about from showing that this construction can always be realized. Given a pseudospherical surface, a constant θ, and a unit vector $v_0 \in TM_{p_0}$, not a principal direction, then there exists a unique surface M^* and a pseudospherical congruence $\ell : M \to M^*$ with constant θ such that $\overline{p_0 p_0^*} = \sin\theta\, v_0$. Analytically this is equivalent to the statement that if q is a solution of the sine-Gordon equation, then the following overdetermined system of ordinary differential equations is solvable for q^*:

$$\begin{cases} q_x^* &= q_x + 4s\sin\left(\frac{q^*+q}{2}\right), \\ q_t^* &= -q_t + \frac{1}{s}\sin\left(\frac{q^*-q}{2}\right), \end{cases} \qquad (BT_{q,s})$$

where $s = \csc\theta - \cot\theta$. Moreover, a solution, q^* is again a solution of the sine-Gordon equation.

Bäcklund's theorem, but for $\theta = \frac{\pi}{2}$, was already known to Bianchi [1879 Bi]. In fact this case had been discovered even earlier (1870) in a different form by Ribaucour ([1870 Ri]). Note that if $X(u, v, \tau)$ is the Ribaucour triply orthogonal system defined above and τ is a fixed constant, then the map $\ell : M \to P_\tau$ defined by $\ell(f(u, v)) = X(u, v, \tau)$ is a pseudospherical congruence with $\theta = \pi/2$. However, Ribaucour did not realize that his theorem gave a method for constructing new pseudospherical surfaces from a given one.

The classic permutability Theorem for Bäcklund transformations is due to Bianchi (cf. [1909 Ei]). Given two pseudospherical congruences $\ell_i : M_0 \to M_i$ with angles θ_i respectively and $\sin\theta_1^2 \neq \sin\theta_2^2$, then there exist an algebraic construction of a unique surface M_3, and two pseudospherical congruences $\tilde{\ell}_i : M_i \to M_3$, such that $\tilde{\ell}_2\ell_1 = \tilde{\ell}_1\ell_2$. The analytic reformulation of this permutabiliy formula is the following. Suppose q is a solution of the sine-Gordon equaton and q_1, q_2 are two solutions of the above system of ordinary differential equations $(BT)_{q,s}$ with constants $s = s_1, s_2$ respectively, where $s_i = \csc\theta_i - \cot\theta_i$. The Bianchi permutability theorem gives a third solution to the sine-Gordon equation

$$q_3 = q + 4\tan^{-1}\left(\frac{s_1 + s_2}{s_1 - s_2}\tan\left(\frac{q_1 - q_2}{4}\right)\right).$$

What are now called Darboux transformations were discovered by Darboux during his investigation of Liouville metrics. A metric $ds^2 = A(x, y)(dx^2 + dy^2)$ is Liouville if there is a coordinate system (u, v) such that ds^2 is of the form

$$ds^2 = (f(u) - g(v))(du^2 + dv^2)$$

for some f and g of one varible. The classical geometers were interested in such metrics at least in part because Liouville had shown that all geodesics on such surfaces can be obtained by quadratures. The question of deciding whether a metric ds^2 is Liouville led to the study of the following special second order linear partial differential equation

$$w_{xy} = (f(x + y) - g(x - y))w.$$

Darboux was led to look for transformations of Hill's operators in the process of separating variables in this equation. The original analytic version of Darboux transformation ([1882 Da] and [1889 Da] Chap. 9) is the following: Let q be a smooth function of one variable, λ_0 a constant, and suppose that f satisfies $f'' = (q + \lambda_0)f$. Set

$$q^\sharp = f(f^{-1})'' - \lambda_0.$$

If $y(x, \lambda)$ is the general solution of the Hills operator with potential q:

$$y'' = (q + \lambda)y,$$

then $z = y' - (f/f')y$ is the general solution of the Hills operator with potential $q^\sharp$:

$$z'' = (q^\sharp + \lambda)z.$$

Next, suppose that we factor

$$D^2 - q - \lambda_0 = (D + v)(D - v),$$

In other words, suppose that v satisfies $v_x + v^2 = q + \lambda_0$. (Here $D = d/dx$.) Choose f so that $f'/f = v$. Then

$$(D - v)(D + v) = D^2 - q^\sharp - \lambda_0.$$

This Darboux theorem gives an algebraic algorithm (without quadrature) to transform general solutions of $D^2 - q - \lambda$ to those of $D^2 - q^\sharp - \lambda$.

Examples where the classical geometry appears in the modern theory of integrable systems are common and easily found. For example, orthogonal systems and their Combescure transformations arise in the local Hamiltonian theory of hydrodynamics of weakly deformed soliton lattices, and are known as the Whitham equations. The latter is the work of Dubrovin and Novikov ([1983 DN]) and Tsarëv ([1991 Ts]). An Egorof metric that is homogeneous of degree m, i.e., $\phi_{x_i}(rx) = r^m \phi_{x_i}(x)$ for all i, gives rise to a Frobenius manifold in the sense of Dubrovin, and such metrics give solutions to the WDVV equation in the study of Gromov-Witten invariants ([1991 W], [1995 RT], [1996 Du]). One of the best known examples is perhaps the Darboux transformations. If $q(x, t)$ is a solution of KdV then the Hills operators with potential $q(\cdot, t)$ are isospectral. It follows that the Darboux transformations of the Hills operators induce transformations on the space of solutions of KdV. This is a critical observation due to Adler and Moser ([1978 AM]) and Deift ([[1978 De]).

III. Many of the topics mentioned in the previous paragraphs will reappear in the papers in this volume, including the classical differential geometry. It is easy to understand the importance of the subject of integrable systems in modern mathematics once one notes that all of the papers except the last three mention either in passing or as fundamental motivation applications to constructions which are relevant in string theory and supersymmetric quantum field theory. Techniques from integrable systems, despite their limitations in classical geometry and applications, have provided one among the successful tools for rigorous mathematical interpretation of ideas in quantum field theory. It should be emphasized that it is mainly geometric ideas about classical geometric moduli spaces which can be treated. These spaces arise only as vacua for the field theories, but the special properties of the various field theories induce geometry on the moduli spaces which is not the usual Riemannian or symplectic geometry familiar to differential geometers. Because of the nature of the questions, the smooth part of the moduli space is not necessarily compact, so local algebraic results reminiscent of the classical 19th century have relevance if a more complete global analysis is not feasible.

Nigel Hitchin's paper "Integrable systems in Riemannian Geometry" starts with a discussion of what integrability means, and of the contrast between this century and the previous one. He goes on to discuss three problems familiar to differential geometers: constant mean curvature surfaces (tori) in R^3, Einstein metrics with certain symmetries in four dimensions, and hyper-Kähler manifolds. We meet loop groups in the section on tori, the flat Egorov metrics studied by Darboux in the section on Einstein manfiolds, and a moduli space (for a sypersymmetric quantum field theory) which has one of the special geometries, the hyper-Kähler one. The paper gives three different, beautiful illustrations of reduction of partial differential equations to finite dimensional flows and a nice explanation of "algebraically integrable".

The paper "Seiberg-Witten Integrable Systems" by Ron Donagi deals entirely with finite dimensional algebraically integrable systems which arise directly from a supersymmetric quantum field theory. The integrable system comes from the study of Higg's bundles on curves, and the paper is the only one to attempt a useful discussion of how an integrable system can relate to high energy physics. A host of references to the physics literature are given for the ambitious geometer, and a certain amount of algebraic geometry is needed to follow the details.

The paper "Five Lectures on Soliton Equations" by Edward Frenkel is motivated by the study of deformations of conformal field theories, and affine Lie algebras play a major role. The basic theory for the KdV equation, which is usally presented in terms of isospectral flows for a second order Hill's operator, is rederived for extensions of modified KdV, a system which is contained in the Non-linear Schrodinger hierarchy. This is particularly helpful in understanding the generality with which the algebraic constructions for KdV can be carried over to other classes of integrable systems and their Poisson brackets.

We are fortunate to include two papers by Boris Dubrovin which are most closely related to the classical geometry, but which in addition have important applications in two dimensional topological quantum field theory. In the paper "Differential Geometry of Moduli Spaces and Applications to Soliton Equations and Topological Conformal Field Theory", the relationship among Egorov metrics, n-wave equations, flat metrics on moduli spaces of curves with marked points and topological conformal field theories is discussed. The moduli space in the second paper "Differential Geometry of the Space of Orbits of a Coxeter Group" is a space of orbits, and carries the natural structure of a Frobenius manifold, a coordinate free description of the WDVV (Witten-Dijkgraaf, Verlinde, Verlinde) equations.

The emphasis in the paper "Symplectic Forms in the Theory of Solitons" by Igor Krichever and Duong Phong lies on the construction of a universal symplectic form, which can be specialized to many specific cases. We meet, for the first time, global questions in the identification of scattering data for the initial data given in two dimensions for the KP equation. A description of the WKB or Whitham flows on the moduli spaces of exact solutions leads into a discussion of leafs of the various moduli spaces of curves and applications to the topological Landau-Ginsburg models treated by Dubrovin. The paper includes applications to the Seiberg-Witten moduli spaces discussed by Donagi. The careful discussion of the analytic versus algebraic constructions is particularly helpful to geometers in the habit of thinking in terms of actual solutions to partial differential equations.

The final recent paper in the collection, "Poisson Actions and Scattering Theory" is our own. While partially motivated by questions from modern physics, it represents an attempt to reconcile the local algebraic methods with an understanding of global inverse scattering theory on the line. Reality conditions consistent with a compact group permit this, but restrict the number of integrable systems to whcih the global theory is applicable. We also give an explanation of Bäcklund (or Darboux) transformations in terms of the action of a rational subgroup of a loop group. The implication is that not all constructions are best described in terms of Lie groups and infinitesimal actions, especially if one is interested in global questions.

We have reprinted two influential papers which where originally published in 1984 and 1985. The paper "Loop groups and Equations of KdV type" by Graeme Segal and George Wilson is an important introduction to both subjects. A description of the KP hierarchy as commuting flows on an infinite dimensional Grassman manifold was originally due to Sato ([1981 S], [1983 DJKM]), but the work of Segal and Wilson is written up with a geometeric emphasis very much appreciated by some of us. We argued for the inclusion of the seminal paper "Scattering and Inverse Scattering for First Order Systems" by Richard Beals and Ronald Coifman. The algebraic geometry needed to treat the periodic theory is referred to by many authors in this volume, but it seems important for geometers to also have some understanding of analytic issues involved in constructing solutions to equations on non-compact domains.

Bibliography

[1813 Du] Dupin, *Dévelopements de Géométrie*, 1813.

[1844 Ru] Russell, J.S., *Report on Waves*, 14th Mtg. of the British Assoc. for the Advance. of Science, John Murray, London (1844), 311–390 + 57 plates.

[1859 La] Lamé, *Lecons sur les coordonnées curvilignes et leurs diverses applications*, 1859, Paris.

[1866 Da] Darboux, G., *Sur les surfaces orthogonales*, Bull. Soci. Philomathique (1866), 16–.

[1867 Co] Combescure, E., *Sur les déterminants fonctionnels et les coordonnées curvilignes*, Ann. de l'Ëcole Normale **4** (1867), 102–122.

[1870 Ri] Ribaucour, *Note sur les développées des surfaces*, Comptes Rendus **70** (1870), 330–333.

[1871 Bo] Boussinesq, M.J., *Théorie de l'intumescence appelée onde solitaire ou de translation se propageant dans un canal rectangulaire*, C. R. Acad. Sc. Paris **72** (1871), 755–759.

[1875 M] Moutard, *Ricerche sulle superficie a curvatura e sulle elicoidi*, Ann. Scuola normale superiore di Pisa **2** (1879), 285–.

[1879 Bi] Bianchi, *Ricerche sulle superficie a curvatura e sulle elicoidi*, Ann. Scuola normale superiore di Pisa **2** (1879), 285–.

[1879 Li] Lie, S., *Ueber Flächen deren Krümmungsradien durch eine Relation Verknüptt sind*, Archiv Math. Naturvidenskab **4** (1879), 510–.

[1882 Da] Darboux, G., *Sur une proposition relative aux équations linéarires*, Comptes Rendus **94** (1882), 1456–.

[1883 Ba] Bäcklund, A.V., *Concerning surfaces with constant negative curvature*, New Era Printing Co., Lancaster, PA, original 1883, translation 1905.

[1889 Da] G. Darboux, *Leçons sur la théorie générale des surfaces, v. II*, Gauthier-Villars, Paris, 1889.

[1894 Da] G. Darboux, *Leçons sur la théorie générale des surfaces, v. III*, Gauthier-Villars, Paris, 1894.

[1895 KdV] Koteweg, D.J., de Vries, G., *On the change of form of long waves advancing in a rectangular canal, and on a new type of long stationary waves*, Philos. Mag. Ser. 5 **39** (1895), 422–443.

[1897 Da] G. Darboux, *Lecons systémes orthogonaux et les coordonnées curvilignes*, Gauthier-Villars, Paris, 1897.

[1906 dR] Da Rios, *Sul moto d'un liquido indefinito con un filetto vorticoso di forma qualunque (on the motion of an unbounded liquid with a vortex filament of any shape)*, Rend. Circ. Mat. Palermo **22** (1906), 117–135.

[1909 Ei] Eisenhart, L.P., *A treatise on the differential geometry of curves and surfaces*, Ginn, 1909.

[1910 Tz] Tzitzeica, G., *Sur une nouvelle classe des surfaces*, C.R. Acad. Sci. Paris **150** (1910), 955–956.

[1919 Ca] Cartan, E., *Sur les variétés de courbure constante d'un espace euclidien ou non-euclidien*, Bull. Soc. Math. France **47** (1919), 132–208.

[1945 Ca] Cartan, É, *Les systèmes différentiels extèrieurs et leurs applications géométriques*, Hermann, Paris, 1945.

[1955 FPU] Fermi, E., Pasta, J., Ulam, S., *Los Alamos Report* **LA1940** (1955).

[1965 W] Whitham, G.B., *Non-linear dispersive waves*, Proc. Royal Soc. London **A. 139** (1965), 283–291.

[1965 ZK] Zabusky, N.J., Kruskal, M.D., *Interaction of solitons in a collisionless plasma and the recurrence of initial states*, Physics Rev. Lett. **15** (1965), 240–243.

[1967 GGKM] Gardner, C.S., Greene, J.M., Kruskal, M.D., Miura, R.M., *Method for solving the Korteweg-de Vries equation*, Physics Rev. Lett. **19** (1967), 1095–1097.

[1968 La] Lax, P.D., *Integrals of nonlinear equations of evolution and solitary waves*, Comm. Pure. Appl. Math. **31** (1968) 467–490.

[1971 Ha] Hasimoto, H. *Motion of a vortex filament and its relation to elastica*, J. Phys. Soc. Jap. **31** (1971) 293–.

[1971 ZF] Zakharov, V.E., Faddeev, L.D., *Korteweg-de Vries equation, a completely integrable Hamiltonian system*, Func. Anal. Appl. **5** (1971), 280–287.

[1972 ZS] Zakharov, V.E., Shabat, A.B., *Exact theory of two-dimensional self-focusing and one-dimensional interaction of waves in non-linear media*, Sov. Phys. JETP **34** (1972), 62–69.

[1973 AKNS] Ablowitz, M.J., Kaup, D.J., Newell, A.C. and Segur, H., *Method for solving the sine-Gordon equation*, Phy. Rev. Lett. **30** (1973), 1262–1264.

[1973 ZMa] Zakharov, V.E., Manakov, S.V., *On resonant interaction of wave packets in non-linear media*, JETP Letters **18** (1973) 243–247.

[1974 Wh] Whitham,G., *Linear and non-linear waves*, Wiley, 1974.

[1974 ZMa] Zakharov, V.E., Manakov, S.V., *The theory of resonant interaction of wave packets in non-linear media*, Sov. Phys. JETP **42** (1974), 842–850.

[1974 FPU] Fermi, E., Pasta, J., Ulam, S., *Studies of Nonlinear Problems. I*, Lectures in Applied Math. **15** (1974), 143–155.

[1975 Mo] Moser, J., *Three integrable Hamiltonian systems connected with isospectral deformations*, Adv. Math. **16** (1975), 197–220.

[1976 GDi] Gel'fand, I.M., Dikii, L.A., *Fractional Powers of Operators and Hamiltonian Systems*, Funkcional'nyi Analiz i ego Prilozenjia **10** (1976).

[1977 Kr] Krichever, I.M., *Methods of algebraic geometry in the theory of non-linear equations*, Russian Math. Surveys **32** (1977), 185–213.

[1978 AM] Adler, M., Moser, J., *On a class of polynomials connected with the Korteweg-deVries equation*, Comm. Math. Phys. **61** (1978), 1–30.

[1978 De] Deift, P., *Applications of a commutation formula*, Duke. Math. J. **45** (1978), 267–310.

[1978 ZMi] Zakharov, V.E., Mikhailov, A.V., *Relativistically invariant 2-dimensional models of field theory which are integrable by means of the inverse scattering problem method*, Soviet Physics JETP **47** (1978), 1017–1027.

[1979 Ad] Adler, M., *On a trace functional for formal pseudo-differential operators and the symplectic structure of the Korteweg-de Vries Type Equations*, Invent. Math. **50** (1979), 219–248.

[1979 FRS] Frenkel, I.E., Reiman, A.G., Semenov-Tian-Shansky, M.A., *Graded Lie algebras and completely integrable dynamical systems*, Soviet Math. Dokl. **20** (1979), 811–814.

[1979 Ko] Kostant, B., *The solution to a generalized Toda lattice and representation theory*, Adv. Math. **34** (1979), 195–338.

[1979 ZS] Zakharov, V.E., Shabat, A.B., *Integration of non-linear equations of mathematical physics by the inverse scattering method, II*, Funct. Anal. Appl. **13** (1979), 166–174.

[1980 AdM] Adler, M., van Moerbeke, P., *Completely integrable systems, Euclidean Lie algebras and curves*, Adv. Math. **38** (1980), 267–317.

[1980 CT] Chern, S.S., Terng, C.L., *An analogue of Bäcklund's theorem in affine geometry*, Rocky Mountain J. Math. **10** (1980), 105–124.

[1980 FFL] Flaschke, H., Forrest, M.G. and McLaughlin, D., *Multiphase averaging and the inverse spectral solution of the Korteveg-deVries equation*, Comm. Pure Appl. Math **33** (1980), 739.

[1980 Mo] Moser, J., *Various aspects of integrable Hamiltonian Systems*, Progress in Mathematics **8** (1980), 233–289, Birkhauser, Boston.

[1980 RS] Reyman, A.G., Semenov-Tian-Shansky, *Current algebras and non-linear partial differential equations*, Sov. Math., Dokl. **21** (1980), 630–634.

[1980 Sy] Symes, W.W., *Systems of Toda type, Inverse spectral problems, and representation theory*, Inventiones Math. **59** (1980), 13–51.

[1980 TT] Tenenblat, K., Terng, C.L., *Bäcklund's theorem for n-dimensional submanifolds of* R^{2n-1}, Ann. Math. **111** (1980), 477–490.

[1981 Ch] Cherednik, I.V., *Integration of the equation of a two dimensional chiral field and of its quantum analogue*, Soviet J. Nucl. Physics **33** (1981), 144–145.

[1981a DJKM] Date, E., Jimbo, M., Kashiwara, M., Miwa, T., *Transfromation groups for soliton eqautions I. The* τ*-function of the KP equation*, Proc. Japan Acad. Ser. A **57** (1981), 342–347.

[1981 DS] Drinfel'd, V.G., and Sokolov, V.V., *Equations of Kortweg-de Vries type and simple Lie algebras*, Dokl. Akad. Nauk SSSR **258** (1981) 11–16, (Trans. as Soviet Math. Dokl. 23, 457-462).

[1981 KW] Kuperschmidt, B.A., Wilson, G., *Modifying Lax equations and the second Hamiltonian structure*, Invent. Math. **62** (1981), 403–436.

[1981 S] Sato,M., *Soliton equations as dynamical systems on infinite dimensional Grassman manifolds*, RIMS Kokyuroku **439** (1982), 30–40.

[1981 Wi] Wilson, G., *The modified Lax equations and two dimensional Toda lattice equations associated with simple Lie algebras*, Ergodic Theory and Dynamical Systems I **30** (1981), 361–380.

[1983 Dr] Drinfel'd, V.G., *Hamiltonian structures on Lie groups, Lie bialgebras and the geometrical meaning of Yang-Baxter equations*, Sov. Math. Doklady **27** (1983), 68–71.

[1983 DN] Dubrovin, B.A., Novikov, S.P., *Hamiltonian formalism of one dimensional systems of hydrodynamic type, and the Bogolyubov-Whitham averaging method*, Soviet Math. Dokl. **27** (1983), 781–785.

[1984 AD] Arbarello, E. and DeConcini, C., *On a set of equations Characterizing Riemann matrices*, Ann. of Math. **120** (1984), 119–140.

[1984 BC] Beals, R., Coifman, R.R., *Scattering and inverse scattering for first order systems*, Commun. Pure Appl. Math. **37** (1984), 39–90.

[1984 Do] Donaldson, S., *Nahm's equations and the classification of monopoles*, Commun. Math. Physics **96** (1984), 387–407.

[1984 DN] Dubrovin, B.A., Novikov, S.P., *On Poisson brackets of hydrodynamic type*, Soviet Math. Dokl. **30** (1984), 294–297.

[1984 NMPZ] Novikov, S., Manakov, S., Pitaevskii, L.B. Zakharov, V.E., *Theory of Solitons*, Plenum. New York, 1984.

[1985 BC] Beals, R., Coifman, R.R., *Inverse scattering and evolution equations*, Commun. Pure Appl. Math. **38** (1985), 29–42.

[1985 SW] Segal, G., Wilson, G., *Loop groups and equations of KdV type*, Publ. Math. IHES **61** (1985), 5–65.

[1987 ABT] Ablowitz, M.J., Beals, R., Tenenblat, K., *On the solution of the generalized wave and generalized sine-Gordon equations*, Stud. Appl. Math. **74** (1987), 177–203.

[1987 Hi] Hitchin, N., *Stable bundles and integrable systems*, Duke Math. Journal **54** (1987), 91–1114.

[1989 PS] Pinkall, U., Sterling, I., *On the classification of constant mean curvature tori*, Ann. of Math. **130** (1989), 407–451.

[1989 U] Uhlenbeck, K., *Harmonic maps into Lie group (classical solutions of the Chiral model)*, J. Differential Geometry **30** (1989), 1–50.

[1991 AC] Ablowitz, M.J., Clarkson, P.A., *Cambridge University*, 1991.

[1991 Bo] Bobenko, A.I., *All constant mean curvature tori in R^3, S^3, H^3 in terms of theta functions*, Math. Ann. **290** (1991), 209–245.

[1991 Ri] Ricca, R.L., *Rediscovery of Da Rios equations*, Nature **352** (1991), 561–562.

[1991 Ts] Tsarëv, S.P., *The geometry of Hamiltonian systems of hydrodynamic type. The generalized Hodograph method*, Math. USSR Izvestiya **37** (1991), 397–419.

[1991 W] Witten, E., *Two dimensional gravity and intersection theory on moduli space*, Survey in Differential Geometry **1** (1991), 243–310s.

[1992 Hi] Hitchin, N., *Hyper-Kähler manifolds*, Astérisque **206** (1992), 137–166.

[1993 BFPP] Burstall, F.E., Ferus, D., Pedit, F., Pinkall, U., *Harmonic tori in symmetric spaces and commuting Hamiltonian systems on loop algebras*, Annals of Math. **138** (1993), 173–212.

[1994 KM] Kontsevich, M., Manin, Y., *Gromov-Witten classes, quantum cohomology, and enumerative geometry*, Comm. Math. Phys. **164** (1994), 625–562.

[1994 RT] Ruan, Y.G., Tian, G., *A mathematical theory of quantum cohomology*, Math. res. Lett. **1** (1994), 269–278.

[1995 RT] Ruan, Y.G., Tian, G., *A mathematical theory of quantum cohomology*, J. Differential Geom. **42** (1995), 259–367.

[1996 DW] Donagi, R., Witten, E., *Supersymmetric Yang-Mills theory and integrable systems*, Nucl. Phys. **B.460** (1996), 299–334.

[1996 D] Dubrovin, B.A., *Geometry of 2D topological field theories*, Integrable systems and quantum groups (1996), 299–334, Lecture notes in Math., v. 1620, Springer-Verlag.

[1997 BS] Bobenko, A.I., Schief, W.K., *Affine spheres: discretization via duality relations* (to appear).

[1997 F] Freed, D, *Special Kähler manifold* (to appear).

[1997 Pa] Palais, R.S., *The symmetries of solitons*, Bulletin of the AMS **34** (1997), 339–403.

Integrable systems in Riemannian geometry

Nigel Hitchin*
Department of Pure Mathematics
and Mathematical Statistics
University of Cambridge
16 Mill Lane
Cambridge CB2 1SB
UK

1 Introduction

What is an integrable system?

There is no precise formulation which will encompass all the mathematical phenomena covered by this term, but we have a host of examples: most mathematicians would agree that any definition of an integrable system should include:

- completely integrable Hamiltonian systems
- systems linearized on an abelian variety
- Painlevé equations

All these have the characteristic property (in Ward's definition of integrability [60]) that "their solutions can in principle be constructed explicitly". Even that vague statement begs the question, for we know that a general solution to a Painlevé equation involves new transcendental functions. What we mean by an integrable system is that there is a systematic method of describing all

*Current address: Mathematical Institute, 24–29 St Giles, Oxford OX1 3LB, UK.

solutions. In an individual problem, we simply have to select the particular one whose constants of integration match the problem.

That sounds like the 19th century approach to differential equations, and it is true that even the modern questions in Riemannian geometry which are amenable to attack using integrable systems frequently involve a journey into the last century to get going. There could not be a stronger contrast between the 20th century use of existence theorems for partial differential equations to discover deep differential geometric results and the method of integrable systems in producing explicit answers, but existence theorems will never give us all the information we want. Nor, as we shall see, does explicitness, when we find it, always answer the interesting questions. What integrability can do is to reveal truths and patterns which are hidden in the existence theory.

There are perhaps three contrasting approaches to mathematics which distinguish the last two centuries:

- local versus global
- explicitness versus existence
- special cases versus unifying structure

I hope to show that, even in those areas of concern to our predecessors, the modern approach as applied to integrable systems has much to commend it.

Riemannian geometry is not *a priori* concerned with abelian varieties or symplectic manifolds—the stuff that integrable systems are made on. The natural questions we ask are those concerning curvature—properties of the extrinsic curvature of manifolds in Euclidean space, for example, or special properties of the Riemann tensor of an abstract manifold. In this paper we shall concentrate on three areas of Riemannian geometry where integrable systems have been used over the past few years:

- Tori of constant mean curvature
- 4-dimensional Einstein manifolds
- Hyperkähler metrics

The first topic is one where the literature is most extensive, and the subject best understood. Whereas ten years ago, one could have argued that the theory of algebraic curves played an important role in the theory of minimal surfaces and harmonic maps, this was essentially based on versions of the twistor construction. In particular, the algebraic curve was the surface itself, with its conformal structure. The integrable system approach to minimal tori is very different (although now integrated into the twistor viewpoint too), and although algebraic curves come in to play, they do so in a more subtle manner: it is a subspace of the Jacobian which models the surface itself, with its linear structure. We give here an account of two approaches to the problem. One is a direct construction of the algebraic curve from holonomy considerations. The other, which is more developed in the literature, concerns the production of a rather special object on the torus—a polynomial Killing field. We show in Section 2 how these two points of view are related by bringing in a correspondence which lies at the heart of many applications of integrable systems. This is the relationship between a line bundle on a curve which is expressed as a covering of the projective line, and a matrix of polynomials. Systems which are linearized on the Jacobian of a curve are usually of this form, and the creation of polynomial Killing fields is one example.

The second topic, treated in Section 3, is broader, but the usefulness of integrable systems in addressing the open problems is less well understood. The specific problem we consider concerns the case of self-dual Einstein manifolds with a three-dimensional group of isometries. Here the aspect of integrability which enters is the isomonodromic deformation problem, as manifested by Painlevé's sixth equation. There are other problems in Riemannian geometry, both classical and modern, where the essential question is to solve a Painlevé equation, but this particular one allows us to produce a complete solution to the original question. In doing so, we can write down some explicit complete non-Kählerian Einstein metrics on the four-dimensional ball. We emphasize here the non-Kähler nature, since existence theorems for Einstein metrics are few and far between in the absence of a complex structure.

In Section 4, we survey the central role of the self-dual Yang-Mills equations as an overarching structure yielding integrable equations in Riemannian geometry as special cases. This is a curious fact, but the role of these equations in explaining integrability is ever-present. We show in particular how the two systems considered above are derived from forms of the Yang-Mills equations, and then study a third system—Nahm's equations—which play a central role in the latter part of this paper.

The first four sections survey material which has already been published by various authors. Section 5 is a new set of results in which the methods of integrable systems are used to provide concrete information about families of metrics which are of interest to both differential geometers and physicists—*hyperkähler metrics.* Many of the examples of such metrics—Kronheimer's metrics on the cotangent bundle, or a coadjoint orbit of, a complex semi-simple Lie group—or monopole moduli spaces—are based on solutions to Nahm's equations. We give a formula for the Kähler potential of those metrics within these families which carry circle actions of a particular type. The formula can be expressed in terms of the data used to construct a solution to Nahm's equations: an algebraic curve and a line bundle on it. In fact it is the logarithmic derivative of the Riemann theta function which provides the essential component of this. In many respects this is the analogue in Riemannian geometry of the formula for the potential in the KdV equation in terms of the τ-function, perhaps the most familiar place in the conventional literature on integrable systems where theta functions arise. Within the general formula we work out some specific examples to compare with alternative methods. These include the Eguchi-Hanson metric and the $\mathcal{L}^2$ metric on the moduli space $\mathcal{M}_k^0$ of centred $SU(2)$-monopoles.

The monopole moduli space and its Riemannian metric has attracted much attention recently as a testing ground for duality theories in physics [53]. Our formula for the Kähler potential ϕ of a given complex structure is the sum of two terms.

$$\phi = \frac{4}{(N+1)(N+2)} \frac{\vartheta^{(N+2)}(0)}{\vartheta^{(N)}(0)} - \frac{1}{3}Q(u,u)$$

The first, which is rotationally invariant, is fundamentally associated to the spectral curve of the monopole: it is defined by the expansion of the Riemann theta function about a distinguished point. The second term, while also expressible in terms of the spectral curve, can also be read off from the monopole itself. It depends on a direction u in $\mathbf{R}^3$, which corresponds to the complex structure in the hyperkähler family. The term $-Q(u,u)/4$ is the coefficient of r^{-3} in the asymptotic expansion of the length of the Higgs field along a radius in the direction u.

2 Tori of constant mean curvature

2.1 Background

The classical problem here is one of determining the shape of a soap bubble—the constant pressure difference between the inside and the outside is translated into the condition on a surface in $\mathbf{R}^3$ that it should have constant mean curvature. Generalizations of this include such surfaces in S^3, hyperbolic space or higher dimensional manifolds.

A.D. Alexandrov [2], in the 1950's, extended the local results of Jellet [32] a hundred years earlier to show that the only constant mean curvature surface *embedded* in $\mathbf{R}^3$ was a round sphere, thus satisfactorily explaining the prevalence of round soap bubbles, but the question remained open as to whether higher genus surfaces could be *immersed* with constant mean curvature.. This problem, in the case of the torus, is attributed to Hopf, and was solved in 1984 by Wente [61]. He found a particular immersed torus which has constant mean curvature. Wente's is a purely 20th century proof—an existence theorem for the sinh-Gordon equation,

$$\omega_{xx} + \omega_{yy} = -\sinh \omega$$

but the subsequent development of the subject, as described by Melko and Sterling [43] is noteworthy.

Wente's existence theorem provoked Abresch to make a computerized numerical analysis of the solution to create pictures of the torus. He noticed, on plotting the curvature lines, that each line of one family appeared to be planar. Taking this as an ansatz, he reduced the partial differential equation to ordinary differential equations solvable by elliptic functions and gave an explicit analytic solution [1]. Ironically, Wente had seen the reformulation of the problem into the sinh-Gordon equation in Eisenhart's classic textbook [20]. Had he read a few pages further, he would have encountered the 19th century version of Abresch's analytical solution [58].

This cautionary tale does not of course reveal the general solution to the problem, which has now developed into a major area of research. The main points of view (see [21] for various articles on the subject) are due to Pinkall and Sterling et al, whose approach can be expressed in terms of Hamiltonian systems and loop groups, Bobenko, who sees it in terms of the much studied finite-gap solutions of the sinh-Gordon equation and the corresponding Baker-Akhiezer function, and the author [25] whose approach is influenced by twistor theory. These viewpoints all embed the specific problem in a bigger structure, and they are all related. In the next sections I want to make the relationship between two of these a little more precise, since it involves a construction which appears in many of the interfaces between integrable systems and differential geometry.

2.2 Harmonic maps

All approaches to constant mean curvature surfaces set the problem in the wider context of harmonic maps of surfaces into Lie groups. It is an old result [50] that the Gauss map of a constant mean curvature surface in $\mathbf{R}^3$ is a harmonic map to S^2. Embedding S^2 totally geodesically as an equator in S^3 we get a harmonic map into $S^3 \cong SU(2)$. Once we are in the 3-sphere we can also think of minimal surfaces in the sphere, which are other examples of harmonic maps. Generalizing, we may consider a Riemann surface M and a harmonic map $f : M \to G$ into a compact semisimple Lie group, with its biinvariant metric.

The equations for a harmonic map have a particularly accessible form. If we trivialize the tangent bundle by left translation, this defines a flat connection ∇_L with trivial holonomy, and by right translation another trivial connection ∇_R. Since the tangent bundle has structure group G, we may equally regard these as connections on a principal G-bundle. The map f relates the two trivializations: $f^{-1}df = \nabla_L - \nabla_R$. However, instead of this formalism we consider

$$\begin{aligned} \nabla &= \frac{1}{2}(\nabla_R + \nabla_L) \\ \phi &= \frac{1}{2}(\nabla_R - \nabla_L) \end{aligned}$$

to define a pair of objects: a connection ∇ and a section $\phi \in \Omega^1(M; \operatorname{ad} P)$. Using the conformal structure, we write $\phi = \Phi - \Phi^*$ where $\Phi \in \Omega^{1,0}(M; \operatorname{ad} P \otimes \mathbf{C})$ is of type $(1,0)$. Although we write $-\Phi^*$ which makes sense only for the unitary group, for a general Lie group we interpret this as the application of the anti-involution on $\Omega^{1,0}(M; \operatorname{ad} P \otimes \mathbf{C})$ defined by the real structure on the complex Lie algebra which gives the compact real form. We then have the following:

Proposition 2.1. *If f is a harmonic map the G^c connection*

$$\nabla + \zeta\Phi - \zeta^{-1}\Phi^*$$

is flat for all $\zeta \in \mathbf{C}^*$. Conversely if ∇, Φ define a flat connection for all ζ which has trivial holonomy for $\zeta = \pm 1$, then it arises from a harmonic map.

The historical origins of this method are unclear: in some respects it is the classical notion of an associated surface, but in the context of integrable systems it seems to begin with Pohlmeyer [49].

2.3 The loop group approach

The approach initiated by Pinkall and Sterling provides a formulation in terms of loop groups and algebras, as studied for example by Burstall et al [8], who we follow in this section. Here one works in the trivial covariant constant gauge at $\zeta = 1$ for the connection $\nabla + \zeta\Phi - \zeta^{-1}\Phi^*$, the connection ∇_R. The connection matrix can be written as

$$A_\zeta = (1-\zeta)\alpha - (1-\zeta^{-1})\alpha^*$$

for some Lie-algebra valued $(1,0)$-form α on the Riemann surface M. The harmonic map f is then defined by the covariant constant trivialization at $\zeta = -1$, the connection ∇_L,

$$f^{-1}df = 2(\alpha - \alpha^*) \tag{2.1}$$

If M is a torus, passing to the universal covering $\mathbf{C}$, we can write $\alpha = adz$, and then comparing coefficients of ζ in the flatness condition gives the equation for f to be harmonic:

$$\frac{\partial a}{\partial \bar{z}} = -[a, a^*] \tag{2.2}$$

The essential point of the loop group approach is to see these equations as the consequence of another equation with more variables, but with a simpler form. One considers a finite Laurent series which vanishes at $\zeta = 1$ (a "polynomial Killing field")

$$\xi = \sum_{|n|\leq d} (1-\zeta^n)\xi_n$$

where the ξ_n are $\mathfrak{g}^c$-valued functions with $\xi_{-n} = -\xi_n^*$ satisfying the equation

$$d\xi + [(1-\zeta)adz - (1-\zeta^{-1})a^*d\bar{z}, \xi] = 0 \tag{2.3}$$

or equivalently

$$\frac{\partial \xi}{\partial z} = [\xi, (1-\zeta)a] \tag{2.4}$$

$$\frac{\partial \xi}{\partial \bar{z}} = -[\xi, (1-\zeta^{-1})a^*] \tag{2.5}$$

and where $a = \xi_d$. Expanding out yields equation (2.2) as one of the many terms.

The principal advantage of using this approach is its integrability in finite-dimensional terms. The highest order coefficient of ξ commutes with a, so a solution of (2.4), (2.5) will still be a Laurent polynomial of the same degree: we can thus rewrite the problem in terms of integrating vector fields on the finite-dimensional vector space Ω_d of Laurent polynomials in the Lie algebra $\mathfrak{g}$ of degree d. These are the two vector fields X_1, X_2 defined by the equations (2.4,2.5)

$$X_1 + iX_2 = Z = [\xi, (1-\zeta)\xi_d] \tag{2.6}$$

They can be simultaneously integrated if they commute, and this happens if and only if $[Z, \bar{Z}] = 0$. But this is true iff

$$\begin{aligned}&[\,[\xi,(1-\zeta^{-1})\xi_{-d}],(1-\zeta)\xi_d] + [\xi,(1-\zeta)[\xi,(1-\zeta^{-1})\xi_{-d}]_d]\\ =&[\,[\xi,(1-\zeta)\xi_d],(1-\zeta^{-1})\xi_{-d}] + [\xi,(1-\zeta^{-1})[\xi,(1-\zeta)\xi_d]_{-d}]\end{aligned}$$

or equivalently

$$\begin{aligned}&(1-\zeta)(1-\zeta^{-1})[\,[\xi,\xi_{-d}],\xi_d] + (1-\zeta)[\xi,[\xi_d,\xi_{-d}]\,]\\ =&(1-\zeta)(1-\zeta^{-1})[\,[\xi,\xi_d],\xi_{-d}] + (1-\zeta^{-1})[\xi,[\xi_{-d},\xi_d]\,]\end{aligned}$$

and verification of this statement follows immediately on writing $(1-\zeta)(1-\zeta^{-1}) = (1-\zeta) + (1-\zeta^{-1})$ and using the Jacobi identity.

The specific form of this approach has certain benefits. Any vector field of the form $X = [\xi, B(\xi)]$ on a Lie algebra (a *Lax pair*) has the property that invariant functions are conserved along the flow. In our case, the algebra is the loop algebra $\Omega\mathfrak{g}$, and we are restricting to the finite-dimensional subspace Ω_d. Any invariant function on $\mathfrak{g}$ then defines a polynomial, all of whose coefficients are constants of integration. In fact, these constants fit into a convenient Hamiltonian formalism for the system, but we shall not go into this here.

A particular consequence of the existence of these conserved quantities is obtained by applying the Killing form. The constant coefficient in the resulting polynomial is the expression

$$\sum_n \langle \xi_n, \xi_{-n} \rangle$$

which, since $\xi_n = -\xi^*_{-n}$, defines an inner product on Ω_d. By compactness the vector fields can be integrated to give a solution to (2.4),(2.5) on the whole of $\mathbf{R}^2$.

A slightly stronger consequence of a Lax form is that the flow is tangential to an orbit of the Lie group, so that (2.6) implies for the complex vector field Z that ξ lies in the orbit of ξ_d under the complex Lie group. In particular, $a = \xi_d : \mathbf{R}^2 \to \mathfrak{g}^c$ itself maps to a fixed orbit.

As it stands, this method is an ansatz designed to produce solutions of the equations on $\mathbf{R}^2$. We shall not go into the Hamiltonian formalism which helps to solve it, involving r-matrices, and the Adler-Kostant-Symes method but (see [9]) it is well-documented.

For the problem at hand, maps of a torus, we need to find doubly periodic solutions. The aim is thus to find the general solution and then choose the right constants of integration to give us maps of a torus. More importantly, what is needed is a proof that the method has general applicability: in particular we need to show that *any* harmonic map from a torus has the property that a lies in a fixed orbit, and then find a polynomial Killing field. Furthermore, if this is true, we need to interpret the equation geometrically in terms of the original problem of harmonic maps to G. The situation here is now quite well understood. In the next section I essentially follow [9].

2.4 Jacobi fields

Suppose we have a harmonic map from a torus to G, then (2.2) shows that evaluating any invariant polynomial on a gives a holomorphic function on the torus, which is constant by compactness. In the simplest case where a is principal semi-simple (has distinct eigenvalues) at some point it follows that it is principal semi-simple everywhere and since all invariant polynomials have the same value, lies in a single complex adjoint orbit. If we make this simplifying assumption for th present, then we see that one condition for the method to apply is satisfied, and this is a consequence of compactness of the torus. It will be compactness again which gives the existence of a polynomial Killing field, but this involves a more complex argument.

The equation (2.3) simply says that ξ is a covariant constant section of the adjoint bundle $\mathrm{ad}\, P \otimes \mathbf{C}$ with respect to the flat connection $d + A_\zeta$. The flatness of the connection for all ζ implies the existence of a local infinite Laurent series solution, but what we need is a polynomial. We begin by finding a semi-infinite series

$$x = \sum_0^\infty x_n \zeta^{-n}$$

This provides a formal solution if the following recurrence relation holds:

$$dx_{n-1} + [\alpha - \alpha^*, x_{n-1}] - [\alpha, x_n] + [\alpha^*, x_{n-2}] = 0 \tag{2.7}$$

We focus attention on the $(1,0)$ part:

$$\frac{\partial x_{n-1}}{\partial z} + [a, x_{n-1}] - [a, x_n] = 0 \tag{2.8}$$

If we begin with $x_0 = a$, we can solve this recursively so long as $\partial x_{n-1}/\partial z$ lies in the image of $\mathrm{ad}(a)$. Since, as we have seen, a lies in a single orbit, $\partial x_0/\partial z = \partial a/\partial z$ is tangential to the orbit and so is indeed in the image of ad(a). To proceed, one engages in a two-step recursion process modifying x_n by a suitable term in ker ad(a) [8]. If a is doubly periodic, so is each x_n, which is thus defined on the torus. We therefore get a formal solution x to

$$d'x + [A_\zeta^{1,0}, x] = 0$$

Attending to the $(0,1)$ part, we note that since $d + A_\zeta$ is flat, $\partial/\partial z + (1-\zeta)a$ and $\partial/\partial\bar{z} - (1-\zeta^{-1})a^*$ commute, so that

$$\partial x/\partial\bar{z} - [(1-\zeta^{-1})a^*, \xi] = \sum_0^\infty \tilde{x}_n \zeta^{-n}$$

is another solution to (2.8). In this case, we have $\tilde{x}_0 = \partial a/\partial\bar{z} - [a^*, a]$ which vanishes by (2.2). Thus, by recurrence, the solution determined above satisfies the full recurrence equation and gives a formal solution to $dx + [A_\zeta, x] = 0$.

The coefficients of this series have an interpretation on the torus itself, which helps to prove finiteness of the series. They are (complex) *Jacobi fields*—solutions of the linearization of the harmonic map equation. To see this, recall from (2.1) that the actual harmonic map f is defined by

$$f^{-1}df = 2(adz - a^*d\bar{z})$$

An infinitesimal deformation $\dot{f}$ of the map $f : M \to G$ defines a tangent vector to G along the torus which we represent as the Lie algebra-valued function $\psi = f^{-1}\dot{f}$. Differentiating the above equation along the deformation gives

$$d\psi + 2[\alpha - \alpha^*, \psi] = 2(\dot{\alpha} - \dot{\alpha}^*)$$

We have the harmonic map equation $d''\alpha - [\alpha^*, \alpha] = 0$, and differentiating this along the deformation we get

$$d''\dot{\alpha} - [\dot{\alpha}^*, \alpha] - [\alpha^*, \dot{\alpha}] = 0$$

Writing $\beta = \dot{\alpha}$ and $\gamma = \dot{\alpha}^*$, a complex solution ψ of the Jacobi equation is equivalent to a solution of the three equations

$$\begin{aligned} d'\psi + 2[\alpha, \psi] &= 2\beta \\ d''\psi - 2[\alpha^*, \psi] &= -2\gamma \\ d''\beta - [\gamma, \alpha] - [\alpha^*, \beta] &= 0 \end{aligned}$$

We want to see that $\psi = x_n$ solves these equations. Now from (2.7),

$$\begin{aligned} d'x_{n-1} + [\alpha, x_{n-1}] - [\alpha, x_n] &= 0 \\ d''x_{n-1} - [\alpha^*, x_{n-1}] + [\alpha^*, x_{n-2}] &= 0 \end{aligned}$$

Setting $2\beta = [\alpha, x_n + x_{n-1}]$ and $2\gamma = [\alpha^*, x_{n-2} + x_{n-1}]$ we obtain the first two of the Jacobi equations. As for the third, we have, using $d''\alpha = [\alpha^*, \alpha]$,

$$\begin{aligned} 2d''\beta &= [\,[\alpha^*, \alpha], x_n + x_{n-1}] - [\alpha, d''(x_n + x_{n-1})] \\ &= [\,[\alpha^*, \alpha], x_n + x_{n-1}] - [\alpha, [\alpha^*, x_n]\,] + [\alpha, [\alpha^*, x_{n-1}]\,] \\ &\qquad\qquad\qquad\qquad - [\alpha, [\alpha^*, x_{n-1}]\,] + [\alpha, [\alpha^*, x_{n-2}]\,] \\ &= 2[\gamma, \alpha] + 2[\alpha^*, \beta] \end{aligned}$$

using the Jacobi identity.

The coefficients x_n of the formal solution therefore each belong to a fixed vector space of Lie-algebra-valued functions - the space of Jacobi fields. Moreover, since the linearization of the harmonic map equations is elliptic, by compactness of the torus this space is finite-dimensional. This is the crux of the finiteness result from this point of view. To reduce x from a series to a polynomial requires one more step.

We consider the polynomials

$$p_k = \sum_{n=0}^{k} \zeta^{k-n} x_n$$

obtained from the first $(k+1)$ terms of $\zeta^k x$. Since $dx + [A_\zeta, x] = 0$, we have $d(\zeta^k x) + [A_\zeta, \zeta^k x] = 0$, and consideration of the polynomial part gives

$$dp_k + [A_\zeta, p_k] = [\alpha, x_{k+1}] \tag{2.9}$$

We want to show that x_k vanishes for k large enough. Suppose not, then we can find polynomials $p_1, \ldots, p_N$ spanning a space of greater dimension than the vector space J of Jacobi fields. But then, since each x_{k+1} is an element of J, it follows from (2.9) that there is a polynomial $q(\zeta) = \sum_0^N \lambda_k p_k$ such that $dq + [A_\zeta, q] = 0$. Without loss of generality, we can assume that the constant coefficient q_0 is non-zero. But from (2.7) with $n = 0$ $q_0 = ha$ for some function h, and from the same formula with $n = 1$, h must be holomorphic and hence constant on the torus. We can thus scale q to be obtained from the same recurrence relation with the same initial condition as x. Since, moreover, the other coefficients of q are multiples of x_k for $k > 0$, they are in $\mathrm{Im}(\mathrm{ad}(a))$ and so q coincides with x, showing that the x_k do indeed vanish for $k > N$. It is now a simple step using reality and dividing by a power of ζ to obtain the required polynomial Killing field.

This is the loop group approach. We now give the alternative viewpoint which involves the geometry of algebraic curves.

2.5 Hyperelliptic curves

The details of this construction have only been worked out for $G = SU(2)$ [25], so we restrict ourselves to this case and consider, for comparison with the previous method, the situation where a is semisimple. The construction starts with an algebraic curve S defined by the equation

$$\eta^2 = P(\zeta)$$

where $P(\zeta)$ is a polynomial of degree $2p+2$ in ζ such that $\bar{P}(\zeta) = \bar{\zeta}^{2p+2}P(\bar{\zeta}^{-1})$. There must be no roots of P at 0 or ∞ nor on the unit circle. The curve is a branched double covering $\pi : S \to \mathbf{P}^1$—a hyperelliptic curve.

We then take two meromorphic differentials $\theta, \tilde{\theta}$ with double poles over 0 and ∞ but with residue zero, anti-invariant with respect to the hyperelliptic involution $\eta \mapsto -\eta$ and satisfying some reality conditions.

To relate this to the problem of harmonic maps of a torus, we have to find the torus in this mass of algebraic data. The differentials have expansions near $\zeta = 0$

$$\theta = \lambda_{-2}\frac{d\zeta}{\zeta^2} + \lambda_0 + \dots \qquad \tilde{\theta} = \tilde{\lambda}_{-2}\frac{d\zeta}{\zeta^2} + \tilde{\lambda}_0 + \dots$$

and since their residue is zero, the principal parts are determined by the coefficients $\lambda_{-2}, \tilde{\lambda}_{-2}$.

We write as usual $\mathcal{O}(d)$ for the pull-back to S of the line bundle of degree d on $\mathbf{P}^1$, then the poles of the differentials occur on the divisor $D = \pi^{-1}\{0, \infty\}$ of $\mathcal{O}(2)$. The principal parts $[\theta], [\tilde{\theta}]$ are then, invariantly speaking, elements of $H^0(D, \mathcal{O}(2))$ and the exact cohomology sequence for the natural sequence of sheaves

$$0 \to \mathcal{O} \to \mathcal{O}(2) \to \mathcal{O}(2)_D \to 0$$

gives a coboundary map

$$\delta : H^0(D, \mathcal{O}(2)) \to H^1(S, \mathcal{O}) \tag{2.10}$$

The image of the principal parts of the differentials then spans a (real) 2-dimensional subspace U. This is going to be the *universal covering* of our 2-torus.

The aim is to construct, for each ζ, a flat connection on a certain rank 2 vector bundle over U, and show that this is of the form of Proposition 1. Now each $x \in U \subseteq H^1(S, \mathcal{O})$ defines by exponentiation (and choice of Poincaré bundle) a line bundle L_x of degree zero over S. The *vector bundle* arises by choosing a fixed line bundle E of degree $p+1$ on S (with some reality property we will not go into here) and defining the fibre at $x \in U$ to be

$$V_x = H^0(S, E \otimes L_x).$$

The *connection* on the bundle V is defined by means of parallel translation, and this involves the interpretation of the cohomology classes in the image of the map δ in (2.10). The coboundary construction of the cohomology class means that after exponentiating, the line bundle $L_x = \exp(\delta(a[\theta] + b[\tilde{\theta}]))$ is naturally trivial outside D. The trivialization extends to the whole of S only if we multiply by

$$\exp(\pm\zeta^{-1}(a\lambda_{-2} + b\tilde{\lambda}_{-2})) \tag{2.11}$$

in a neighbourhood of $\pi^{-1}(0)$ and analogous expressions at ∞. For $x, y \in U$ the ratio of these trivializations gives a non-vanishing section P_{xy} of $L_y \otimes L_x^*$ outside D. Now if the line bundle $E \otimes L_x(-1)$ of degree $p-1$ is non-special (i.e. $H^0(S, E \otimes L_x(-1)) = H^1(S, E \otimes L_x(-1)) = 0$), it is easy to see that restricting sections of $E \otimes L_x$ to $D_\zeta = \pi^{-1}(\zeta)$ is an isomorphism, so we have

$$V_x = H^0(S, E \otimes L_x) \cong H^0(D_\zeta, E \otimes L_x) \tag{2.12}$$

Furthermore, if $\zeta \neq 0, \infty$, then multiplication by P_{xy} defines an isomorphism

$$H^0(D_\zeta, E \otimes L_x) \cong H^0(D_\zeta, E \otimes L_y) \tag{2.13}$$

Putting (2.12) and (2.13) together, we have our definition of parallel translation $\Pi_{xy} : V_x \to V_y$. It is clearly independent of the path, and so for each $\zeta \neq 0, \infty$ we have a flat connection on V.

As ζ approaches 0, then (2.11) shows that the connection matrix acquires a simple pole in ζ. A similar consideration at $\zeta = \infty$ shows that the connection is of the form

$$\nabla + \zeta\Phi - \zeta^{-1}\Phi^*$$

as required.

This, in brief, is the construction. We start with the hyperelliptic curve, and for each line bundle E satisfying suitable reality constraints (which actually guarantee it is non-special—see [25]) we obtain a harmonic map from $U = \mathbf{R}^2$ to $SU(2)$. In order to obtain a map of the torus we need the connection to descend to a quotient of the vector space U, and for this we need $U \in H^1(S, \mathcal{O})$ to intersect $H^1(S, \mathbf{Z})$ in a lattice. An equivalent way of saying this is to insist that the periods of the differentials θ and $\tilde{\theta}$ should lie in $2\pi i\mathbf{Z}$, and this of course imposes severe constraints on the curve.

Even if those constraints are satisfied, we only get a harmonic map if the holonomy is trivial at $\zeta = \pm 1$. In fact, given that θ has periods in $2\pi i\mathbf{Z}$, we can write $\theta = dh/h$ for some holomorphic function on $S\backslash\pi^{-1}\{0, \infty\}$, and this 2-valued function on $\mathbf{P}^1\backslash\{0, \infty\}$ can, by examining the construction above, be seen to be the eigenvalue of the holonomy of the flat connection $\nabla + \zeta\Phi - \zeta^{-1}\Phi^*$ around one generator of the fundamental group of the torus. Similarly $\tilde{\theta}$ gives the holonomy for the other generator. The existence of a harmonic map thus requires this extra constraint, which can also be written in terms of periods of differentials using the reciprocity formula [25].

The algebraic curve method provides another construction in finite-dimensional terms, of harmonic maps from the torus. As with the loop group method, we need to prove its generality—that to every harmonic map from a torus to $SU(2)$ there exists a hyperelliptic curve.

2.6 Spectral curves

Let $f : M \to SU(2)$ be a harmonic map of a 2-torus. Where do we find an algebraic curve? The key idea is to consider the *holonomy* of the flat connection $\nabla + \zeta\Phi - \zeta^{-1}\Phi^*$. Since $\pi_1(M) = \mathbf{Z} \oplus \mathbf{Z}$, we have two holonomy matrices for each $\zeta \neq 0, \infty$ and so holomorphic functions $H, \tilde{H} : \mathbf{C}^* \to SL(2, \mathbf{C})$ which *commute.*

As we saw in the previous section, if the hyperelliptic curve construction applies, then the eigenvalues of the holonomy are single-valued holomorphic functions on $S \backslash \pi^{-1}\{0, \infty\}$, so to construct S we need to consider the eigenvalues

$$h = \frac{\operatorname{tr} H \pm \sqrt{(\operatorname{tr} H)^2 - 4}}{2}$$

This function has branch points where $(\operatorname{tr} H)^2 - 4$ has odd zeros, and the essential point for producing an algebraic curve is to show that there are only finitely many such points. As with the previous method, this finiteness will depend on solutions of an elliptic equation on the compact manifold M. In fact, we have to rule out first the case where the holonomy is trivial for all ζ, but this can rather readily be shown to correspond to the harmonic map defined by a holomorphic or antiholomorphic map to $\mathbf{P}^1$.

The important point to note here is that if $(\operatorname{tr} H)^2 - 4$ has an odd zero at $\zeta = \zeta_0$, so does $(\operatorname{tr} \tilde{H})^2 - 4$. To prove this, we work in the field K of fractions of the convergent power series in $(\zeta - \zeta_0)$. Then H and $\tilde{H}$ are 2×2 matrices with entries in the field. If $(\operatorname{tr} H)^2 - 4$ has an even zero, the eigenvalues of H are in K and are distinct since ζ_0 is an isolated zero. But since $\tilde{H}$ commutes with H, the eigenvectors are also eigenvectors of $\tilde{H}$ which thus has eigenvalues in K, and hence $(\operatorname{tr} \tilde{H})^2 - 4$ also has an even zero.

A consequence of this is that at an odd zero of $(\operatorname{tr} H)^2 - 4$, the eigenvalues of both H and $\tilde{H}$ are ± 1, and since they commute, there is a common eigenvector with eigenvalue 1 for some choice of $\pm H, \pm \tilde{H}$. We now interpret this fact in terms of flat connections. It means that after possibly tensoring with a flat unitary line bundle with holonomy ± 1, we have a global solution s to the equation

$$\nabla s + \zeta_0 \Phi s - \zeta_0^{-1} \Phi^* s = 0$$

and in particular to the elliptic equation

$$\nabla^{1,0} s + \zeta_0 \Phi s = 0$$

For $\|\zeta\| \leq 1$, $\nabla^{1,0} + \zeta\Phi$ is a holomorphic family of elliptic operators of index zero which therefore has a *determinant* which is a holomorphic function of ζ.

Being holomorphic, it vanishes at a finite number of points or identically. If the latter, then for $\zeta = e^{i\theta}$, the connection is flat and unitary and we can use a standard Weitzenböck argument (as in the theory of stable bundles) to deduce from $\nabla^{1,0}s + e^{i\theta}\Phi s = 0$ that s is covariant constant and thus the holonomy trivial. Since this is true for all θ, we are back in the trivial holonomy case. So there must be only finitely many odd zeros of $(\operatorname{tr} H)^2 - 4$ in the unit disc. Arguing similarly with the (0,1) part we get only finitely many outside the disc.

This is the essential finiteness property, which as the interested reader will find in [25], leads to the fact that any harmonic map from a torus to $SU(2)$ can be constructed from a (possibly singular) hyperelliptic curve—the *spectral curve*—in the manner of the previous section.

We have seen here two rather different methods of integrating the equations for a harmonic map of the torus. One is based on a polynomial with values in the Lie algebra, the other on line bundles over an algebraic curve. The two are in fact closely connected, and the link is provided by the following well-known result, to be found, for example in [5].

2.7 A basic result in integrable systems

Many classical (and not so classical) integrable systems can be linearized on the Jacobian of an algebraic curve. The fundamental idea behind this is the link between a line bundle over a certain curve and a matrix of polynomials. Applications to integrable systems concern the evolution of those matrices as the class of the line bundle follows a straight line on the Jacobian.

As usual, let $\mathcal{O}(d)$ be the line bundle of degree d over $\mathbf{P}^1$. We consider an element $A \in H^0(\mathbf{P}^1, \mathcal{O}(d)) \otimes \mathfrak{gl}(k)$, so that in terms of an affine coordinate ζ on $\mathbf{P}^1$, $A(\zeta)$ is simply a polynomial of degree d with coefficients which are $k \times k$ matrices. Let η denote the tautological section of $\pi^*\mathcal{O}(d)$ over the total space of $\mathcal{O}(d)$ and $S \subset \mathcal{O}(d)$ the curve defined by $\det(\eta - A(\zeta)) = 0$, the *spectral curve* of A. We then have the theorem [5]:

Theorem 1. *Suppose S is smooth, and let X be the space of all $B \in H^0(\mathbf{P}^1, \mathcal{O}(d)) \otimes \mathfrak{gl}(k)$ with spectral curve S. Then $PGL(k, \mathbf{C})$ acts freely on X by conjugation and the quotient can be identified with $J^{g-1}(S)\backslash\Theta$.*

Proof: In algebro-geometric language, a line bundle L on S is equivalent to a vector bundle $V = \pi_* L$ with the structure of a $\pi_*\mathcal{O}$-module. But this is the same thing as a homomorphism

$$A : V \to V(d)$$

satisfying the equation $P(A, \zeta) = 0$ where $\det(\eta - A(\zeta)) = P(\eta, \zeta)$. Since S is smooth and in particular irreducible, then by the Cayley-Hamilton theorem A has characteristic polynomial P. Now L is not contained in the theta divisor if and only if

$$H^0(S, L) = H^1(S, L) = 0$$

and from the functorial properties of the direct image, this is equivalent to

$$H^0(\mathbf{P}^1, V) = H^1(\mathbf{P}^1, V) = 0.$$

But from the Birkhoff-Grothendieck classification of bundles on the projective line, this means that $V \cong \mathcal{O}^k(-1)$, and so $\mathrm{Hom}(V,V) \cong \mathrm{Hom}(\mathcal{O}^k, \mathcal{O}^k)$ and we can thus interpret $A \in H^0(\mathbf{P}^1, \mathcal{O}(d)) \otimes \mathfrak{gl}(k)$

Remarks:
1. Note that the genus g of S is

$$g = \frac{1}{2}(k-1)(dk-2) \tag{2.14}$$

2. As described in [5], the assumption of smoothness is by no means necessary in the proof: if S is irreducible and reduced then we can repeat the argument using torsion-free rank 1 sheaves, and in the general case using invertible sheaves so long as A is regular (i.e. it has a k-dimensional space of commuting matrices) at each point.

The slick algebraic proof perhaps disguises the meaning of the correspondence, so let us spell it out. The matrix A has a single-valued eigenvalue η not on $\mathbf{P}^1$, but on the covering S. Over a point $\zeta \in \mathbf{P}^1$, the fibre V_ζ is by definition $H^0(D_\zeta, L)$ where D_ζ is the divisor $\pi^{-1}(\zeta)$. At a generic point the fibre consists of k distinct points $p_1, \ldots, p_k$ and we can find a basis of sections $s_1, \ldots, s_k$ with $s_i(p_j) = 0$ if $i \neq j$. This is a basis of eigenvectors of $A(\zeta)$.

Let us now compare the two methods of solving the harmonic map equations for a 2-torus to $SU(2)$. On the one hand, the loop group method produces a polynomial Killing field

$$\xi = \sum_{-d}^{d} (1 - \zeta^n)\xi_n$$

which we may regard as a section of $\mathcal{O}(2d) \otimes \mathfrak{sl}(2, \mathbf{C})$. Its coefficients depend on a point of the torus M. It satisfies the differential equation

$$d\xi + [A_\zeta, \xi] = 0$$

On the other hand, the spectral curve approach produces a hyperelliptic curve S with equation $\eta^2 = P(\zeta)$. Since P is a section of $\mathcal{O}(2p+2)$, the curve S naturally lies inside $\mathcal{O}(p+1)$. Moreover the points x of the torus in the construction correspond to line bundles $E \otimes L_x$ on S. It seems plausible that the two points of view may be linked by taking $d = p+1$ and $k = 2$. This is indeed so:

Proposition 2.2. *Let f be a harmonic map of a torus M to $SU(2)$. Then, if the spectral curve S is smooth, the polynomial Killing field at the point $x \in M$ is obtained by the procedure of Theorem 1 from the line bundle $L = EL_x(-1)$ over S.*

Proof: Recall that the construction of the vector bundle V over the torus gave

$$V_x \cong H^0(D_\zeta, EL_x) = \pi_*(EL_x)_\zeta = (\pi_*L)(1)_\zeta$$

so that $A(\zeta)$ acts naturally as an automorphism of V. We have to prove that it is covariant constant, that is commutes with parallel translation. But as we saw, parallel translation is defined by multiplying $H^0(D_\zeta, EL_x)$ by the section P_{xy} of $L_x^* L_y$. If $s \in H^0(D_\zeta, EL_x)$ vanishes at all but one point in D_ζ, then so does $P_{xy}s$ hence parallel translation preserves the eigenspaces of $A(\zeta)$. Equivalently A commutes with parallel translation and is therefore covariant constant for each ζ. It therefore satisfies the differential equation, and thus agrees, up to a constant, with ξ.

It follows that, given the polynomial Killing field, we obtain the spectral curve from its characteristic polynomial, and given the curve, Theorem 1 produces the Killing field. Seen in this light, the spectral curve construction (at least in the smooth case) clearly works for $G = SU(n)$ for general n.

2.8 Successes and failures

The two methods have different advantages, and it is clear that a synthesis of the two is the best way for further progress. On the one hand, our criteria of integrability are best satisfied by the algebraic curve approach, for here we have a general form of solution, the constants of integration being essentially the coefficients of the curve and the parameters of a point on its Jacobian. Our particularly relevant geometrical problem of harmonic maps to the torus involves choosing those coefficients to satisfy constraints (which are incidentally transcendental in nature and not at all easy to put into effect). On the other hand, it is most effective for the group $SU(2)$, or at best $SU(n)$, since the direct image of a line bundle constructs a vector bundle. For the other classical groups, the Jacobian is replaced by a Prym variety (see [24]), but for a general Lie group one has to consider more general abelian varieties as in [15]. By contrast, the Lie-algebraic formulation of the loop-group approach does not distinguish the linear from non-linear groups.

The generality of the methods is approached in different ways. In the spectral curve approach [25], the case where a is semisimple or nilpotent are treated side-by-side with no essential difference in method. The case of trivial holonomy was simply set aside to be dealt with by different methods. In geometrical terms nilpotency of a corresponds to a conformal harmonic map to $SU(2)$—so its image is a minimal surface—and the trivial holonomy case to a conformal map to S^2 This contrasts somewhat with the loop-group approach where the essential finiteness result seems to require semi-simplicity. Nevertheless, the loop-group approach seems to be more effective for higher rank groups.

One issue which arises in higher rank is the question of how one can incorporate other constructions of harmonic maps, those which factor through holomorphic maps to flag manifolds—the superminimal surfaces, for example—into the integrable system approach. Here the loop-group formalism really comes into its own (see [10], [38]).

What, then, putting both methods together, are the successes?

- There is now a reasonably general method for constructing harmonic tori. Under precisely stated conditions, *any* harmonic torus can be constructed by a particular ansatz.

- The methods allow one to construct explicit new examples, Wente's torus being but the first.

- Some non-existence results ensue due to parameter counting: in particular there are constraints on the conformal structure of a torus to be minimally immersed in S^3 [25], whereas Bryant showed using superminimal surfaces that it can always be minimally immersed in S^4.

- An unexpected feature is the essential presence of deformations. In the loop-group approach we obtain non-trivial Jacobi fields as part of the ansatz, but perhaps more importantly the algebraic curve approach shows that at least some of these integrate to give deformations through harmonic maps. This consists of the choice of the line bundle E on the spectral curve. If the polynomial P is of degree $2p + 2$, then curve S is of genus p, so its Jacobian is p-dimensional. The constraints for a harmonic map, severe though they are, are independent of the choice of E, so given one harmonic map, it possesses a p-dimensional family of deformations.

And what are the failures?

- The methods are restricted to the 2-torus (or, as in [8] to pluriharmonic maps of a complex torus of higher dimension). If M is a surface of higher genus, then of course we can still consider the holonomy of the flat connection $\nabla + \zeta\Phi - \zeta^{-1}\Phi^*$ around generators of the fundamental group, but it is not clear how much help that is since the group is now non-abelian. In particular, it was the existence of a commuting holonomy matrix in the torus case which gave finiteness for the branch points. More seriously, it is difficult to see how the surface itself, whose universal covering is now no longer a vector space, can appear in the context of integrable systems linearized on a Jacobian.

- Despite having full control of the parameters, it is very difficult to read off information of a differential-geometric nature from the algebraic-geometric origins. There are formulas for energy in terms of the expansions of the differentials [25], but even estimating the size of the energy from the hyperelliptic curve is currently impossible.

- Some of the longstanding questions in the subject remain unanswered, for example the conjecture of Hsiang and Lawson that the only *embedded* minimal torus in S^3 is the Clifford torus. All that the method offers in this direction, since the Clifford torus corresponds to the case $p = 0$, is the invitation to prove that an embedded torus has no deformations.

3 Four-dimensional Einstein manifolds

3.1 Background

Since the days of relativity, the search for 4-dimensional solutions to the Einstein equations $R_{ij} = \Lambda g_{ij}$ in both positive definite and Lorentzian signature has been a driving force in differential geometry. As a result of Yau's proof of the Calabi conjecture, and subsequent developments by Yau, Tian and others since, the 20th century approach of global existence theorems has produced many examples, but always restricted to the Kähler case. For a compact non locally-homogeneous example which is not Kähler, we only have the Page metric (see [6]) on the non-trivial 2-sphere bundle over S^2.

The non-compact case is less amenable to analysis, but offers more opportunities for constructions. Here again, both analysis and explicit construction are easier in the Kähler case. I want to indicate now how solutions of Painlevé's sixth equation can enter into the construction of some complete *non-Kählerian* Einstein metrics of negative scalar curvature on the unit ball in $\mathbf{R}^4$, providing deformations of both the hyperbolic metric and the Bergmann metric. The details are in [26].

The Painlevé equations are well-known not to be solvable in terms of "known functions", and require the introduction of the so-called Painlevé transcendants in general. Fortuitously, the problem discussed here does not need these and can be solved explicitly in terms of theta functions. The reason, as we shall see, is fundamentally associated with the sense in which these equations can be thought of as being integrable.

The starting point for this construction is the work of Tod [57], who directly approached the question of finding Einstein metrics in four dimensions which have self-dual Weyl tensor, and are invariant under the action of $SU(2)$. He showed that the conformal structure could be written, using the standard basis $\sigma_1, \sigma_2, \sigma_3$ of left-invariant 1-forms on $SU(2)$, in the form:

$$g_0 = \frac{ds^2}{s(1-s)} + \frac{\sigma_1^2}{\Omega_1^2} + \frac{(1-s)\sigma_2^2}{\Omega_2^2} + \frac{s\sigma_3^2}{\Omega_3^2}$$

where the functions Ω_i satisfy the differential equations:

$$\begin{aligned} \Omega_1' &= -\frac{\Omega_2\Omega_3}{s(1-s)} \\ \Omega_2' &= -\frac{\Omega_3\Omega_1}{s} \\ \Omega_3' &= -\frac{\Omega_1\Omega_2}{1-s}. \end{aligned} \tag{3.1}$$

and the value of $\Omega_1^2 - \Omega_2^2 - \Omega_3^2$, which is a constant as a consequence of (3.1), is $-1/4$. This is not the metric itself. The Einstein metric is $g = e^{2u}g_0$ and has scalar curvature 4Λ where

$$-4\Lambda e^{2u} = \frac{8s\Omega_1^2\Omega_2^2\Omega_3^2 + 2\Omega_1\Omega_2\Omega_3(s(\Omega_1^2+\Omega_2^2) - (1-4\Omega_3^2)(\Omega_2^2 - (1-s)\Omega_1^2))}{(s\Omega_1\Omega_2 + 2\Omega_3(\Omega_2^2 - (1-s)\Omega_1^2))^2} \tag{3.2}$$

This equation is one whose integrability we shall consider. Again, as is characteristic of the subject, its history goes back beyond Einstein and his equations to the 19th century study of *orthogonal coordinates.*

3.2 Orthogonal coordinates

Consider a metric given locally by

$$g = \frac{\partial \phi}{\partial x_1}dx_1^2 + \cdots + \frac{\partial \phi}{\partial x_n}dx_n^2$$

Clearly $x_1, \ldots, x_n$ are orthogonal coordinates, so if the metric is *flat* we have an orthogonal coordinate system on $\mathbf{R}^n$. Flatness leads to a nonlinear equation for ϕ, which can be put in various forms. Metrics of this type were studied by Darboux and Egorov (see [14, Chapter VIII], and [18]) at the turn of the century. In more recent times, such metrics arise in Dubrovin's theory [17] of Frobenius manifolds with its connections to conformal field theories and quantum cohomology. Such metrics have extra properties:

- $d\phi$ is covariant constant
- ϕ is homogeneous as a function of $x_1, \ldots, x_n$

The first condition is that the 1-form

$$d\phi = \frac{\partial \phi}{\partial x_1}dx_1 + \cdots + \frac{\partial \phi}{\partial x_n}dx_n$$

is covariant constant, which means that, using the metric, its dual vector field

$$X = \frac{\partial}{\partial x_1} + \cdots + \frac{\partial}{\partial x_n}$$

is covariant constant. In particular X is an infinitesimal isometry. It generates the $\mathbf{R}$-action $x_i \mapsto x_i + t$. The homogeneity condition clearly gives an action of $\mathbf{R}^*$ by conformal transformations, so we are considering a problem of orthogonal coordinates invariant under the two symmetries:

$$x_i \mapsto x_i + t$$
$$x_i \mapsto \lambda x_i$$

In positive definite signature, the degree of homogeneity of the coefficients of the metric must be zero, because X being covariant constant implies that $g(X, X)$ is a non-zero constant. Thus the coefficients are functions of $\mathbf{R}^n$ invariant under this two-parameter group. In three dimensions, this means that they are functions of the single variable

$$s = \frac{x_1 - x_2}{x_3 - x_2}$$

and this is where the differential equation (3.1) enters:

Proposition 3.1. *Let $\Omega_1, \Omega_2, \Omega_3$ be functions of $s = (x_1 - x_2)/(x_3 - x_2)$, then the metric*

$$g = -\Omega_1^{-2} dx_1^2 + \Omega_2^{-2} dx_2^2 + \Omega_3^{-2} dx_3^2$$

is flat iff $\Omega_1, \Omega_2, \Omega_3$ satisfy equation (3.1).

Darboux's 1910 book [14] on orthogonal coordinates has more pages than Besse's comprehensive 1987 book [6] on Einstein manifolds, which is perhaps an indicator of the importance of the subject at the time, but in some sense we are seeing here another manifestation of the same piece of mathematics in a different context. We face the same problem, though: how to solve the differential equation (3.1).

3.3 Isomonodromic deformations

The equation (3.1) is integrable because it can be reduced to the geometrical problem of isomonodromic deformations, which we describe briefly next.

Consider a meromorphic connection on a trivial bundle over $\mathbf{P}^1$ with connection form (in an affine coordinate z)

$$A = \sum_{i=1}^{n} \frac{A_i dz}{z - z_i}$$

An isomonodromic deformation $A_i(z_1, \ldots, z_n)$ for $(z_1, \ldots, z_n) \in U \subset \mathbf{C}^n$ is a family of such connections with constant holonomy (up to conjugation). It necessarily satisfies the Schlesinger equation [40]:

$$dA_i + \sum_{j \neq i} [A_i, A_j] \frac{dz_i - dz_j}{z_i - z_j} = 0. \tag{3.3}$$

Suppose we have four points $z_1, \dots, z_4$. By a projective transformation we can make these points $0, 1, s, \infty$. Then

$$A(z) = \frac{A_1}{z} + \frac{A_2}{z-1} + \frac{A_3}{z-s}$$

and Schlesinger's equation becomes:

$$\begin{aligned} \frac{dA_1}{ds} &= \frac{[A_3, A_1]}{s} \\ \frac{dA_2}{ds} &= \frac{[A_3, A_2]}{s-1} \\ \frac{dA_3}{ds} &= \frac{[A_1, A_3]}{s} + \frac{[A_2, A_3]}{s-1} \end{aligned} \tag{3.4}$$

where the last equation is equivalent to

$$A_1 + A_2 + A_3 = -A_4 = \text{const.}$$

Assume that the A_i are 2×2 matrices of trace 0. Note that from Schlesinger's equation (3.3),

$$d(\operatorname{tr} A_i^2) = -2 \sum_{j \neq i} \operatorname{tr}(A_i[A_i, A_j]) \frac{dz_i - dz_j}{z_i - z_j} = 0,$$

so that each $\operatorname{tr} A_i^2$ is independent of z_1, z_2, z_3, z_4, and is a constant of integration. By its very origin, the full constants of integration of this equation consist of the prescription of the holonomy: a representation of the fundamental group of the 4-punctured sphere in $SL(2, \mathbf{C})$. Since this group is a free group on three generators (taken to be loops passing once around three of the punctures), these constants are essentially the choice of a triple of elements M_1, M_2, M_3 in $SL(2, \mathbf{C})$ modulo conjugation. When the eigenvalues of A_j do not differ by an integer, the holonomy around a small loop surrounding the pole z_j is conjugate to

$$M_j = \exp(2\pi i A_j)$$

so that if $\operatorname{tr} A_j^2 = k$,

$$\operatorname{tr} M_j = 2\cos(\pi\sqrt{2k}) \tag{3.5}$$

The constancy of k is thus just part of the full constants of motion—invariants of the holonomy.

From the point of view of integrable systems, we have here the general solution. A specific one is determined by the choice of a holonomy representation. That is precisely what we shall do when we relate the equations for the Ω_i to isomonodromic deformations.

To make that relationship, we consider solutions to the isomonodromic deformation problem for which the holonomies for small curves surrounding the poles are all conjugate. This means that the residues A_i satisfy

$$\operatorname{tr} A_1^2 = \operatorname{tr} A_2^2 = \operatorname{tr} A_3^2 = \operatorname{tr}(A_1 + A_2 + A_3)^2 = k \tag{3.6}$$

Now if we set

$$\Omega_1^2 = -(k + \operatorname{tr} A_1 A_2) \quad \Omega_2^2 = (k + \operatorname{tr} A_2 A_3) \quad \Omega_3^2 = (k + \operatorname{tr} A_3 A_1) \tag{3.7}$$

The Schlesinger equation for isomonodromic deformations (3.4) shows that

$$\begin{aligned} 2\Omega_1 \frac{d\Omega_1}{ds} &= \frac{\operatorname{tr}([A_1, A_2]A_3)}{s(s-1)} \\ 2\Omega_2 \frac{d\Omega_2}{ds} &= -\frac{\operatorname{tr}([A_1, A_2]A_3)}{s} \\ 2\Omega_3 \frac{d\Omega_3}{ds} &= \frac{\operatorname{tr}([A_1, A_2]A_3)}{s-1} \end{aligned} \tag{3.8}$$

On the other hand, in the Lie algebra of $SL(2, \mathbf{C})$, we have the identity

$$(\operatorname{tr}([A_1, A_2]A_3))^2 = -2 \det \begin{pmatrix} \operatorname{tr} A_1^2 & \operatorname{tr} A_1 A_2 & \operatorname{tr} A_3 A_1 \\ \operatorname{tr} A_1 A_2 & \operatorname{tr} A_2^2 & \operatorname{tr} A_2 A_3 \\ \operatorname{tr} A_3 A_1 & \operatorname{tr} A_2 A_3 & \operatorname{tr} A_3^2 \end{pmatrix}$$

Expanding the determinant and using (3.6), we see from this that $\operatorname{tr}([A_1, A_2]A_3)$ is a distinguished square root of $4\Omega_1^2\Omega_2^2\Omega_3^2$, which gives the required equations (3.1):

$$\begin{aligned} \Omega_1' &= -\frac{\Omega_2\Omega_3}{s(1-s)} \\ \Omega_2' &= -\frac{\Omega_3\Omega_1}{s} \\ \Omega_3' &= -\frac{\Omega_1\Omega_2}{1-s}. \end{aligned}$$

Does this approach give explicit solutions? In some ways, the answer is no, for a simple substitution

$$\frac{dy}{dx} = \frac{y(y-1)(y-x)}{x(x-1)} \left(2z - \frac{1}{2y} - \frac{1}{2(y-1)} + \frac{1}{2(y-x)}\right)$$

(which defines the auxiliary variable z), and

$$\begin{aligned} \Omega_1^2 &= \frac{(y-x)^2 y(y-1)}{x(1-x)} \left(z - \frac{1}{2(y-1)}\right)\left(z - \frac{1}{2y}\right) \\ \Omega_2^2 &= \frac{y^2(y-1)(y-x)}{x} \left(z - \frac{1}{2(y-x)}\right)\left(z - \frac{1}{2(y-1)}\right) \\ \Omega_3^2 &= \frac{(y-1)^2 y(y-x)}{(1-x)} \left(z - \frac{1}{2y}\right)\left(z - \frac{1}{2(y-x)}\right). \end{aligned}$$

reduces the general isomonodromic deformation problem for four points to a Painlevé equation:

$$\frac{d^2y}{dx^2} = 1/2\left(\frac{1}{y}+\frac{1}{y-1}+\frac{1}{y-x}\right)\left(\frac{dy}{dx}\right)^2 - \left(\frac{1}{x}+\frac{1}{x-1}+\frac{1}{y-x}\right)\frac{dy}{dx} + \frac{y(y-1)(y-x)}{x^2(x-1)^2}\left(\alpha+\beta\frac{x}{y^2}+\gamma\frac{x-1}{(y-1)^2}+\delta\frac{x(x-1)}{(y-x)^2}\right)$$

and the general solutions of such equations are known to involve new transcendental functions. However, the specific problem on Einstein manifolds we began with concerned finding solutions to this equation with $\Omega_1^2 - \Omega_2^2 - \Omega_3^2 = -1/4$. This has a natural interpretation in terms of the holonomy, for

$$\Omega_1^2 - \Omega_2^2 - \Omega_3^2 = -3k - \operatorname{tr} A_1A_2 - \operatorname{tr} A_2A_3 - \operatorname{tr} A_3A_1 = -2k$$

and so we must take $k = 1/8$. From (3.5) this means that the holonomy around each pole has eigenvalues $\pm i$ and so is of order 4. We can use this fact to give explicit solutions.

The basic idea is that by taking a double covering of $\mathbf{P}^1$ branched over the four poles of the meromorphic connection, we pull back the connection to one on an elliptic curve E. This time the eigenvalues of the holonomy around each pole are ± 1 and since the connection has holonomy in $SL(2,\mathbf{C})$, this is a multiple of the identity. In $PSL(2,\mathbf{C})$ this is trivial, and so the holonomy extends to a representation of the fundamental group of the elliptic curve, which is abelian. We can now rewrite the connection in terms of meromorphic connections on line bundles and the holonomy is then defined by the periods of meromorphic differentials on the elliptic curve E. This allows us to find explicit solutions of the isomonodromic deformation problem, and consequently (see [26] for details) explicit forms for Einstein metrics. For information, the general solution to this particular Painlevé equation, where the coefficients in the equation are $\alpha = 1/8, \beta = -1/8, \gamma = 1/8, \delta = 3/8$, is

$$y(x) = \frac{\vartheta_1'''(0)}{3\pi^2\vartheta_4^4(0)\vartheta_1'(0)} + \frac{1}{3}\left(1+\frac{\vartheta_3^4(0)}{\vartheta_4^4(0)}\right) + \frac{\vartheta_1'''(\nu)\vartheta_1(\nu) - 2\vartheta_1''(\nu)\vartheta_1'(\nu) + 4\pi i c_1(\vartheta_1''(\nu)\vartheta(\nu) - \vartheta_1'^2(\nu))}{2\pi^2\vartheta_4^4(0)\vartheta_1(\nu)(\vartheta_1'(\nu) + 2\pi i c_1\vartheta_1(\nu))}$$

where $\nu = c_1\tau + c_2$ and $x = \vartheta_3^4(0)/\vartheta_4^4(0)$. It is then a matter of using estimates for theta functions to describe global properties of the resulting Einstein metrics in terms of the two constants of integration c_1 and c_2.

This is a classical approach to the problem, but just as with the case of harmonic tori, it gives some new examples and illuminates the area in a number of ways.

3.4 Metrics on the ball

The *complete* metrics produced this way are of different forms. There are two obvious ones—the well-known (and unique) *compact* self-dual Einstein manifolds S^4 and CP^2 with positive scalar curvature. The isomonodromic problem they correspond to has finite holonomy group: the binary dihedral group for a triangle and a square respectively (see [27]). The three matrices M_1, M_2, M_3 cover three reflections in the dihedral group.

In the case of zero scalar curvature the metric on the moduli space of 2-monopoles [3] is the unique complete metric and corresponds to the holonomy representation

$$M_1 = \begin{pmatrix} -i & i \\ 0 & i \end{pmatrix}, \quad M_2 = \begin{pmatrix} -i & i \\ 0 & i \end{pmatrix}, \quad M_3 = \begin{pmatrix} -i & 0 \\ 0 & i \end{pmatrix}.$$

The rest are metrics of negative scalar curvature, and correspond to representations of the following form:

$$M_1 = \begin{pmatrix} 0 & \lambda \\ -\lambda^{-1} & 0 \end{pmatrix}, \quad M_2 = \begin{pmatrix} 0 & \lambda e^{i\theta} \\ -\lambda^{-1} e^{-i\theta} & 0 \end{pmatrix}, \quad M_3 = \begin{pmatrix} 0 & e^{i\theta} \\ -e^{-i\theta} & 0 \end{pmatrix} \tag{3.9}$$

They live naturally on the unit ball in $\mathbf{R}^4$ and fall into two types. The first consists of taking a general value for λ and θ. It has the following properties:

- The conformal structure extends over the boundary 3-sphere, and induces there a left-invariant conformal structure on $SU(2)$. If we add in the hyperbolic metric and Pedersen's metrics [48], then we have a two-parameter family of self-dual Einstein metrics which induce *any* left-invariant conformal structure on the 3-sphere. Thus these conformal structures are, in the language of LeBrun [36] of "positive frequency". There are obstructions in general to the positive frequency condition, involving the eta-invariant of the boundary (see [28]).

- The metrics approach the hyperbolic metric near the boundary of the ball, but not fast enough to be covered by the rigidity theorem of Min-Oo [44] which would force it to be the hyperbolic metric itself. It is interesting to note that Min-Oo's approach is based on Witten's proof of the positive mass theorem, concerning zero scalar curvature. The Taub-NUT metric (see e.g. [6]) is a self-dual Einstein metric on $\mathbf{R}^4$ with zero scalar curvature which does not fall within the scope of Witten's theorem. Our metrics are in some sense hyperbolic analogues of this metric. Indeed, the Pedersen metrics (which have an extra degree of symmetry) can be obtained by a quaternionic-Kähler quotient construction which generalizes the hyperkähler quotient which yields the Taub-NUT metric.

- There are no quotients of these metrics by discrete groups: the central fixed point of the $SU(2)$ is distinguished by its local geometry.

- The full family is parametrized by the left-invariant conformal structures on the 3-sphere.

The second family corresponds to taking a special value $\theta = 0$ in the holonomy (3.9). It has the following features:

- One of the coefficients of σ_i^2 in the metric decays much faster than the others as the boundary 3-sphere is approached and so the conformal structure of the metric does not extend over the sphere. In the limit it induces a degenerate conformal structure, which we can think of as a CR-structure on the boundary. If we add in the Bergmann metric on the unit disc in $\mathbf{C}^2$, then we get a family which realizes every left-invariant CR-structure on S^3.
- Although the Bergmann metric is a Kähler metric, this is not so for any other member of the family.
- There are no quotients by discrete groups.
- The full family is parametrized by the left-invariant CR-structures on the 3-sphere. In fact, if $\mu\sigma_1^2 + \sigma_2^2$ represents this structure, with $\mu < 1$, we take the holonomy with
$$\lambda = \pm\sqrt{\frac{1+\mu}{1-\mu}}$$

4 Integrability and self-duality

4.1 Background

We remarked initially that among the distinguishing characteristics of 20th century mathematics is the goal of setting up unifying structures linking together disparate areas. The most noteworthy case for us concerns integrable systems and their relationship with the self-dual Yang-Mills equations. For a thorough account of this, we refer to the book of Mason and Woodhouse [42]. The essential idea is to see integrable systems as *dimensional reductions* of solutions to the equations
$$F_A = *F_A$$
for the curvature F_A of a connection A on a principal G-bundle P over $\mathbf{R}^4$. In fact, it is unwise to rely on the Euclidean signature alone. We may consider complex connections, or more commonly real ones which are self-dual in signature $(2, 2)$ (recall that there are no real self-dual 2-forms in Lorentzian signature since in that case $*^2 = -1$).

By "dimensional reduction" we mean that we take solutions which are invariant under a group H of diffeomorphisms of $\mathbf{R}^4$ which preserve the equations. Since the Hodge star operator is conformally invariant in the middle degree, this must be a group of conformal transformations, i.e. a subgroup of $SO(5, 1)$ for

Euclidean signature, or $SO(3,3)$ in the other case. If $\dim H = d < 4$, then the self-duality equations can be reinterpreted as equations in a $(4-d)$-dimensional quotient space, which in the case of translations consists of another vector space, but where the induced metric may be degenerate.

The equations on the quotient are not for just a connection, but involve extra data—the "Higgs fields". These arise, invariantly speaking, from the fact that H-invariance of the connection only has meaning if we have a lifting of the action of H from $\mathbf{R}^4$ to the principal bundle P. If X is a vector field on $\mathbf{R}^4$ generated by the action of H, then for any section s of a vector bundle V associated to P there is another section, the Lie derivative $\mathfrak{L}_X s$. If we have a connection, we can also define the covariant derivative $\nabla_X s$ and the difference $(\nabla_X - \mathfrak{L}_X)s$ is a zero-order operator: an endomorphism of V. If the connection is H-invariant, then this endomorphism is defined on the quotient space, and is called a Higgs field. Thus for a d-dimensional group H, we have d Higgs fields.

To relate the corresponding coupled equations for connections and Higgs fields to known integrable systems, two processes are involved. In the first place, since systems are not necessarily written in gauge-theoretical terms, some gauge choices have to be made. Secondly, a curious phenomenon takes place with a reduction to two dimensions: the equations become, after a suitable reinterpretation, conformally invariant. This has nothing to do with the conformal invariance in four dimensions, and is much stronger since there is an infinite dimensional pseudogroup of such transformations in two dimensions. In the case that the quotient metric is degenerate, this is the pseudogroup of Galilean transformations. What it means is that a reduction to a relatively few canonical models can be achieved.

In this manner, choosing G to be an appropriate real form of $SL(2,\mathbf{C})$, the KdV equation and non-linear Schrödinger equations appear as dimensional reductions, as pointed out by Mason and Sparling [41].

We have considered so far two examples of integrable systems arising in Riemannian geometry: harmonic maps of a surface and the isomonodromic deformation problem corresponding to Painlevé's sixth equation. We can see these now as dimensional reductions of the Yang-Mills equations.

4.2 Two examples

We begin with $\mathbf{R}^4$ with metric $g = dx_1^2 + dx_2^2 - dx_3^2 - dx_4^2$ and and volume form $dx_1 \wedge dx_2 \wedge dx_3 \wedge dx_4$. The Hodge star operator is then:

$$\begin{aligned} *dx_1 \wedge dx_2 &= dx_3 \wedge dx_4 \\ *dx_2 \wedge dx_3 &= -dx_4 \wedge dx_2 \\ *dx_1 \wedge dx_4 &= -dx_2 \wedge dx_3 \end{aligned}$$

We take for the group H the 2-dimensional group of translations

$$(x_1, x_2, x_3, x_4) \mapsto (x_1, x_2, x_3 + a_1, x_4 + a_2)$$

for $(a_1, a_2) \in \mathbf{R}^2$. The quotient space is $\mathbf{R}^2$ with Euclidean metric $dx_1^2 + dx_2^2$ and we have two Higgs fields $\phi_1(x_1, x_2)$ and $\phi_2(x_1, x_2)$. In these terms, the self-dual connection has connection form

$$A_1 dx_1 + A_2 dx_2 + \phi_1 dx_3 + \phi_2 dx_4$$

Equating the three anti-self-dual coefficients of the curvature to zero gives the three equations:

$$\begin{aligned} F_{12} &= [\phi_1, \phi_2] \\ \nabla_1 \phi_1 &= \nabla_2 \phi_2 \\ \nabla_1 \phi_2 &= -\nabla_2 \phi_1 \end{aligned}$$

Putting $\Phi = (\phi_1 + i\phi_2)d(x_1 + ix_2)$, we obtain the equations

$$\nabla^{0,1}\Phi = 0$$
$$F = [\Phi, \Phi^*]$$

for a harmonic map in the form of Proposition 1.

For the next example (following [42]) we pass to complex coordinates, setting $z = x_1 + ix_2, \tilde{z} = x_1 - ix_2, w = x_3 + ix_4, \tilde{w} = x_3 - ix_4$, so that the metric is given by

$$g = dz\, d\tilde{z} - dw\, d\tilde{w}.$$

In this case, the anti-self-dual 2-forms have as basis $dz \wedge dw, d\tilde{z} \wedge d\tilde{w}, dz \wedge d\tilde{z} - dw \wedge d\tilde{w}$.

Consider the 3-dimensional group $H = \mathbf{C}^* \times \mathbf{C}^* \times \mathbf{C}^*$ acting by conformal transformations as follows:

$$(z, \tilde{z}, w, \tilde{w}) \mapsto (\lambda\nu^{-1} z, \mu\tilde{z}, \lambda w, \mu\nu^{-1}\tilde{w})$$

The quotient space is 1-dimensional. Since $s = z\tilde{z}/w\tilde{w}$ is invariant under the group action, we can take it to be a parameter on the quotient. Now introduce coordinates

$$\begin{aligned} p &= -\log w \\ q &= -\log \tilde{z} \\ r &= \log(\tilde{w}/\tilde{z}) \end{aligned}$$

Under the group action these transform as

$$(p, q, r) \mapsto (p - \log \lambda, q - \log \mu, r - \log \nu)$$

and so an H-invariant connection defines Higgs fields P, Q, R which are Lie-algebra valued functions of s. In one dimension, a gauge transformation locally trivializes any connection, so there is a gauge for which the connection form is

$$Pdp + Qdq + Rdr.$$

In these coordinates the anti-self-dual 2-forms are spanned by

$$ds \wedge dp + s dr \wedge dp, \quad dq \wedge dr, \quad (s-1) dp \wedge dq + dp \wedge dr - ds \wedge dq$$

and the curvature is

$$P' ds \wedge dp + Q' ds \wedge dq + R' ds \wedge dr + [P,Q] dp \wedge dq + [Q,R] dq \wedge dr + [R,P] dr \wedge dp$$

For self-duality of the connection, the product with the anti-self-dual 2-forms must vanish and this leads to the three equations

$$\begin{aligned} P' &= 0 \\ Q' &= \frac{1}{s}[R,Q] \\ R' &= \frac{1}{(s-1)}[R,P] + \frac{1}{s(s-1)}[R,Q] \end{aligned}$$

and taking

$$\begin{aligned} P &= -A_1 - A_2 - A_3 \\ Q &= A_1 \\ R &= A_3 \end{aligned}$$

we obtain Schlesinger's equation in the form (3.4). So Schlesinger's equation, and the isomonodromic deformation problem for four singular points, is a dimensional reduction of the self-dual Yang-Mills equations.

The two concrete applications of integrable systems to problems in Riemannian geometry which we have considered thus arise in a natural way by choosing a group H of conformal transformations and studying the self-dual Yang-Mills equations invariant under H, for differing gauge groups G. Much more can be said, in particular with regard to the twistor methods of solving the equations, but that will take us too far afield. Suffice it to note that the indeterminate ζ in the flat connection $\nabla + \zeta\Phi - \zeta^{-1}\Phi^*$ for a harmonic map is essentially a complex parameter on a twistor line. Many of the standard, inverse scattering methods of solving integrable systems have a reinterpretation in twistor terms (see [42]).

The one feature which does emerge from this general point of view, is that we can't expect all of the interesting problems in Riemannian geometry to succumb to the method of integrable systems. As Ward has pointed out in [59], the self-dual Yang-Mills equations have the "Painlevé property" whereas the full Yang-Mills equations do not. By analogy, it would be surprising if the full Einstein equations could be solved by any integrable system method. In four dimensions, as we have seen, self-duality may lead to integrability, and we shall see later a very direct relationship between certain integrable systems and the construction of hyperkähler metrics in higher dimensions. For this, though, we need to study another dimensional reduction, that of Nahm's equations.

4.3 Nahm's equations

Take $\mathbf{R}^4$ with positive definite Euclidean metric $dx_0^2 + dx_1^2 + dx_2^2 + dx_3^2$ and consider the 3-dimensional group H of translations of the form

$$(x_0, x_1, x_2, x_3) \mapsto (x_0, x_1 + a_1, x_2 + a_2, x_3 + a_3)$$

An invariant connection now gives three Higgs fields T_1, T_2, T_3, functions of x_0, and, as in the previous example, trivializing the connection in one dimension, the connection form can be written $T_1 dx_1 + T_2 dx_2 + T_3 dx_3$. The curvature of the connection is

$$\begin{aligned} T_1' dx_0 \wedge dx_1 + T_2' dx_0 \wedge dx_2 + T_3' dx_0 \wedge dx_3 \\ + [T_1, T_2] dx_1 \wedge dx_2 + [T_2, T_3] dx_2 \wedge dx_3 + [T_3, T_1] dx_3 \wedge dx_1 \end{aligned}$$

The self-dual Yang-Mills equations now become *Nahm's equations*

$$\begin{aligned} T_1' &= [T_2, T_3] \\ T_2' &= [T_3, T_1] \\ T_3' &= [T_1, T_2] \end{aligned}$$

Since these are obtained by the action of a three-dimensional group of translations, it is not surprising that there is a close relationship to the equations for a harmonic map, where H was a two-dimensional translation group. The only difference is in the signature on the metric on $\mathbf{R}^4$. In fact, harmonic maps of a torus which are S^1-invariant reduce to the very similar equations

$$\begin{aligned} T_1' &= [T_2, T_3] \\ T_2' &= -[T_3, T_1] \\ T_3' &= -[T_1, T_2] \end{aligned}$$

which also arise in the theory of variations of Hodge structure [52]. Given that we can linearize the equations for harmonic maps of a torus on the Jacobian of a curve, it is not surprising that Nahm's equations can be too. We write, with an indeterminate ζ,

$$A(\zeta) = (T_2 + iT_3) - 2iT_1\zeta + (T_2 - iT_3)\zeta^2 \tag{4.1}$$

so that $A \in H^0(\mathbf{P}^1, \mathcal{O}(2)) \otimes \mathfrak{g}$. When $G = U(n)$, Proposition 1 tells us that, modulo overall conjugation, this corresponds to a line bundle L on the spectral curve $S \subset \mathcal{O}(2)$ defined by $\det(\eta - A(\zeta)) = 0$.

If we now set A_+ to be the polynomial part of $A\zeta^{-1}$, Nahm's equations become equivalent to the Lax pair (putting $s = x_0$)

$$\frac{dA}{ds} = [A, A_+]$$

As a consequence of the above Lax form, the spectral curve remains the same, and the line bundle evolves along a curve in the Jacobian, which in fact is a straight line. This is very similar to the case of harmonic maps from a torus, but there the points $\zeta = 0, \infty$ were distinguished. In the present case, because A is only quadratic in ζ, these points play no particular role. The direction in the Jacobian in which the straight line evolution takes place is determined by a canonical element in $H^1(S, \mathcal{O})$, (the tangent space to the Jacobian at any point). We take the canonical generator x of $H^1(\mathbf{P}^1, K) \cong H^1(\mathbf{P}^1, \mathcal{O}(-2))$ and the tautological element $\eta \in H^0(\mathcal{O}(2), \pi^*\mathcal{O}(2))$ and define $\eta x \in H^1(\mathcal{O}(2), \mathcal{O})$. Restricting ηx to $S \subset \mathcal{O}(2)$ gives a canonical element in $H^1(S, \mathcal{O})$.

The principal result (see e.g.[22]) is that (T_1, T_2, T_3) satisfy Nahm's equations if and only if the line bundle L_s evolves in a straight line on the Jacobian in the distinguished direction ηx.

Nahm's equations originated in the study of magnetic monopoles, but they have become a means of constructing concrete Einstein metrics in higher dimensions than four. We shall use the integrable systems approach to study these in some detail, and find explicit formulae.

5 Hyperkähler metrics

5.1 Background

It is now 20 years since Yau's proof of the Calabi conjecture. This theorem provided a great many compact manifolds satisfying the Einstein equation $R_{ij} = 0$. Given a compact Kähler manifold with first Chern class zero, the theorem asserts the existence of an essentially unique Kähler metric cohomologous to the initial one, but with zero Ricci tensor. The first examples, of K3 surfaces, are also the first examples of *hyperkähler manifolds*—Riemannian manifolds whose holonomy is contained in $Sp(n) \subseteq SU(2n)$, and, as pointed out in [4], a minor extension of Yau's theorem can be used to prove the existence of a hyperkähler metric on any compact Kähler manifold with a non-degenerate holomorphic 2-form. In particular, the Hilbert scheme $X^{[n]}$ of 0-cycles of length n on a K3 surface or an abelian surface X has a hyperkähler metric.

These existence proofs are impressive, especially given the state of affairs 25 years ago, when no complete non-trivial Ricci-flat metric was known to exist. They have provided a great source of information for algebraic geometers, in particular for studying moduli problems, but there are simple questions which they cannot hope to answer. For example, is the hyperkähler metric on the Hilbert scheme in any way locally defined by that on the K3 surface itself?

The existence theorems work best in the compact situation, and despite more than 15 years of study, the situation in the non-compact case—trying to put a Ricci-flat Kähler metric on the complement of an anticanonical divisor—is still not fully understood (see [56]). On the other hand, some construction methods have arisen over the last few years to give plenty of explicit hyperkähler metrics on noncompact manifolds. These are of three types:

- twistor space methods
- finite-dimensional hyperkähler quotients
- infinite-dimensional hyperkähler quotients

Sometimes all three methods can be used to give a single metric, for example the ubiquitous Taub-NUT metric on $\mathbf{R}^4$, whose twistor construction is given in [6] (Chapter 13) is expressed as a quotient of $\mathbf{C}^2 \times \mathbf{C}^2$ in [6] (Addendum E), and has recently appeared in the context of duality as the natural metric obtained by an infinite-dimensional quotient construction on a certain family of $SU(3)$ monopoles [37].

Among the infinite-dimensional quotient constructions is a series of metrics defined on coadjoint orbits of complex semisimple Lie groups. These ideas were initiated by Kronheimer [35] and developed more fully by Biquard [7] and Kovalev [33]. The construction makes use of Nahm's equations, as do other metrics of interest. In [34] Nahm's equations are used to put complete hyperkähler metrics on the total space of the cotangent bundle to a complex Lie group. Finally, Nakajima's result [45] show that the natural metrics on the moduli spaces of $SU(2)$ monopoles on $\mathbf{R}^3$ as described in [3] may also be obtained from the Nahm matrices which are used to construct the monopole. A more recent result of Nakajima and Takahasi [46], [55] shows that this applies also to the $SU(3)$ monopole metrics studied by Dancer [11].

As we have seen, Nahm's equations are solvable in terms of a linear flow on the Jacobian of a curve, and we might hope to be able to write down the hyperkähler metric explicitly using the data of the curve. In some sense, as we shall see, this is the case.

5.2 Hyperkähler quotients and Nahm's equations

Recall that a hyperkähler manifold is a Kähler manifold M with a nondegenerate covariant constant holomorphic 2-form ω^c. The real and imaginary parts of ω^c together with the Kähler form constitute a triple of closed 2-forms $\omega_1, \omega_2, \omega_3$, each one symplectic and satisfying some algebraic constraints. These constraints can be summed up as saying that the stabilizer of all three at each point is conjugate to the quaternionic unitary group $Sp(n) \subset GL(4n, \mathbf{R})$. Each form $\omega_1, \omega_2, \omega_3$ is the Kähler form of a complex structure I, J, K and these generate an action of the quaternions on the tangent bundle.

If G is a Lie group acting on M, preserving all three 2-forms, we have three moment maps, which can be collected into a single function:

$$\mu : M \to \mathfrak{g} \otimes \mathbf{R}^3$$

If $\mu^{-1}(0)$ is smooth, then the induced metric is G-invariant and descends to the quotient. The hyperkähler quotient construction [23] consists of the observation that this quotient metric is again hyperkähler. For most purposes, the initial space M is taken to be a flat quaternionic vector space, so that in this case the hyperkähler metric on the quotient is simply induced from the restriction of a Euclidean metric to a submanifold. This is quite explicit, except for the fact that the non-linear algebraic equation $\mu(m) = 0$ may not be easy to solve.

A standard class of examples can be obtained by taking a compact semisimple Lie group G, setting
$$M = \mathfrak{g} \otimes \mathbf{H}$$
and taking the adjoint action of G. The moment map equations are then

$$[A_0, A_1] = [A_2, A_3], \quad [A_0, A_2] = [A_3, A_1], \quad [A_0, A_3] = [A_1, A_2]$$

where $A \in M$ is $A = A_0 + iA_1 + jA_2 + kA_3$. An infinite-dimensional version of this is to consider the interval $[0,1]$ and a connection $A_0 = d/ds + B_0(s)$ on a trivial G-bundle, with A_1, A_2, A_3 replaced by Higgs fields $B_i : [0,1] \to \mathfrak{g}$. We can then consider the infinite-dimensional quaternionic affine space $\mathcal{A}$ of differential operators of the form

$$\frac{d}{ds} + B_0 + iB_1 + jB_2 + kB_3$$

as a hyperkähler manifold, using the $\mathcal{L}^2$ inner product. The appropriate group is now the infinite-dimensional gauge group $\mathcal{G}_0^0$ of smooth functions $g : [0,1] \to G$ such that $g(0) = g(1) = 1$. The moment map equations for this action read

$$\begin{aligned} B_1' + [B_0, B_1] &= [B_2, B_3] \\ B_2' + [B_0, B_2] &= [B_3, B_1] \\ B_3' + [B_0, B_3] &= [B_1, B_2] \end{aligned}$$

Formally speaking, we expect a hyperkähler metric to be induced on the quotient space. The appropriate analysis was carried out in [34] and gives a complete hyperkähler metric on the cotangent bundle of the complex group G^c. A related paper which describes properties of these metrics is [13].

The identification of the hyperkähler quotient as a cotangent bundle proceeds as follows: define
$$\alpha = B_0 - iB_1 \qquad \beta = B_2 + iB_3$$
and let $f : [0,1] \to G^c$ be the solution to the equation

$$\frac{df}{ds} = f\alpha \tag{5.1}$$

satisfying the initial condition $f(1) = 1$. Consider the map defined by

$$\psi(\alpha, \beta) = (f(0)^{-1}, \beta(1)) \tag{5.2}$$

Then because the group $\mathcal{G}_0^0$ consists of functions vanishing at the end-points, the map ψ is easily seen to be defined on the quotient, and as shown in Kronheimer's paper [34] (where the arguments are modelled on those of Donaldson [16]), this gives a diffeomorphism to $G^c \times \mathfrak{g}^c \cong T^*G^c$.

Now there is a unique gauge transformation $g : [0,1] \to G$ with $g(0) = 1$ such that

$$\frac{dg}{ds} = -B_0 g$$

and after applying this and putting $T_i = \mathrm{Ad}(g)B_i$, the moment map equations for B_i become Nahm's equations, so that in principle we only have to solve Nahm's equations to determine the zero set of the moment map and hence the metric.

The other uses of Nahm's equations to give hyperkähler metrics depend on different boundary conditions for the Nahm matrices, and we shall deal with these separately later.

5.3 Kähler potentials

It is easy to ask for an explicit form of a metric, but less easy to decide in what form one would really like it. When we ask for explicitness and receive it, it may not be what we really wanted, since the questions we pose initially may not be readily answered by using some complicated expression in transcendental functions. The examples of metrics on the ball in Section 3 are borderline in this respect: it is just possible to determine global behaviour of the metric, but it involves a mixture of expansions of theta functions—well-known because of their long lineage—and consequences of the differential equations they satisfy. So what would be a good answer for a Kähler metric on a manifold? Perhaps the simplest is to find a *Kähler potential*—a locally-defined function ϕ such that the Kähler form is expressible as

$$i\partial\bar{\partial}\phi = \omega$$

This is just a single function on the manifold, and as a consequence has an interpretation independent of coordinates. It disguises the fact that one needs to know the holomorphic coordinates as well in order to write down the metric, but it may well be that some properties of the manifold can be deduced from the potential itself. This is what we shall do for the metrics constructed from solutions of Nahm's equations. It turns out that there is a natural global Kähler potential for one of the complex structures of the hyperkähler family. This is a consequence of the following result [23]

Proposition 5.1. *Let M be a hyperkähler manifold with Kähler forms $\omega_1, \omega_2, \omega_3$ and suppose X is a vector field on M such that*

$$\mathcal{L}_X\omega_1 = 0, \quad \mathcal{L}_X\omega_2 = \omega_3, \quad \mathcal{L}_X\omega_3 = -\omega_2$$

Let μ be the moment map for X with respect to ω_1, then 2μ is a Kähler potential for the complex structure J.

Proof: By definition, μ satisfies $d\mu = \iota(X)\omega_1$. Using the complex structure J, we have

$$d\mu(JY) = (\partial_J\mu + \bar{\partial}_J\mu)(JY) = i(\partial_J\mu - \bar{\partial}_J\mu)(Y)$$

But we also have

$$d\mu(JY) = \iota(X)\omega_1(JY) = g(IX, JY) = g(KX, Y) = \omega_3(X, Y)$$

Hence

$$\iota(X)\omega_3 = i(\partial_J\mu - \bar{\partial}_J\mu)$$

and so

$$-2i\partial_J\bar{\partial}_J\mu = d(\iota(X)\omega_3) = \mathfrak{L}_X\omega_3 = -\omega_2$$

giving, as required,

$$\omega_2 = 2i\partial_J\bar{\partial}_J\mu$$

Remarks:
1. Note that so long as the moment map μ is globally defined (and this will certainly be true if $b_1(M) = 0$), so is the Kähler potential. This has serious implications for the complex structure J. In particular, since the Kähler form is cohomologically trivial, there can be no compact complex subvarieties.
2. A circle action which acts non-trivially on the 3-dimensional space of covariant constant 2-forms spanned by $\omega_1, \omega_2, \omega_3$ will, after some orthogonal change of basis, always be of the above form. The vector field X generating it is normalized by the conditions $\mathfrak{L}_X\omega_2 = \omega_3, \quad \mathfrak{L}_X\omega_3 = -\omega_2$.
3. It is clear from the symmetry of the problem that the complex structure J is not determined by the circle action, and any complex structure $\cos\theta J + \sin\theta K$ orthogonal to I in the 2-sphere of all complex structures will share the same Kähler potential 2μ.

In the case of Nahm's equations, there is an obvious action of $SO(3)$ on the space of (B_0, B_1, B_2, B_3) given by

$$B_0 \mapsto B_0, \quad B_i \mapsto \sum_i^3 P_{ij}B_j$$

where $P \in SO(3)$. This rotates the Kähler forms, as does the action which descends to the quotient. Choosing the $SO(2)$ subgroup which leaves fixed the Kähler form ω_1 then gives an action which differentiates to a vector field X of precisely the nature of Proposition 5.1. Thus Kronheimer's metric on T^*G^c has a globally defined Kähler potential for one of the complex structures. In fact, since the $SO(3)$ action acts transitively on the hyperkähler complex structures, these are all equivalent to the standard one.

The chosen circle action acts trivially on $\alpha = B_0 - iB_1$ and takes $\beta = B_2 + iB_3$ to $e^{i\theta}\beta$. In terms of the parametrization (5.2) above, this is just scalar multiplication by $e^{i\theta}$ in the fibres of the cotangent bundle.

Now on the affine space $\mathcal{A}$, the vector field X generated by the circle action is given by

$$X = jB_3 - kB_2$$

On a Kähler manifold, the moment map μ for the Hamiltonian vector field X satisfies $\operatorname{grad}\mu = IX$, but since

$$IX = jB_2 + kB_3$$

we see that for $A \in \mathcal{A}$,

$$\mu(A) = \frac{1}{2}\int_0^1 \langle B_2, B_2\rangle + \langle B_3, B_3\rangle ds$$

This descends to the quotient, and so defines a potential for the metric there.

The inner product on the Lie algebra is Ad-invariant, so we may equally write μ as a function on the space of solutions (T_1, T_2, T_3) of Nahm's equations:

$$\mu(A) = \frac{1}{2}\int_0^1 \langle T_2, T_2\rangle + \langle T_3, T_3\rangle ds \qquad (5.3)$$

The challenge now is to express this in terms of the data which yields the solution to Nahm's equations for the group $G = SU(n)$: a family of line bundles over an algebraic curve.

5.4 Theta functions

We consider the integrand in the formula (5.3) in the case that $G = SU(n)$. With the invariant inner product $\langle U, U\rangle = \operatorname{tr} UU^* = -\operatorname{tr} U^2$, this can be expressed in terms of

$$\operatorname{tr}(T_2^2 + T_3^2).$$

Since this does not involve the adjoint, it makes sense for arbitrary $T_i \in \mathfrak{sl}(k, \mathbf{C})$.

Consider then what $\operatorname{tr}(T_2^2 + T_3^2)$ expresses: a conjugation-invariant function on the space of triples (T_1, T_2, T_3) of matrices of trace zero. If we put these together, as in (4.1)

$$\begin{aligned} A(\zeta) &= (T_2 + iT_3) - 2iT_1\zeta + (T_2 - iT_3)\zeta^2 \\ &= A_0 + A_1\zeta + A_2\zeta^2 \end{aligned}$$

we obtain an element of $H^0(\mathbf{P}^1, \mathcal{O}(2)) \otimes \mathfrak{gl}(k)$. According to Theorem 1, a dense open set in the space of triples modulo conjugation corresponds to the choice of a non-singular curve S

$$\det(\eta - A(\zeta)) = \eta^k + a_1(\zeta)\eta^{k-1} + a_2(\zeta)\eta^{k-2} \ldots + a_k(\zeta) = 0$$

in $\mathcal{O}(2)$ and a point in the Jacobian of line bundles of degree $g-1$, not in the theta divisor Θ. Since $A(\zeta)$ has trace zero, in fact the coefficient $a_1(\zeta)$ vanishes. From the algebraic nature of the correspondence, $\operatorname{tr}(T_2^2+T_3^2)$ is a meromorphic function on $J^{g-1}(S)$ having a pole on Θ, and our task is to determine what it is.

There are quadratic functions of $(T_1.T_2.T_3)$ which are *constant* on the torus: the coefficients of the polynomial $a_2(\zeta)$ in the formula for the spectral curve. In fact,

$$\begin{aligned} a_2(\zeta) &= -\frac{1}{2}\operatorname{tr} A(\zeta)^2 \\ &= c_0 + c_1\zeta + c_2\zeta^2 + c_3\zeta^3 + c_4\zeta^4 \end{aligned}$$

so that

$$\begin{aligned} c_0 + c_4 &= \operatorname{tr}(T_3^2) - \operatorname{tr}(T_2^2) \\ c_0 - c_4 &= -2i \operatorname{tr} T_2T_3 \\ c_1 + c_3 &= 2i \operatorname{tr} T_1T_2 \\ c_1 - c_3 &= -2 \operatorname{tr} T_3T_1 \\ c_2 &= 2\operatorname{tr}(T_1^2) - \operatorname{tr}(T_2^2) - \operatorname{tr}(T_3^2) \end{aligned} \tag{5.4}$$

It is clear that $\operatorname{tr}(T_2^2+T_3^2)$ is not in the space spanned by these coefficients, and indeed it is not constant on the Jacobian. Nevertheless, it is more natural to ask for the more symmetrical invariant

$$\Delta = \operatorname{tr}(T_1^2+T_2^2+T_3^2)$$

and we can then obtain

$$\operatorname{tr}(T_2^2+T_3^2) = \frac{1}{3}(2\Delta - c_2) \tag{5.5}$$

Before we give the answer, recall some basic features of $J^{g-1}(S)$. It has a distinguished linear vector field $\mathcal{X} = \eta x \in H^1(S, \mathcal{O})$. ¿From (2.14), the genus of the spectral curve is $g = (k-1)^2$, so $\pi^*\mathcal{O}(k-2)$, which has degree $k(k-2) = (k-1)^2 - 1 = g-1$ is a distinguished point, which we can use to identify with $J(S)$. Furthermore, it is easy to see [22] that $K_S \cong \pi^*\mathcal{O}(2k-4)$, so the distinguished origin is a theta-characteristic: a line bundle L such that $L^2 \cong K$. Finally, since $\mathcal{O}(k-2)$ has $k-1$ sections on $\mathbf{P}^1$ (and usually even more on S itself), the distinguished point always lies on the theta divisor, and if $k > 2$ it is a singular point. With these preliminaries, we can state:

Theorem 2. *Let (T_1, T_2, T_2) be a triple of trace-free $k \times k$ matrices for which the spectral curve S is smooth. Then*

$$\Delta = \operatorname{tr}(T_1^2+T_2^2+T_3^2) = \frac{3}{2}\frac{d^2}{ds^2}\log\vartheta + c$$

where ϑ is the Riemann theta function translated by a half-period, d/ds denotes the derivative along the vector field $\mathcal{X}$, and the constant c is given by

$$c = -\frac{3}{(N+2)(N+1)} \frac{\vartheta^{(N+2)}(0)}{\vartheta^{(N)}(0)}$$

with $N = k(k^2-1)/6$.

Proof: First recall that the Riemann theta function is a holomorphic function on the universal covering $\mathbf{C}^g$ of $J(S)$. We may write

$$\vartheta(z) = \sum_{m \in \mathbf{Z}^g} \exp[\pi i(\langle Bm, m\rangle + 2\langle z, m\rangle)]$$

where A_i, B_i are the integrals of the holomorphic differentials on the curve over a canonical basis, and B_i are the columns of the matrix B. The Riemann theta function depends on the choice of canonical basis, which in turn determines a theta characteristic, providing an isomorphism $J(S) \cong J^{g-1}(S)$. This isomorphism carries the theta divisor $\Theta \subset J^{g-1}$ to the zero set of the theta function ϑ. In our geometrical problem, we have another theta characteristic, independent of the choice of canonical basis, and the difference is a half-period. We shall continue to denote this translate by ϑ.

The theta function satisfies the basic properties:

$$\vartheta(z + A_i) = \vartheta(z) \tag{5.6}$$

$$\vartheta(z + n_1 B_1 + \cdots + n_g B_g) = \exp[-\pi i(\langle Bn, n\rangle + 2\langle z, n\rangle)]\vartheta(z) \tag{5.7}$$

It is a consequence of these relations that

$$\frac{\partial^2}{\partial z_i \partial z_j} \log \vartheta$$

is invariant under translation by the lattice generated by A_i, B_i and hence is a meromorphic function on $J(S)$. In particular, so is

$$\frac{d^2}{ds^2} \log \vartheta = \sum_{i,j} \mathcal{X}_i \mathcal{X}_j \frac{\partial^2}{\partial z_i \partial z_j} \log \vartheta$$

Both Δ and $(\log \vartheta)''$ are meromorphic with poles along the theta-divisor. The strategy of proof will be to show that, taking an appropriate multiple of Δ, the principal parts coincide, so the difference is holomorphic and constant. A calculation at the distinguished origin will then evaluate the constant. To analyze the pole, we shall use Nahm's equations.

We begin by considering a generic smooth point of Θ where the vector field $\mathcal{X}$ is transversal. From [5], the holomorphic structure on the vector bundle $V(1)$

jumps from $\mathcal{O}^k$ to $\mathcal{O}(1) \oplus \mathcal{O}(-1) \oplus \mathcal{O}^{k-2}$. We need to investigate the behaviour of $A(\zeta)$ near this point. It acquires a singularity clearly from Theorem 1, but its precise nature requires further investigation.

In fact, in [22] (§5), it is shown that at the origin in $J(S)$, a solution to Nahm's equations acquires a simple pole and thus does so at a general point of Θ. Expanding, we have

$$T_i(s) = \frac{\rho_i}{s} + \sigma_i + \tau_i s \dots$$

and applying Nahm's equations about any simple pole gives the three sets of relations:

$$-\rho_1 = [\rho_2, \rho_3] \tag{5.8}$$
$$0 = [\rho_2, \sigma_3] + [\sigma_2, \rho_3] \tag{5.9}$$
$$\tau_1 = [\tau_2, \rho_3] + [\sigma_2, \sigma_3] + [\rho_2, \tau_3] \tag{5.10}$$

and similar expressions obtained by cyclic permutation of the indices. From (5.8), the residues ρ_i define a k-dimensional representation of $SL(2, \mathbf{C})$. At the origin, the purpose of the long argument in [22] is to show that this is the unique irreducible k-dimensional representation, a fact we shall need at a later stage.

Consider then a solution of Nahm's equations which acquires a pole as $s \to 0$, for which the line bundle L_s approaches a *smooth* point of Θ. The construction of Theorem 1 always yields a vector bundle V_s on $\mathbf{P}^1$ and a homomorphism $A_s : V_s \to V_s(2)$. The bundle $V_s(1)$ is trivial for $s \neq 0$. At $s = 0$, $V_0(1) \cong \mathcal{O}(1) \oplus \mathcal{O}(-1) \oplus \mathcal{O}^{k-2}$, and then $V_0(1)$ still has a k-dimensional space of sections, just as in the trivial case. It follows that we can choose a basis $v_1(\zeta, s), \dots, v_k(\zeta, s)$ of sections of V_s, holomorphic for s in a neighbourhood of 0. For $s \neq 0$ these span the fibre at each point $\zeta \in \mathbf{P}^1$, but for $s = 0$, they all lie in the subbundle $\mathcal{O}(1) \oplus \mathcal{O}^{k-2}$. Now since the Nahm matrices acquire a simple pole,

$$A_s(\zeta) v_i(\zeta, s) = \sum_j \frac{R(\zeta)_{ji} v_j(\zeta, s)}{s} + \dots$$

and so since this is finite at $s = 0$,

$$\sum_j R(\zeta)_{ji} v_j(\zeta, 0) = 0$$

But $v_1(\zeta, 0), \dots, v_k(\zeta, 0)$ span a $(k-1)$-dimensional subspace, so the rank of $R(\zeta)$ is 1. The residues of the Nahm matrices thus define a representation for which

$$R(\zeta) = (\rho_2 + i\rho_3) - 2i\rho_1\zeta + (\rho_2 - i\rho_3)\zeta^2$$

has rank 1. But on the irreducible representation of dimension n this has rank $n - 1$, so the representation is the sum of a $(k-2)$-dimensional trivial representation and the irreducible 2-dimensional one. The residues thus only have a pole on the 2-dimensional component, and this, up to conjugacy, is

$$\rho_1 = -\frac{1}{2}\begin{pmatrix} i & 0 \\ 0 & -i \end{pmatrix} \quad \rho_2 = -\frac{1}{2}\begin{pmatrix} 0 & 1 \\ -1 & 0 \end{pmatrix} \quad \rho_3 = -\frac{1}{2}\begin{pmatrix} 0 & i \\ i & 0 \end{pmatrix}$$

It follows that

$$\Delta(s) = \operatorname{tr}(T_1^2 + T_2^2 + T_3^2)(s) = -\frac{3}{2s^2} + \frac{c_{-1}}{s} + \ldots \tag{5.11}$$

We now need to determine the constant

$$c_{-1} = 2\sum_i \operatorname{tr}(\rho_i \sigma_i)$$

Note that from Nahm's equations (5.9), we have

$$[\rho_2, \sigma_3] + [\sigma_2, \rho_3] = 0 \quad \text{etc.}$$

and so

$$[\rho_1, [\rho_2, \sigma_3]] = [\rho_1, [\rho_3, \sigma_2]]$$

But from the Jacobi identity,

$$\begin{aligned}[\rho_1, [\rho_2, \sigma_3]] &= -[\sigma_3, [\rho_1, \rho_2]] - [\rho_2, [\sigma_3, \rho_1]] \\ &= [\sigma_3, \rho_3] + [\rho_2, [\rho_1, \sigma_3]]\end{aligned}$$

Doing a similar calculation on the right hand side, we obtain

$$\begin{aligned}[\sigma_3, \rho_3] + [\sigma_2, \rho_2] &= -[\rho_2, [\rho_1, \sigma_3]] + [\rho_3, [\rho_2, \sigma_1]] \\ &= -[\rho_2, [\rho_3, \sigma_1]] + [\rho_3, [\rho_2, \sigma_1]] \\ &= -[\sigma_1, \rho_1]\end{aligned}$$

As a consequence of this relation

$$\begin{aligned}0 &= [\rho_1, [\rho_1, \sigma_1]] + [\rho_1, [\rho_2, \sigma_2]] + [\rho_1, [\rho_3, \sigma_3]] \\ &= [\rho_1, [\rho_1, \sigma_1]] - [\rho_2, [\sigma_2, \rho_1]] - [\sigma_2, [\rho_1, \rho_2]] - [\rho_3, [\sigma_3, \rho_1]] - [\sigma_3, [\rho_1, \rho_3]] \\ &= [\rho_1, [\rho_1, \sigma_1]] - [\rho_2, [\sigma_1, \rho_2]] + [\sigma_2, \rho_3] - [\rho_3, [\sigma_1, \rho_3]] - [\sigma_3, \rho_2] \\ &= ((\operatorname{ad}\rho_1)^2 + (\operatorname{ad}\rho_2)^2 + (\operatorname{ad}\rho_3)^2)\sigma_1\end{aligned}$$

Consider for a moment the case of any simple pole for Nahm's equations. The residues ρ_1, ρ_2, ρ_3 give $\mathbf{C}^k$ the structure of a representation space for $SL(2, \mathbf{C})$. We decompose $\mathbf{C}^k = \bigoplus_n E_n$ where E_n is an irreducible representation space. Since from the calculation above, $((\operatorname{ad}\rho_1)^2 + (\operatorname{ad}\rho_2)^2 + (\operatorname{ad}\rho_3)^2)\sigma_j = 0$, σ_j commutes with each ρ_i. Its component in each $\operatorname{Hom}(E_n, E_n)$ is thus a scalar by irreducibility. It follows that $\mathrm{tr}(\rho_i\sigma_i) = 0$ since $\mathrm{tr}(\rho_i) = 0$. Hence the coefficient of s^{-1} vanishes and

$$\Delta(s) = -\frac{3}{2s^2} + c_0 + c_1 s + \dots \tag{5.12}$$

Now compare this with the Riemann theta function ϑ. We know that this has a simple zero at a smooth point on Θ, so in terms of the parameter s, we have an expansion

$$\vartheta(s) = s(a_0 + a_1 s + a_2 s^2 + \dots)$$

so

$$\frac{d^2}{ds^2}\log\vartheta = -\frac{1}{s^2} + \frac{2a_0a_2 - a_1^2}{a_0^2} + \dots$$

Hence from (5.12),

$$\Delta - \frac{3}{2}\frac{d^2}{ds^2}\log\vartheta$$

is holomorphic on $J(S)$ and so is equal to a constant, depending on the curve S.

To evaluate the constant, we focus attention on the origin. Here, as proved in [22], the residues ρ_i define the k-dimensional irreducible representation of $SL(2, \mathbf{C})$. Thus $\mathbf{C}^k = E_1$, and by the argument above each σ_i is a scalar. However, $\mathrm{tr}\,\sigma_i = 0$ so

$$\sigma_i = 0$$

Now consider the constant term

$$\sum_i (2\,\mathrm{tr}(\rho_i\tau_i) + \mathrm{tr}(\sigma_i)^2) = 2\sum_i \mathrm{tr}(\rho_i\tau_i)$$

in the expansion of Δ at the origin. From (5.10), with $\sigma_i = 0$, we have

$$\tau_1 = [\tau_2, \rho_3] + [\rho_2, \tau_3]$$

and so

$$\begin{aligned}\mathrm{tr}(\rho_1\tau_1) &= \mathrm{tr}(\rho_1[\tau_2, \rho_3]) + \mathrm{tr}(\rho_1[\rho_2, \tau_3]) \\ &= \mathrm{tr}([\rho_3, \rho_1]\tau_2) + \mathrm{tr}([\rho_1, \rho_2]\tau_3) \\ &= -\,\mathrm{tr}(\rho_2\tau_2) - \mathrm{tr}(\rho_3\tau_3)\end{aligned}$$

from the definition of representation (5.8). Consequently

$$\sum_i \mathrm{tr}(\rho_i\tau_i) = 0$$

and Δ has no constant term at the origin.

Compare this with the expansion of ϑ at the origin. Since this is a singular point of the theta-divisor, ϑ vanishes to high order N. Since the origin is a theta-characteristic, our translate ϑ of the theta function is either odd or even and so

$$\vartheta(s) = s^N(a_0 + a_2 s^2 + \dots)$$

where

$$a_0 = \frac{1}{N!}\vartheta^{(N)}(0), \quad a_2 = \frac{1}{(N+2)!}\vartheta^{(N+2)}(0)$$

hence

$$\frac{d^2}{ds^2}\log\vartheta(s) = -\frac{N}{s} + \frac{2\vartheta^{(N+2)}(0)}{(N+1)(N+2)\vartheta^{(N)}(0)} + \dots$$

Now for the k-dimensional irreducible representation,

$$\sum_i \mathrm{tr}(\rho_i^2) = -\frac{k(k^2-1)}{4}$$

so near the origin,

$$\Delta(s) = -\frac{k(k^2-1)}{4s^2} + \dots$$

and

$$\frac{3}{2}\frac{d^2}{ds^2}\log\vartheta = -\frac{3N}{2s^2} + \dots$$

Hence since

$$\Delta - \frac{3}{2}\frac{d^2}{ds^2}\log\vartheta$$

is constant, we must have $N = k(k^2-1)/6$ and finally

$$\Delta = \frac{3}{2}\frac{d^2}{ds^2}\log\vartheta - \frac{3}{(N+1)(N+2)}\frac{\vartheta^{(N+2)}(0)}{\vartheta^{(N)}(0)}$$

as required.

Example:
When $k = 2$, the curve S is elliptic, and we can identify $J(S)$ with S itself. The theta divisor is a single point, which we take as the origin, and the translate of the theta function which vanishes at the origin is traditionally called ϑ_1, and regarded as a function of $\nu = u/2\omega_1$. The classical formula for the Weierstrass zeta function

$$\zeta(u) = \frac{\eta_1 u}{\omega_1} + \frac{1}{2\omega_1}\frac{\vartheta_1'(\nu)}{\vartheta_1(\nu)} \tag{5.13}$$

shows that

$$-\wp(u) = \zeta'(u) = \frac{\eta_1}{\omega_1} + \frac{d^2}{du^2}\log\vartheta_1$$

and as is well known, the Weierstrass $\wp$-function has an expansion at the origin

$$\wp(u) = \frac{1}{u^2} + c_2 u^2 + \dots$$

Thus there is no constant term for $\wp$. Now $u = \kappa s$ for some constant κ. This constant is a function of the coefficients of the spectral curve: it expresses the canonical vector field $\mathcal{X}$ in terms of the Weierstrass vector field d/du. As a consequence of the theorem, we can say that

$$\operatorname{tr}(T_1^2 + T_2^2 + T_3^2)(s) = -\frac{3}{2}\kappa^2 \wp(\kappa s) \tag{5.14}$$

The constant κ can be evaluated by referring to Hurtubise's calculation [29]. The spectral curve can be reduced by a rotation to the canonical form

$$\eta^2 = r_1\zeta^3 - r_2\zeta^2 - r_1\zeta$$

with r_1, r_2 real and $r_1 \geq 0$. In this form $\kappa = r_1^{1/2}$.

If we return to the question of Kähler potentials, then Theorem 2 provides an answer for Kronheimer's metrics on T^*G^c. From Proposition 5.1 and (5.3) and (5.5) we have

$$\begin{aligned} \phi = 2\mu(A) &= -\int_0^1 \operatorname{tr}(T_2^2 + T_3^2)ds \\ &= -\int_0^1 \frac{1}{3}(2\Delta - c_2) \end{aligned}$$

so from Theorem 2, we have

$$\phi = \int_0^1 -\frac{d^2}{ds^2}(\log\theta)ds + \frac{2\vartheta^{N+2}(0)}{(N+1)(N+2)\vartheta^N(0)} + \frac{c_2}{3}$$

Before we state this as a theorem, we need to understand better the role of the coefficient c_2 in the formula. As it stands, the formula for the Kähler potential is the sum of two terms. The first, involving the theta function, depends only on the modulus of the curve, and so is invariant under the $SO(3)$ action

$$B_0 \mapsto B_0, \quad B_i \mapsto \sum_i^3 P_{ij}B_j$$

The second term consists essentially of the coefficient of ζ^2 in the quartic polynomial $a_2(\zeta)$ appearing in the formula for the spectral curve. This is picked out by the action of the circle subgroup in $SO(3)$ which preserves the complex structure I. Invariantly speaking, $a_2(\zeta)$ lies in the vector space $H^0(\mathbf{P}^1, \mathcal{O}(4))$. The circle action generates the vector field $X = \zeta d/d\zeta$, a holomorphic section of the tangent bundle $\mathcal{O}(2)$, and so $X^2 \in H^0(\mathbf{P}^1, \mathcal{O}(4))$. But the space $H^0(\mathbf{P}^1, \mathcal{O}(4))$ is an irreducible representation space for $SL(2, \mathbf{C})$ and has an invariant inner product. If $a \in H^0(\mathbf{P}^1, \mathcal{O}(4))$ is written as

$$a = c_0 + c_1\zeta + c_2\zeta^2 + c_3\zeta^3 + c_4\zeta^4$$

then the inner product is

$$\langle a, a\rangle = 12c_0c_4 - 3c_1c_3 + c_2^2$$

From this point of view, the coefficient c_2 has the invariant meaning

$$c_2 = \langle a_2, X^2\rangle$$

We are aiming to give a formula for the Kähler potential of the complex structure J, but our formula seems to involve the holomorphic vector field X which vanishes at I, and I is not canonically associated to J. There are circle actions in $SO(3)$ fixing *any* complex structure orthogonal to J. We cannot expect a formula for the Kähler potential to be independent of the complex structure I, because as remarked above (following Proposition 5.1), it is already independent of the complex structure orthogonal to I and putting both facts together, we would have an $SO(3)$-invariant potential which is not the case.

The explanation, of course, is that Kähler potentials for a given complex structure are not uniquely defined: we may add on any pluriharmonic function—a function f satisfying

$$\partial\bar{\partial} f = 0.$$

With this in mind, consider the effect of rotating the complex structure I by a circle action preserving J. A rotation of θ replaces c_2 by

$$c_2 + \frac{3}{2}\sin^2\theta(c_0 + c_4 - c_2) - \frac{3}{2}\sin\theta\cos\theta(c_1 - c_3) \tag{5.15}$$

Note from (5.4) that

$$2\,\mathrm{tr}(T_3 + iT_1)^2 = c_0 + c_4 - c_2 - 2i(c_1 - c_3)$$

and $\mathrm{tr}(T_3 + iT_1)^2 = \mathrm{tr}(B_3 + iB_1)^2$ is holomorphic relative to the complex structure J. Its real and imaginary parts $c_0 + c_4 - c_2$ and $c_1 - c_3$ are thus pluriharmonic. Hence from (5.15) any one of these choices gives the same Kähler metric.

The vector field Y conjugate to X which fixes J is

$$Y = \frac{1}{2i}(\zeta^2 + 1)\frac{d}{d\zeta}$$

and this gives

$$\langle a_2, Y^2\rangle = -\frac{3}{2}(c_0 + c_4 - c_2) - 2c_2$$

so that

$$\partial_J \bar{\partial}_J c_2 = -\frac{1}{2}\partial_J \bar{\partial}_J \langle a_2, Y^2 \rangle \tag{5.16}$$

Thus $-\langle a_2, Y^2 \rangle/2$ provides an alternative to the c_2 term in the Kähler potential, which is now invariantly defined by the complex structure J under consideration.

Perhaps a better way to express this is to use the identification of the 5-dimensional irreducible representation space of $SU(2)$ which here appears as $H^0(\mathbf{P}^1, \mathcal{O}(4))$ as the space of 3×3 symmetric matrices of trace zero, in other words to define from $q \in H^0(\mathbf{P}^1, \mathcal{O}(4))$ a quadratic form Q on the 3-dimensional space $H^0(\mathbf{P}^1, \mathcal{O}(2))$. This is easily achieved by defining for $x \in H^0(\mathbf{P}^1, \mathcal{O}(2))$

$$Q(x, x) = \langle q, x^2 \rangle$$

From this point of view, we think of a complex structure in the hyperkähler family as being a point $u \in S^2 \subset \mathbf{R}^3$ and then (5.16) becomes

$$\partial\bar{\partial} c_2 = -\frac{1}{2}\partial\bar{\partial}\langle a_2, Y^2 \rangle = -\frac{1}{2}\partial\bar{\partial} Q(u, u)$$

We can now formulate the theorem in a more natural form:

Proposition 5.2. *Let $u \in S^2$ be a complex structure of the hyperkähler family on $T^*SL(k, \mathbf{C})$, then the Kähler potential ϕ is given by the formula*

$$\phi = \frac{\vartheta'(a)}{\vartheta(a)} - \frac{\vartheta'(b)}{\vartheta(b)} + \frac{2\vartheta^{(N+2)}(0)}{(N+1)(N+2)\vartheta^{(N)}(0)} - \frac{1}{6}Q(u, u)$$

where the spectral curve S given by $\eta^k + a_2(\zeta)\eta^{k-2} + \ldots + a_k(\zeta) = 0$, Q is the quadratic form defined by the coefficient a_2, the points $a, b \in J(S)$ are the line bundles corresponding to the triple (T_1, T_2, T_3) at $s = 0$ and $s = 1$ respectively, and ϑ is a translate by a half-period of the Riemann theta function on $J(S)$.

Remark: Although this is an explicit formula, it is only useful in the parametrization of the space $T^*SL(k, \mathbf{C})$ by the integrable system approach. The data involves firstly $(k+1)^2 - 4$ parameters for the coefficients of the spectral curve (these are real because the curve satisfies reality conditions), and secondly $(k-1)^2$ real parameters for the points a on the Jacobian, the initial point for the flow of the vector field $\mathcal{X}$. These provide

$$(k+1)^2 - 4 + (k-1)^2 = 2(k^2 - 1)$$

parameters. The full space has real dimension $4(k^2 - 1)$ since $\dim SU(k) = k^2 - 1$, but there is a free isometric action of $SU(k) \times SU(k)$ given by the two quotient groups $\mathcal{G}_0/\mathcal{G}_0^0$ and $\mathcal{G}^0/\mathcal{G}_0^0$ where $\mathcal{G}_0$, $\mathcal{G}^0$ are the groups of gauge transformations $g : [0,1] \to G$ for which $g(0) = 1$ or $g(1) = 1$ respectively (see [13]), and this provides the extra degrees of freedom. Since $SU(k) \times SU(k)$ preserves the metric and all three complex structures, the Kähler potential naturally exists on the quotient.

A more satisfying answer would result if we could relate these coordinates to the holomorphic parametrization given by (5.2). This, however, involves solving a supplementary linear differential equation $df/ds = f\alpha$ to determine f, and even in the case of $SL(2, \mathbf{C})$, where solutions to Nahm's equations can be explicitly written down with elliptic functions, this is not a practical prospect. Instead we can consider another class of metrics obtained by solutions to Nahm's equations: those on *coadjoint orbits.*

5.5 Hyperkähler metrics on coadjoint orbits

Kronheimer's use of Nahm's equations to construct hyperkähler metrics on coadjoint orbits of a complex semi-simple Lie group [35] has been extended from the semi-simple or nilpotent orbit case which he originally considered by Biquard and Kovalev [7],[33], but we shall restrict ourselves here to the semi-simple case. It is the same basic set-up as above, except that one studies solutions of Nahm's equations on a semi-infinite instead of finite interval. We also make a special choice of boundary condition to ensure the existence of a circle action, for only this will give us our Kähler potential by the above method.

Choose a point ξ in the Lie algebra of the maximal torus of a compact semi-simple group G, and consider solutions to Nahm's equations on the interval $(-\infty, 0]$ with boundary condition

$$(T_1, T_2, T_3) \to (\mathrm{Ad}(g)\xi, 0, 0)$$

for some $g \in G$. Because of this choice, we have the same circle action as before: $(T_1, T_2 + iT_3) \mapsto (T_1, e^{i\theta}(T_2 + iT_3))$. The Lax form of Nahm's equations implies

$$(T_1 + iT_2)' = [iT_3, T_1 + iT_2]$$

so that for all $s \in (-\infty, 0]$, $(T_1 + iT_2)(s)$ lies in the G^c-orbit of ξ. Kronheimer showed in [35] that

$$(T_1, T_2, T_3) \mapsto (T_1 + iT_2)(0)$$

identifies the moduli space of solutions diffeomorphically with the adjoint ($\cong$ coadjoint) orbit of ξ.

Now for $G = SU(k)$, the spectral curve $\det(\eta - A(\zeta, s)) = 0$ is independent of s. Letting $s \to -\infty$, $(T_1, T_2, T_3) \to (\mathrm{Ad}(g)\xi, 0, 0)$ so we have

$$\begin{aligned}\det(\eta - A(\zeta, s)) &= \det(\eta - ((T_2 + iT_3) - 2iT_1\zeta + (T_2 - iT_3)\zeta^2)) \\ &= \det(\eta + 2i\xi\zeta) \\ &= (\eta - \lambda_1\zeta)(\eta - \lambda_2\zeta)\dots(\eta - \lambda_k\zeta)\end{aligned}$$

and the spectral curve is reducible to a union of projective lines, all meeting at $\zeta = 0$ and $\zeta = \infty$. Biquard's analysis [7] shows that any solution of Nahm's equations arising from this spectral curve also satisfies the boundary conditions at $s = -\infty$, and also shows that the Kähler potential defined by

$$\phi = -\int_{-\infty}^{0} \mathrm{tr}(T_2^2 + T_3^2)ds \tag{5.17}$$

is finite.

What is required then is to study such solutions from the point of view of integrable systems, involving line bundles and the construction of triples from curves of the above type. Such a study was made in the paper of Santa Cruz [51]. We would like to apply Theorem 2 to obtain a description in terms of theta functions, but that was only stated for a non-singular spectral curve S. One might hope in general that the statement of the theorem would hold for these curves too. The line bundles on the degenerate spectral curve S can be described as in [51] in terms of transition functions and from that, a means of determining those triples (T_1, T_2, T_3) for which $A(\zeta)$ is regular (in the Lie group sense) for all ζ is provided. As Santa Cruz points out, this leads to explicit determinantal formulas for the theta divisor, and rationality in terms of $e^{\lambda_i s}$ for the integrand $\mathrm{tr}(T_2^2 + T_3^2)$ in the Kähler potential. With such concrete forms for each side it seems likely that the theorem would still hold.

The case $k = 2$ is somewhat easier, for then the spectral curve consists of two rational curves meeting transversely at two points, and this is a situation where degeneration methods tell us what the theta divisor should be [39]. The metric is the well known Eguchi-Hanson metric [19]. This is more commonly described as being defined on the cotangent bundle of the projective line, but this is an exceptional complex structure among the hyperkähler family. The general one is that of an affine quadric in $\mathbf{C}^3$—a semisimple orbit of $SL(2, \mathbf{C})$.

Consider then the spectral curve

$$\eta^2 - \lambda^2\zeta^2 = (\eta - \lambda\zeta)(\eta + \lambda\zeta) = 0$$

since ξ lies in the Lie algebra of $SU(2)$ and $\pm\lambda$ are the eigenvalues of $2i\xi$, λ is real. The curve has genus $(k-1)^2 = 1$ and any line bundle of degree zero is obtained as $\exp(u\eta x)$ for the canonical element $\eta x \in H^1(S, \mathcal{O})$. Setting U_0 to be the complement of $\zeta = \infty$ and U_∞ the complement of U_0, this is the line bundle with transition function

$$\exp(u\eta/\zeta).$$

On the component $\eta = \lambda\zeta$, this can be trivialized by the constant functions $e^{\lambda u}$ on U_0 and 1 on U_∞, and on the other component $\eta = -\lambda\zeta$ by the functions $e^{-\lambda u}$ and 1. For these ordinary singular points the bundle is trivial if and only if the two trivializations agree at $\zeta = 0$ and $\zeta = \infty$, and this is true if and only if

$$\exp(\lambda u) = \exp(-\lambda u)$$

The Jacobian of S is thus $\mathbf{C}/(\pi i/\lambda)\mathbf{Z}$, and following [39], the theta function is

$$\vartheta(u) = \sinh(\lambda u)$$

According to Theorem 2, we have

$$\begin{aligned}\Delta &= \frac{3}{2}\frac{d^2}{du^2}\log\vartheta - \frac{\vartheta'''(0)}{2\vartheta'(0)} \\ &= -\frac{3}{2}\lambda^2 \operatorname{cosech}^2 \lambda u - \frac{1}{2}\lambda^2\end{aligned}$$

Now the spectral curve has the equation $\eta^2 - \lambda^2\zeta^2 = 0$, so the coefficient $a_2(\zeta)$ is given by

$$a_2(\zeta) = -\lambda^2\zeta^2$$

and the constant c_2 is $-\lambda^2$. Consequently, we have from (5.5)

$$\begin{aligned}\operatorname{tr}(T_2^2 + T_3^2) &= \frac{1}{3}(2\Delta - c_2) \\ &= -\lambda^2 \operatorname{cosech}^2 \lambda u - \frac{1}{3}\lambda^2 + \frac{1}{3}\lambda^2 \\ &= -\lambda^2 \operatorname{cosech}^2 \lambda u\end{aligned}$$

The solution to Nahm's equations for $s \in (-\infty, 0]$ is derived from a linear flow along the Jacobian using the vector field $\mathcal{X}$ and this corresponds to setting $u = s - a$. Thus at $s = 0$, $u = -a$. Since the solution must be non-singular for $s \in (-\infty, 0]$, the theta divisor $u = 0$ must not be in this interval, so $a > 0$. From (5.17) we can now evaluate the Kähler potential as

$$\begin{aligned}\phi &= -\int_{-\infty}^{0} \operatorname{tr}(T_2^2 + T_3^2)ds \\ &= \int_{-\infty}^{-a} \lambda^2 \operatorname{cosech}^2(\lambda u)du \\ &= \lambda(1 - \coth\lambda a)\end{aligned}$$

Now we must relate the parameter a to the complex coordinates of the coadjoint orbit. This is the more difficult part in higher rank groups, but it is purely algebraic: there is no extra differential equation to solve. Recall that a solution to Nahm's equations defines the matrix

$$X = (T_1 + iT_2)(0)$$

on the orbit of ξ. Note that

$$\operatorname{tr} XX^* = -\operatorname{tr}(T_1^2 + T_2^2) = -\operatorname{tr}(T_2^2 + T_3^2) + \operatorname{tr}(T_3^2 - T_1^2)$$

but from the coefficients of $a_2(\zeta) = -\lambda^2\zeta^2$ (5.4) we see that $\operatorname{tr} T_3^2 = \operatorname{tr} T_2^2$ and $2\operatorname{tr} T_1^2 - \operatorname{tr} T_2^2 - \operatorname{tr} T_3^2 = -\lambda^2$ and hence

$$\operatorname{tr}(T_3^2 - T_1^2) = \frac{1}{2}\lambda^2$$

so that now

$$\operatorname{tr} XX^* = \lambda^2 \operatorname{cosech}^2 \lambda a + \frac{1}{2}\lambda^2$$

This now provides the final formula for the potential in terms of the coadjoint orbit:

$$\phi(X) = \lambda + \sqrt{(\operatorname{tr} XX^* + \lambda^2/2)}$$

Remark: This formula has been encountered before. In [51], Santa Cruz uses conjugates of the Nahm matrices explicitly derived from the line bundle approach to give the integrand $\operatorname{tr}(T_2^2 + T_3^2)$. Stenzel [54] obtains the same expression by using the $SU(2)$ symmetry to reduce the problem to an ordinary differential equation.

5.6 Monopole moduli spaces

Nahm's equations were originally produced in order to solve another set of differential equations: the Bogomolny equations. These are dimensional reductions of the self-dual Yang-Mills equations by the group H of translations

$$(x_0, x_1, x_2, x_3) \mapsto (x_0 + a, x_1, x_2, x_3)$$

The quotient space is $\mathbf{R}^3$ with the Euclidean metric and we have a G-connection A and a single Higgs field $-\phi$. The equations are then

$$\nabla\phi = *F_A$$

A solution to these equations with the boundary conditions that the curvature F_A is $\mathcal{L}^2$ is called a magnetic monopole. The boundary conditions and equations imply that as $r \to \infty$, ϕ tends to a particular orbit in $\mathfrak{g}$. Let $G = SU(n)$, then up to conjugation the Higgs field has an asymptotic expansion in a radial direction

$$\phi = i \operatorname{diag}(\lambda_1, \ldots, \lambda_n) - \frac{i}{2r} \operatorname{diag}(k_1, \ldots, k_n) + \ldots$$

where for topological reasons $k_1, \ldots, k_n$ are integers.

In 1981, W. Nahm proposed a construction of $SU(2)$ monopoles by performing a non-linear Fourier transform to reduce the Bogomolny equations to the ordinary differential equations which have now become known as Nahm's equations. We assume then that $n = 2$, and

$$\phi = i \operatorname{diag}(\lambda, -\lambda) - \frac{i}{2r} \operatorname{diag}(k, -k) + \ldots$$

The interpretation of the integer k is as the first Chern class of the $i\lambda$ eigenspace bundle of ϕ at large distances.

The formalism for the Nahm transform is rather like that of the hyperkähler quotient construction for

$$A = A_0 + iA_1 + jA_2 + kA_3$$

in Section 5.2. In $\mathbf{R}^4$ the formally written operator

$$D = \nabla_0 + i\nabla_1 + j\nabla_2 + k\nabla_3$$

can be thought of as the Dirac operator coupled to the vector bundle with connection. It has a dimensional reduction to three dimensions

$$D = -\phi + i\nabla_1 + j\nabla_2 + k\nabla_3$$

which is also a Dirac operator, now with the zero-order Higgs term ϕ. One considers an eigenvalue problem for this operator, showing that for $s \in (-\lambda, \lambda)$ the space of $\mathcal{L}^2$ solutions ψ of

$$(D - is)\psi = 0$$

is of dimension k. For varying s this space is a vector bundle of rank k inside the space of $\mathcal{L}^2$ functions. Orthogonal projection then defines a connection $d/ds + B_0$ on the vector bundle over the interval $(-\lambda, \lambda)$ and orthogonal projection of the operations of multiplication $\psi \mapsto x_i\psi$ defines three Higgs fields $B_1(s)$. One then shows (cf. [22]) that, gauging B_0 to zero, the matrices $B_i = T_i$ satisfy Nahm's equations. They acquire poles at the end points of the interval whose residues are irreducible k-dimensional representations of $SL(2, \mathbf{C})$.

It is for this reason that so much attention in [22] was expended on this singular behaviour, but which was also of some considerable use in our proof of Theorem 2.

The return journey, from a solution to Nahm's equations to an $SU(2)$ monopole, involves the same procedure, considering the $\mathcal{L}^2$ solutions to the equation $(\hat{D} - ix)\psi = 0$ on the interval, where

$$\hat{D} = \frac{d}{ds} + iT_1 + jT_2 + kT_3$$

and $x = x_1 i + x_2 j + x_3 k$ is a "quaternionic eigenvalue".

The Fourier transform analogy enables one to prove a "Plancherel formula" (see [45]) that the natural $\mathcal{L}^2$ metrics from both points of view coincide. For physical reasons (see [3]), the metric from the Bogomolny equation point of view is the most fundamental, but for calculation it is the Nahm equation metric which is most accessible. Nakajima and Takahasi [46], [55] have studied not only the case of $SU(2)$, but also $SU(n)$ where the $\lambda_2 = \lambda_3 = \cdots = \lambda_n$ and

$k_2 = k_3 = \cdots = k_n$. This means that the Higgs field breaks the symmetry from $SU(n)$ to $U(n-1)$ at infinity. The Chern class of the line bundle at infinity for Nakajima must be $n-1$ which we call the charge of the $SU(n)$ monopole. Dancer [11] made a close study of this case where $n = 3$. At present these are the cases where the Plancherel formula is known and where the monopole metric is the natural metric on the moduli space of solutions to Nahm's equations. Let us consider this in more detail, first in the more studied $SU(2)$ case.

5.7 $SU(2)$ monopoles

The hyperkähler quotient setting of Section 5.2 needs to be modified for the case of $SU(2)$ monopoles since the Nahm matrices are singular at the end-points of the interval. The standard normalization here, achievable by rescaling the metric on $\mathbf{R}^3$, is to take the eigenvalues of the Higgs field at infinity to be $\pm i$. The interval for Nahm's equations then has length 2 and it is convenient to take it, as in [22], to be $[0, 2]$. We consider solutions T_1, T_2, T_3 to Nahm's equations on this interval, for which $\operatorname{tr} T_i = 0$, (this being equivalent to centering the monopole [3]) and which have simple poles at the endpoints whose residue defines an irreducible representation of $SL(2, \mathbf{C})$. The circle action

$$(T_2 + iT_3) \mapsto e^{i\theta}(T_2 + iT_3)$$

is well-defined on this space, but the $\mathcal{L}^2$ metric is not, since the residues may vary within a conjugacy class. We therefore have to adopt the point of view of fixing the residues at the poles. This necessitates the reintroduction of the connection matrix B_0. We thus consider the space $\mathcal{A}$ of operators $d/ds + B_0 + iB_1 + jB_2 + kB_3$ with $B_i : (0, 2) \to \mathfrak{su}(k)$ on a rank k complex vector bundle over the interval $[0, 2]$ where at $s = 0$, B_0 is smooth and for $i > 0$,

$$B_i = \frac{\rho_i}{s} + \dots$$

for a *fixed* irreducible representation defined by ρ_i. At $s = 2$ we have the same behaviour:

$$B_i = \frac{\rho_i}{s-2} + \dots$$

Tangent vectors (A_0, A_1, A_2, A_3) to this space are then smooth at the end-points, and using the group $\mathcal{G}_0^0$ of smooth maps $g : [0, 2] \to SU(k)$ for which $g(0) = g(1) = 1$, and a little analysis, we obtain a hyperkähler metric on the space $\mathcal{B}$ of solutions to the equations

$$\begin{aligned} B_1' + [B_0, B_1] &= [B_2, B_3] \\ B_2' + [B_0, B_2] &= [B_3, B_1] \\ B_3' + [B_0, B_3] &= [B_1, B_2] \end{aligned} \tag{5.18}$$

modulo the action of the gauge group $\mathcal{G}_0^0$.

This is our metric, but to find the Kähler potential we need the circle action. Having fixed the residues, this is less easy to describe, because it involves a compensating gauge transformation outside $\mathcal{G}_0^0$. The potential only depends on the infinitesimal version of the action, and this is represented by a vector field

$$(\psi' + [B_0, \psi], [B_1, \psi], B_3 + [B_2, \psi], -B_2 + [B_3, \psi])$$

This is a vector field on the space $\mathcal{B}$ in the infinite-dimensional flat space $\mathcal{A}$: we are using the linear structure of the ambient space to write down tangent vectors. It must be smooth at the end-points, so

$$\psi(0) = \psi(2) = -\rho_1.$$

The Kähler potential is defined in terms of the moment map μ for this vector field, which uses the symplectic form on the quotient. But the quotient construction tells us that its pull-back to $\mathcal{B}$ is the restriction of the constant symplectic form on $\mathcal{A}$:

$$\int_0^2 [-\operatorname{tr}(A_0\tilde{A}_1) + \operatorname{tr}(A_1\tilde{A}_0) + \operatorname{tr}(A_2\tilde{A}_3) - \operatorname{tr}(A_3\tilde{A}_2)]ds$$

The moment map thus satisfies

$$\begin{aligned} d\mu(A) = \int_0^2 & -\operatorname{tr}(A_0[B_1, \psi]) + \operatorname{tr}(A_1(\psi' + [B_0, \psi])) + \operatorname{tr}(A_2(-B_2 + [B_3, \psi])) \\ & - \operatorname{tr}(A_3(B_3 + [B_2, \psi]))ds \\ = \int_0^2 & -\operatorname{tr}([A_0, B_1]\psi) + \operatorname{tr}(A_1\psi') - \operatorname{tr}([B_0, A_1]\psi) - \operatorname{tr}(A_2B_2) \\ & + \operatorname{tr}([A_2, B_3]\psi) - \operatorname{tr}(A_3B_3) + \operatorname{tr}([B_2, A_3]\psi)ds \end{aligned}$$

On the other hand A is tangent to $\mathcal{B}$, so A satisfies the linearization of the equations (5.18), so in particular

$$A_1' + [A_0, B_1] + [B_0, A_1] = [A_2, B_3] + [B_2, A_3]$$

Substituting in the formula for $d\mu$ then gives

$$\begin{aligned} d\mu(A) &= \int_0^2 \operatorname{tr}(A_1'\psi) + \operatorname{tr}(A_1'\psi)ds - \int_0^2 \operatorname{tr}(A_2B_2 + A_3B_3)ds \\ &= [\operatorname{tr}(A_1\psi)]_0^2 - \int_0^2 \operatorname{tr}(A_2B_2 + A_3B_3)ds \end{aligned}$$

Now as we saw in the proof of Theorem 2, at an irreducible representation, the Nahm matrix has an expansion

$$T_i = \frac{\rho_i}{s} + \tau_i s + \dots$$

so the conjugate B_i by a smooth gauge transformation behaves like

$$B_i = \frac{\rho_i}{s} + [\varphi, \rho_i] + \dots$$

hence

$$\operatorname{tr}(B_i \rho_i) = -\frac{N}{2s} + b_1 s + \dots$$

Thus, differentiating with respect to a parameter, a tangent vector A to $\mathcal{B}$ has the property

$$A_i(0) = 0$$

and similarly at $s = 2$, which tells us that the boundary term vanishes and

$$d\mu(A) = -\int_0^2 \operatorname{tr}(A_2 B_2 + A_3 B_3) ds$$

The integrand is conjugation-invariant, so we can write this in terms of Nahm matrices

$$d\mu(\dot{T}) = -\int_0^2 \operatorname{tr}(T_2 \dot{T}_2 + T_3 \dot{T}_3) ds \tag{5.19}$$

To find the potential, we would therefore like to make sense of the divergent integral

$$-\frac{1}{2}\int_0^2 \operatorname{tr}(T_2^2 + T_3^2) ds$$

Since again there is no constant term in the expansion of T_i at an irreducible pole, we have

$$\operatorname{tr}(T_i^2) = -\frac{N}{2s^2} + 2\operatorname{tr}(\rho_i \tau_i) + \dots$$

the most obvious function μ that will satisfy (5.19) is thus

$$\mu(T) = -\frac{1}{2}\int_0^2 \left(\operatorname{tr}(T_2^2 + T_3^2) + \frac{N}{s^2} + \frac{N}{(s-2)^2}\right) ds \tag{5.20}$$

We can cast this in another form if we recall how Nahm's equations are to be solved with a linear flow in the direction of the canonical vector field $\mathcal{X}$ on the Jacobian of the spectral curve. As we saw, a pole which defines an irreducible representation corresponds to the flow passing through the origin of $J(S)$, so an irreducible pole at both ends of the interval means that the flow is periodic. (This is the $L^2 \cong \mathcal{O}$ condition on the spectral curve—see [3]). Thus the function

$$\operatorname{tr}(T_2^2 + T_3^2)$$

is periodic in s of period 2.

Now let $f(s)$ be a meromorphic function with an expansion about $s = 0$ of

$$f(s) = \frac{a_{-2}}{s^2} + a_0 + a_1 s + \dots$$

and Γ_ϵ a contour consisting of an interval $[a, b]$ indented around the origin with a semicircular contour of radius ϵ. The contour integral is the limit as $\epsilon \to 0$ of

$$\int_a^{-\epsilon} f(s)\,ds + \int_\epsilon^b f(s)\,ds - \frac{2a_{-2}}{\epsilon}$$

Comparing with (5.20) and using the periodicity, we can take μ to be the integral of $-\operatorname{tr}(T_2^2 + T_3^2)/2$ around a closed contour in the Jacobian which is an orbit of the flow missing the origin.

Now return to Theorem 2. We saw that

$$\operatorname{tr}(T_2^2 + T_3^2) = \frac{1}{3}(2\Delta - c_2)$$

and

$$\Delta = \frac{3}{2}\frac{d^2}{ds^2}\log\vartheta + c$$

so apart from the constants c and c_2, the Kähler potential is obtained by integrating $(\log\vartheta)''$ around a closed cycle. This would be zero if $(\log\vartheta)'$ were single valued, but (5.7) shows that this is not so. However (5.6) gives

$$\vartheta(z + A_i) = \vartheta(z)$$

so if we choose a canonical basis so that our cycle is in the linear span of the A-cycles, then the integral is indeed zero. We thus obtain a formula for the Kähler potential:

$$\phi = \frac{4}{(N+1)(N+2)}\frac{\vartheta^{(N+2)}(0)}{\vartheta^{(N)}(0)} - \frac{1}{3}Q(u,u) \tag{5.21}$$

In the case of monopoles the term $Q(u, u)$ has a direct interpretation in terms of the Higgs field ϕ. For this, we have to use Hurtubise's approach to the asymptotic Higgs field.

Recall that the boundary conditions of a monopole imply an expansion

$$|\phi| = 1 - \frac{k}{2r} + O(r^{-2})$$

Hurtubise shows in [30] that this extends to a complete asymptotic expansion which defines a harmonic function, corresponding to some distribution of charge. Moreover, this can be calculated from the spectral curve.

If we consider the expansion of $|\phi|$ along the ray $(r, 0, 0)$, then we set $\eta = -2r\zeta = R\zeta$ in the equation of the spectral curve

$$\eta^k + a_2(\zeta)\eta^{k-2} + \cdots + a_k(\zeta)$$

to obtain a polynomial of degree $2k$ in ζ whose coefficents are functions of R. As $R \to \infty$, the $2k$ roots of this equation separate unambiguously into two groups: k which tend to 0 and k which tend to ∞. If $\zeta_1, \ldots, \zeta_k$ are the first group, then Hurtubise's formula is

$$|\phi| \sim 1 - \frac{\partial}{\partial R}\log(\zeta_1\zeta_2\ldots\zeta_k)$$

Making the substitution in the equation of the spectral curve, we obtain

$$\zeta^k + a_2(\zeta)\frac{\zeta^{k-2}}{R^2} + \cdots + \frac{a_k(\zeta)}{R^k} = 0 \tag{5.22}$$

Now each root ζ_i has an asymptotic expansion

$$\zeta_i = \frac{\alpha_i}{R} + \ldots$$

and it follows that each term of the form $c\zeta^{k+m}$ for $m \geq 1$ in (5.22) decays at least as fast as $R^{-(k+3)}$. Collecting together the terms of the form $c\zeta^n$, for $n \leq k$ we have a polynomial of degree k such that for each $\zeta = \zeta_i$,

$$(1 + c_2R^{-2})\zeta^k + \cdots + z_0R^{-k} = O(R^{-(k+3)})$$

where $a_2(\zeta) = c_0 + c_1\zeta + c_2\zeta^2 + \cdots + c_4\zeta^4$ and $a_k(\zeta) = z_0 + z_1\zeta + \cdots + z_{2k}\zeta^{2k}$. Thus the product of the roots satisfies

$$\zeta_1\zeta_2\ldots\zeta_k = (-1)^k\frac{z_0R^{-k}}{(1 + c_2R^{-2})} + O(R^{-(k+3)})$$

and so

$$\frac{\partial}{\partial R}\log(\zeta_1\zeta_2\ldots\zeta_k) = -\frac{k}{R} - \frac{2c_2}{R^3} + \ldots$$

Replacing R by $-2r$, we have

$$|\phi| \sim 1 - \frac{\partial}{\partial R}\log(\zeta_1\zeta_2\ldots\zeta_k) = 1 - \frac{k}{2r} - \frac{c_2}{4r^3} + \ldots$$

In our parametrization, the unit direction $u = (1, 0, 0)$ corresponds to $\zeta = 0$, and so as in 5.4, $c_2 = \langle a_2, X^2\rangle = Q(u, u)$ The asymptotic expansion in the direction u is thus

$$|\phi| = 1 - \frac{k}{2r} - \frac{Q(u,u)}{4r^3} + \ldots$$

We thus have the following theorem about the moduli space metric:

Theorem 3. *Let $\mathcal{M}_k^0$ denote the moduli space of centred $SU(2)$ monopoles on $\mathbf{R}^3$ and let $u \in S^2$ be a complex structure of the hyperkähler family on $\mathcal{M}_k^0$. Then the Kähler potential for the natural $\mathcal{L}^2$ metric in this complex structure is*

$$\phi = \frac{4}{(N+1)(N+2)} \frac{\vartheta^{(N+2)}(0)}{\vartheta^{(N)}(0)} - \frac{1}{3} Q(u,u)$$

where ϑ is a translate of the Riemann theta function of the spectral curve and $-Q(u,u)/4$ is the coefficient of r^{-3} in the asymptotic expansion of the length of the Higgs field in the radial direction u.

As with the metric on the cotangent bundle of the group, to express this potential in terms of the natural complex coordinates of $\mathcal{M}_k^0$, considered as a space of rational functions [15] requires the solution of a linear differential equation whose coefficients are Nahm matrices. This clearly places constraints on explicitness. Even the coordinates determined by the integrable system approach are complicated to determine because of the constraints. The periodicity condition imposes $g = (k-1)^2$ conditions on the coefficients of the spectral curve which are $5 + 7 + (2k+1) = (k+1)^2 - 4$ in number, giving the moduli space $\mathcal{M}_k^0$ dimension $4k - 4$.

Example:
Consider the 2-monopole moduli space, where the metric is fully described in [3]. Here, $k = 2$, the curve S is elliptic, and the periodicity of the Nahm flow for $s \mapsto s + 2$ requires us in standard Weierstrassian coordinates to put

$$s = u/\omega_1 = 2\nu$$

In this case $N = 1$ and

$$\frac{4}{(N+1)(N+2)} \frac{\vartheta^{(N+2)}(0)}{\vartheta^{(N)}(0)} = \frac{2}{3} \frac{\vartheta^{(3)}(0)}{\vartheta^{(1)}(0)} = \frac{1}{6} \frac{\vartheta'''(0)}{\vartheta'(0)}$$

using differentiation with respect to ν. From the the standard formula in elliptic functions (cf. (5.13)), we have

$$2\eta_1\omega_1 = -\frac{1}{6} \frac{\vartheta'''(0)}{\vartheta'(0)} \tag{5.23}$$

so that the $SO(3)$-invariant term in the potential is

$$-2\eta_1\omega_1$$

In [47], D.Olivier gives a derivation of this potential, directly from the formula in [3]

$$g = \beta^2\gamma^2\delta^2 \frac{(dk^2)^2}{(4k^2k'^2K^2)^2} + \beta^2\sigma_1^2 + \gamma^2\sigma_2^2 + \delta^2\sigma_3^2$$

where

$$\beta\gamma = -K^2(k'^2 + u)$$
$$\gamma\delta = K^2(k^2 - u)$$
$$\beta\delta = -K^2 u$$

and

$$u = \frac{E}{K} - k'^2$$

All these expressions use the Legendre notation for complete elliptic integrals. For Olivier, the Kähler potential is

$$\phi = \frac{\beta\gamma + \gamma\delta + \delta\beta}{4} - J$$

where J is direction-dependent. After changing from Legendrian formulae to Weierstrassian ones, this corresponds (up to a constant multiple) to the formula here.

5.8 $SU(n)$ monopoles

The study of $SU(n)$ monopoles is more complicated because of the choice of boundary conditions

$$\phi = i\,\mathrm{diag}(\lambda_1, \ldots, \lambda_n) - \frac{i}{2r}\,\mathrm{diag}(k_1, \ldots, k_n) + \ldots$$

for the Higgs field. The most studied case is that of maximal symmetry breaking: where the eigenvalues at infinity are distinct. Thus the structure group $SU(n)$ is reduced to its maximal torus at infinity. Here there is a construction involving Nahm's equations [31]. In this case one solves the equations on a sequence of intervals corresponding to vertices of the Dynkin diagram. It seems reasonable to believe that the same methods advanced here may work in that situation, but the full correspondence, including the Plancherel formula, has not been worked out. We restrict ourselves instead to a special case of minimal symmetry breaking:

$$\phi = i\,\mathrm{diag}(n-1, -1\ldots, -1) - \frac{i}{2r}\,\mathrm{diag}(n-1, -1, \ldots, -1) + \ldots$$

where the group is reduced from $SU(n)$ to $U(n-1)$ at infinity. Here the work of Nakajima, Takahasi [46] and Dancer [11], [12] provide a fuller picture.

The arguments involving the Dirac operator can be extended in this case so that there is an $\mathcal{L}^2$ nullspace of dimension $k = n - 1$ for s between the two distinct eigenvalues k and -1 of ϕ at infinity, or normalizing so that the origin is one end-point, for $s \in (0, k+1)$. Nahm matrices can then be defined just as in the $SU(2)$ case, but with the boundary condition that there is a simple pole at $s = 0$ whose residue defines the irreducible k-dimensional representation of $SL(2, \mathbf{C})$. At the other end-point $s = k + 1$, the matrices are finite.

Using matrices (B_0, B_1, B_2, B_3) as above, we can define a hyperkähler metric on the moduli space of solutions to Nahm's equations, and by the Plancherel theorem this is the natural metric on the moduli space of monopoles with this structure. There is, as with all monopole metrics, an action of $SO(3)$ rotating the complex structures and in this case an extra action of the group $\mathcal{G}_0/\mathcal{G}_0^0 \cong SU(k)$ where $\mathcal{G}_0$ is the group of smooth maps $g : [0, k+1] \to SU(k)$ with $g(0) = 1$. Using the integrable system approach we can count dimensions, since solving Nahm's equations with these boundary conditions consist of starting at the origin and following the flow for a time $t = k + 1$. We therefore have $(k+1)^2 - 4$ parameters for the spectral curve and $k^2 - 1 = \dim SU(k)$ for the gauge action giving a moduli space M_k of dimension $(k+1)^2 - 4 + k^2 - 1 = 2(k^2 + k - 2)$.

As above, we remove the singularity at $s = 0$ by defining

$$\mu(T) = -\frac{1}{2}\int_0^{k+1}\left(\operatorname{tr}(T_2^2 + T_3^2) + \frac{N}{s^2}\right) ds$$

and using

$$\operatorname{tr}(T_2^2 + T_3^2) = \frac{1}{3}(2\Delta - c_2)$$

with

$$\Delta = \frac{3}{2}\frac{d^2}{ds^2}\log\vartheta + c$$

This gives the potential

$$\phi = -\frac{\vartheta'(k+1)}{\vartheta(k+1)} + \frac{k(k-1)}{6} + \frac{2(k+1)}{(N+2)(N+1)}\frac{\vartheta^{(N+2)}(0)}{\vartheta^{(N)}(0)} + \frac{k+1}{3}c_2$$

Example:
For $k = 2$ (the case considered by Dancer) we can again express this using Weierstrass functions. We obtain from (5.14)

$$\Delta(s) = -\frac{3}{2}\kappa^2\wp(\kappa s)$$

and so using the Weierstrass zeta-function and ignoring the constant term,

$$\phi = -\kappa\zeta(3\kappa) + c_2 \tag{5.24}$$

Bibliography

[1] U. Abresch, *Constant mean curvature tori in terms of elliptic functions*, J. reine angew. Math. **374** (1987), 169–192.

[2] A.D. Alexandrov, *Uniqueness theorems for surfaces in the large*, Amer. Math. Soc. Trans. (Series 2) **21** (1958), 412–416.

[3] M.F. Atiyah and N.J. Hitchin, "The Geometry and Dynamics of Magnetic Monopoles", Princeton University Press, Princeton (1988).

[4] A. Beauville, *Variétés Kählériennes dont la 1ére classe de Chern est nulle*, J. Diff. Geometry **18** (1983), 755–782.

[5] A. Beauville, *Jacobiennes des courbes spectrales et systèmes hamiltoniens complètement intégrables*, Acta Math. **164** (1990), 211–235.

[6] A.L. Besse, *Einstein Manifolds*, Springer Verlag, Berlin-Heidelberg (1987).

[7] O. Biquard, *Sur les équations de Nahm et la structure de Poisson des algèbres de Lie semi-simples complexes*, Math. Ann. **304** (1996), 253–276.

[8] F.E. Burstall, D. Ferus, F. Pedit and U. Pinkall, *Harmonic tori in symmetric spaces and commuting Hamiltonian systems on loop algebras*, Annals of Math. **138** (1993), 173–212.

[9] F.E. Burstall and F. Pedit, *Harmonic maps via Adler-Kostant-Symes theory* in "Harmonic maps and integrable systems", 221–272, (eds A.P. Fordy & J.C. Wood), Vieweg, Wiesbaden (1994).

[10] F.E. Burstall, *Harmonic tori in spheres and complex projective spaces*, J. Reine Angew. Math. **469** (1995), 149–177.

[11] A.S. Dancer, *Nahm's equations and hyperkähler geometry*, Commun. Math. Phys. **158** (1993), 545–568.

[12] A.S. Dancer, *A family of hyperkähler manifolds*, Quart. J. Math. Oxford 45. (1994), 463–478.

[13] A.S. Dancer and A.F. Swann, *Hyperkähler metrics associated to compact Lie groups*, Math. Proc. Camb. Phil. Soc.,**120** (1996), 61–69.

[14] G. Darboux, "Leçons sur les Systèmes Orthogonaux" (2nd edition), Gauthier-Villars, Paris (1910).

[15] R. Donagi and E. Markman, *Spectral covers, algebraically completely integrable Hamiltonian systems, and moduli of bundles*, in "Integrable systems and quantum groups (Montecatini Terme, 1993)" 1–119, Lecture Notes in Math. **1620**, Springer, Berlin (1996).

[16] S.K. Donaldson, *Nahm's equations and the classification of monopoles*, Commun. Math. Phys. **96** (1984), 387–407.

[17] B. Dubrovin, *Geometry of 2D topological field theories*, in "Integrable systems and quantum groups (Montecatini Terme, 1993)", 120–348, Lecture Notes in Math. **1620**, Springer, Berlin, 1996.

[18] D.-Th. Egorov, *Sur les systèmes orthogonaux admettant un groupe continu de transformations de Combescure*, Comptes rendus Acad. Sci. Paris **131** 668 (1900), **132** 174 (1901).

[19] T. Eguchi and A.J. Hanson, *Asymptotically flat self-dual solutions to Euclidean gravity* Phys. Lett. **74B** (1978) 249–251.

[20] L.P. Eisenhart, "A treatise on the differential geometry of curves and surfaces", Dover, New York (1909).

[21] A.P. Fordy and J.C. Wood (eds), "Harmonic maps and integrable systems", Vieweg, Wiesbaden (1994).

[22] N.J. Hitchin, *On the construction of monopoles*, Commun. Math. Phys. **89** (1983), 145–190.

[23] N.J. Hitchin, A. Karlhede, U. Lindström and M. Roček, *Hyperkähler metrics and supersymmetry*, Commun. Math. Phys. **108** (1987), 535–589.

[24] N.J. Hitchin, *Stable bundles and integrable systems*, Duke Math. Jour. **54** (1987) 91–114.

[25] N.J. Hitchin, *Harmonic maps from a 2-torus to the 3-sphere*, J. Differential Geometry **31** (1990), 627–710.

[26] N.J. Hitchin, *Twistor spaces, Einstein metrics and isomonodromic deformations*, J. Differential Geometry **42** (1995), 30–112.

[27] N.J. Hitchin, *A new family of Einstein metrics*, in "Manifolds and Geometry (Pisa 1993)" (eds P. de Bartolomeis, F. Tricerri and E. Vesentini), Symposia Mathematica **XXXVI**, Cambridge University Press, Cambridge (1996).

[28] N.J. Hitchin, *Einstein metrics and the eta-invariant*, Bollettino U.M.I. **10**, Suppl.fasc. (1996).

[29] J. Hurtubise, *$SU(2)$ monopoles of charge 2*, Commun. Math. Phys. **92** (1983), 195–202.

[30] J. Hurtubise, *The asymptotic Higgs field of a monopole*, Commun. Math. Phys. **97** (1985), 381–389.

[31] J. Hurtubise and M.K. Murray, *On the construction of monopoles for the classical groups*, Commun. Math. Phys. **122** (1989), 35–89.

[32] J.H. Jellet, *Sur la surface dont la courbure moyenne est constante*, J. Maths. Pures. Appl. **18** (1853), 163–167.

[33] A.G. Kovalev, *Nahm's equations and complex adjoint orbits*, Quart. J. Math. Oxford **47** (1996), 41–58.

[34] P.B. Kronheimer, *A hyperkähler structure on the cotangent bundle of a complex Lie group*, MSRI preprint (1988).

[35] P.B. Kronheimer, *A hyperkählerian structure on coadjoint orbits of a semisimple Lie group*, J. London Math. Soc. **42** (1990) 193–208.

[36] C. LeBrun, *On complete quaternionic-Kähler manifolds*, Duke Math. Journal **63** (1991) 723–743.

[37] K. Lee, E.J. Weinberg and P. Yi, *Electromagnetic duality and $SU(3)$ monopoles*, Phys. Lett. B **376** (1996) 97–102.

[38] I. McIntosh, *A construction of all non-isotropic harmonic tori in complex projective space*, Internat. J. Math. **6** (1995) 831–879.

[39] H.P. McKean, *Theta functions, solitons and singular curves*, in "Partial differential equations and geometry", Proc. Park City Conf. 1977, Lect. Notes Pure Appl. Math. **48** (1979) 237–254 Marcel Dekker, New York.

[40] B. Malgrange, *Sur les deformations isomonodromiques I. Singularités régulières*, in "Mathématique et Physique", Séminaire de l'Ecole Normale Supérieure 1979–1982, Progress in Mathematics **37**, Birkhäuser, Boston (1983), 401–426.

[41] L.J. Mason and G.A.J. Sparling, *Nonlinear Schrödinger and Korteweg de Vries are reductions of self-dual Yang-Mills*, Phys. Lett. **A137** (1989), 29–33.

[42] L.J. Mason and N.M.J. Woodhouse, "Integrability, self-duality, and twistor theory", Oxford University Press, Oxford (1996).

[43] M. Melko and I. Sterling, *Integrable systems, harmonic maps and the classical theory of surfaces*, in "Harmonic maps and integrable systems", 129–144, (eds A.P. Fordy & J.C. Wood), Vieweg, Wiesbaden (1994).

[44] M. Min-Oo, *Scalar curvature rigidity of asymptotically hyperbolic spin manifolds*, Math. Ann. **285**, (1989), 527–539.

[45] H. Nakajima, *Monopoles and Nahm's equations*, in " Einstein metrics and Yang-Mills connections. Proceedings of the 27th Taniguchi international symposium, held at Sanda, Japan, December 6–11, 1990" (eds T. Mabuchi et al) Lect. Notes Pure Appl. Math. **145**, Marcel Dekker, (New York) (1993).

[46] H. Nakajima and M. Takahasi, (in preparation).

[47] D.Olivier, *Complex coordinates and Kähler potential for the Atiyah-Hitchin metric*, Gen. Relativity Gravitation **23** (1991), 1349–1362.

[48] H. Pedersen, *Einstein metrics, spinning top motions and monopoles*, Math. Ann. **274** (1986) 35–59.

[49] K. Pohlmeyer, *Integrable Hamiltonian systems and interactions through quadratic constraints*, Commun. Math. Phys. **46** (1976), 207–221.

[50] E.A. Ruh and J. Vilms, *The tension field of the Gauss map*, Trans. Amer. Math. Soc. **149** (1970), 569–573.

[51] S. d'Amorim Santa-Cruz, *Twistor geometry for hyperkähler metrics on complex coadjoint orbits*, Ann. of Global Analysis and Geometry **15** (1997), 361–377.

[52] W. Schmid, *Variation of Hodge structure: the singularities of the period mapping*, Invent. Math. **22** (1973), 211–319.

[53] A. Sen, *Dyon-monopole bound states, self-dual harmonic forms on the multi-monopole moduli space, and* $SL(2, \mathbf{Z})$*-invariance of string theory*, Phys. Lett. **B 329** (1994), 217–221.

[54] M.B. Stenzel, *Ricci-flat metrics on the complexification of a compact rank one symmetric space*, Manuscr. Math. **80** (1993), 151–163.

[55] M. Takahasi, *Nahm's equation, rational maps from* CP^1 *to* CP^k*, and monopoles*, Master's Thesis, Tôhoku University, (1996) (in Japanese).

[56] G. Tian and S.-T. Yau, *Complete Kähler manifolds with zero Ricci curvature. II.* Invent. Math. **106** (1991), 27–60.

[57] K.P. Tod, *Self-dual Einstein metrics from the Painlevé VI equation*, Phys. Lett. **A 190** (1994), 221–224.

[58] M. Voretzsch, *Untersuchung einer speziellen Fläche constanter mittlerer Krümmung bei welcher die eine der beiden Schaaren der Krümmungslinien von ebenen Curven gebildet wird*, Dissertation, Universität Göttingen (1883).

[59] R.S. Ward, *The Painlevé property for the self-dual gauge field equations*, Phys. Lett. **A102** (1984), 279–282.

[60] R.S. Ward, *Integrable systems in twistor theory* in "Twistors in mathematics and physics", 246–259, (eds T.J. Bailey & R.J. Baston) London Math. Soc. Lecture Note Ser., 156, Cambridge Univ. Press, Cambridge, (1990).

[61] H. Wente, *Counterexample to a conjecture of H. Hopf*, Pacific J. Math. **121** (1986), 193–243.

[62] S.-T. Yau, *On Calabi's conjecture and some new results in algebraic geometry*, Proc. Nat. Acad. Sci. USA **74** (1977), 1798–1799.

Seiberg-Witten Integrable Systems

Ron Y. Donagi*†

Department of Mathemtics
University of Pennsylvania
Philadelphia, PA 19104-6395

Introduction

While many mathematicians are aware of the monopole equations [W] and of the impact they have had in topology, probably not very many are as familiar with the works [SW1, SW2] of Seiberg and Witten on which the monopole equations are based. The appearance of these works was a major breakthrough in Quantum Field Theory (QFT), and has to a large extent revolutionized and transformed the subject. The actual achievement in [SW1] was an *exact* solution to a problem in QFT (pure $N = 2$ supersymmetric Yang-Mills in four dimensions with gauge group SU(2)) which could traditionally be solved only approximately, by computing the first few terms in a perturbative expansion. The solution is close to algebraic geometry both in its details (involving families of elliptic curves) and in its general approach, which is to use global properties (monodromy, symmetries) of a physical theory to uniquely characterize a geometric model of the theory. This work opened the doors to solutions, along similar lines, of many specific quantum theories, some of which are reviewed here. It also led directly to the ongoing revolution in string theory involving string dualities and the appearance of higher dimensional objects (branes) on more or less equal footing with the strings themselves.

*Work partially supported by NSF grant DMS95-03249, and by grants from the Univesity of Pennsylvania Research Foundation and The Harmon Duncombe Foundation.

†Current address: School of Mathematics, Institute for Advanced Study, Princeton, NJ 08540

The purpose of these notes is to try to make this exciting and rapidly evolving chapter of high energy physics accessible to mathematicians, particularly to algebraic geometers. I have had to make some choices: I decided to focus on the original work and its further development within QFT, thus omitting the spectacular (but well documented) applications to topology and strings. Even so, it was impossible to render with mathematical rigor the enormous amount of background material needed from the physics. I opted instead, in §1, for the telling of one uniform but very imprecise story of the physics. My hope is that it will afford, to a mathematical reader with no prior knowledge of the physics, at least a general sense of the physical theory, of how the different parts interact, and of what some of the basic concepts mean. The remaining, mathematical, sections of these notes discuss what is now known (and what is still missing) about Seiberg-Witten type solutions of four-dimensional $N = 2$ supersymmetric Yang-Mills theories in algebro-geometric terms, with the emphasis on an interpretation via integrable systems.

Section 1 starts with Maxwell's classical description of electromagnetism. We arrive at our central object of study, the duality in super Yang-Mills theory, after discussing separately the three directions in which Maxwell's theory needs to be extended: nonabelian gauge groups, quantization, and supersymmetry. We are then ready for an overview of the Seiberg-Witten solution, [SW1], emphasizing its algebro-geometric elements. The section concludes with a discussion of the other theories which could be exactly solvable by algebro-geometric techniques. These theories occupy us in the remainder of these notes.

Algebraically integrable systems are quickly reviewed in §2. The upshot is that all the structures arising in the Seiberg-Witten physics of §1 are naturally realized here, so algebraically integrable systems are the natural context for describing the physics (cf. [DW], [F]). We also clarify the role of the Seiberg-Witten differential, which occurs naturally in any integrable system of *spectral* type. One important feature of this differential, the linear dependence of its residues on the mass parameters, turns out to be part of a very general phenomenon. We state and outline a proof of an apparently new result, which is a complex analogue of the theorem of Duistermaat-Heckman [DH] on variation of the symplectic structure in a family of symplectic quotients. The complex case is somewhat subtler than the compact version due to the presence of non-semisimple (e.g. nilpotent) orbits.

Section 3 is mostly an attempt to rewrite the solution [DW] of the SU(n) theory with adjoint matter, with an emphasis on the math rather than the physics. We review the work of Markman [Mn] and others on the integrable systems of meromorphic Higgs bundles; among these systems we pick the particular subsystem we need. The system (i.e. its spectral curves) is described explicitly, then compared to the elliptic soliton description [T, TV]. A major part of [DW] conisted of "tests" showing that this system accurately reproduces physical properties of the theory; we describe some of these "tests" here.

In the final section we discuss the other major class of integrable systems which is known to solve the pure $N = 2$ SYM theories: the periodic Toda lattices [MW1]. In particular, we compare the various forms assumed by the spectral curves. We then move to a common generalization of both systems, the elliptic Calogero-Moser systems, which were proposed in [Mc] as a solution of the theory with adjoint matter and arbitrary gauge group. This seems quite promising, but it still lacks a satisfactory algebro-geometric description. We also mention briefly some other theories where at least the family of spectral curves is known, and conclude by mentioning some recent ideas in string theory which exhibit the QFT integrable systems discussed here as limits of stringy systems.

It is a pleasure to acknowledge pleasant and useful conversations, about some of the topics covered here, with D. Freed, L. Jeffrey, A. Kirillov, D. Lowe, N. Nekrasov, R. Plesser, N. Seiberg, S. Singer, R. Sjammar, J. Weitsman, and E. Witten, as well as helpful comments on the manuscript by D. Freed, A. Grassi, A. Ksir, N. Nekrasov, M. Verbitsky and E. Witten.

1 Some Physics Background

1.1 Electromagnetism

The electric field E and the magnetic field B in $\mathbb{R}^3$ are governed by Maxwell's equations, of T-shirt fame. Think of E and B, respectively, as a 1-form and a 2-form on $\mathbb{R}^3$. Both are time-dependent, so we combine them into the electromagnetic field $F := E \wedge dt + B$, a 2-form on $\mathbb{R}^4$. Maxwell's equations in vacuo are:

$$dF = 0 \qquad\qquad d * F = 0.$$

Here $*$ is the Hodge star operator on forms on $\mathbb{R}^4$, determined by the Minkowski metric $g = dx^2 + dy^2 + dz^2 - c^2dt^2$. (The speed of light c is usually set to 1 by adjusting units.)

The same equations make sense for an unknown 2-form F on any (pseudo) Riemannian manifold (X, g). They are clearly invariant under the group of global isometries of (X, g). When dim $X = 4$, the equations are also invariant under the duality $F \mapsto *F$. When $X = \mathbb{R}^4$ with the Minkowski metric, this duality interchanges the electric and magnetic fields. The group of isometries in this case is the Poincare group, an extension of the Lorentz group $\mathrm{SO}(3,1)$ of rotations by the group $(\mathbb{R}^4, +)$ of translations.

Geometrically, we fix a $\mathrm{U}(1)$-bundle L on X and consider a connection A on L as a new variable. The first equation $dF = 0$ is interpreted as saying that (locally) $F = dA$ is the curvature. The remaining equation then is $d * dA = 0$. The group of isometries of the triple (X, g, L) acts on the connections and preserves the equation. This is an extension of the group of those isometries of (X, g) which preserve L by the *gauge group* of C^∞ maps from X to $\mathrm{U}(1)$. The duality is still there, but is now hidden.

Now impose an external field j. (In $\mathbb{R}^4$, j is a 3-form whose $dx \wedge dy \wedge dz$ component is the charge density, while the other 3 components give the current.)

The equations become

$$dF = 0, \qquad d * F = j,$$

and duality is lost. Geometrically, once we think of F as curvature, the first equation becomes a tautology (the Bianchi identity), while the second is an extra condition which may or may not hold. Physically, we explain this asymmetry as due to the absence of magnetic charges ("monopoles").

Dirac proposed that some magnetic charges can be introduced by changing the topology, or allowing singularities. For example, a monopole at the origin of $\mathbb{R}^3$ is represented by a U(1)-bundle on $(\mathbb{R}^3_{x,y,z} \setminus 0) \times \mathbb{R}_t$. Such bundles are characterized topologically by their Chern class $c_1 \in \mathbb{Z}$, and this c_1 is interpreted as the magnetic charge at the origin. Indeed, the "magnetic charge" at the origin is just the total magnetic flux across a small sphere S^2 around the origin in $\mathbb{R}^3$; this is the integral over S^2 of B, which equals the integral of F, which is c_1. We can extend F to $\mathbb{R}^4$ as a distribution, but now dF is the δ-function $c_1.\delta_{x,y,z}$ instead of $dF = 0$. This is not yet sufficient to restore the symmetry to Maxwell's equations: the electric charge is arbitrary, but the magnetic charge is allowed only in discrete "quanta". This does raise the possibility, though, that complete symmetry could be restored in a quantum theory, where, as in the real world, electric charge also comes in discrete units. Indeed, Dirac discovered that in quantum mechanics, where an electrically charged particle is represented by a wave function satisfying Schrödinger's equation, the electric charge e of one particle and the magnetic charge g of another (in appropriate units) satisfy $eg \in \mathbb{Z}$. This has the curious interpretation that all electric charges are automatically quantized as soon as one magnetic monopole exists somewhere in the universe.

In [MO], Montonen and Olive conjectured that in certain supersymmetric Yang-Mills quantum field theories the electric-magnetic duality is indeed restored. This involves an extension of the basic picture in three separate directions: replacing the "abelian" Maxwell equations by the non-abelian Yang-Mills; quantizing the classical theory; and adding supersymmetry to the Poincaré invariance.

1.2 Yang-Mills theory

Yang-Mills theory is quite familiar to mathematicians. It involves replacing the structure group U(1) by an arbitrary reductive group G. We thus have a connection A on some G-bundle V over our fourfold X, the curvature is now the ad(V)-valued two-form $dA + \frac{1}{2}[A, A]$, and the equations are as before,$dF = 0$, $d * F = 0$, except that the covariant derivative $d = d_A$ on ad(V) now replaces the exterior derivative d on the trivial bundle ad(L).

These equations can be, and usually are, considered as variational equations for the Lagrangian

$$L(A) := \int_X \mathrm{Tr}(F \wedge *F),$$

where the trace Tr can be any non-degenerate invariant pairing on the Lie algebra $\mathfrak{g}$ of G. This Lagrangian is considered as a functional on an appropriate space of connections A (if X is not compact, we must impose some decay condition at infinity for the integral to converge). The equation $dF = 0$ is again the Bianchi identity, while $d * F = 0$ is the equation of the critical locus of L. Either from the Lagrangian description or directly from the equations, it is clear that the symmetry group includes global isometries of X which lift to V ("the Poincare group") as well as local diffeomorphisms of the bundle ("the gauge group") and, in dimension 4, the duality $F \mapsto *F$. In the original model of Yang and Mills, the group G was SU(2), while in QCD it is SU(3). For any "realistic" model of all of particle physics, G would need to contain at least $\mathrm{U}(1) \times \mathrm{SU}(2) \times \mathrm{SU}(3)$.

1.3 Quantization

Quantization is an art form which, when applied to classical physical theories, yields predictions of subatomic behavior which are in spectacular agreement with experiments. The quantization of the classical mechanics of particles leads to quantum mechanics, which can be described, in a Hamiltonian approach, via operators and rays in Hilbert space, or in a Lagrangian formulation, via path integrals. Quantization of continuous fields leads to QFT: quantum field theory. It has a Hamiltonian formulation, where the Hilbert space is replaced by the "larger" Fock space, as well as a Lagrangian formulation, based on averaging the action (= exp of the Lagrangian) over "all possible histories" via path integrals. These are generally ill-defined integrals over infinite-dimensional spaces. Nevertheless, there exist powerful tools for expanding them perturbatively, as power series in some "small" parameters such as Planck's constant, around a point corresponding to a "free" limit where the integral becomes Gaussian and can be assigned a value. (Actually, in modern QFT the "small parameters" used are often taken to be dimensionless coupling constants rather than Planck's constant.) The coefficients of the expansion are finite dimensional integrals encoded combinatorially as Feynman diagrams. These power series typically diverge and need to be "regularized" and "renormalized". These processes are somewhat akin to the way we make the Weierstrass $\wp$-function converge, by subtracting a correction term from each summand in an infinite series: a first attempt to write down a doubly periodic meromorphic function in terms of an infinite series of functions leads, unfortunately, to an everywhere divergent series; nevertheless we can obtain a series which does converge, and indeed to a doubly periodic meromorphic function, by subtracting an appropriate constant from each term. A physicist might say that we "cancelled the infinity" of the original series by subtracting "an infinite constant". In any case, it is some of these renormalized values which are confirmed to amazing accuracy by experiment, providing physicists with unshakable confidence in the validity of the technique.

For a physicist, a "theory", either classical or quantum, is usually specified by its Lagrangian L, a functional given by integration over the underlying space X of a Lagrangian density $\mathfrak{L}$, which is a local expression involving the values and derivatives of a given collection of fields. The classical theory considers extrema of L, while the quantum theory averages the action over all possible fields. The largest weight is thus still assigned to the extrema of L, but all "quantum fluctuations" around the extrema are now included. The Lagrangian density typically consists of a quadratic part, plus higher order terms which in a perturbative approach are interpreted as small perturbations. In a theory consisting of scalar fields φ_i, the quadratic term might be $\sum_i(|d\varphi_i|^2 + m_i^2\varphi_i^2)$, where m_i are the masses. A perturbation might include terms like $g_I\varphi^I$, where $I = (i_1, \ldots, i_k)$ is a multiindex, and g_I is a (supposedly small) *coupling constant* giving the strength of an interaction involving the k fields $\varphi_i, i \in I$. (Other types of theories might involve also some fields which are vectors or spinors with respect to the Poincare group.) A typical quantity to be computed is the n-point function

$$\langle x_1, \ldots, x_n \rangle := \int \varphi(x_1) \ldots \varphi(x_n) \exp{(iL(\varphi))} D\varphi,$$

where integration is over the infinite-dimensional space of all φ's, with respect to a (non-existent) measure $D\varphi$, and $L(\varphi) := \int_X \mathfrak{L}(\varphi)$.

A formal approach to this integral is to interpret it as a power series in the coupling constants. Terms of the expansion are described by *Feynman diagrams.* The types of edges in a diagram correspond to the types of fields in the theory. The vertex types correspond to the multiindices I with non-zero g_I. A φ^4 term thus corresponds to a vertex where four φ edges meet (this may represent two incoming particles which collide and scatter away), while a $\varphi\psi\bar{\psi}$ vertex can represent an electron ψ annihilating a positron $\bar{\psi}$ to produce a photon φ. The *Feynman rules* assign to each diagram a 4ℓ dimensional integral, where ℓ is the number of loops, i.e. first Betti number, of the diagram, and 4 is the dimension of space-time: the integration is over all ways of assigning a momentum (a 4-vector) to each oriented edge, subject to Kirchhoff's law of null-sum at each vertex. The integrand for each diagram is the product over all edges of a basic function (*the propagator*) of the momentum assigned to that edge, times a factor involving the coupling constant g_I at each vertex of type I, and an overall numerical factor accounting for graph automorphisms. The n-point function, or rather its Fourier tranform, is then expanded as the sum of all possible diagram integrals with n external legs.

One problem is that some of these Feynman integrals may be divergent. Of course, even if they converge individually, their sum may not. So the individual integrals need to be *regularized*, and the entire theory needs to be *renormalized.* A theory is said to be *superrenormalizable* if all divergences in it disappear after a finite number of its divergent diagrams are made finite by adjusting some of the parameters. In a *renormalizable* theory there are infinitely many divergent diagrams, but they can all be renormalized by iteratively adjusting the values of just a finite collection of parameters. There is an easily verifiable necessary condition for renormalizability: each field φ carries a certain dimensionality $[\varphi]$. (This is measured in powers of the mass; time and distance are converted to mass-inverse by setting the speed of light c and Planck's constant h to 1.) Now L, which occurs as an exponent, must have no dimensionality (it is a pure number, independent of units), so each term in $\mathcal{L}$ must have dimensionality 4. This determines the dimensionality of each g_I. The obvious necessary condition for renormalizability is that $[g_I] \geq 0$ for all I: otherwise, the insertion of a vertex of type I in an already divergent diagram will give a new diagram with divergence which is worse. Similarly, a necessary condition for superrenormalizability is that $[g_I] > 0$ for all I, since otherwise insertions into one divergent diagram would yield others with equally bad divergence. These conditions serve to point out the very few candidates for renormalizable theories. Actually proving renormalizability in each case tends to be much harder.

For regularization, we introduce an auxiliary variable Λ so that the integral can be interpreted as (the "limit" of) a function of Λ going to infinity with Λ. For example, Λ can be a momentum cutoff, meaning that the integral is carried out only over the ball of momenta p satisfying $|p| \leq \Lambda$. (Or better, use some approximation of unity by smooth compactly supported cutoff functions.) Another popular variation is dimensional regularization: one writes down the expression for a given integral in a d-dimensional space, and observes that the answer is an analytic function in d for sufficiently small d, typically acquiring a singularity at the relevant dimension $d = 4$. The analogue of Λ in this version is $\exp\left(\frac{1}{4-d}\right)$.

The modern approach to renormalization is based on the *renormalization group flow*. Very crudely, one might illustrate this process as follows. The integral for each diagram D translates into a function $f_D(m, g, p, \Lambda)$ of the masses, coupling constants, external momenta and cutoff (or whatever else we used for Λ). If the number of divergent diagrams is finite (and $\leq$ the number of parameters in L) then the simultaneous level sets of the corresponding f_D will be (or will contain) a family of curves on which Λ is unbounded. So we may hope to find a flow parametrized by Λ along these curves:

$$m = m(\Lambda), \; g = g(\Lambda), \; p = p(\Lambda).$$

In a general renormalizable theory there may be infinitely many f_D, but each may be modified (by a quantity which is bounded as a function of Λ) so there is still a flow tangent to all of them. Instead of attempting to fix the parameters at their "bare" values and taking Λ to infinity, renormalization is accomplished by flowing along this renormalization group flow. The effective, physically measurable value of the parameters thus varies with the scale Λ. Similar phenomena are rather common in statistical physics. For example, the electric charge of a particle in a dielectric appears to be reduced, or *screened*, with distance, as a result of the alignment of the dielectric's charges around it. In QFT this occurs even in vacuum, because of the quantum fluctuations.

A famous series of experiments at the Stanford Linear Accelerator showed that, when bombarded by very high energy electrons, a proton behaves as if it is made of three separate subparticles (quarks) which move freely within it. Yet at low energies (or equivalently, as viewed on a larger distance scale), the quarks are confined within the proton. A major success of QFT was its interpretation of these results in terms of the varying coupling constant g for the strong force. The relevant theory (QCD: quantum chromodynamics) is *asymptotically free*, which means that $g \to 0$ as $\Lambda \to \infty$, and $g \to \infty$ as $\Lambda \to 0$. Thus at high energy scale Λ, the quarks behave as if $g = 0$, i.e. as in the free theory, while at large distance $\Lambda \to 0$ the force becomes stronger and is presumably sufficient to confine the quarks. (This last point is not rigorously proved; Seiberg and Witten obtained a model for confinement in the supersymmetric analogue.)

An often computable quantity which describes the behavior of a coupling constant is its *beta function*, $\beta := \Lambda \partial g / \partial \Lambda$. An asymptotically free theory corresponds to $\beta < 0$ for small values of g. In the boundary case where β is identically 0, the theory is scale-invariant.

An additional complication is involved in the quantization of theories which are (that is, their Lagrangian is) gauge-invariant. The problem with gauge-invariant theories stems from the degeneracy of their Lagrangian. In classical gauge theory, the classical solutions (extrema of the action) are not isolated, but come in entire gauge-group orbits. The equivalence of the Lagrangian and Hamiltonian descriptions breaks down: this equivalence is based on a Legendre transform, which makes sense only near isolated points. In the quantized theory, the path integral needs to be taken not over all paths, but over gauge-equivalence classes. This is seen already for the free (unperturbed) theory: the value of a Gaussian integral is given by the inverse determinant of the operator in the exponent; degeneracy (or gauge invariance) implies that this determinant vanishes, and needs to be replaced by a determinant on an appropriate transversal slice or quotient. For quantized Yang-Mills, or gauge, theories, a good reference is [FS]. The reward for handling these additional complications is a collection of theories which are renormalizable and sometimes asymptotically free: the beta-function of the basic theory (corresponding to Maxwell's in vacuo) is negative. The coupling to matter increases beta, but it remains negative when the number of quarks is small enough.

1.4 Supersymmetry

Supermathematics is the habit of adding the prefix "super" to ordinary, commutative objects, to denote their sign-commutative generalization. Thus, a super space is a locally ringed topological space whose structure sheaf $\mathcal{O} = \mathcal{O}_0 \oplus \mathcal{O}_1$ is $\mathbb{Z}_2$ graded, $\mathcal{O}_0$ is central, and multiplication $\mathcal{O}_1 \times \mathcal{O}_1 \to \mathcal{O}_0$ is skew symmetric. Affine super space $\mathbb{R}^{m,n}$ is the topologcal space $\mathbb{R}^m$ with the sheaf $C^\infty(\mathbb{R}^m) \otimes \Lambda^\cdot(\mathbb{R}^n)$ and the $\mathbb{Z}_2$ grading coming from the $\mathbb{Z}$-grading of the exterior algebra. An (m,n)-dimensional super manifold is a super space which is locally isomorphic to $\mathbb{R}^{m,n}$ (The isomorphisms are, of course, isomorphisms of $\mathbb{Z}_2$-graded algebras, and need not preserve any $\mathbb{Z}$-grading. A typical automorphism of $\mathbb{R}^{1,2}$, with even coordinate x and odd coordinates θ_1, θ_2, may send x

to $x + \theta_1\theta_2$. So the identification of $\mathcal{O}$ as C^∞ tensor exterior algebras is not preserved under coordinate change.) All of calculus extends to super manifolds. The only non-obvious point is that the transformation law for differential forms involves the determinant of the transformation in the even variables, but the inverse determinant of the transformation in the odd variables, and a hybrid—the *Berezinian*—for a general transformation mixing the parities. The resulting operation of *integration* looks, in the odd directions, more like usual *differentiation*, i.e., in local coordinates $x_1, \ldots, x_m, \theta_1, \ldots, \theta_n$ it reads off the coefficient of the top odd part, $\prod_{i=1}^n d\theta_i$.

The infinitesimal symmetries of a super manifold are described by a super Lie algebra: a $\mathbb{Z}_2$-graded algebra $\mathfrak{g} = \mathfrak{g}_0 \oplus \mathfrak{g}_1$ with a $[\ ,\ ]$ operation which is sign-antisymmetric and satisfies the only reasonable version of the Jacobi identity:

$$\sum_{i=1}^{3} (-1)^{x_{i-1}x_{i+1}} [X_{i-1}, [X_i, X_{i+1}]] = 0,$$

where the X_i ($i \in \mathbb{Z}_3$) are in $\mathfrak{g}_{x_i}$. In particular $\mathfrak{g}_0$ is an ordinary Lie algebra, and $\mathfrak{g}_1$ is a $\mathfrak{g}_0$-module with a $\mathfrak{g}_0$-valued symmetric bilinear form $[\ ,\ ]$ with respect to which $\mathfrak{g}_0$ acts as derivations:

$$[X, [\theta_1, \theta_2]] = [[X, \theta_1], \theta_2] + [\theta_1, [X, \theta_2]],$$

for $X \in \mathfrak{g}_0$, $\theta_i \in \mathfrak{g}_1$. The basic example is the Poincare super algebra $\mathbf{p} = \mathbf{p}_0 + \mathbf{p}_1$, whose even part $\mathbf{p}_0$ is the usual Poincare algebra, and whose odd part $\mathbf{p}_1$ is the spinor module S of the Lorentz group $\mathrm{SO}(3,1)$. (The Poincare group P_0 acts, via its quotient $\mathrm{SO}(3,1)$, on $\mathbf{p}_1$; this induces the infinitesimal action of $\mathbf{p}_0$.) The pairing $\gamma\colon \mathbf{p}_1 \times \mathbf{p}_1 \to \mathbf{p}_0$ is Clifford multiplication, with image the translation subgroup of $\mathbf{p}_0$. A physicist would write everything in coordinates, so γ becomes the collection of Pauli or Dirac matrices (depending on whether real or complex forms are used). The N-extended super Poincaré algebra has the same $\mathbf{p}_0$, but $\mathbf{p}_1$ is replaced by $S^{\oplus N}$, the sum of N orthogonal copies of the spinors.

A supersymmetric ("SUSY") space is the super analogue of a homogeneous space for the super Poincaré algebra or its N-extended version. An affine example is $(V, \mathcal{O})$, where V is a vector space with non-degenerate quadratic form (of signature $(3,1)$, in our case), and $\mathcal{O}$ is given in terms of the spinor module $S = S(V)$ as $C^\infty(V) \otimes \Lambda^{\cdot}(S^{\oplus N})$. For more details on SUSY spaces, we refer to [Be], where they are defined to be super manifolds modelled on the above affine example.

An $N = 1$ SUSY theory is a theory (i.e. a Lagrangian involving certain fields) with an infinitesimal action of the super Poincaré algebra. Likewise, a theory "with N supersymmetries" has an action of the N-extended Poincaré super algebra. These theories can often, though not always, be described in terms of a super Poincaré invariant Lagrangian involving superfields on a SUSY space. The ordinary Lagrangian, on ordinary space, is recovered by Berezin integration along all the odd directions. The classic reference on supersymmetry is [WB]. A possibly more friendly introduction is [Fr], and the most geometrical version is [Be]. The first two references include a study of the supersymmetric version of Yang-Mills theory.

In physics, there are two types of particles: bosons, which combine with each other freely, and fermions, whose combinations are restricted by Pauli's exclusion principle. Mathematically, the distinction is roughly this: consider a collection of n particles with distinct physical properties. If H_i is the quantum Hilbert space of states of the ith particle, then the space of states of the entire collection is $\otimes_{i=1}^{n} H_i$. Now if instead we have n identical particles, there are fewer distinct states of the ensemble: for bosons we get $\mathrm{Sym}^n H$, while for fermions, $\Lambda^n H$. In four dimensional theories, bosonic particles ordinarily live in representations of the Poincare group with integer spin, while fermions have half-integer (that is, non-integer) spin. (This is the "spin-statistics" theorem.) Most "matter" particles (electrons, protons, quarks) are fermions; photons are bosons.

At a somewhat superficial level, the motivation for introducing supersymmetry is to be able to treat the two types of particles uniformly. A complementary and deeper reason is based on the no-go theorem of Coleman and Mandula (cf. [Fr], [WB]). This says that the most general algebra of symmetries of a class of quantum theories satisfying some rather reasonable non-degeneracy assumptions (one of these is non-freeness, another is the existence of massive particles) is of the form $\mathbf{p}_0 \oplus \mathfrak{g}$, where $\mathbf{p}_0$ is the Poincaré algebra of space-time isometries, and $\mathfrak{g}$ is the gauge algebra. The point is that this is a direct sum (of algebras), so there can be no non-trivial mixing of the two symmetries. Since symmetries are equivalent to conserved quantities, this implies for instance that any conserved quantity ("current") other than energy-momentum and angular momentum (which come from $\mathbf{p}_0$) must transform as a scalar under $\mathbf{p}_0$. This is unsatisfactory, since some free theories (e.g. of one real scalar field plus one real vector field, [Fr]; such theories are not covered by the Coleman-Mandula theorem) do admit conserved vectors and tensors, which should survive under small perturbation.

The tasks of mixing the two particle types and of avoiding the Coleman-Mandula restriction can fortunately be accomplished simultaneously through the introduction of super Lie algebras: even elements preserve particle types, odd elements exchange them. The even part of the algebra obeys Coleman-Mandula, but there is room for odd symmetries and the corresponding odd, or spinorial, conserved quantities. A theorem of Haag, Sohnius and Lopuszanski classifies the possible super Lie algebras of symmetries of physically interesting theories: there are the N-extended super Poincare algebras, as well as some central extensions and variants including an (even) gauge algebra $\mathfrak{g}$ acting on the odd sector.

1.5 $N = 2$ Super Yang-Mills

The setting for Montonen-Olive duality [MO], for the work of Seiberg-Witten [SW1, SW2], and for the development discussed in later sections, is $N = 2$ supersymmetric Yang-Mills theory (SYM) in four-dimensional space. The Lagrangian for this theory is usually written in terms of super fields on an affine SUSY space [WB], making the super Poincare invariance evident. When written out explicitly in terms of its component fields, the Lagrangian is quite complicated. It starts with a purely bosonic term analogous to the usual Yang-Mills Lagrangian $\int \mathrm{Tr}(F \wedge *F)$. Additional terms, involving additional fields, are required for the $N = 2$ supersymmetry. The fields involved are grouped into $N = 1$ multiplets (representations of the $N = 1$ super Poincare algebra), which in turn combine into $N = 2$ multiplets.

Such a theory depends, of course, on the choice of a compact gauge group G. For a given G there is the pure $N = 2$ theory, analogous to Maxwell's equations in vacuo, containing only those fields and terms in the Lagrangian required for supersymmetry. There are also various theories in which some combination of additional fermionic ("matter") particles is thrown in. We will return to these shortly.

Each SYM theory has a moduli space B of vacuum states (= eigenstates of the Hamiltonian or energy operator corresponding to the lowest eigenvalue). This is of course a consequence of the degeneracy, or of the gauge invariance, of the Lagrangian: a non-degenerate Lagrangian should correspond to a unique vacuum, and an ordinary degeneracy is obtained when two eigenvalues happen to coincide, so there are two (or more) independent vacua. The situation in gauge theory is that there is a continuum B of independent vacua. The complex dimension of B is the rank r of G, and we can in fact describe a coordinate system (u_i), $i = 1, \dots, r$, on it. First, the classical analogue: the classical potential is

$$V(\varphi) = \frac{1}{g^2} \mathrm{Tr}[\varphi, \varphi^\dagger]^2,$$

where g is a coupling constant (a parameter in the Lagrangian), and φ is a Higgs field—a complex field transforming as a scalar under Poincare, and in the adjoint representation of G. Since $[\varphi, \varphi^\dagger]$ is self-adjoint, the minima occur precisely when φ commutes with $\varphi^\dagger$. This locus is G-invariant: it is the union of the G-orbits of elements in the Cartan subalgebra $\mathbf{t}$ of the complexified Lie algebra $\mathfrak{g}_{\mathbb{C}}$. The quotient by G is thus $B = \mathbf{t}/W$, where W is the Weyl group. For example, when $G = \mathrm{SU}(n)$ we have $\mathbf{t} \approx \mathbb{C}^{n-1}$, and the quotient is again $\mathbb{C}^{n-1}$, with coordinates $u_2 = \mathrm{Tr}\ \varphi^2, \dots, u_n = \mathrm{Tr}\ \varphi^n$. (The reader will no doubt recognize this as a (real) symplectic reduction: the action of G on $\mathfrak{g}_{\mathbb{C}}$ is Hamiltonian with respect to the natural translation-invariant symplectic structure induced on the real vector space underlying $\mathfrak{g}_{\mathbb{C}}$ by the imaginary part of the Killing form. The moment map sends φ to $\mathrm{Tr}[\varphi, \varphi^\dagger]$, and the symplectic quotient is $\mathfrak{g}_{\mathbb{C}}//G = B = \mathbf{t}/W$. We will run into the complex analogue of symplectic quotients in the discussion of linearity in the next section.)

Now the constraints imposed by $N = 2$ supersymmetry (i.e. the difficulty of writing down a supersymmetric Lagrangian) imply that the quantum moduli space remains the same B, as an abstract algebraic variety, so it still has the coordinates u_i. But we will see that just about all other features of B change in going from the classical to the quantum: for example, this applies to the Kähler metric on it, the locus of singularities (of the metric), the flat structure, and the relation of the u coordinates to the natural fields in the theory.

A central quantity in any SYM theory is the electric charge a, which is the vector of eigenvalues of the Higgs field φ. It lives in the Cartan, $\mathbf{t}$. For SU(2) this is a complex number, as is u. We can take φ to have eigenvalues $\pm a/2$, and then $u = \mathrm{Tr}\ \varphi^2 = \frac{a^2}{2}$, in the classical theory. In the quantum theory, this relationship still holds approximately as $u \to \infty$, by the asymptotic freedom of the theory. But for finite u we will see that the relationship is very different.

A crucial property of these theories is that their low-energy (read: real-world observable) behavior can be described in terms of only a finite number of parameters—the coordinates u_i on the quantum moduli space B. This description is via a "low energy effective Lagrangian" $\mathcal{L}_{\text{eff}}$, obtained from the full Lagrangian through a process of averaging over all heavy degrees of freedom. So all measurable quantities become (locally) functions on B. In particular, this applies to the components $a_i = a_i(u)$ of the electric charge. Supersymmetry implies that these functions are holomorphic, and we will use them as alternative local coordinates on B.

While the existence of $\mathcal{L}_{\text{eff}}$ is guaranteed by the general theory, its actual computation can be extremely difficult. The supersymmetry allows $\mathcal{L}_{\text{eff}}$ to be expressed locally in terms of a single holomorphic function $\mathcal{F} = \mathcal{F}(a)$, the prepotential. (This is due to Seiberg, cf. [SW1], (2.7) and (2.11). Note that we need $\mathcal{F}$ as a function of a, not of u.) All other quantities in the theory can then be expressed in terms of $\mathcal{F}$ and the coordinates a. This includes the vector

$$a^D = \frac{d\mathcal{F}}{da},$$

whose physical significance will appear shortly; the matrix

$$\tau(a) = \frac{d^2\mathcal{F}}{da^2} = \frac{da^D}{da},$$

which is a complexified coupling constant (for SU(2), the low-energy effective values of the real coupling constant g (encountered in the potential $V(\varphi)$) and of a phase angle θ are given by $\tau = \frac{\theta}{2\pi} + \frac{4\pi i}{g^2}$. In particular, $\mathcal{F}(a)$ must be such that $\mathrm{Im}(\tau) > 0$); the Kähler metric

$$ds^2 = \mathrm{Im}(\tau\, da\, d\bar{a}) = \mathrm{Im}(da^D\, d\bar{a})$$

which was mentioned above; and its Kähler potential

$$K = K(a, \bar{a}) = \mathrm{Im}\left(\sum_i a_i^D \bar{a}_i\right) = \mathrm{Im}\left(\sum_i \frac{d\mathcal{F}}{da_i}\bar{a}_i\right).$$

The effective Lagrangian itself is given as a sum of kinetic and potential parts, each a Berezin integral over an $N = 1$ SUSY affine space of dimensions (4,4) and (4,2), respectively:

$$\mathcal{L}_{\text{eff}} = \frac{1}{4\pi}\int K(\Phi, \bar{\Phi})d^4\eta + \frac{1}{8\pi}\operatorname{Im}\int \tau(\Phi)|W|^2 d^2\eta.$$

Here Φ and W are the two $N = 1$ multiplets which together make up the $N = 2$ multiplet containing the Higgs field φ. (φ and a fermion ψ are the fields in the "chiral multiplet" Φ. The "vector multiplet" W contains another fermion λ and the gauge field A. It can be thought of as the field strength, $N = 1$ super analogue of the curvature F in ordinary YM.) The integrands involve an analytic function (τ) or the real part of one (K); this function is to be applied formally to the super fields Φ, W and their conjugates. The integration over the four (or two) odd space directions η is interpreted via Berezin's recipe as a derivative, or as the reading off of the relevant coefficients. (The second term is actually written as an integral over a (4,2)-dimensional "chiral" quotient of the full (4,4)-dimensional SUSY space, in which two of the odd directions have been reduced.)

What is missing is an explicit expression for $\mathcal{F}$. Traditionally, $\mathcal{F}$ was written as an infinite sum involving its classical ("tree level") value, the one-loop correction which is logarithmic in a (higher loop contributions were known to vanish), and an infinite sum of mysterious instanton corrections involving negative powers of a. The breakthrough came when Seiberg and Witten succeeded, in [SW1], in computing $\mathcal{F}$ (and hence $\mathcal{L}_{\text{eff}}$, ds^2, etc.) exactly, for the pure $N = 2$ theory with $G = \mathrm{SU}(2)$. Their solution is based on the global properties of the theory.

1.6 Duality

The original conjecture of Montonen and Olive [MO] was that the low energy theory should be invariant under a duality exchanging electric and magnetic charges, and also exchanging the coupling constant g with its inverse. In other words: electric charge in the strongly coupled theory (large g) behaves like magnetic charge at weak coupling, and vice versa. Seiberg and Witten found that a modified version of this duality holds in pure $N = 2$ SYM for SU(2). They applied the super version of the Hodge $*$ operator to the Lagrangian $\mathcal{L}_{\text{eff}}$, added some variables (Lagrange multipliers) and integrated out others, and ended up with a new Lagrangian $\mathcal{L}^D_{\text{eff}}$ which is equivalent to the original, and has the same form, but in terms of a new quantity $a^D := \frac{d\mathcal{F}}{da}$:

$$\mathcal{L}^D_{\text{eff}}(a^D) = \mathcal{L}_{\text{eff}}(a).$$

The same transformation sends a^D to $-a = -\frac{d\mathcal{F}}{da^D}$. (This must be so, since the first term in $\mathcal{L}_{\text{eff}}$ changes sign when a, a^D are simply exchanged.) The *physical meaning* of a^D is therefore that of the dual, or magnetic, charge. As to the coupling constants, the defining relations

$$\tau = \frac{da^D}{da}, \quad \tau^D = \frac{d(-a)}{da^D}$$

together with the chain rule, imply that the actual duality transformation is

$$\tau^D = \tau^D(a^D) = -\tau(a)^{-1}.$$

This specializes to the Montonen-Olive transformation

$$g^D = g^{-1}$$

when $\theta = 0$, but not otherwise.

There is another, more elementary, transformation which preserves the Lagrangian: the θ angle is defined only modulo $2\pi\mathbb{Z}$, so τ is defined modulo $\mathbb{Z}$. There is therefore a transformation which fixes a and sends $\tau \mapsto \tau + 1$. Since $\tau = \frac{da^D}{da}$, this requires that a^D goes to $a^D + a$.

The two transformations can be represented by the matrices

$$S = \begin{pmatrix} 0 & 1 \\ -1 & 0 \end{pmatrix} \qquad T = \begin{pmatrix} 1 & 1 \\ 0 & 1 \end{pmatrix},$$

acting linearly on the 2-vector $\binom{a^D}{a}$ and fractional-linearly on τ. Since S and T generate $\mathrm{SL}(2,\mathbb{Z})$,we see that the theory has an $\mathrm{SL}(2,\mathbb{Z})$ action. (For other groups, $\mathrm{SL}(2,\mathbb{Z})$ is replaced by a subgroup of finite index in $\mathrm{Sp}(2r,\mathbb{Z})$.)

1.7 The Seiberg-Witten solution

The foregoing suggests that what lives intrinsically over a generic $u \in B$ is not the electric charge $a(u)$ but the unimodular lattice $\mathbb{Z}a(u)+\mathbb{Z}a^D(u)$ of all charges (electric, magnetic, and "dyonic", or mixed). As u varies, we get a $\mathbb{Z}^2$ local system V over B minus some singular locus, together with a homomorphism to (the trivial bundle with fiber) $\mathbb{C}$. Dually, this homomorphism is equivalent to a holomorphic section $\binom{a^D}{a}$ of the vector bundle $V_{\mathbb{C}}$ (i.e. of the sheaf $V \otimes \mathcal{O}_{B \setminus \mathrm{sing}}$). The plan is to identify the local system (variation of Hodge structure, really) V, and then to fix the section $\binom{a^D}{a}$, which in turn will determine the prepotential $\mathcal{F}$ (up to constant).

Seiberg and Witten identify V by finding its monodromy. Near $u = \infty$ we know that $u \sim \frac{1}{2}a^2$, so the monodromy around ∞ sends $a \mapsto -a$. The known non-trivial logarithmic (one-loop) term in the perturbative expansion of $\mathcal{F}$ then fixes the shift in a^D: the monodromy is

$$M_\infty = \begin{pmatrix} -1 & 2 \\ 0 & -1 \end{pmatrix}.$$

In particular, the local system is non trivial, so τ is a non-constant holomorphic function on the universal cover of $B \setminus \mathrm{sing}$, with values in $\mathbb{H}$. The singular locus must therefore contain at least two points (in addition to ∞). From a physical argument, Seiberg and Witten deduce that the system is as simple as possible: exactly three singularities, which can be relocated to $u = \pm 1$ and ∞ by an automorphism. Duality instructs that the finite monodromies should be conjugate to $-M_\infty$. Seiberg and Witten label things so that a^D vanishes at $u = 1$, yielding the monodromy

$$M_1 = \begin{pmatrix} 1 & 0 \\ -2 & 1 \end{pmatrix}$$

which then determines M_{-1}.

The monodromies generate the level-2 congruence subgroup $\Gamma(2) \subset \mathrm{SL}(2, \mathbb{Z})$, and the local system itself can now be identified with the fiber cohomology of the universal level-2 elliptic curve

$$E_u\colon\ y^2 = (x+1)(x-1)(x-u)$$

over the u-line $\mathbb{C} \setminus \{\pm 1\}$.

The complexification $V_{\mathbb{C}}$ has a global trivialization in terms of the holomorphic 1-form $\lambda_1 = \frac{dx}{y}$ and the residueless meromorphic form $\lambda_2 = \frac{x\,dx}{y}$. We can also fix an integral homology basis, say $\gamma =$ loop around the branch points $0, 1$, and $\gamma^D =$ loop around $1, u$. To determine $a(u)$ and $a^D(u)$, we use the condition that $\mathrm{Im}(\tau) > 0$, where $\tau(u) := \frac{da^D/du}{da/du}$. One geometric solution is to take $\tau(u)$ as the period

$$\tau_u := \frac{\oint_{\gamma^D} \lambda_1}{\oint_\gamma \lambda_1}$$

of the elliptic curve E_u; rigidity implies that this is in fact the only solution. We then find that, up to a multiplicative constant, a and a^D are the periods over γ, γ^D of the meromorphic 1-form

$$\lambda = \frac{y\,dx}{x^2 - 1} = \frac{(x-u)dx}{y} = \lambda_2 - u\lambda_1.$$

1.8 Adding matter

In a quantum field theory, a matter particle is represented by a fermion field ψ which, in the case of a gauge theory, needs to be in some representation of G. It is added (or: coupled) to the theory by adding to the YM Lagrangian a term corresponding to the Lagrangian of the free particle (this includes a parameter, the particle mass m), and another term corresponding to its interaction with the YM field. In SYM the field needs to be in a representation ρ of G, tensored with an $N = 2$ multiplet, to preserve the SUSY. (The relevant multiplet, which is built from the fermion ψ, is called a *hypermultiplet*; we have already encountered the other basic type, the $N = 2$ *vector multiplet*, which contained the scalar Higgs field φ.) If the number of particles added is not too large, the theory remains asymptotically free ($\beta < 0$) or becomes scale invariant ($\beta = 0$), and in either case retains many properties of the pure SYM theory. There still is an r-dimensional moduli space B. At $u \in B$ there are r electric charges $a_i(u)$ and r dual, magnetic, charges, $a_i^D(u)$, each charge living in the Cartan $\mathbf{t} \approx \mathbb{C}^r$ as before. Duality is modified somewhat: instead of $\mathrm{Sp}(2r, \mathbb{Z})$ transformations acting on $\binom{a^D}{a}$, one gets affine symplectic transformations of the form

$$\begin{pmatrix} a^D \\ a \end{pmatrix} \longmapsto R \begin{pmatrix} a^D \\ a \end{pmatrix} + \sum_{i=1}^{N_f} m_i \begin{pmatrix} n_i^D \\ n_i \end{pmatrix}$$

where $R \in \mathrm{Sp}(2r, \mathbb{Z})$, the m_i are the masses of the N_f particles added, and n_i, n_i^D are integral $r \times r$ matrices. ("N_f" stands for number of *flavors*. The number of *colors* N_c is the dimension of the representation space of ρ. The terminology comes from quantum chromodynamics (QCD), a gauge theory with $G = \mathrm{SU}(3)$.)

It is reasonable to guess that this structure can be implemented by a family of r-dimensional abelian varieties A_u over $u \in B \smallsetminus (\text{sing})$: The $r \times 2r$ period matrix of A_u in some integral homology basis γ^D, γ is $\left(\frac{da^D(u)}{du}, \frac{da(u)}{du}\right)$. (The polarization on the complex torus corresponding to the lattice generated by columns of this matrix is given by the Dirac quantization condition.) On the total space of the family there is a meromorphic 1-form λ, the *Seiberg-Witten differential.* There are N_f divisors D_i along which λ acquires a pole with constant residue $\frac{m_i}{2\pi i}$. The charges a^D, a can then be recovered as the periods of λ over γ^D, γ. These have precisely the ambiguity needed: monodromy of the cycles γ^D, γ acts as $\mathrm{Sp}(2r, \mathbb{Z})$, while moving such a cycle across the divisor D_i changes the period by an integer multiple of m_i. The masses m_i acquire a cohomological interpretation: the 2-form $\sigma = d\lambda$ is holomorphic (in fact, it is a holomorphic symplectic form) on the total space, and its de Rham cohomology class can be expressed in terms of the fundamental classes of the D_i:

$$[\sigma] = \sum m_i [D_i].$$

In particular, as we vary the masses, we obtain a family of algebraic symplectic manifolds in which the symplectic class varies *linearly* in the parameters.

In the case of scale-invariant theories ($\beta = 0$), the theory and therefore also its geometric model depend on one parameter in addition to the masses. This is τ_{cl}, the classical limit ($u \to \infty$) of the quantum coupling constant $\tau = \tau(u)$. In asymptotically free theories ($\beta < 0$) we have

$$\lim_{u \to \infty} \tau(u) = \infty,$$

and instead of τ the theory depends on a scale parameter Λ. In the [SW1] solution, this Λ is set to 1 for convenience.

In [SW2] Seiberg and Witten show that the above geometric model indeed leads to an exact solution of the various SU(2) SYM theories with matter. For each of these, the solution requires a family parametrized by the vector

$m = (m_i)$ of masses (and by the additional parameter τ or Λ, which we omit from notation), of elliptic surfaces

$$X_m \to B \approx \mathbb{C}$$

with meromorphic 1-forms λ_m whose residues equal $\frac{m}{2\pi i}$. The solution satisfies numerous consistency checks (the way one theory is known to flow to another in the limit where some parameter, such as a mass, goes to 0 or ∞, is reflected in corresponding degenerations of one geometric model to another) and has all the internal symmetry of each of the physical theories.

The number of particles which can be added to a SYM theory is determined as follows. The Killing form on the reductive Lie algebra $\mathfrak{g}$ determines an element c_2 in the center of the universal enveloping algebra $\mathrm{U}(\mathfrak{g})$. By Schur's lemma, c_2 acts on the total space of each irreducible representation ρ as multiplication by some scalar $c_2(\rho)$. (This c_2 is closely related to the second chern class of a bundle: Let V, V' be two vector bundles on a surface associated via representations ρ, ρ' to the same G-bundle. Then

$$\frac{c_2(\rho')}{c_2(\rho)} = \frac{c_2(V')}{c_2(V)},$$

where c_2 on the right denotes Chern classes. This can be used to give another definition of c_2 of a representation.) In any case, the β function of the SYM theory with particles in representations ρ_i, $i = 1, \ldots, N_f$, is

$$\beta = c_2(\mathrm{ad}) - \sum_i c_2(\rho_i)$$

where ad is the adjoint representation. For the classical algebras we can adjust the scaling of the Killing form so that $c_2(F) = 1$ where F is the fundamental representation. With this normalization, $c_2(\mathrm{ad}) = 2h^\vee_{\mathfrak{g}}$, where $h^\vee_{\mathfrak{g}}$ is the dual Coxeter number. For $\mathrm{SU}(n)$ we have $h^\vee_{\mathfrak{g}} = n$, so the SU(2) theory, for example, remains asymptotically free with$N_f = 0, 1, 2$ or 3 particles in the fundamental representation F, and becomes scale invariant for $N_f = 4$ fundamental particles or for one particle in the adjoint representation. These are precisely the cases studied and solved in [SW2].

It is natural to ask whether the four dimensional, $N = 2$ SYM theory with arbitrary gauge groups G and collection of particles ρ_i with $\beta \leq 0$ can be similarly solved in geometric terms via families of abelian varieties and meromorphic one-forms. This question will occupy us in the remainder of this work.

2 Why integrable systems

2.1 Algebraically integrable systems

An algebraically integrable system consists of a complex algebraic manifold X, an everywhere non-degenerate, closed holomorphic $(2, 0)$-form σ on X, and a morphism $\pi : X \to B$ whose general fiber is a (polarized) abelian subvariety $X_b := \pi^{-1}(b)$ which is Lagrangian with respect to σ. In practice, the object which is naturally given is often an affine open $X_0 \subset X$, the fibers being affine subsets of abelian varieties. This is a straightforward algebraic analogue of the (real, C^∞) notion of integrable system in classical mechanics. As there, any function H ("Hamiltonian") on the base B determines a 1-form $d\pi^* H$ on X, and when contracted with the symplectic form σ it gives a vector field v_H on X, tangent to the fibers of π. The pullbacks $\pi^* H_1, \pi^* H_2$ of two functions on B Poisson-commute, and in particular the corresponding vector fields commute.

Locally on the base of an algebraically integrable system there is a natural holomorphic flat structure, depending on a discrete choice. The fiber homology $H_1(X_b, \mathbb{Z})$ has a non-degenerate skew-symmetric pairing (induced by the polarization). The choice involved is that of a Lagrangian subspace $V_{\mathbb{Z}}^*$ for this pairing. This is often achieved by specifying a symplectic basis $\alpha_1, \ldots, \alpha_g, \beta_1, \ldots, \beta_g$ of $H_1(X_b, \mathbb{Z})$, and taking the subspace $V_{\mathbb{Z}}^*$ spanned by the α cycles. This choice needs to be made for one $b \in B$, and (by Gauss-Manin) is propagated to nearby fibers. Integration over these α-cycles gives a local trivialization of the relative cotangent bundle $\pi_* \omega_X$, and the symplectic form σ converts this to a flat structure on (the tangent bundle of) B, modelled on $V := \mathrm{Hom}(V_{\mathbb{Z}}^*, \mathbb{C})$.

Locally on the flat base B, we can ask what kind of data is required to specify an algebraically integrable system inducing the given flat structure. We think of the family of Abelian varieties, plus choice of α and β cycles, as specified by its period map

$$p : B \to \mathbb{H}_g$$

to the Siegel half space $\mathbb{H}_g$ of $g \times g$ symmetric complex matrices with positive imaginary part. The total space $\pi : X \to B$ is then retrieved as pullback of the universal abelian variety $\mathfrak{X}_g \to \mathbb{H}_g$. The question then is to understand the condition on p for there to be a symplectic form on X for which π is a Lagrangian fibration. The fixing of the flat structure is achieved by thinking of $\mathbb{H}_g$ as open subset of $\mathrm{Sym}^2 V^*$, where V is the tangent space to B (at any $b \in B$). The derivative of the period map is

$$dp_b \in \mathrm{Hom}(V, \mathrm{Sym}^2 V^*) = V^* \otimes \mathrm{Sym}^2 V^*.$$

A straightforward computation done in [DM2] shows that the condition is that dp_b lives in the subspace $\mathrm{Sym}^3 V^*$, for all $b \in B$. So an integrable system with base B is specified by a field of cubics on the tangent bundle of B.

This field of cubics is subject, of course, to a strong integrability condition. In flat local coordinates z_i on B, let $(p_{ij}(z))$ be the period matrix and $(c_{ijk}(z))$ the field of cubics. Then equality of mixed partials

$$\frac{\partial p_{ij}}{\partial z_k} = c_{ijk} = \frac{\partial p_{ik}}{\partial z_j}$$

implies the existence of functions w_i (locally) on B such that

$$p_{ij} = \frac{\partial w_i}{\partial z_j}.$$

These again have symmetric partials,

$$\frac{\partial w_i}{\partial z_j} = p_{ij} = \frac{\partial w_j}{\partial z_i},$$

so there is a single function $\mathcal{F}$ on B, called the *prepotential*, satisfying:

$$\begin{aligned} w_i &= \frac{\partial \mathcal{F}}{\partial z_i} \\ p_{ij} &= \frac{\partial^2 \mathcal{F}}{\partial z_i \partial z_j} \\ c_{ijk} &= \frac{\partial^3 \mathcal{F}}{\partial z_i \partial z_j \partial z_k} \end{aligned}$$

The prepotential determines also a Kähler potential on B

$$K := \operatorname{Im}\left(\sum_i w_i \bar{z}_i\right) = \operatorname{Im}\left(\sum_i \frac{\partial \mathcal{F}}{\partial z_i} \bar{z}_i\right)$$

and its associated Kähler form

$$\frac{i}{2}\partial\bar{\partial}K = \frac{i}{2}\sum_{i,j} \operatorname{Im}(p_{ij}) dz_i \wedge d\bar{z}_j.$$

Equivalently, this Kähler (1,1)-form on B is obtained by integration over the g-dimensional fibers X_b of the $(g+1, g+1)$-form $\sigma \wedge \bar{\sigma} \wedge t^{g-1}$ on X, where t is any $(1,1)$-form on X whose restrictions to fibers give the polarization.

Conversely, given B with coordinates z_i and a holomorphic function $\mathcal{F}$, the above formulas produce an algebraically integrable system over the open subset of B where $\operatorname{Im}\left(\frac{\partial^2 \mathcal{F}}{\partial z_i \partial z_j}\right) > 0$: the family of abelian varieties is determined by its periods p_{ij}, and the symplectic form is given by the cubic c_{ijk}. The slightly weaker notion of an analytically integrable system (just replace abelian varieties by polarized complex tori) is recovered over the open subset where $\operatorname{Im}\left(\frac{\partial^2 \mathcal{F}}{\partial z_i \partial z_j}\right)$ is invertible, cf. [DM2]. The Kähler "metric" in this case is still non-degenerate, but possibly indefinite.

We note that the w_i play a role dual to that of the z_i: the z_i give the flat structure on B determined by integration over the α cycles, while the w_i give the flat structure determined by integration over the β cycles. We also note that, up to this point, the z_i as well as the w_i have been determined only up to arbitrary additive constants. In other words, what we have are the vector fields $\partial/\partial z_i, \partial/\partial w_i$, or the 1-forms dz_i, dw_i. We will discuss below the additional choice used in the SW setup to fix these additive constants.

Similar structures have arisen in various more specialized contexts, e.g. in the WDVV equations of conformal field theory and on moduli spaces of Calabi-Yau threefolds, as well as in the Seiberg-Witten setup described in §1. They often go under the title of "special geometry". The point is that these specialized contexts are not really needed: the structure of special geometry on B (consisting of a covering by open subsets on which an appropriate prepotential, dual flat holomorphic coordinate systems, and a Kähler metric are defined), arises naturally on the base of any algebraically integrable system. The converse is also true, locally, as we have just seen. Globally, though, there is an integrability constraint: the monodromy action (say on the coordinates z_i, w_i) in an integrable system involves integral symplectic transformations plus (complex) translations, while special geometry allows the symplectic transformation to be real instead of integral. We refer the reader to [F] for a very clear differential-geometric discussion of special geometry and its relationship to integrable systems.

2.2 Seiberg-Witten Differentials

In the Seiberg-Witten picture of super Yang-Mills, there are the charges a_i and a_i^D which are described locally as holomorphic functions on the quantum moduli space B. There is less ambiguity in choosing these a_i, a_i^D than in the coordinates z_i, w_i on the base of an algebraically integrable system: the group of their linear combinations $\sum n_i a_i + n_i^D a_i^D$ (with integer coefficients n_i) should be uniquely determined in the pure SYM case, and determined modulo the masses of the added particles in general. A natural way to get functions with precisely such ambiguity on the base of an integrable system $\pi : X \to B$ is to choose a meromorphic differential 1-form λ on X and let a_i, a_i^D be its periods over a set of 1-cycles representing α_i, β_i respectively and avoiding the poles of λ. We want a_i, a_i^D to be a possible choice of the coordinates z_i, w_i, which are defined up to translation (once we have fixed the α_i, β_i). In other words, we want

$$da_i = dz_i, \quad da_i^D = dw_i.$$

Since dz_i, dw_i were defined as the contraction of α_i, β_i respectively with the symplectic form σ, the condition becomes simply:

$$d\lambda = \sigma.$$

The behavior of the singularities of λ is constrained by this condition. When restricted to a general fiber X_b, λ will have poles along the union of some irreducible divisors $D_{b,j}$. One constraint is that the residue Res_j of λ along $D_{b,j}$ is (locally in B) independent of b. A *Seiberg-Witten differential* is a meromorphic 1-form λ satisfying $d\lambda = \sigma$ and $\mathrm{Res}_j = m_j$, the mass of the j-th particle. On an algebraically integrable system with a specified Seiberg-Witten differential, the local coordinates z_i, w_i can be specified as the charges a_i, a_i^D, with just the right ambiguity.

One way to describe an integrable system (though by no means the only way!) is in terms of a family $C \to B$ of spectral curves C_b. In the simplest situation, the abelian varieties X_b are just the Jacobians $J(C_b)$. More generally,

X_b may be realized as the Prym of C_b with respect to an involution, or as the generalized Prym with respect to a correspondence, cf. [D1]. An example of such a system, due to Mukai [Mu], is obtained from the Jacobians of a complete linear system $C \to B$ of curves on a symplectic surface $(S, \bar{\sigma})$. The symplectic form σ on X then corresponds via Abel-Jacobi to the pullback to C of $\bar{\sigma}$. Mukai studied the case where S is a K3 surface. Taking $S = T^*E$ to be the cotangent bundle of some curve E yields Hitchin's system on E. One obtains many more examples by allowing S to have a symplectic form $\bar{\sigma}_0$ defined only on an open subset $S_0 \subset S$, and imposing appropriate restrictions on the intersection of the spectral curves C_b with $S - S_0$. For example, we can take S to be the total space of the line bundle $\omega_E(D)$ for some effective divisor D on E, to recover Markman's system [Mn] parametrizing meromorphic Higgs bundles on E. Note that this S has a natural meromorphic 1-form $\bar{\lambda}$, given away from D (i.e. where S can be identified with T^*E) as the action 1-form $\bar{\lambda} = p\,dq$, whose differential is $d\bar{\lambda} = \bar{\sigma}_0$. It follows that the "tautological" 1-form on C (pullback of $\bar{\lambda}$) induces the Seiberg-Witten differential λ on the integrable system $X = J(C/B) \to B$. The system discussed in §3 is of this form.

2.3 Linearity: complexified Duistermaat-Heckman

One feature of the Seiberg-Witten picture which we have not yet discussed is the linear dependence of the symplectic form on the particle masses. From one point of view, this linearity follows from existence of the Seiberg-Witten differential: the equation $\sigma = d\lambda$ implies that the cohomology class of σ is a fixed linear combination of the residues of λ, which are just the masses. On the other hand, linearity can also be interpreted as arising from a complex analogue of the theorem of Duistermaat-Heckman [DH].

Let (X, σ) be a complex symplectic manifold (i.e. the symplectic form σ is holomorphic, of type (2,0)) equipped with a Hamiltonian action of a complex reductive group G. The Hamiltonian property of the action means that there is a moment map $\mu : X \to \mathfrak{g}^*$ to the dual of the Lie algebra. Ignoring bad loci, we will assume (or pretend) that there is a good non-singular quotient X/G. Let $G_a \subset G$ be the stabilizer of $a \in \mathfrak{g}^*$ under the coadjoint action. The symplectic quotient X_a (or $X//_aG$) is defined as

$$X_a := \mu^{-1}(a)/G_a = \mu^{-1}(\mathcal{O}_a)/G,$$

where $\mathcal{O}_a$ is the (coadjoint) orbit of a. It is again a complex symplectic manifold, with form σ_a. As a ranges over orbits of a fixed "type" in $\mathfrak{g}^*$, the topology of X_a remains locally constant, so cohomologies of nearby X_a's of fixed type can be identified. The complex Duistermaat-Heckman theorem says that, under this identification, the cohomology class $[\sigma_a] \in H^2(X_a)$ varies linearly with a.

Consider first the abelian case where $G = T$ is a torus $T = \mathrm{Hom}(\Lambda, \mathbb{C}^*)$. The moment map then factors

$$X \to X/T \to \mathbf{t}^*,$$

and we assume both maps are bundles, at least over some open $\mathbf{t}_0^* \subset \mathbf{t}^*$. The T-bundle $X \to X/T$ has first chern class

$$c_1 \in H^2(X/T, \Lambda^*).$$

For each a in $\mathbf{t}_0^*$, this restricts to a class (still denoted by c_1) in $H^2(X_a, \Lambda^*)$. The theorem in this case says that for $a, a_0 \in \mathbf{t}_0^* = \Lambda \otimes \mathbb{C}$,

$$[\sigma_a] - [\sigma_{a_0}] = \langle c_1, a - a_0 \rangle.$$

Because of possible monodromy of the fibers of X/T over $\mathbf{t}_0^*$, both sides may be multivalued, so we should think of the linearity as being local in a.

Next, let G be an arbitrary reductive group, and consider reduction at a regular semisimple element a. This means that the stabilizer G_a is a maximal torus $T \subset G$. The coadjoint orbits of semisimple elements are parametrized by the quotient $\mathbf{t}^*/W$ of the dual Cartan by the Weyl group. Let $\mathfrak{g}_0^*, \mathbf{t}_0^*$ denote the open subsets of $\mathfrak{g}^*, \mathbf{t}^*$ respectively parametrizing regular semisimple elements. The parameter space for regular semisimple coadjoint orbits is then

$$\mathbf{t}_0^*/W \approx \mathfrak{g}_0^*/G.$$

The moment map in this nonabelian situation no longer factors. The closest we get to factorization is the commutative diagram:

$$\begin{array}{ccccc} & & X_0 & & \\ & \swarrow & & \searrow^{\mu} & \\ X_0/G & & & & \mathfrak{g}_0^* \\ & \searrow & & \swarrow & \\ & & \mathbf{t}_0^*/W = \mathfrak{g}_0^*/G & & \end{array}$$

with $X_0 := \mu^{-1}(\mathfrak{g}_0^*)$. Somewhat surprisingly, it is easy to reduce this situation to the abelian case, by considering $Y := \mu^{-1}(\mathbf{t}^*)$. The symplectic form of X restricts to a symplectic form on Y, and the T-action on Y is Hamiltonian with respect to this form, so there is a moment map $\mu_T : Y \to \mathbf{t}^*$ compatible with μ via the inclusions $Y \subset X$ and $\mathbf{t}^* \subset \mathfrak{g}^*$. Now for $a \in \mathbf{t}_0^*$ there is an obvious identification of symplectic manifolds $X//_aG \approx Y//_aT$, so the dependence of the symplectic class $[\sigma_a]$ of $X//_aG$ on $a \in t_0^*$ is again locally linear. (In the original case with G compact, the multivaluedness is replaced by discontinuities across walls, resulting in $[\sigma_a]$ being linear on each Weyl chamber).

What happens when a is allowed to degenerate? If G is compact, a remains semisimple (though not regular), and there is a fairly obvious linearity for all such a in a given open face $\mathbf{f}_0$ of the Weyl chamber, i.e. a face $\mathbf{f}$ minus its proper subfaces. In the complex reductive case, there are more possibilities: the most natural orbit which corresponds to an a in a complexified open face $\mathbf{f}_0$ consists of elements which are not semisimple but regular, in the sense that their centralizers have the minimal possible dimension. The closure of this orbit consists of a finite number of other orbits. The smallest of these orbits (the unique closed orbit, which is in the closure of each of the others) consists of the semisimple elements. Although we will not encounter non-regular elements in the integrable systems we need in the following section, it seems worthwhile to point out that linearity is indeed a general phenomenon applicable to all strata.

We identify $\mathfrak{g}, \mathbf{t}$ with $\mathfrak{g}^*, \mathbf{t}^*$ via an invariant form. Consider an arbitrary stratum $S \subset \mathfrak{g}$ corresponding to a face $\mathbf{f}_0 \subset \mathbf{f} \subset \mathbf{t}$. S is a bundle over (a finite quotient of the complexified) $\mathbf{f}_0$, and the fibers are (co)adjoint orbits of a fixed type. This bundle has local sections, which can be taken of the form $n + \mathbf{f}_0$, where n is some fixed nilpotent in the reductive subalgebra $\mathbf{l} = \mathbf{l}_\mathbf{f}$ (= common centralizer of all elements of $\mathbf{f}_0$). All elements $a = n + t$ of this section have the same centralizer $\mathbf{c}$ in $\mathfrak{g}$. ($\mathbf{c}$ is the centralizer in $\mathbf{l}$ of n). Since $\mathbf{f}$ is the center of $\mathbf{l}$, it is contained in $\mathbf{c}$, which actually decomposes as $\mathbf{c} = \mathbf{f} + \mathbf{d}$ with $\mathbf{d}$ semisimple. Let $D \subset G$ be the subgroup corresponding to $\mathbf{d}$. The reduction to the abelian case now proceeds as follows: We observe that $\mu^{-1}(n + \mathbf{f}_0)/D$ is symplectic, and has a residual Hamiltonian action of the subtorus $F \subset T$ whose algebra is $\mathbf{f}$. Let μ_F be the moment map to $\mathbf{f}$ (which we identify with $\mathbf{f}^*$). Then for $a = n + t$ as above, we have an isomorphism of symplectic manifolds:

$$X//_aG \approx \mu_F^{-1}(t)/F = (\mu^{-1}(n + \mathbf{f}_0)/D)//_tF.$$

Since the symplectic form on the right varies linearly with t (which parametrizes $a = n + t$), we conclude that $[\sigma_a]$ varies linearly with a, as claimed.

In a sense, there is linearity even as we degenerate from one stratum to another. We illustrate this phenomenon with the following example.

The left action of G on itself lifts to a Hamiltonian action of G on $X := T^*G \approx G \times \mathfrak{g}^*$. The quotient map $X \to X/G$ is

$$\begin{aligned} G \times \mathfrak{g}^* &\to \mathfrak{g}^* \\ (g, a) &\mapsto a, \end{aligned}$$

while the moment map sends

$$(g, a) \mapsto \operatorname{ad}_g^* a.$$

The symplectic quotients

$$X//_aG = \mathcal{O}_a$$

are the coadjoint orbits. These fit, as discussed above, into a finite number of continuous families, each indexed by the W-quotient of some open face. Inside each family we expect linearity. As we move between families, the cohomology (and even the dimension) jumps. Yet there is a way to express linearity uniformly for all $a \in \mathfrak{g}^*$. Let $G_a, \mathfrak{g}_a$ be the stabilizers of a in $G, \mathfrak{g}$ respectively. Then a restricts to $\overline{a} \in \mathfrak{g}_a^*$ which is G_a-invariant, so it gives an element of $(\mathfrak{g}_a^*)^{G_a}$, a bi-invariant 1-formon G_a. This gives a class in $H^1(G_a, \mathbb{C})$, which is then sent to $H^2(\mathcal{O}_a, \mathbb{C})$ by the transgression map (= the coboundary map $E_2^{01} \to E_2^{20}$

in the Leray spectral sequence of the fibration $G \to \mathcal{O}_a$, with fiber G_a, where $E_2^{pq} = H^p(\mathcal{O}_a, H^q(G_a, \mathbb{C}))$). The image is seen to be $[\sigma_a]$, and the construction is obviously linear on each $\mathbf{f}_0 \subset \mathbf{f} \subset \mathbf{t}$, since the stabilizers are actually constant there.

3 Which integrable system

We have seen that all the structures on the quantum moduli space of $N = 2$ SYM predicted by Seiberg and Witten can be realized in terms of algebraically integrable systems. The features of "special geometry" occur in any algebraically integrable system. The distinguished electric and magnetic charges appear as soon as we fix a Seiberg-Witten differential, and perhaps the most natural way to get this is as tautological 1-form on a family of curves contained in the total space of a line bundle $\omega_E(D)$ of meromorphic differentials on some fixed curve E. Since a quantum theory with $\beta = 0$ depends on a parameter $\tau = \tau_{cl}$, it is natural to take E to be the elliptic curve corresponding to this τ. Finally, the linearity properties suggest that our family of systems, parametrized by the masses, could arise as the family of symplectic reductions of a larger symplectic manifold.

There remains the technicality of identifying a specific integrable system for each type of theory, and testing that these theories have reasonable physical properties. In this section we first give several descriptions of the system [DW] which solves the SU(n) theory with one adjoint, and then describe some of the tests which confirm the choice.

3.1 Meromorphic Higgs bundles

Fix a curve E, an effective divisor D on it, and an integer $n \geq 2$. The total space of the system is the moduli space X of equivalence classes of pairs (V, φ), where V is a rank-n vector bundle on E with trivial determinant, $\varphi : V \to V \otimes \omega_E(D)$ is a traceless endomorphism with values in $\omega_E(D)$, and the pairs are subject to an appropriate semistability condition. The base is

$$B := \oplus_{i=2}^{n} H^0(E, (\omega_E(D))^{\otimes i}),$$

and the map $\pi : X \to B$ sends a pair (V, φ) to (the sequence of coefficients of) the characteristic polynomial

$$\det(t \cdot I - \varphi).$$

(It is best to think of t as a fiber coordinate taking values in $\omega_E(D)$.)

The generic fibers of π are abelian varieties. Consider the surface S which is the total space of $\omega_E(D)$. The points $b = (b_2, \ldots, b_n) \in B$ parametrize a linear system of curves $C_b \subset S$, given by the equation $t^n + b_2 t^{n-2} + \cdots + b_n = 0$. Let λ

be the tautological meromorphic 1-form on S. Then the pair (V, φ) determines a sheaf $L := \mathrm{coker}(\varphi - \lambda \cdot Id)$ on S whose support is the curve C_b. In fact, for (V, φ) generic, L is a line bundle on C_b. Thus we have a map $\pi^{-1}(b) \to \mathrm{Pic}(C_b)$. If we were working with $\mathrm{GL}(n)$-bundles rather than $\mathrm{SL}(n)$-bundles (remove the restrictions $\det V \approx \mathcal{O}$, $\mathrm{Tr}(\varphi) = 0$), this would be an isomorphism of $\pi^{-1}(b)$ with a component of $\mathrm{Pic}(C_b)$. The restrictions mean that we get instead a translate of $\mathrm{Prym}(C_b/E)$.

This construction, modelled on Hitchin's system [Hi], works much more generally, cf. [D2]. In particular, the structure group $\mathrm{SL}(n)$ can be replaced by any reductive G. The total space X now parametrizes pairs (V, φ) with V a G-bundle over E, and φ a section of $\mathrm{ad}(V) \otimes \omega_E(D)$. The base is replaced by

$$B := \oplus_{i=1}^{r} H^0(E, (\omega_E(D))^{\otimes d_i}),$$

where r is the rank of G and d_i are the degrees of the basic invariant polynomials on the Lie algebra $\mathfrak{g}$. Once we choose a representation ρ of G, we get a family of spectral curves $\{C_{\rho,b} \mid b \in B\}$ in S as before.

As shown in [D2], [Fa] and further references therein, the generic fiber of π can be identified as (a translate of) a certain distinguished abelian variety $\mathrm{Prym}(C_b)$ which is an isogeny summand in the Jacobian $J(C_{\rho,b})$ for each representation ρ. The most natural way to view this is to abandon the spectral curves $C_{\rho,b}$ altogether, in favor of the cameral cover $C_b \to E$, a Galois cover with group W, the Weyl group of G. This is the pullback to E, via the classifying map $E \dashrightarrow \mathfrak{t}/W \approx \mathfrak{g}/G$ determined locally in E by the point $b \in B$, of the W-cover $\mathfrak{t} \to \mathfrak{t}/W$. ($\mathfrak{t}$ is the Cartan subalgebra, and the classifying map is really a map from the total space of $(\omega_E(D))^*$ to $\mathfrak{t}/W$. The local form above depends on the choice of a local section of the line bundle $(\omega_E(D))^*$, so is really defined only modulo homotheties). Now $J(C_b)$ decomposes according to the irreducible representations of the Galois group W, and the distinguished Prym, $\mathrm{Prym}(C_b)$, is the component corresponding to the action of W on the weights. (The question of the dependence of the spectral Jacobian on ρ was raised in [AM] and studied, in the case of the Toda system, in [Ml]. A solution, under some conditions, was given in [Ka]. The case $E = \mathbb{P}^1$ was analyzed in [AHH] and [B], and the general case is in [D2], [Me]). In this section we will use only the $\mathrm{SL}(n)$ system, postponing the appearance of other groups to §4.

The total space X is not symplectic, but has a Poisson structure [B], [Mn]. In general, a Poisson structure on an algebraic manifold X is a Lie bracket operation on $\mathcal{O}_X$ satisfying the properties of the usual Poisson bracket. Equivalently, it can be given by a section α of $\wedge^2 T_X$ satisfying an integrability condition. If α is non-degenerate on some open set X_0, then its inverse $\sigma = \alpha^{-1}$ is a two-form on X, and the integrability condition says that σ is closed, i.e. σ then gives a symplectic structure. In general, a Poisson manifold admits a natural foliation by submanifolds (whose conormal bundles consist of the null spaces of α) which inherit a symplectic structure. A typical example is $\mathfrak{g}^*$, which has the natural (Kirillov-Kostant) Poisson bracket, the symplectic leaves being precisely the coadjoint orbits. More generally, a Hamiltonian G-action on a symplectic manifold Y induces a Poisson structue on $X = Y/G$. This is the ordinary quotient; the symplectic quotients $X_a := Y//_a G$, for $a \in \mathfrak{g}^*$, are recovered as the symplectic leaves of X. (The Kirillov-Kostant Poisson structure on $\mathfrak{g}^*$ is recovered when we take $Y := T^*G$).

Markman's construction of the Poisson structure is based on the moduli space $\mathcal{M}_D$ of "bundles with a level-D structure" on E : $\mathcal{M}_D$ parametrizes equivalence classes of stable pairs (V, η) where

$$\eta : V_D \stackrel{\sim}{\rightarrow} \longrightarrow \mathcal{O}_D^n$$

is a trivialization of the fibers of V at points of D (sending the given volume element of V_D to the standard one on $\mathcal{O}_D^n$, since our structure group is $\mathrm{SL}(n)$). Consider the cotangent bundle $Y := T^*\mathcal{M}_D$. By elementary deformation theory, its points are (equivalence classes of) triples (V, η, φ) with $(V, \eta) \in \mathcal{M}_D$ and $(V, \varphi) \in X$, a meromorphic Higgs bundle with poles on D. The symmetry group G is the product of copies of $\mathrm{SL}(n)$, one for each point of D. (More precisely, $G = \{g \in \mathrm{Aut}_D(\mathcal{O}_D^n) \mid \det g = 1\}$, allowing for possibly non-reduced D.) The G action on $\mathcal{M}_D$ (with quotient $\mathcal{M} = \mathcal{M}_0$, the moduli of stable bundles on E) lifts to a Hamiltonian action on Y:

$$g : (V, \eta, \varphi) \mapsto (V, g \circ \eta, \varphi),$$

with quotient $X = Y/G$, so we get a Poisson structure on X. The moment map

$$\mu : Y \to \mathfrak{g}^* \approx \mathfrak{g}$$

sends (V, η, φ) to φ^η, the element of $\mathrm{End}_D(\mathcal{O}_D^n)$ which is conjugate via η to the residue $\mathrm{Res}_D \varphi \in \mathrm{End}_D(V_D)$. The symplectic leaf X_a, for $a \in \mathfrak{g}$, can therefore be identified as the moduli space of pairs $(V, \varphi) \in X$ such that $\mathrm{Res}_D \varphi$ is conjugate to the given a. In particular, each abelian variety $\pi^{-1}(b) \approx \mathrm{Prym}(C_b/E)$ is contained in one symplectic leaf X_a, and is a Lagrangian subvariety there.

A number of special cases of this construction are well known. When $D = 0$, we get Hitchin's system [Hi] on $T^*\mathcal{M}$. When E has genus 0, we get the polynomial matrices system studied in [AHH] and [B]. For more details and examples, see [Mn] or [DM1].

For the SYM theory with adjoint matter, we take E to be the elliptic curve whose τ parameter (in the halfplane $\mathbb{H}$) is the coupling constant of the theory. We are looking for a 1-parameter family of integrable systems (parametrized by mass m), each having an $(n-1)$-dimensional base and $(n-1)$-dimensional fibers. One checks easily that $\dim(X_a)$ equals the dimension of the adjoint orbit of a; so we are looking for orbits of the smallest possible dimension $2(n-1)$. There is essentially only one 1-parameter family of possibilities: D consists of a single point $\infty \in E$, and a is the conjugacy class of the diagonal matrix with eigenvalues $m(1, 1, \ldots, 1, 1-n)$. This conjugacy class consists of all diagonalizable $\mathrm{SL}(n)$ matrices A with

$$\mathrm{rank}(A - m \cdot Id) \leq 1.$$

(As $m \to 0$ there is one further, nilpotent, orbit of the right dimension: the conjugacy class of the elementary matrix e_{12}. But this turns out to give a trivial integrable system, i.e. the same as for $a = 0$: all the spectral curves are completely reducible).

Our system can be described rather explicitly. The moduli space of degree 0, rank n vector bundles on E can be identified with the symmetric product $S^n E$: each equivalence class contains a unique decomposable bundle $V = \oplus_{i=1}^{n} L_i$, with $L_i = \mathcal{O}_E(q_i - \infty) \in \mathrm{Pic}^0 E$. The moduli of $\mathrm{SL}(n)$-bundles is the fiber over 0 of the Abel-Jacobi map $S^n E \to E$. It is a projective space

$$\left\{ q_1 + \cdots + q_n \,\middle|\, \sum q_i = n\infty \right\} = \mathbb{P}H^0(E, \mathcal{O}(n\infty)) \approx \mathbb{P}^{n-1}.$$

In terms of the decomposition $V = \oplus L_i$, any Higgs field φ on V can be written as a matrix with entries

$$\varphi_{ij} \in \Gamma(L_j \otimes L_i^{-1}(\infty)).$$

Let $A := \mathrm{Res}_\infty \varphi$. Its diagonal elements are all zeroes, and $A - m \cdot Id$ has rank ≤ 1, so A is conjugate, via a diagonal matrix, to m times $e = \sum_{i \neq j} e_{ij}$. This means that each (V, φ) in our system is equivalent to a pair (V, φ) with $\mathrm{Res}_\infty(\varphi_{ij}) = m$ for $i \neq j$. For a given V, this condition uniquely determines the φ_{ij} (as long as $L_i \neq L_i$). The remaining free parameters are the coordinates q_i of $V = \oplus \mathcal{O}(q_i - \infty)$ and the diagonal entries $p_i = \varphi_{ii}$. Away from the diagonals $q_i = q_j$, these (q_i, p_i) can be identified with the canonical coordinates on $T^*\mathcal{M}$.

3.2 The spectral curves

The family of spectral curves C_b, for the fundamental, n-dimensional representation of $\mathrm{SL}(n)$, can be described as a linear system on the compactification $\overline{S} = \mathbb{P}(\omega_E(\infty) \oplus \mathcal{O}_E)$ of the surface S, or on various birationally equivalent surfaces. More precisely, it forms an affine open subspace of this linear system, the complement consisting of curves which contain the section at infinity, $Z_\infty := \overline{S} - S$, with some multiplicity $k > 0$. By removing this multiple component, we recover the spectral curves for the analogous problem with n replaced by $n - k$. (If A is an endomorphism of a vector space V, and W is an invariant subspace, then

$$\mathrm{rank}_W(A - m \cdot Id) \leq \mathrm{rank}_V(A - m \cdot Id),$$

so the conjugacy class restricts correctly.)

We can write down an explicit equation

$$p_n(t) = t^n + \sum_{k=2}^{n} f_k t^{n-k} = 0$$

for the spectral curves. Here t is a coordinate on the fibers of $\omega_E(\infty)$, and $f_k \in \Gamma(\mathcal{O}_E(k\infty))$. (We identify $\mathcal{O}_E \approx \omega_E$). Once we have such a p_n for each

n, the previous remark and the tracelessness condition imply that the general spectral curve has equation

$$C_u : p_n + \sum_{\ell=0}^{n-2} u_\ell p_\ell = 0,$$

with arbitrary constants u_ℓ which can therefore be interpreted as coordinates on the quantum moduli-space, the base of the integrable system.

The condition on p_n which specifies the right conjugacy class at infinity is that when we substitute $t = t' + \xi^{-1}$, where ξ is a local coordinate on E at ∞, we get a function of t' and ξ with at most first order poles in ξ. Write the equation of E in Weierstrass form,

$$E : y^2 = x^3 + bx - c.$$

The local coordinate ξ determines a sequence $x_k, k = 2, 3, \dots$, of rational functions on E which are regular on $E - \infty$ and such that near ∞,

$$x_k - \xi^{-k} = O(\xi^{-1}).$$

We may as well use the local coordinate $\xi = x^{-1/2}$, and then the x_k can be read off the Taylor expansions:

$$y = \xi^{-3}(1+b\xi^4 - c\xi^6)^{1/2} = \xi^{-3}(1+\frac{b}{2}\xi^4 - \frac{c}{2}\xi^6 - \frac{b^2}{8}\xi^8 + \frac{bc}{4}\xi^{10} - \frac{2c^2-b^3}{16}\xi^{12} + \dots)$$

and

$$\begin{aligned} y^{-1} &= \xi^3(1+b\xi^4 - c\xi^6)^{-1/2} \\ &= \xi^3(1 - \frac{b}{2}\xi^4 + \frac{c}{2}\xi^6 + \frac{3b^2}{8}\xi^8 - \frac{3bc}{4}\xi^{10} + \frac{6c^2-5b^3}{16}\xi^{12} + \dots). \end{aligned}$$

Explicitly, we get:

$$\begin{aligned} x_2 &= x \\ x_3 &= y \\ x_4 &= x^2 \\ x_5 &= xy \\ x_6 &= x^3 \\ x_7 &= x^2y - \frac{b}{2}y \\ x_8 &= x^4 \\ x_9 &= x^3y - \frac{b}{2}xy + \frac{c}{2}y \\ x_{10} &= x^5 \\ x_{11} &= x^4y - \frac{b}{2}x^2y + \frac{c}{2}xy + \frac{3b^2}{8}y \\ x_{12} &= x^6, \end{aligned}$$

and so on. In terms of this basis, one checks by an easy induction argument that

$$p_n = t^n - \sum_{k=2}^{n} (-1)^k (k-1) \binom{n}{k} x_k t^{n-k},$$

so the first few polynomials are:

$$\begin{aligned}
&1 \\
&t \\
&t^2 - x \\
&t^3 - 3xt + 2y \\
&t^4 - 6xt^2 + 8yt - 3x^2 \\
&t^5 - 10xt^3 + 20yt^2 - 15x^2t + 4xy \\
&t^6 - 15xt^4 + 40yt^3 - 45x^2t^2 + 24xyt - 5x^3 \\
&t^7 - 21xt^5 + 70yt^4 - 105x^2t^3 + 84xyt^2 - 35x^3t + 6\left(x^2y - \frac{b}{2}y\right).
\end{aligned}$$

This is a bit cleaner than the formulas in [DW], mostly because we set the sum of the roots in the Weierstrass equation of E to 0. A more-or-less equivalent solution was given already in [T], where the generating function for the p_n was written in terms of theta functions and exponentials on E. A slightly different (but equivalent) set of polynomials was given in [IM1].

3.3 Elliptic solitons

It is curious to note that the same system arises in several rather different contexts. Here is a brief outline of one of these, following the work of Treibich and Verdier [T], [TV].

Finite dimensional solutions of the KP hierarchy (= infinite dimensional integrable system) are well understood. They are parametrized by Krichever data, consisting of a curve C, a point $p \in C$, a line bundle L, and local trivializations. Such data determine a point in the total space of the hierarchy with the property that the infinite collection of commuting KP flows emanating from the point sweep out only a finite dimensional space, isomorphic to $J(C)$. The curve C, the point p and the trivializations remain fixed, while L moves freely in $J(C)$.

An elliptic soliton is a finite dimensional solution of KP in which the first flow, KP_1, already closes up, i.e. evolves on an elliptic curve inside the possibly larger Jacobian $J(C)$. The direction of KP_1, for points coming from Krichever data, is the tangent vector at p to the Abel-Jacobi image of C in $J(C)$. So the condition is that $J(C)$ should contain an elliptic curve E which is tangent at $AJ(p)$ to the Abel-Jacobi image $AJ(C)$.

Start with an n-sheeted branched cover $\pi : C \to E$ and a point $p \in E$ with local coordinate ξ. Let $\pi^*(p) = q_0 + \cdots + q_{n-1}$, and let $\xi_i := \xi \circ \pi$ near q_i. The tangent vector $t_i := \partial/\partial\xi_i$ lives naturally in $\Gamma(\mathcal{O}_{q_i}(q_i))$. Its image

$$(AJ)_* t_i \in T_{AJ(q_i)} J(C)$$

can then be identified with the image of t_i under the coboundary map δ for the sequence

$$0 \to \mathcal{O}_C \to \mathcal{O}_C(q_i) \to \mathcal{O}_{q_i}(q_i) \to 0.$$

Similarly, the tangent at $\pi^* p$ to the image $\pi^* E$ of E in $J(C)$ is given by $\sum_{i=0}^{n-1} \delta(t_i) = \delta(\sum_{i=0}^{n-1} t_i)$. The tangency condition then says that $\delta(\sum_{i=0}^{n-1} t_i)$ is proportional to $\delta(t_0)$, and therefore equal (by the residue theorem) to $\delta(nt_0)$. This happens if and only if there is some $f \in \Gamma(\mathcal{O}(\pi^* p))$ with

$$\mathrm{Res}_{q_i}(f \pi^* dz) = \begin{cases} 1, & 1 \leq i \leq n-1 \\ 1-n, & i = 0 \end{cases}$$

where dz is a non-zero closed 1-form on E. This is equivalent to saying that $\pi : C \to E$ is one of our spectral curves, and $f \pi^* dz$ becomes the Seiberg-Witten differential.

3.4 Tests

What makes this description of SYM with adjoint matter convincing for a physicist, is that it can be "tested" by working out various limits and special cases, and that it yields results consistent with the "predictions" based on the expected behavior of the physical system. We outline some of these tests.

3.5 Consistency with Seiberg-Witten

For $n = 2$, with given E and m, our system produces a 1-parameter family of double covers $C_u \to E$. Each C_u has genus 2, and the integrable system is the family of 1-dimensional Pryms. Write the equation of E as $y^2 = \prod_{i=1}^{3}(x - e_i) = x^3 + bx - c$, with $\sum e_i = 0$ as before. The cover C_u is given explicitly in terms of our polynomials p_i as:

$$0 = p_2 + up_0 = t^2 - x + u.$$

Its Galois group over the $\mathbb{P}^1$ with coordinate x is $\mathbb{Z}_2 \times \mathbb{Z}_2$, with involutions $\alpha : y \mapsto -y$, $, \beta : t \mapsto -t$, and $\alpha\beta$. The quotients are a rational curve (with coordinate t); the original E; and another elliptic curve E_u, respectively.

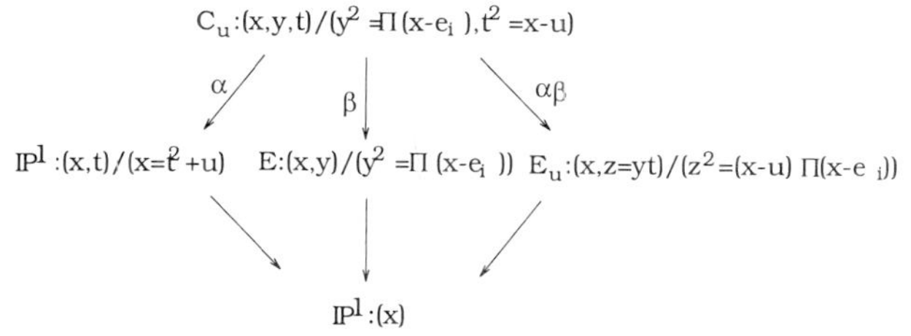

The Prym can then be identified with E_u.

The original Seiberg-Witten solution is given directly as the family of elliptic curves E'_u, double covers of $\mathbb{P}^1$ branched at $e_i u + e_i^2$ $(i = 1, 2, 3)$ and ∞ ([SW2], eq. (16.17)). These are indeed isomorphic to our E_u: the transformation

$$x \mapsto \frac{(u^2 + b)x - c}{u - x}$$

takes the e_i and u to $e_i u + e_i^2$ and ∞, respectively.

3.6 Mass to zero: the $N = 4$ limit

The limit $m \to 0$ of $N = 2$ SYM with gauge group G and adjoint matter is known to acquire a higher form of supersymmetry: it is the $N = 4$ SYM theory, with the same gauge group. (A physicist might rather say that two of the four supersymmetries of $N = 4$ SYM are broken when the theory is perturbed by the addition of a mass term to the Lagrangian). The extra symmetry of the $N = 4$ theory means that "everything" (i.e. the ingredients of the special geometry package encountered in §2: the quantum moduli space, the Kähler metric, the prepotential...) can be computed explicitly. The outcome matches perfectly to the corresponding quantities for Hitchin's system [Hi], i.e. the system of holomorphic (no poles) Higgs bundles on E. (This computation is in §2.3 of [DW]). So the physics predicts that our system should specialize, as $m \to 0$, to Hitchin's for SU(n), which it of course does. (This is not really an honest "test", since we started out searching for deformations of Hitchin's system, so this limit was built in from the beginning).

3.7 τ to ∞: the flow to pure $N = 2$

One of the most interesting limits occurs as both m and τ go to infinity simultaneously, with τ proportional to $\log(m)$. Equivalently, after an obvious rescaling, we can keep $m = 1$ and let the elliptic curve degenerate ($\tau \to \infty$) while adjusting the coefficients u_ℓ in the spectral curves C_u.

The physics of this limit is easy to work out. Essentially, the mass going to infinity means that in this limit the particle is so heavy that it cannot be affected by the rest of the system; we get the pure N=2 theory, with no mass.

This theory was solved in [AF] and [KLTY]. The solution does not involve τ, as we saw in the section on adding matter: instead of τ having a limit (which was there called τ_{cl}), it now goes to ∞. The spectral curves found in [AF] and [KLTY] have the explicit equation:

$$C'_b : w^2 = (z^n + b_2 z^{n-2} + \cdots + b_n)^2 + 1$$

in the (z, w)-plane, where the coefficients b_k play a role analogous to our u_ℓ.

In [DW], we show that these curves are indeed recovered as the appropriate limits of the spectral curves for the theory with adjoint mass. This is done by substituting the $\tau \to \infty$ limit into the explicit equations obtained above for the C_u, and performing the relevant rescalings. Instead of repeating the derivation here, we will describe the geometry of this degeneration.

The pure N=2 curve C'_b is the fibre product over $\mathbb{P}^1$ (with coordinate v) of the rational two-sheeted cover $w^2 = v^2 + 1$ and the (again rational, with coordinate z) n-sheeted cover $v = z^n + b_2 z^{n-2} + \cdots + b_n$. In the adjoint theory, the analogue of the double cover is just the elliptic curve $E \to \mathbb{P}^1$ (which becomes rational in the limit $\tau \to \infty$), and C_u is again an n-sheeted branched cover of E, but it is no longer the fiber product of E with any n-sheeted cover of $\mathbb{P}^1$. The degree 2 map $\psi : C \to \mathbb{P}^1_z$ and the degree n map $\mathbb{P}^1_z \to \mathbb{P}^1_v$ arise only in the limit: C becomes hyperelliptic.

The type of corank-1 degeneration involved in going from C_u to C'_b can be seen more generally. Consider a degree n branched cover $\pi : C \to E$ where E is elliptic and C has genus n. Up to isogeny, the Jacobian decomposes $J(C) \approx P \times E$, where P is the Prym. There are the Abel-Jacobi and Abel-Prym maps:

$$AJ : C \to J(C)$$
$$AP : C \to P$$

and their derivatives

$$\phi : C \to \mathbb{P}^{n-1}$$
$$\psi : C \to \mathbb{P}^{n-2}.$$

Here ϕ is the canonical map and ψ is the composition of ϕ with linear projection from the point $0 \in \mathbb{P}^{n-1}$ determined by the isogeny.

In the degeneration, we want each of E, C to acquire a node: the singular curves are E_0, C_0, with normalizations E', C', and singular points $p_0 \in E_0, q_0 \in C_0$ with branches $p_i, q_i, i = 1, 2$ in the normalizations. The Jacobians $J(C), J(E)$ become non-compact: $J(E_0)$ is a C^*, while $J(C_0)$ is a C^*-extension of $J(C')$; but the Prym P remains compact. This type of degeneration occurs when $\pi' : C' \to E'$ is totally ramified above each of the p_i: ${\pi'}^{-1}(p_i) = nq_i$. In this case we still have the Abel-Jacobi and Abel-Prym maps, going to the compactified Jacobians. The derivative ϕ is still the canonical map of the singular C_0; its image has a node at $0 = \phi(q_0)$. On the other hand, ψ is no longer a morphism on C_0; its restriction $\psi' : C' \to \mathbb{P}^{n-2}$ is now the canonical map of C'.

Within the $2n-2$-dimensional family of all covers $\pi : C \to E$ we restrict attention to the n-dimensional family of our spectral curves. Here there exist points $p \in E, q \in C$ such that $AJ(C)$ at q is tangent to $\pi^*(E)$ at p. This is equivalent to saying that $AP : C \to P(C/E)$ is ramified at q, or that the linear system giving ψ has a base point at q, or that $\phi(q) = 0$.

Combining these two effects, we find that when the corank-1 degeneration described above occurs within the family of spectral curves, the two branches $\phi(q_1), \phi(q_2)$ are tangent at 0. Equivalently, the canonical map ψ' satisfies $\psi(q_1) = \psi(q_2)$; so C' is hyperelliptic, ψ is two-to-one, and $\mathcal{O}(p_1 + p_2)$ is the hyperelliptic bundle. It follows immediately that the two maps, of degrees 2 and n respectively, from C' to $\mathbb{P}^1$ fit in a cartesian diagram. The corank-1 degeneration of the spectral curves (for the adjoint theory) therefore reproduces exactly the spectral curves of the pure $N = 2$ theory.

3.8 Higgs to ∞: symmetry breaking

This time we fix both the curve E and the mass m, and instead we flow to the theory whose gauge group is some Levi subgroup $G \subset \mathrm{SU}(n)$. Fix a diagonal matrix ψ, and consider the flow sending a Higgs bundle (V, φ) in our system to (V, φ_s) where $\varphi_s := \varphi + s\psi$. (As before we decompose $V = \oplus L_i$ and let ψ act diagonally. The flow is well-defined only on the $n!$-sheeted cover of the system on which this decomposition exists). The physics predicts that the limit of the $\mathrm{SU}(n)$ integrable system as $s \to \infty$ along this flow should be the G-system, where $G \subset \mathrm{SU}(n)$ is the centralizer of ψ.

For generic ψ, with distinct eigenvalues, this is clear: After rescaling we have $\varphi_s \sim \varphi'_s := \psi + \frac{1}{s}\varphi$, so the limiting spectral curve is $\det(t - \psi) = 0$, which consists of n copies of E meeting at ∞. The Prym is then the product $J(E)^{(n-1)}$, independent of the initial (V, φ). This trivial system corresponds to the Abelian structure group $\mathrm{U}(1)^{n-1}$ which is the maximal torus of $\mathrm{SU}(n)$ and the centralizer of ψ.

In general, ψ will have, say, k distinct eigenvalues with multiplicities $n_1, \ldots, n_k$, $\sum n_i = n$. The centralizer is

$$G = \mathrm{U}(1)^{k-1} \times \prod_{i=1}^{k} \mathrm{SU}(n_i).$$

Now the limit of the spectral curves of φ'_s is non-reduced, consisting of k components, the ith with multiplicity n_i. What we need is the limit of the Jacobians. This clearly has a factor (up to isogeny) from each component, and by rescaling around the ith eigenvalue ψ_i of ψ (i.e. setting $t - \psi_i = s^{-1}\tilde{t}$) we see that this factor is the Jacobian of the spectral curve

$$\det(\tilde{t} - \varphi_i),$$

where φ_i is the $n_i \times n_i$ dimensional ith block of φ. The claim is then that this is one of the $\mathrm{SU}(n_i)$ spectral curves, i.e. that the residue at ∞ of the ith block is conjugate to $m \cdot \mathrm{diag}(1, 1, \ldots, 1, 1 - n_i)$. An equivalent condition is: $\mathrm{rank}(\varphi_i - m) \leq 1$, and this condition is clearly inherited by any block in φ, as required.

3.9 Singularities

Much of the effort in [DW] was devoted to analyzing the discriminant, the hypersurface $D \subset B$ over which the spectral curves are singular. Physically, that is where all the interesting phenomena occur, and where many predictions can be made. For instance, as $\tau \to \infty$ ("at weak coupling"), the limit of D should be reducible: this is a classical limit, where the mass spectrum can be analyzed directly from the classical Lagrangian. D is the locus where some mass vanishes. This mass can belong to either of the two types (vector or hyper) of $N = 2$ SUSY multiplets mentioned in section 1.8 ("Adding matter"), and this gives rise to the two components. In fact, one knows the explicit equations in B for these components. A Maple assisted computation, using the explicit equations obtained above for the spectral curves, confirmed the reducibility and produced the precise predicted factors for SU(3).

For any τ (and any n), the most interesting locus in B is the $(n-1)$-fold self-intersection of D, a finite set. Physically, this corresponds to the massive vacua, where the (low energy) gauge group is broken to its maximal torus by monopole condensation. A classification result due to 't Hooft predicts that the possible massive vacua should correspond to index n subgroups of $\mathbb{Z}_n \times \mathbb{Z}_n$.

The corresponding result can be obtained directly from the integrable system. The general spectral curve has genus n, and we are imposing the maximal number, $n - 1$, of nodes. The normalized spectral curve is then of genus 1, and must be an unramified n-sheeted cover of the base E. These covers are of course classified by the subgroups as above. Conversely, given any n-sheeted cover $\pi : E' \to E$ and a point ∞' above $\infty \in E$, there is a unique meromorphic 1-form on E' with residue m at $\pi^{-1}(\infty) \backslash \infty'$ and residue $(1-n)m$ at ∞' (modulo the constant form $\pi^* dz$), hence a unique map of E' to the total space of $\omega_E(\infty)$ whose image is a spectral curve, as required. (Changing the choice of ∞' amounts to a deck transformation, leaving the spectral curve unchanged).

We also analyzed, assisted with some massive and miraculous-looking Maple computations, the locus of cusps in the SU(3) theory, again confirming some rather delicate predictions coming from the physics. It seems that there is some very beautiful geometry still waiting to be understood in the structure of the discriminant for the SU(n) theories.

4 Other integrable systems

Having studied in some detail the system which solves one particular family of SYM theories, we proceed to consider what is known for some of the other families. Our discussion will include,

- The pure $N = 2$ SYM for arbitrary gauge group G, and its relation to the Toda systems.
- The situation when fundamental matter is added to the pure theory.
- The theory with adjoint matter for arbitrary G, and its relation to the elliptic Calogero-Moser system.
- A few words about the integrable systems arising in string theory.

4.1 Pure $N = 2$ SYM

The solutions by Seiberg and Witten of the pure $N = 2$ SYM theory [SW1] and the theories with fundamental or adjoint matter [SW2] led rapidly to the creation of a whole literature of solutions to specific SYM theories. The general pattern was: postulate the general form of a family of curves which builds-in some of the expected symmetries of the theory, then adjust coefficients to enforce all other available information: further symmetries, limiting theories which were already known, internal consistencies. All cases which were worked out this way, e.g. [KLTY], [AF], [DS], [HO], [APS], [BL], [AS], [H], [AAG] and further references therein, were based on postulating curves which are hyperelliptic or closely related to hyperelliptics.

It was observed in [GKMMM] that the curves obtained in [KLTY], [AF] for the pure $G = \mathrm{SU}(n)$ theory are precisely the spectral curves of the periodic Toda system. The analogue for all semisimple groups was found in [MW], which we follow here.

4.2 The Toda system

A simple way to describe the Toda system is via the construction of Adler-Kostant-Symes. Start with a decomposition $\mathfrak{g} = \mathbf{a} \oplus \mathbf{b}$ of a Lie algebra $\mathfrak{g}$ as vector space direct sum of two of its subalgebras $\mathbf{a}, \mathbf{b}$. This induces a dual decomposition $\mathfrak{g}^* = \mathbf{a}^* \oplus \mathbf{b}^*$, an injection $i := \mathbf{b}^* \hookrightarrow \mathfrak{g}^*$, and a projection $\pi : \mathfrak{g} \twoheadrightarrow \mathbf{b}$. Let $\mathcal{O}_x \subset \mathbf{b}^*$ denote the coadjoint orbit of some $x \in \mathbf{b}^*$. The image $i(\mathcal{O}_x) \subset \mathfrak{g}^*$ has, in general, nothing to do with the coadjoint orbit $\mathcal{O}_{i(x)}$ of $i(x)$ in $\mathfrak{g}^*$. So i^* of a G-invariant function f on $\mathfrak{g}^*$ can be a complicated, non-constant function on $\mathcal{O}_x$. The AKS claim is that any two such, i^*f_1 and i^*f_2, Poisson commute with respect to the natural (Kirillov-Kostant) symplectic structure $\sigma_{\mathcal{O}_x}$ on $\mathcal{O}_x$. Indeed, let $v_j \in \mathfrak{g}$ correspond to the differential $df_j|_{i(x)} \in \mathfrak{g}^{**} = \mathfrak{g}$, so i^*df_j is given by $\pi(v_j)$. G-invariance of f_j implies that v_j is conormal to $\mathcal{O}_{i(x)}$, i.e. that

$$0 = \langle i(x), [\mathfrak{g}, v_j] \rangle.$$

We set $a_j := v_j - \pi(v_j) \in \mathbf{a}$, and compute:

$$\begin{aligned}
\{i^*f_1, i^*f_2\}_{\mathcal{O}_x} &= \sigma_{\mathcal{O}_x}(\pi v_1, \pi v_2) \\
&= \langle x, [\pi(v_1), \pi(v_2)]_{\mathbf{b}} \rangle \\
&= \langle i(x), [\pi(v_1), \pi(v_2)]_{\mathfrak{g}} \rangle \\
&= \langle i(x), [v_1, \pi(v_2)]_{\mathfrak{g}} \rangle - \langle i(x), [a_1, \pi(v_2)]_{\mathfrak{g}} \rangle \\
&= \langle i(x), [v_1, \pi(v_2)] \rangle - \langle i(x), [a_1, v_2] \rangle + \langle i(x), [a_1, a_2] \rangle.
\end{aligned}$$

We have just seen that the first two terms vanish when f_1, f_2 are G-invariant, and the third term vanishes since it pairs $i(x) \in \mathbf{b}^* = \mathbf{a}^\perp$ with $[a_1, a_2] \in \mathbf{a}$. This proves AKS.

A shifted version of AKS is very convenient in applications. Replace the inclusion $i : \mathbf{b}^* \hookrightarrow \mathfrak{g}^*$ by its shifted version $\epsilon + i$, where ϵ is an element of $\mathfrak{g}^*$ which annihilates $[\mathbf{a}, \mathbf{a}]$ and $[\mathbf{b}, \mathbf{b}]$. The above argument goes through verbatim. For example, take $\mathfrak{g} = \mathbf{sl}(n)$, $\mathbf{a} =$ strictly lower triangular matrices, $\mathbf{b} =$ upper triangulars, and (after identifying $\mathfrak{g}$ with $\mathfrak{g}^*$, so that $\mathbf{b}^*$ becomes the lower triangulars):

$$\epsilon = \begin{pmatrix} 0 & 1 & & & \\ & 0 & & & \\ & & \ddots & & \\ & & & 0 & 1 \\ & & & & 0 \end{pmatrix}, \quad x = \begin{pmatrix} 0 & & & & \\ 1 & 0 & & & \\ & & \ddots & & \\ & & & 0 & \\ & & & 1 & 0 \end{pmatrix}.$$

The orbit $\mathcal{O}_x \subset \mathbf{b}^*$ consists of all matrices of the form:

$$\begin{pmatrix} b_1 & & 0 \\ a_1 & \ddots & \\ & \ddots & \\ 0 & & a_{n-1} b_n \end{pmatrix}$$

with $a_j \neq 0$ and $\sum b_j = 0$. The shifted orbit consists of tridiagonal matrices:

$$\begin{pmatrix} b_1 & 1 & & & 0 \\ a_1 & \cdot & \cdot & & \\ & \cdot & \cdot & \cdot & \\ & & \cdot & \cdot & 1 \\ 0 & & & a_{n-1} & b_n \end{pmatrix}.$$

This is a $2(n-1)$-dimensional symplectic manifold (X, σ), isomorphic as symplectic manifold to the cotangent bundle of the maximal torus $T \approx (\mathbb{C}^*)^{n-1}$. The restriction to it of the symmetric polynomials of $\mathrm{SL}(n)$ (= coefficients of the characteristic polynomial) gives $n-1$ independent, commuting Hamiltonians. This is the classical Toda system.

An equivalent system is obtained if ϵ is replaced by $m\cdot\epsilon$ with $m \in \mathbb{C}^*$. When $m = 0$, the Hamiltonians become the symmetric functions of the b_i. We get the cotangent bundle X of $(\mathbb{C}^*)^{n-1}$, fibered by its moment map over $(\mathbb{C})^{n-1}$. Another variant is to take $\epsilon = 0$, but replace $\mathbf{a}$ by the algebra $\mathbf{so}(n)$ of skew symmetric matrices. Now $\mathbf{b}^*$ sits in $\mathbf{sl}(n)$ as the symmetric matrices, and the orbit consists of symmetric tridiagonal matrices

$$\begin{pmatrix} b_1 & c_1 & & & \\ c_1 & b_2 & & & \\ & & \ddots & & \\ & & & b_{n-1} & c_{n-1} \\ & & & c_{n-1} & b_n \end{pmatrix}.$$

This is conjugate to the previous matrix with $a_i = c_i^2$, so we obtain a new version $(\tilde{X}, \tilde{\sigma})$ of Toda which is just a 2^{n-1}-sheeted cover of the previous, basic version, (X, σ).

The Toda system for a semisimple Lie algebra $\mathfrak{g}$ is obtained with $\mathbf{b}$ a Borel subalgebra, $\mathbf{a}$ the nilpotent radical of the opposite Borel, $x = \sum e_\alpha$ and $\epsilon = \sum e_{-\alpha}$, where the sums are over the simple roots α with respect to $\mathbf{b}$, and the $e_{\pm\alpha}$ are normalized root vectors. The total space can again be identified with the cotangent bundle of the maximal torus, a typical element having the form $\sum_{\alpha \text{ simple}}(b_\alpha h_\alpha + a_\alpha e_\alpha + e_{-\alpha})$ with respect to a standard basis $h_\alpha, e_{\pm\alpha}$.

4.3 Periodic Toda

The periodic Toda systems are obtained when the semisimple algebra $\mathfrak{g}$ is replaced by a loop algebra $\mathfrak{g}^{(1)}$. This has an extended Dynkin diagram consisting of the simple roots α of $\mathfrak{g}$ plus the affine root α_0, corresponding to the highest root of $\mathfrak{g}$. The general element of the periodic Toda system can thus be written as

$$\sum_\alpha (b_\alpha h_\alpha + a_\alpha e_\alpha + e_{-\alpha}) + z e_{\alpha_0} + z^{-1} a_0 e_{-\alpha_0},$$

where z is the loop variable, and the product $a_0 \prod_\alpha a_\alpha = \mu$ is a fixed non-zero constant. (For $\mathfrak{g} = \mathbf{sl}(n)$, this is:

$$\begin{pmatrix} b_1 & 1 & & a_0 z^{-1} \\ a_1 & \ddots & \ddots & \\ & \ddots & \ddots & 1 \\ z & & a_{n-1} & b_n \end{pmatrix},$$

with $\prod a_i = \mu, \sum b_j = 0$).

As in the case of Higgs bundles, the spectral curves $C_{\rho,b}$ depend on the choice of a representation ρ of G. These Toda curves and their relationships were first studied in [Ml] and [MS], with some clarifications added in [MW1].

Each $C_{\rho,b}$ is a branched cover of the z-line $\mathbb{P}^1_z$. The latter plays the role for periodic Toda which the base curve E played for the meromorphic Higgs system in §2. The cameral cover in this situation has a strikingly simple description. The classifying map $\mathbb{P}^1_z \dashrightarrow \mathbf{t}/W$, which is usually only locally defined, is now a morphism for $z \neq 0, \infty$. It factors through the double cover

$$\begin{aligned} \mathbb{P}^1_z &\to \mathbb{P}^1_w \\ z &\mapsto w = z + \mu z^{-1} \end{aligned}$$

which maps $\mathbb{P}^1_z \backslash \{0, \infty\}$ to $\mathbb{A}^1_w = \mathbb{P}^1_w \backslash \{\infty\}$, followed by an affine linear map

$$\mathbb{A}^1_w \mapsto \mathbf{t}/W$$

whose image in $\mathbf{t}/W$ is a straight line pointing in the direction of the invariant polynomial of highest degree, which is equal to the Coxeter number $h_{\mathfrak{g}}$ of $\mathfrak{g}$. More precisely, the algebra of polynomial functions on $\mathbf{t}/W$ is isomorphic to the algebra of invariant polynomials on $\mathbf{t}$, which is filtered by the degree. The equations of the image of $\mathbb{A}^1_w$ are obtained by fixing the values of all polynomials on $\mathbf{t}/W$ whose pullback to $\mathbf{t}$ has degree $< h_{\mathfrak{g}}$. The induced W-cover of $\mathbb{P}^1_w$ then has $r = \text{rank}(\mathfrak{g})$ branch points in $\mathbb{A}^1_w$, and one at $w = \infty$. The choice of a Borel labels the finite branch points w_α by the simple roots α in such a way that the monodromy at w_α is the reflection corresponding to α. The monodromy at ∞ is then the product of these reflections, the Coxeter element. The Cameral cover is the pullback to $\mathbb{P}^1_z$. It can be described as the W-cover obtained from the trivial cover $W \times \mathbb{P}^1_z$ by making $r+1$ cuts labelled by the extended set of roots $\{\alpha\} \cup \{\alpha_0\}$, each running from a point z_α to μz_α^{-1}, and pasting across each cut according to the corresponding element of W.

In addition to the loop algebras $\mathfrak{g}^{(1)}$, there are twisted versions $\mathfrak{g}^{(k)}$ where k is the order of an automorphism of the Dynkin diagram of $\mathfrak{g}$, and each of these also gives rise to a periodic Toda system. The basic observation of Martinec and Warner [MW1] is that each of the known families of pure $N = 2$ SYM curves arises as spectral curves for one of these twisted periodic Toda systems. The specific identification involves Langlands duality (Olive-Montonen duality, to a physicist) exchanging the direction of arrows in the extended diagram. The (untwisted) extended diagrams for types A, D, E are self-dual, yielding ordinary Toda systems. The remaining cases yield twisted algebras and systems:

$$(B_r^{(1)})^\vee = A_{2r-1}^{(2)}, (C_r^{(1)})^\vee = D_{r+1}^{(2)}, (F_4^{(1)})^\vee = E_6^{(2)}, (G_2^{(1)})^\vee = D_4^{(3)}.$$

We will now describe explicitly the spectral curves for some of these systems.

4.4 The spectral curves

For the classical groups we can write the equation of the Toda spectral curves $C_{F,u}$ for the fundamental representation F as

$$w = z + \mu z^{-1}, x^\epsilon w + p = 0$$

where

$$p = p(x) = x^{n+1} + u_2 x^{n-1} + \cdots + u_{n+1}$$

for type A_n ($G = \mathrm{SL}(n+1)$), and

$$p = p(x^2) = x^{2n} + u_2 x^{2n-2+} \ldots + u_{2n}$$

for types B_n, C_n, and D_n, and the u_ℓ are coordinates on the base. Here $\epsilon = 0$ for A_n, B_n, C_n and $\epsilon = 2$ for D_n. In general, $\deg_x(p)$ is the dimension of the representation, while the w term modifies the coefficient $u_{h_{\mathfrak{g}}}$ corresponding to the invariant of highest degree. For A_n, B_n, C_n, D_n, these Coxeter numbers are $n+1, 2n, 2n, 2n-2$. (The B_n case is slightly different since 0 is a weight of F. So the actual "spectral" curve has equation $x(w + p(x)) = 0$. We may safely ignore the x factor, since the distinguished Prym lives in the Jacobian of the other factor).

For A_n, the fundamental curve has genus n, so its Jacobian is the distinguished Prym. It can be described geometrically as the fiber product over $\mathbb{P}^1_w$ of the double cover $w = z + \mu z^{-1}$, branched at $w = \pm 2\sqrt{\mu}$, and the degree $n+1$ cover $w = -p(x)$, with total ramification over $w = \infty$ and n simple branch points at the images of the roots of $p'(x) = 0$. The substitution $y = 2z + p(x)/2$, $4\mu = \Lambda^{2(n+1)}$ converts it to

$$y^2 = p^2(x) - \Lambda^{2(n+1)},$$

the form obtained in [AF] and [KLTY]. (Here we have reinstated the scale parameter Λ of the asymptotically free theory. Previously this was set to 1).

The other spectral curves for A_n are easily obtained from the fundamental one. The curve for $\wedge^2 F$ is of degree $\frac{n(n+1)}{2}$ over the z line, and of genus $n(n-1)+1-[\frac{n+1}{2}]$. The obvious correspondence between it and $C_{F,u}$ produces a copy of $J(C_{F,u})$ in $J(C_{\wedge^2 F,u})$. The A_3 case is described in [Ml], [MS]: The tetragenal curve $C_{F,u}$ of genus 3 gives rise via Recillas' trigonal construction (cf. [D3]) to the degree 6, genus 5 curve $C_{\wedge^2 F,u}$, together with an involution whose quotient is a trigonal curve C' of genus 2. The Jacobian of $C_{F,u}$ is isogenous to $\mathrm{Prym}(C_{\wedge^2 F,u}/C')$. (Actually, the degree 6 and 3 curves produced by the trigonal construction have genera 7 and 4 respectively, but each has 2 nodes, over $z = 0$ and $z = \infty$; the curves $C_{\wedge^2 F,u}$ and C', of genera 5 and 3, are their respective desingularizations). A beautiful picture of the $g = 11$ curve $C_{\wedge^2 F,u}$ for A_4 is given in [MW1]. In contrast to these small curves, there is the gigantic cameral cover, of degree $n!$ over $\mathbb{P}^1_z$ and of genus $1+\frac{n-1}{2(n+1)}(n+2)!$, whose Jacobian contains those of all (irreducible components of) spectral covers.

For $D_n, C_{F,u}$ has equation $x^2 w + p(x^2) = 0$. This depends on n parameters, but has genus $2n-1$, too big. The quotient C' by the involution $x \mapsto -x$ is hyperelliptic of genus $n-1$; the n-dimensional piece is $\mathrm{Prym}(C_{F,u}/C')$, and it can be realized as the Jacobian of a curve C'', the quotient of $C_{F,u}$ by the involution $x \mapsto -x, z \mapsto \mu z^{-1}$. The substitution $y := x^2(z - \mu z^{-1})$, $4\mu = \Lambda^{4(n-1)}$ takes the equation of $C_{F,u}$ to

$$y^2 = p^2(x^2) - \Lambda^{4(n-1)} x^4,$$

the form of the SO($2n$) SYM curve obtained in [BL]. In terms of $s = xy, t = x^2$, the quotient C'' becomes:

$$s^2 = tp^2(t) - \Lambda^{4(n-1)}t^3.$$

For type C_n, the fundamental Toda curve $C_{F,u}$ has equation $w + p(x^2) = 0, w = z + \mu z^{-1}$. The substitution $y = 2z + p(x^2)$ changes this to

$$y^2 = p^2(x^2) - 4\mu.$$

Just as in the D_n case, $C_{F,u}$ has automorphism group $\mathbb{Z}_2 \times \mathbb{Z}_2$ over the $t = x^2$ line:

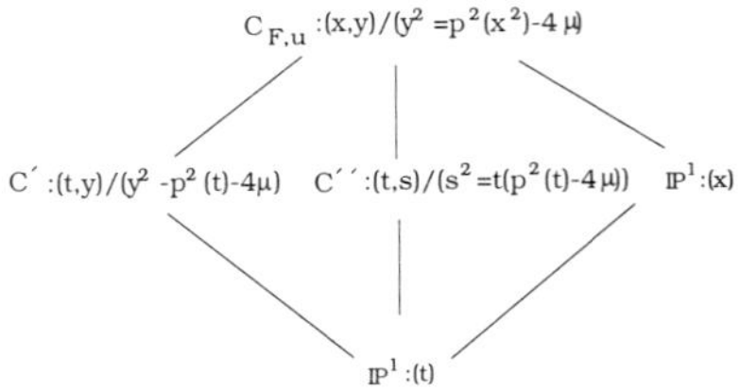

and the n-dimensional piece is $J(C'')$. But this is *not* the right abelian variety for the SYM theory. Rather, we should consider the Toda curves for $(C_n^{(1)})^\vee = D_{n+1}^{(2)}$. These have the form

$$w^2 + p(x^2) = 0, \ w = z + \mu z^{-1}.$$

The coordinates on the three $\mathbb{Z}_2$ quotients by the involutions $x \mapsto \pm x, z \mapsto \pm \mu z^{-1}$ are now:

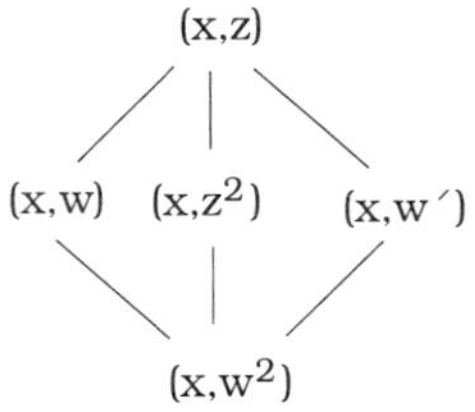

where $w' = w - 2\mu z^{-1} = z - \mu z^{-1}$. The corresponding genera are:

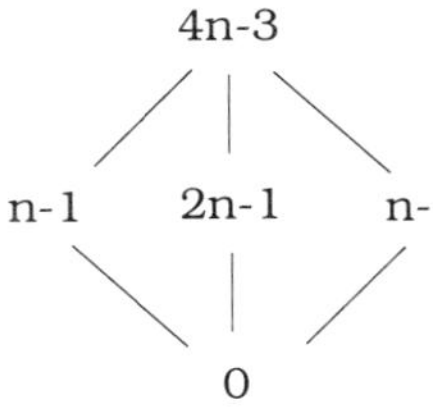

The middle curve has equation $v + \frac{\mu^2}{v} + p(x^2) - 2\mu = 0$, where $v = z^2$. It factors further: in fact, it looks just like the C_n-Toda curve, with z, μ, p replaced by $v, \mu^2, p - 2\mu$ respectively. So its Jacobian has an n-dimensional piece, which is the Jacobian of:

$$s^2 = t((p(t) - 2\mu)^2 - 4\mu^2) = tp(t)(p(t) - 4\mu).$$

This is essentially the C_n-curve obtained in [AS].

For B_n the story is similar, but simpler. The dual Toda curve has equation $x(z + \mu z^{-1}) + p(x^2) = 0$. The substitution $y = 2xz + p(x^2)$, $4\mu = \Lambda^{2(2n-1)}$ converts this to the form obtained in [DS],

$$y^2 = p^2(x^2) - \Lambda^{2(2n-1)} x^2.$$

Again this is of genus $2n - 1$. The genus n quotient is

$$s^2 = t(p^2(t) - \Lambda^{2(2n-1)} t).$$

For G_2, we need Toda curves for $(G_2^{(1)})^\vee = D_4^{(3)}$. These are given [MW1] as:

$$3(z - \mu z^{-1})^2 - x^8 + 2u_2 x^6 - [u_2^2 + 6(z + \mu z^{-1})]x^4 + [u_4 + 2u_2(z + \mu z^{-1})]x^2 = 0.$$

The various quotients involve the functions:

$$\begin{aligned} w &= z + \mu z^{-1} \\ s &= x^2 \\ u &= x(z - \mu z^{-1}) \\ v &= x\left(w - s^2 + \frac{u_2}{3} s\right). \end{aligned}$$

Some quotient curves, indicated by the functions which generate their fields, are:

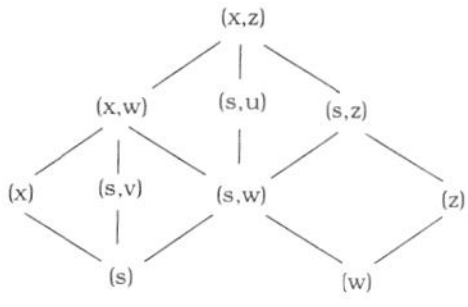

The corresponding genera are:

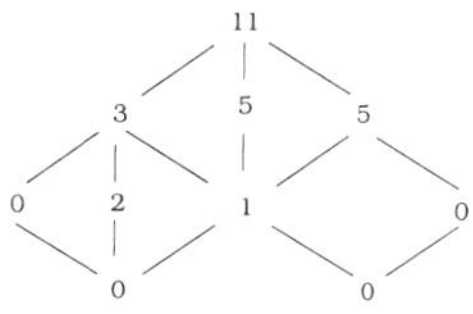

I have not checked whether the genus 2 curve with functions (s, v) is the one corresponding to the G_2 theory. An explicit family of genus 2 curves was proposed in several works (see references in [AAG]) for the G_2 theory, but according to [LPG] this family does not have good physical properties, and does not match the Toda curves.

The E_6 curves were considered in [LW], and turn out to involve some beautiful geometry. The simplest curve is based on the 27-dimensional representation of E_6. The resulting 27-sheeted cover of the z line behaves like the lines on a 1-parameter family of cubic surfaces. Each of the 6 simple roots corresponds to an ordinary double point acquired by the surface, so the local monodromy at each finite branch point is the product of 6 disjoint transpositions. Lerche and Warner study this genus-34 spectral curve, and go on to relate it to an integrable system coming from a string theory compactification on a Calabi-Yau threefold which degenerates to a fibration of the E_6 ALE singularity by a family of cubic surfaces.

4.5 Fundamental matter

The spectral curves for SYM with various gauge groups and numbers of particles ("quarks") in the fundamental representation have been determined by various means. A very incomplete list includes [HO], [APS], [AS], [H], [MN], [KP], [DKP1],[DKP2]. Each of these solutions is supported by substantial evidence, and yet the total picture seems far from complete. No really clear unifying principle is known, and there is no analogue of the Toda integrable system which is responsible for the various curves. There is also a certain level of disagreement among comparable solutions, see the discussion in [MN] and [DKP1]. (The curves involved in those solutions are the same, but the parametrizations differ. An integrable system would of course give a preferred parametrization).

4.6 Adjoint matter

As soon as [MW] and [DW] appeared, each with a class of integrable systems which solves its respective version of SYM, the question arose whether there was a common generalization: an integrable system for SYM with adjoint matter and arbitrary gauge group G, going to the periodic Toda system in the limit as m and τ go together to ∞.

In [Mc], Martinec suggested that the elliptic Calogero-Moser system (cf. [OP]) may provide this common generalization. There is such a system for each semisimple G, and the case $G = \mathrm{SU}(n)$ agrees exactly with the system in [DW]. There is also a direct computation [I1] showing that the A_n and D_n systems can degenerate to the respective Toda systems, and there is a deformed family of integrable systems depending on some additional parameters (the elliptic spin models, [I2]). A way of obtaining the A_n system by symplectic reduction is given in [GN].

One way to try to understand the geometry of these systems is as follows.

Fix an elliptic curve E, a semisimple group G, and its maximal torus T. Let $\mathcal{M}, \mathcal{M}'$ denote the moduli spaces of semistable bundles on E with structue groups G, T respectively. A special feature of genus 1 is that the natural map $\mathcal{M}' \to \mathcal{M}$ is surjective, in fact it is a Galois cover with group W. (It is not true that the structure group of every semistable G-bundle on E can be reduced to T; but every S-equivalence class of semistable bundles contains one whose structure group can be reduced). Now let X be the moduli space of Higgs bundles on E with first order pole at ∞, and let X' be its finite cover induced by $\mathcal{M}' \to \mathcal{M}$. (A point of X' is an equivalence class of pairs (V, φ) where V is a T-bundle on E and φ is a meromorphic Higgs field on $V \otimes_T G$). There is a "moment map"

$$\mu : X' \to \mathfrak{t}^{\perp}/T,$$

where $\mathfrak{t}^{\perp} \subset \mathfrak{g}^*$ is the annihilator of the Cartan $\mathfrak{t}$, and the T-action is coadjoint. Now **all** the fibers of μ are symplectic, of the minimal dimension $2r$ (where $r = \text{rank}(\mathfrak{g})$), and at least for some of them (but certainly not for all!), the restrictions of all the X-Hamiltonians Poisson commute, giving a large family of small integrable systems. (Presumably, this involves an analogue of the AKS proof of integrability of the Toda system). As the elliptic curve degenerates to $\mathbb{P}^1$ with 0 glued to ∞, this construction specializes to the periodic Toda.

The question remains, though: of all the possible T-orbits, which one corresponds to SYM with adjoint matter? A necessary condition seems to be that the T orbit $\mathcal{O}$ needs to be W-invariant; otherwise the limit as $m \to 0$ will not be the Hitchin system $T^*\mathcal{M}$, but its finite cover $T^*\mathcal{M}'$. (Or something in-between, if $\mathcal{O}$ is invariant under a subgroup of W.) For $\text{SU}(n)$, there is a unique such class (up to scalars), namely $\sum_{i \neq j} e_{ij}$. The natural analogue for arbitrary $\mathfrak{g}$ would be $\sum_{\alpha} e_{\alpha}$, the sum over all roots. Unfortunately, this is ill defined: the e_{α} can be normalized only up to sign, and different signs give distinct T orbits. For the orthogonal groups, there does not seem to be any W-invariant T orbit. David Lowe has pointed out [L] that for $Sp(n)$ there is an invariant orbit, namely the T orbit of the defining matrix J. (It is a signed sum of the long root vectors). This does give an integrable system, but a rather boring one: it is the n-fold product of the $\text{SU}(2)$ system, which does not flow to the pure $N = 2$ C_n spectral curves described earlier.

In sum, the elliptic Calogero-Moser systems (or their spin variants) seem to be a promising general class of systems for realizing the SYM theory with adjoint matter for arbitrary gauge group, but at the moment we lack a good geometric understanding of these systems, e.g. via spectral curves, except in the case of $\text{SU}(n)$.

4.7 Stringy integrable systems?

Quantum field theory is a limit of string theory compactified on a Calabi-Yau threefold, and the SW duality can be realized as the limiting form of various string dualities. A series of recent papers, such as [MW2], [KLMVW], [GHL1],

[GHL2], [LW], suggest that much of the QFT structure studied in this paper—the spectral curves, integrable systems, SW differentials and so on—arises as limit of corresponding stringy structures. A possible framework for studying the transition from stringy integrable systems to the SW systems is proposed in [F], where SW-type integrable systems are shown to arise from rigid special geometry, while the stringy type arises from local special geometry. From a purely mathematical point of view, it should be interesting to understand the relations between the SW-type systems, whose basic input is an elliptic curve; the Mukai system on a K3 surface, which degenerates to Hitchin's on a curve ([DEL], [DM1]); and the Calabi-Yau systems constructed in [DM2].

Bibliography

[AAG] M. R. Abolhasani, M. Alishahiha, A. M. Ghezelbash, hep-th 9606043.

[AHH] M. Adams, J. Harnad, J. Hurtubise, *Isospectral Hamiltonian flows in finite and infinite dimensions II: Integration of flows*, Commun. Math. Phys. (1990), 555–585.

[AM] M. Adler, P. van Moerbeke, *Completely integrable systems, Euclidean Lie algebras, and curves*, Advances in Math. **38** (1980), 267–379.

[AF] P. C. Argyres, A. Faraggi, *The vacuum Structure and Spectrum of $N = 2$ Supersymmetric $SU(N)$ Gauge Theory*, Phys. Rev. Lett. **74** (1995), hep-th 9411057.

[APS] P. C. Argyres, R. Plesser, A. D. Shapere, *The Coulomb phase of $N = 2$ supersymmetric QCD*, Phys. Rev. Lett. **75:1699-1702** (1995), no. e-Print Archive: hep-th 9505100.

[AS] P. C. Argyres, A. D. Shapere, *The vacuum structure of $N = 2$ super-QCD with classical gauge groups*, hep-th 9509175.

[B] A. Beauville, *Jacobiennes des courbes spectrales et systemes hamiltoniens completement integrables*, Acta Math. **164** (1990), 211–235.

[Be] J. Bernstein, *Lectures on Supersymmetry, Lectures at IAS, Fall 1996*, Preprint available at http://www.math.ias.edu/drm/QFT.

[Bo] F. Bottacin, *Symplectic geometry on moduli spaces of stable pairs*, Orsay, 1992.

[BL] A. Brandhuber, K. Landsteiner, *On the Monodromies of $N = 2$ Supersymmetric Yang-Mills Theory with Gauge Group $SO(2n)$*, hep-th 9507008.

[DS] U. H. Danielsson, B. Sundborg, *The Moduli Space and Monodromies of $N = 2$ Supersymmetric $SO(2r+1)$ Yang-Mills Theory*, hep-th 9504102.

[DKP1] E. D'Hoker, I. M. Krichever, D. H. Phong, hep-th 9609041.

[DKP2] E. D'Hoker, I. M. Krichever, D. H. Phong, hep-th 9609145.

[D1] R. Donagi *Decomposition of spectral covers*, in Journees de Geometrie Algebrique D'Orsay, Asterisque **218** (1993), 145–175.

[D2] R. Donagi *Spectral Covers* Current Topics in Complex Algebraic Geometry, (H. Clemens and J. Kollar, eds.), MSRI Pub., vol. 28, 1995, pp. 65–86, Alg Geom, 9505009.

[D3] R. Donagi, *The tetragonal construction* Bull. Amer. Math. Soc. (N.S.) **4** (1981), 181–185.

[DEL] R. Donagi, L. Ein and R. Lazarsfeld, *A non-linear deformation of the Hitchin dynamical system*, AlgGeom 9504017.

[DM1] R. Donagi, E. Markman, *Spectral curves, algebraically completely integrable Hamiltonian systems, and moduli of bundles*, R. Donagi et al, Integrable systems and Quantum Groups, LNM 1620, (1996), 1–119 Alg Geom, 9507017.

[DM2] R. Donagi, E. Markman *Cubics, Calabi-Yau threefolds, Integrable Systems and Mirrors*, Israel Math. Conf. Proc., vol. 9, AMS, 1995, AlgGeom 9408004.

[DW] R. Donagi, E. Witten *Supersymmetric Yang-Mills theory and integrable systems*, Nucl. Phys B460 **299** (1996), hep-th, 9510101.

[DH] J. J. Duistermaat, G. J. Heckman *On the Variation in the Cohomology of the Symplectic Form of the Reduced Phase Space*, Invent. Math. **69** (1982), 259.

[FS] L. D. Faddeev, A. A. Slavnov, *Gauge Fields: Introduction to Quantum Theory*, Benjamin/Cummings, 1980.

[Fa] G. Faltings, *Stable G-bundles and Projective Connections*, Jour. Alg. Geo **2** (1993), 507–568.

[F] D. S. Freed, *On special Kähler manifolds*, preprint.

[Fr] P. Freund, Introduction to supersymmetry, *Camb. Monographs on Math. Phys.*, Camb. U. Press, 1986.

[GHL1] C. Gomez, R. Hernandez, E. Lopez, hep-th 9604057.

[GHL2] C. Gomez, R. Hernandez, E. Lopez, hep-th 9608104.

[GKMMM] A. Gorsky, I. Krichever, A. Marshakov, A. Mironov, A. Morozov, hep-th 9505035.

[GN] A. Gorsky, N. Nekrasov, *Elliptic Calogero-Moser system from two-dimensional current algebra*, hep-th 9401021.

[H] A. Hanany, *On the quantum moduli space of vacua of $N = 2$ supersymmetric gauge theories*, hep-th 9509176.

[HO] A. Hanany, Y. Oz, *On the quantum moduli space of vacua of $N = 2$ supersymmetric $SU(N)$ gauge theories*, hep-th 9505075.

[Hi] N. Hitchin, *Stable bundles and integrable systems*, Duke Math. J. **54** (1987), 91–114.

[I1] . I. Inozemtsev, *The finite Toda lattices*, Comm. Math. Phys. **121** (1989), 629–38.

[I2] V. I. Inozemtsev, *Lax representation with spectral parameter on a torus for integrable particle systems*, J. Lett. Math. Phys. **17** (1989), 11–17.

[IM1] H. Itoyama, A. Morozov, hep-th 9511126.

[IM2] H. Itoyama, A. Morozov, hep-th 9512161.

[Ka] V. Kanev, *Spectral curves, simple Lie algebras and Prym-Tjurin varieties*, Proc. Symp. Pure Math. Part I **49** (1989), 627–645.

[KLMVW] A. Klemm, W. Lerche, P. Mayr, C. Vafa, N. Warner, hep-th 9604034.

[KLTY] A. Klemm, W. Lerche, S. Theisen, S. Yankielowicz *Simple Singularities and $N = 2$ Supersymmetric Yang-Mills Theory* Phys. Lett. **169** (1995), no. B344, hep-th 9411048.

[K] I. M. Krichever, hep-th 9205110.

[KP] I. M. Krichever, D. H. Phong, hep-th 9604199.

[LPG] K. Landsteiner, J. M. Pierre, S. B. Giddings, hep-th 9609059.

[LW] W. Lerche, N. P. Warner, hep-th 9608183.

[Mn] E. Markman, *Spectral Curves and Integrable Systems*, Comp. Math. **93** (1994), 255–290.

[Mv] A. Marshakov, hep-th 9607159.

[Mc] E. Martinec, hep-th 9510204.

[MW1] E. Martinec, N. P. Warner *Integrable Systems and Supersymmetric Gauge Theory*, hep-th 9509161.

[MW2] E. Martinec, N. P. Warner, hep-th 9511052.

[Ml] A. McDaniel, *Representations of* $sl(n, \mathbb{C})$ *and the Toda lattice*, Duke Math. J. **56** (1988), 47–99.

[MS] A. McDaniel, L. Smolinsky, *A Lie theoretic Galois Theory for the spectral curves of an integrable system II*, prep. (1994).

[Me] J. Y. Merindol, *Varietes de Prym d'un Revetement Galoisien*, prep. (1993).

[MN] J. A. Minahan, D. Nemeschansky, hep-th 9601059.

[MO] C. Montonen, D. Olive, *Magnetic Monopoles as Gauge Particles?*, Phys. Lett. **72B** (1977), 177.

[Mu] S. Mukai, *Symplectic structure of the moduli space of sheaves on an abelian or K3 surface*, Inv. Math. **77** (1984), 101–116.

[NT] T. Nakatsu, K. Takasaki, hep-th 9509162.

[N] N. Nekrasov, *Holomorphic bundles and many-body systems* hep-th 9503157.

[OP] M. A. Olshanetski, A. M. Perelomov, *Classical integrable finite-dimensional systems related to Lie algebras*, Phys. Rep. **71C** (1981), 313–400.

[SW1] N. Seiberg, E. Witten, *Monopole Condensation, and Confinement in* $N = 2$ *Supersymmetric Yang-Mills Theory*, Nucl. Phys. **B426** (1994), 19, hep-th 9407087.

[SW2] N. Seiberg, E. Witten, *Monopoles, duality and chiral symmetry breaking in* $N = 2$ *supersymmetric QCD*, Nucl. Phys. **B431** (1994), 484–550, hep-th 9408099.

[T] A. Treibich, *Tangential polynomials and elliptic solitons*, Duke Math. J. **59** (1989), 611–627.

[TV] A. Treibich, J. L. Verdier, *Solitons Elliptiques*, The Grothendieck Festschrift III (1990), Birkhäuser, 437–480.

[W] E. Witten, *Monopoles and Four Manifolds*, hep-th/9411102.

[WB] J. Wess, J. Bagger, *Supersymmetry and supergravity*, Princeton U. Press, 1992.

Five Lectures on Soliton Equations

Edward Frenkel
Department of Mathematics
University of California
Berkeley, CA 94720
E-mail: frenkel@math.berkeley.edu

Introduction

In these lectures we review a new approach to soliton equations of KdV type developed by the author together with B. Feigin and B. Enriquez [11, 12, 8].

The KdV equation is the partial differential equation on a function $v = v(z,\tau)$:

$$v_\tau = 6vv_z + v_{zzz}. \tag{0.1}$$

Introduced by Korteweg and de Vries in 1895, it has a long and intriguing history, see, e.g., [26]. We will view this equation as a flow, parametrized by the variable τ, on the space of functions $v(z)$ in the variable z. Perhaps, the most remarkable aspect of the KdV equation is its *complete integrability*, i.e., the existence of infinitely many other flows defined by equations of the type

$$\partial_{\tau_n} = P_n[v, v_z, v_{zz}, \ldots], \qquad n = 3, 4, \ldots, \tag{0.2}$$

which commute with the flow defined by (0.1) (we set $\tau_1 = z$ and $\tau_2 = \tau$). The important fact is that the right hand sides P_n of equations (0.2) are polynomials in v and its derivatives, i.e., *differential polynomials in* v.

This fact allows us to treat the KdV hierarchy (0.2) in the following way. Consider the ring $R = \mathbb{C}[v^{(n)}]_{n\geq 0}$ of polynomials in the variables $v^{(n)}$ (we shall write v for $v^{(0)}$). Let ∂_z be the derivation of R defined by the formula $\partial_z \cdot v^{(n)} = v^{(n+1)}$ (so that $v^{(n)} = \partial_z^n v$). Note that any derivation on R is uniquely determined by its values on $v^{(n)}$'s via the Leibnitz rule.

Any equation of the form $\partial_\tau v = P[v, v_z, \dots]$ gives rise to an evolutionary derivation D of R (i.e., such that commutes with ∂_z) defined by the formula $D \cdot v = P \in R$. The condition of commutativity with ∂_z implies that $D \cdot v^{(n)} = \partial_z^n \cdot P$, so that D can be written as

$$D = \sum_{n \geq 0} (\partial_z^n \cdot P) \frac{\partial}{\partial v^{(n)}}.$$

From this point of view, KdV hierarchy is an infinite set of mutually commuting evolutionary derivations of R. This is the way we will think of the KdV hierarchy and other, more general, hierarchies throughout these lectures.

We will not discuss the analytic issues related to the KdV equation, but will focus on its universal aspects, such as the origin of its integrability. This approach to soliton equations, initiated by I.M. Gelfand and L.A. Dickey [16], can be summarized as follows. A particular choice of functional space F, in which we want to look for solutions (we assume that it is an algebra closed under the action of the derivative ∂_z) gives us a specific realization of the hierarchy. On the other hand, R is nothing but the ring of functions on the space of ∞–jets of functions at a point. Hence a solution $\mathbf{v}(z, \tau)$ in the realization F gives us a family of homomorphisms $r(t) : R \to F$, which send $v^{(n)}$ to $\partial_z^n \mathbf{v}(z, t)$. Thus, $\operatorname{Spec} R$ can be considered as a scheme-like object associated with the KdV hierarchy, while various solutions correspond to various "points" of this "scheme". This way R captures the universal, realization independent properties of the KdV equations.

In these notes we will try to answer the question: where do the infinitely many commuting derivations of KdV hierarchy (and other hierarchies of KdV type) come from? From the outset, they look rather mysterious. But in fact there is a very simple explanation. Very briefly, the answer is as follows:

(i) we can identify the vector space with coordinates $v^{(n)}, n \geq 0$, with a homogeneous space of an infinite-dimensional unipotent algebraic group (a subgroup of a loop group);

(ii) this homogeneous space has an obvious action of an infinite-dimensional abelian Lie algebra by vector fields;

(iii) rewritten in terms of the coordinates $v^{(n)}$, these vector fields become the KdV flows (0.2).

Thus we will show that the KdV hierarchy is encoded in the differential geometry of a loop group.

For technical reasons, this program is easier to realize for a close relative of the KdV hierarchy—the so-called modified KdV (or mKdV) hierarchy; such a hierarchy is associated to an arbitrary affine Kac-Moody algebra. In Lecture 1 we explain (i)–(iii) for the generalized mKdV hierarchies. The reader who is only interested in the main ideas may read just this lecture.

In Lecture 2 we study the mKdV hierarchy in more detail, and write the equations of the hierarchy explicitly in the so-called zero-curvature form. In Lecture 3 we compare our approach to mKdV with the earlier approaches of Drinfeld-Sokolov and Wilson. We remark that in Sects. 2.1–2.2 and Sects. 3.4–3.6 we follow closely [8].

In Lecture 4 we consider the generalized KdV hierarchies from the point of view of our approach and that of the Drinfeld-Sokolov reduction [7]. Some of the results of this lecture (e.g., Theorem 4.1) are new. Finally, in Lecture 5 we discuss the Toda field theories and their connection to the mKdV and KdV hierarchies, following [12]. In particular, we show that the KdV and mKdV hierarchy are hamiltonian, and their hamiltonians are the integrals of motion of the corresponding affine Toda field theory.

The original motivation for the formalism explained in these lectures came from the study of deformations of conformal field theories, where one needs to show that the integrals of motion of affine Toda field theories can be quantized (see [32, 10]). It turns out that, in contrast to other approaches, our formalism is well suited for tackling the quantization problem. In particular, using the results of Lecture 5 and certain results on quantum groups, we were able to prove that the Toda integrals of motion can be quantized [10, 11]. We briefly discuss this at the end of Lecture 5. Recently these quantum integrals found some spectacular applications in two-dimensional quantum field theory [1]. We hope that quantization of our formalism will help elucidate further the connections between differential geometry of integrable systems and quantum field theory.

Acknowledgements. I thank S.-T. Yau for encouraging me to write this review. I am grateful to B. Enriquez and B. Feigin for their collaboration on the subject of these notes. The notes can serve as a summary of a lecture course on soliton theory that I gave at Harvard University in the Spring of 1996. I thank D. Ben-Zvi for letting me use his notes of my lectures, and for useful discussions.

I am indebted to B. Kostant for kindly supplying a proof of Proposition 4.1, which allowed me to simplify the proof of Theorem 4.1 on the structure of the KdV jet space. I also thank L. Feher for pointing out an inaccuracy in [8], which is corrected here.

The author's research was supported by grants from the Packard Foundation and the NSF.

Lecture 1

1.1 Background material on affine algebras

Let $\widetilde{\mathfrak{g}}$ be an affine Kac-Moody algebra. It has generators $e_i, f_i, \alpha_i^\vee, i = 0, \ldots, \ell$, and d, which satisfy the standard relations, see [22]. The Lie algebra $\widetilde{\mathfrak{g}}$ carries a non-degenerate invariant inner product $(\cdot,\cdot)$. One associates to $\widetilde{\mathfrak{g}}$ the labels $a_i, a_i^\vee, i = 0, \ldots, \ell$, the *exponents* $d_i, i = 1, \ldots, \ell$, and the *Coxeter number* h, see [23, 22]. We denote by I the set of all positive integers, which are congruent to the exponents of $\widetilde{\mathfrak{g}}$ modulo h (with multiplicities).

The exponents and the Coxeter numbers are given in a table below. Note that for all affine algebras except $D_{2n}^{(1)}$, each exponent occurs exactly once. In the case of $D_{2n}^{(1)}$, the exponent $2n-1$ has multiplicity 2.

Type	Coxeter number	Exponents
$A_n^{(1)}$	$n+1$	$1, 2, \ldots, n$
$A_{2n}^{(2)}$	$4n+2$	$1, 3, 5, \ldots, 2n-1, 2n+3, \ldots, 4n+1$
$A_{2n-1}^{(2)}$	$4n-2$	$1, 3, 5, \ldots, 4n-3$
$B_n^{(1)}$	$2n$	$1, 3, 5, \ldots, 2n-1$
$C_n^{(1)}$	$2n$	$1, 3, 5, \ldots, 2n-1$
$D_n^{(1)}$	$2n-2$	$1, 3, 5, \ldots, 2n-3, n-1$
$D_{n+1}^{(2)}$	$2n+2$	$1, 3, 5, \ldots, 2n+1$
$D_4^{(3)}$	12	$1, 5, 7, 11$
$E_6^{(1)}$	12	$1, 4, 5, 7, 8, 11$
$E_6^{(2)}$	18	$1, 5, 7, 11, 13, 17$
$E_7^{(1)}$	18	$1, 5, 7, 9, 11, 13, 17$
$E_8^{(1)}$	30	$1, 7, 11, 13, 17, 19, 23, 29$
$F_4^{(1)}$	12	$1, 5, 7, 11$
$G_2^{(1)}$	6	$1, 5$

The elements $e_i, i = 0, \ldots, \ell$, and $f_i, i = 0, \ldots, \ell$, generate the *nilpotent subalgebras* $\mathfrak{n}_+$ and $\mathfrak{n}_-$ of $\widetilde{\mathfrak{g}}$, respectively. The elements $\alpha_i^\vee, i = 0, \ldots, \ell$, and d generate the *Cartan subalgebra* $\widetilde{\mathfrak{h}}$ of $\widetilde{\mathfrak{g}}$. We have: $\widetilde{\mathfrak{g}} = \mathfrak{n}_+ \oplus \mathfrak{b}_-$, where $\mathfrak{b}_- = \mathfrak{n}_- \oplus \widetilde{\mathfrak{h}}$.

The element

$$C = \sum_{i=0}^{l} a_i^\vee \alpha_i^\vee$$

of $\widetilde{\mathfrak{h}}$ is a *central element* of $\widetilde{\mathfrak{g}}$. Let $\mathfrak{g}$ be the quotient of $[\widetilde{\mathfrak{g}}, \widetilde{\mathfrak{g}}]$ by $\mathbb{C}C$. We can identify $\widetilde{\mathfrak{g}}$ with the direct sum $\mathfrak{g} \oplus \mathbb{C}C \oplus \mathbb{C}d$. The Lie algebra $\mathfrak{g}$ has a Cartan decomposition $\mathfrak{g} = \mathfrak{n}_+ \oplus \mathfrak{h} \oplus \mathfrak{n}_-$, where $\mathfrak{h}$ is spanned by $\alpha_i^\vee, i = 1, \ldots, \ell$. Each $x \in \mathfrak{g}$ can be uniquely written as $x_+ + x_-$, where $x_+ \in \mathfrak{n}_+$ and $x_- \in \mathfrak{b}_-$.

There exists a unique element $\rho^\vee \in \widetilde{\mathfrak{h}}$, such that $(\alpha_i, \rho^\vee) = 1, \forall i = 0, \ldots, \ell$, and $(d, \rho^\vee) = 0$. But $\widetilde{\mathfrak{h}}^*$ is isomorphic to $\widetilde{\mathfrak{h}}$ via the non-degenerate inner product $(\cdot, \cdot)$. We shall use the same notation for the image of $\rho^\vee$ in $\widetilde{\mathfrak{h}}$ under this isomorphism. Then $\rho^\vee$ satisfies: $[\rho^\vee, e_i] = e_i, [\rho^\vee, h_i] = 0, [\rho^\vee, f_i] = -f_i, i = 0, \ldots, \ell$. The adjoint action of $\rho^\vee$ on $\mathfrak{g}$ defines the *principal* $\mathbb{Z}$*-gradation* on $\mathfrak{g}$.

Denote by $\overline{\mathfrak{g}}$ the simple Lie algebra of rank ℓ, generated by $e_i, f_i, \alpha_i^\vee, i = 1, \ldots, \ell$ (the Dynkin diagram of $\overline{\mathfrak{g}}$ is obtained from that of $\widetilde{\mathfrak{g}}$ by removing the 0th node). It has the Cartan decomposition $\overline{\mathfrak{g}} = \overline{\mathfrak{n}}_+ \oplus \mathfrak{h} \oplus \overline{\mathfrak{n}}_-$, where $\mathfrak{h}$ is the Cartan subalgebra of $\overline{\mathfrak{g}}$, spanned by $\alpha_i^\vee, i = 1, \ldots, \ell$.

Remark 1.1. Affine algebra $\widetilde{\mathfrak{g}}$ can be identified with the universal central extension of the loop algebra $\mathfrak{g}^0[t, t^{-1}]$, where $\mathfrak{g}^0$ is a simple Lie algebra, or its subalgebra [22]. More precisely, if $\widetilde{\mathfrak{g}}$ is non-twisted, then $\mathfrak{g}^0 = \overline{\mathfrak{g}}$ and $\mathfrak{g} = \overline{\mathfrak{g}}[t, t^{-1}]$, so that $\widetilde{\mathfrak{g}}$ is the universal central extension of $\overline{\mathfrak{g}}[t, t^{-1}] \oplus \mathbb{C}d$:

$$0 \to \mathbb{C}K \to \widetilde{\mathfrak{g}} \to \overline{\mathfrak{g}}[t, t^{-1}] \oplus \mathbb{C}d \to 0.$$

A twisted affine algebra $\widetilde{\mathfrak{g}}$ can be identified with a subalgebra in the universal central extension of $\mathfrak{g}^0[t, t^{-1}] \oplus \mathbb{C}d$. □

1.2 The principal abelian subalgebra

Set

$$p_{-1} = \sum_{i=0}^{\ell} \frac{(\alpha_i, \alpha_i)}{2} f_i,$$

where α_i's are the simple roots of $\mathfrak{g}$, considered as elements of $\mathfrak{h}$ using the inner product. Let $\mathfrak{a}$ be the centralizer of p_{-1} in $\mathfrak{g}$. Recall that we have a principal gradation on $\widetilde{\mathfrak{g}}$, such that $\deg e_i = -\deg f_i = 1, \deg \alpha_i^\vee = \deg d = 0$. Since p_{-1} is a homogeneous element with respect to this gradation, the Lie algebra $\mathfrak{a}$ is a $\mathbb{Z}$-graded subalgebra of $\mathfrak{g}$. In particular, it can be decomposed as $\mathfrak{a} = \mathfrak{a}_+ \oplus \mathfrak{a}_-$, where $\mathfrak{a}_+ = \mathfrak{a} \cap \mathfrak{n}_+$, and $\mathfrak{a}_- = \mathfrak{a} \cap \mathfrak{b}_-$.

Proposition 1.1 ([20], Prop. 3.8).

(1) The Lie algebra $\mathfrak{a}$ is abelian.

(2) The homogeneous component of $\mathfrak{a}_\pm$ of degree i with respect to the principal gradation has dimension equal to the number of occurencies of i in the set $\pm I$.

We call $\mathfrak{a}$ the principal abelian subalgebra of $\mathfrak{g}$. Its pull-back to $[\widetilde{\mathfrak{g}}, \widetilde{\mathfrak{g}}]$ is called the principal Heisenberg subalgebra.

Proposition 1.1 means that for all $\widetilde{\mathfrak{g}}$ except $D_{2n}^{(1)}$, each homogeneous component of $\mathfrak{a}_\pm$ is either zero-dimensional or one-dimensional, and it is one-dimensional if and only if $i \in I$. In the latter case we choose a generator p_i of this component. In the case $\widetilde{\mathfrak{g}} = D_{2n}^{(1)}$, the homogeneous component of degree i congruent to $2n-1$ modulo the Coxeter number $4n-2$ is two-dimensional. We choose two generators of $\mathfrak{a}$, p_i^1 and p_i^2, that span this component.

In particular, we choose

$$p_1 = \sum_{i=0}^{l} a_i e_i.$$

The properties of $\mathfrak{a}$ that are most important to us are described in the following fundamental proposition, due to V. Kac.

Proposition 1.2 ([20], Prop. 3.8).

(1) $\operatorname{Ker}(\operatorname{ad} p_{-1}) = \mathfrak{a}$.

(2) The Lie algebra $\mathfrak{g}$ *has a decomposition*

$$\mathfrak{g} = \mathfrak{a} \oplus \operatorname{Im}(\operatorname{ad} p_{-1}).$$

(3) With respect to the principal gradation, $\operatorname{Im}(\operatorname{ad} p_{-1}) = \oplus_{j\in\mathbb{Z}}\mathfrak{g}_j$, *where* $\dim \mathfrak{g}_j = \ell$, *and the map* $\operatorname{ad} p_{-1} : \mathfrak{g}_j \to \mathfrak{g}_{j-1}$ *is an isomorphism.*

Remark 1.2. Propositions 1.1 and 1.2 can be derived from B. Kostant's results about cyclic elements in simple Lie algebras (see Sect. 6 of [23]). □

Remark 1.3. L. Feher has pointed out to us that Prop. 5 of [8] was stated incorrectly: it is true only if n is relatively prime to h (see, e.g., [6]). However, the case $n = 1$ (given above) is sufficient for the purposes of [8], as well as for us here. □

Examples. Consider first the case of

$$L\mathfrak{sl}_2 = \left\{ \begin{pmatrix} a(t) & b(t) \\ c(t) & d(t) \end{pmatrix}, a(t) + d(t) = 0 \right\}.$$

Then

$$\mathfrak{a} = \operatorname{span}\{p_i\}_{i\,\mathrm{odd}},$$

where

$$p_{2j+1} = \begin{pmatrix} 0 & t^j \\ t^{j+1} & 0 \end{pmatrix}.$$

More generally, for $\mathfrak{g} = L\mathfrak{sl}_N$,

$$\mathfrak{a} = \operatorname{span}\{p_i\}_{i\not\equiv 0 \bmod N},$$

where

$$p_i = (p_1)^i, \qquad p_1 = \begin{pmatrix} 0 & 1 & 0 & \dots & 0 \\ 0 & 0 & 1 & \dots & 0 \\ \dots & \dots & \dots & \dots & \dots \\ 0 & 0 & 0 & \dots & 1 \\ t & 0 & 0 & \dots & 0 \end{pmatrix}.$$

Remark 1.4. It is known that in the non-twisted case maximal abelian subalgebras of $\mathfrak{g}$ are parametrized by the conjugacy classes of the Weyl group of $\overline{\mathfrak{g}}$, see [21]. In particular, $\mathfrak{h}[t,t^{-1}]$ corresponds to the identity class, and $\mathfrak{a}$ corresponds to the class of the *Coxeter element.* □

1.3 The unipotent subgroup

Let $\widehat{\mathfrak{n}}_+$ be the completion of the Lie subalgebra $\mathfrak{n}_+$ of $\mathfrak{g}$ defined as the inverse limit of finite-dimensional nilpotent Lie algebras $\mathfrak{n}_+^{(m)}, m > 0$, where $\mathfrak{n}_+^{(m)} = \mathfrak{n}_+/(\mathfrak{g}^0 \otimes t^m \mathbb{C}[[t]] \cap \mathfrak{n}_+)$. In the non-twisted case, $\mathfrak{n}_+ = \overline{\mathfrak{n}}_+ \oplus \overline{\mathfrak{g}} \otimes t\mathbb{C}[t]$, and so $\widehat{\mathfrak{n}}_+ = \overline{\mathfrak{n}}_+ \oplus \overline{\mathfrak{g}} \otimes t\mathbb{C}((t))$. In general, $\widehat{\mathfrak{n}}_+$ is a Lie subalgebra of $\mathfrak{g}^0((t))$.

Let $N_+^{(m)}$ be the unipotent algebraic group over $\mathbb{C}$ corresponding to $\mathfrak{n}_+^{(m)}$. Clearly, the exponential map $\mathfrak{n}_+^{(m)} \to N_+^{(m)}$ is an isomorphism. We define the prounipotent proalgebraic group N_+ as the inverse limit of $N_+^{(m)}, m > 0$. We have the exponential map $\widehat{\mathfrak{n}}_+ \to N_+$, which is an isomorphism of proalgebraic varieties. Thus, the ring $\mathbb{C}[N_+]$ of regular functions on N_+, is isomorphic (non-canonically) to the ring of polynomials in infinitely many variables. Below we will construct such an isomorphism explicitly, i.e., we will construct explicitly a natural coordinate system on N_+.

1.4 The action of $\mathfrak{g}$

The group N_+ acts on itself from the left and from the right:

$$n \cdot_L g = n^{-1} g, \qquad n \cdot_R g = gn.$$

In addition, it has a remarkable structure, which is the key in soliton theory: N_+ is equipped with an infinitesimal action of the Lie algebra $\mathfrak{g}$ by vector fields. Let $\overline{G}$ be the simply-connected simple algebraic group over $\mathbb{C}$ with Lie algebra $\overline{\mathfrak{g}}$. In the non-twisted case, there exists an ind-group G, whose set of $\mathbb{C}$–points is $\overline{G}((t))$. Thus, G can be viewed as the Lie group of $\mathfrak{g}$. In the twisted case G is defined in a similar way. The group G has subgroups N_+ and B_-, where the latter is the ind-group corresponding to the Lie subalgebra $\mathfrak{b}_-$. We have an open embedding $N_+ \to B_-\backslash G$. Hence the obvious infinitesimal action of $\mathfrak{g}$ on $B_-\backslash G$ from the right can be restricted to N_+.

It is easy to write down the resulting action of $\mathfrak{g}$ on N_+ explicitly. We can faithfully represent N_+ in the Lie algebra of matrices, whose entries are Taylor power series; for instance, N_+ is faithfully represented on $\mathfrak{g}^0[[t]]$. Each Fourier coefficient of such a series defines an algebraic function on N_+, and the ring $\mathbb{C}[N_+]$ is generated by these functions. Hence any derivation of $\mathbb{C}[N_+]$ is uniquely determined by its action on these functions. We can write the latter concisely as follows: $\nu \cdot x = y$, where x is a "test" matrix representing an element of N_+, i.e. its (i,j)th entry is considered as a regular function f_{ij} on N_+, and y is another matrix, whose (i,j)th entry is $\nu \cdot f_{ij} \in \mathbb{C}[N_+]$.

Remark 1.5. One can also take a slightly different point of view: all formulas below make sense for an arbitrary representation of N_+. Hence instead of picking a particular faithful representation of N_+, one can consider all representations simultaneously. The corresponding matrix elements generate $\mathbb{C}[N_+]$. □

For $a \in \mathfrak{g}$, let us denote by a^R the derivation of $\mathbb{C}[N_+]$ corresponding to the right infinitesimal action of a. Also, for $b \in \mathfrak{n}_+$ we denote by b^L the derivation of $\mathbb{C}[N_+]$ corresponding to the left infinitesimal action of b.

In order to simplify notation, we will write below axa^{-1} instead of $\mathrm{Ad}_a(x)$.

Lemma 1.1.

$$a^R \cdot x = (xax^{-1})_+ x, \qquad \forall a \in \overline{\mathfrak{g}}, \tag{1.3}$$

$$b^L \cdot x = -bx, \qquad \forall b \in \mathfrak{n}_+. \tag{1.4}$$

Proof. Consider a one-parameter subgroup $a(\epsilon)$ of G, such that $a(\epsilon) = 1 + \epsilon a + o(\epsilon)$. We have: $x \cdot a(\epsilon) = x + \epsilon xa + o(\epsilon)$. For infinitesimally small ϵ we can factor $x \cdot a(\epsilon)$ into a product $y_- y_+$, where $y_+ = x + \epsilon y_+^{(1)} + o(\epsilon) \in N_+$ and $y_- = 1 + \epsilon y_-^{(1)} \in B_-$. We then find that $y_-^{(1)} x + y_+^{(1)} = xa$, from which we conclude that $y_+^{(1)} = (xax^{-1})_+ x$. This proves formula (1.3). Formula (1.4) is obvious. □

1.5 The main homogeneous space

Now we introduce our main object. Denote by $\widehat{\mathfrak{a}}_+$ the completion of $\mathfrak{a}_+$ in $\widehat{\mathfrak{n}}_+$. Let A_+ be the abelian subgroup of N_+, which is the image of $\widehat{\mathfrak{a}}_+$ in N_+ under the exponential map. Consider the homogeneous space N_+/A_+. Since $\mathfrak{g}$ acts on N_+ infinitesimally from the right, the normalizer of $\widehat{\mathfrak{a}}_+$ in $\mathfrak{g}$ acts on N_+/A_+ infinitesimally from the right. In particular, $\mathfrak{a}_-$ acts on N_+/A_+, and each $p_{-n}, n \in I$, gives rise to a derivation of $\mathbb{C}[N_+/A_+]$, which we denote by p_{-n}^R.

Our goal is to show that the derivations p_{-n}^R are "responsible" for the equations of the mKdV hierarchy.

1.6 The space of jets

Now we change the subject and consider the space $\mathbf{U}$ of ∞–jets of an $\mathfrak{h}$–valued smooth function $\mathbf{u}(z)$ at $z = 0$. Equivalently, this is the space of functions $\mathbf{u}(z) : \mathcal{D} \to \mathfrak{h}$, where $\mathcal{D}$ is the formal disk, $\mathcal{D} = \operatorname{Spec} \mathbb{C}[[z]]$.

We can view $\mathbf{u}(z)$ as a vector $(u_1, \dots, u_\ell)$, where $u_i = (\alpha_i, \mathbf{u}(z))$. The space $\mathbf{U}$ is therefore the inverse limit of the finite-dimensional vector spaces

$$\mathbf{U}_m = \operatorname{span}\{u_i^{(n)}\}_{i=1,\dots,\ell; n=1,\dots,m},$$

where $u_i = (\alpha_i, \mathbf{u}(0))$, and $u_i^{(n)} = \partial_z^n u_i$ (so, the value of $u_i^{(n)}$ on $\mathbf{u}(z)$ is given by $\partial_z^n u_i(0)$). Thus, the ring $\mathbb{C}[\mathbf{U}]$ of regular functions on $\mathbf{U}$ is $\mathbb{C}[u_i^{(n)}]_{i=1,\dots,\ell; n\geq 0}$.

The derivative ∂_z gives rise to a derivation ∂_z of $\mathbb{C}[\mathbf{U}]$, such that $\partial_z \cdot u_i^{(n)} = u_i^{(n+1)}$. We will write u_i for $u_i^{(0)}$.

Introduce a $\mathbb{Z}$-gradation on $\mathbb{C}[\mathbf{U}]$ by setting $\deg u_i^{(n)} = n+1$.

1.7 Main result

Theorem 1.1 ([12],Prop. 4). *There is an isomorphism of rings*

$$\mathbb{C}[N_+/A_+] \simeq \mathbb{C}[\mathbf{U}],$$

under which p_{-1}^R gets identified with ∂_z.

Proof. We will explicitly construct a homomorphism $\mathbb{C}[\mathbf{U}] \to \mathbb{C}[N_+/A_+]$ and then show that it is actually an isomorphism. In order to do that, we have to construct regular functions on N_+/A_+ corresponding to $u_i^{(n)} \in \mathbb{C}[\mathbf{U}], i = 1, \ldots, \ell; n \geq 0$.

Define a function $\widetilde{u}_i$ on N_+ by the formula

$$\widetilde{u}_i(K) = (\alpha_i, K p_{-1} K^{-1}), \qquad K \in N_+. \tag{1.5}$$

This function is invariant under the right action of A_+. Indeed, if $y \in A_+$, then $y p_{-1} y^{-1} = p_{-1}$, and so $\widetilde{u}_i(Ky) = \widetilde{u}_i(K)$. Hence $\widetilde{u}_i$ can be viewed as a regular function on N_+/A_+. Next, we define the functions

$$\widetilde{u}_i^{(n)} = (p_{-1}^R)^n \cdot \widetilde{u}_i, \qquad n > 0,$$

on N_+/A_+. Consider the homomorphism $\mathbb{C}[\mathbf{U}] \to \mathbb{C}[N_+/A_+]$, which sends $u_i^{(n)}$ to $\widetilde{u}_i^{(n)}$.

To prove that this homomorphism is injective, we have to show that the functions $\widetilde{u}_i^{(n)}$ are algebraically independent. We will do that by showing that the values of their differentials at the identity coset $\bar{1} \in N_+/A_+$, $d\widetilde{u}_i|_{\bar{1}}$, are linearly independent. Those are elements of the cotangent space to N_+/A_+ at $\bar{1}$, which is canonically isomorphic to $(\widehat{\mathfrak{n}}_+/\widehat{\mathfrak{a}}_+)^*$. Using Proposition 1.2 and the invariant inner product on $\widetilde{\mathfrak{g}}$, we identify $(\widehat{\mathfrak{n}}_+/\widehat{\mathfrak{a}}_+)^*$ with $(\mathfrak{a}_+)^\perp \cap \mathfrak{n}_-$. By Proposition 1.2, with respect to the principal gradation, $(\mathfrak{a}_+)^\perp = \oplus_{j>0} (\mathfrak{a}_+)^\perp_{-j}$, where $\dim(\mathfrak{a}_+)^\perp_{-j} = \ell$.

Let us first show that the vectors $d\widetilde{u}_i|_{\bar{1}}, i = 1, \ldots, \ell$, form a linear basis in $(\mathfrak{a}_+)^\perp_{-1}$. Indeed, the tangent space to $\bar{1}$ is isomorphic to $\widehat{\mathfrak{n}}_+/\widehat{\mathfrak{a}}_+$, and it is clear that the projections of $e_i, i = 1, \ldots, \ell$, onto $\widehat{\mathfrak{n}}_+/\widehat{\mathfrak{a}}_+$ are linearly independent. Hence it suffices to check that the matrix $[\langle e_i, d\widetilde{u}_i|_{\bar{1}}\rangle]_{1 \leq i,j \leq \ell}$ is non-degenerate. But

$$\langle e_i, d\widetilde{u}_i|_{\bar{1}}\rangle = (e_i^L \cdot \widetilde{u}_j)(\bar{1}),$$

and we find

$$(e_i^L \cdot \widetilde{u}_j)(K) = -(\alpha_j, [e_i, K p_{-1} K^{-1}]) \tag{1.6}$$

$$= -(\alpha_j, [e_i, p_{-1}] + \ldots)$$

$$= -(\alpha_j, \alpha_i) \tag{1.7}$$

(the dots above stand for the terms lying in $\widehat{\mathfrak{n}}_+$, and so their inner product with α_j is 0). Thus, we see that the matrix $[\langle e_i, d\widetilde{u}_i|_{\mathrm{I}}\rangle]_{1\leq i,j\leq \ell}$ coincides with minus the symmetrized Cartan matrix of $\overline{\mathfrak{g}}$, $-[(\alpha_i,\alpha_j)]_{1\leq i,j\leq\ell}$, and hence is non-degenerate. Therefore the vectors $d\widetilde{u}_i|_{\mathrm{I}}, i=1,\dots,\ell$, are linearly independent.

Now, by definition,

$$d\widetilde{u}_i^{(m+1)}|_{\mathrm{I}} = \operatorname{ad} p_{-1}\cdot d\widetilde{u}_i^{(m)}|_{\mathrm{I}}.$$

But according to Proposition 1.2, $\operatorname{ad} p_{-1} : (\mathfrak{a}_+)^{\perp}_{-j} \to (\mathfrak{a}_+)^{\perp}_{-j-1}$ is an isomorphism for all $j>0$. Hence the vectors $d\widetilde{u}_i^{(m)}|_{\mathrm{I}}$ are all linearly independent. Therefore the functions $\widetilde{u}_i^{(n)}, i=1,\dots,\ell, n\geq 0$, are algebraically independent and our homomorphism $\mathbb{C}[\mathbf{U}]\to\mathbb{C}[N_+/A_+]$ is an embedding.

Now let us compare gradations. Consider the derivation $(\rho^\vee)^R$ on N_+. It is clear that for any $v\in\mathfrak{a}_+, [\rho^\vee, v]\in\mathfrak{a}_+$ (see Sect. 1.1). Hence $(\rho^\vee)^R$ is a well-defined derivation on $\mathbb{C}[N_+/A_+]$. We take $-(\rho^\vee)^R$ as the gradation operator on $\mathbb{C}[N_+/A_+]$ (to simplify notation, we will denote it by $-\rho^\vee$).

We have by formula (1.3),

$$\begin{aligned}(-\rho^\vee\cdot\widetilde{u}_i)(K) &= -(\alpha_i,[(K\rho^\vee K^{-1})_+, Kp_{-1}K^{-1})] \\ &= -(\alpha_i,[K\rho^\vee K^{-1}, Kp_{-1}K^{-1})] + (\alpha_i,[(K\rho^\vee K^{-1})_-, Kp_{-1}K^{-1})] \\ &= -(\alpha_i, K[\rho^\vee, p_{-1}]K^{-1}) + ([\alpha_i,(K\rho^\vee K^{-1})_-], Kp_{-1}K^{-1})] \\ &= (\alpha_i, Kp_{-1}K^{-1}) = \widetilde{u}_i(K).\end{aligned}$$

Hence the degree of $\widetilde{u}_i$ equals 1. Since $[-\rho^\vee, p_{-1}] = p_{-1}$, the degree of p_{-1}^R also equals 1, and so $\deg\widetilde{u}_i^{(n)} = n+1$.

On the other nand, the function $u_i^{(n)}$ has degree $n+1$ with respect to the $\mathbb{Z}$–gradation on $\mathbb{C}[\mathbf{U}]$. Hence our homomorphism $\mathbb{C}[\mathbf{U}]\to\mathbb{C}[N_+/A_+]$ is homogeneous of degree 0. To show its surjectivity, it suffices to prove that the characters of $\mathbb{C}[\mathbf{U}]$ and $\mathbb{C}[N_+/A_+]$ coincide. Here by character of a $\mathbb{Z}$–graded vector space V we understand the formal power series

$$\operatorname{ch} V = \sum_{n\in\mathbb{Z}} \dim V_n q^n,$$

where V_n is the homogeneous subspace of V of degree n. Clearly,

$$\operatorname{ch}\mathbb{C}[\mathbf{U}] = \prod_{n>0}(1-q^n)^{-\ell}.$$

Now, it is well-known that as a $\mathfrak{g}$–module, $\mathbb{C}[N_+]$ is isomorphic to the module M_0^* contragradient to the Verma module with highest weight 0 (these modules are defined in Lecture 4). Hence, as an $\mathfrak{n}_+$–module, it is isomorphic to the restricted dual $U(\mathfrak{n}_+)^*$ of $U(\mathfrak{n}_+)$. This implies that $\mathbb{C}[N_+/A_+]$ is isomorphic to the space of $\mathfrak{a}_+$–invariants on $U(\mathfrak{n}_+)^*$.

By the Poincare-Birkhoff-Witt theorem, $U(\mathfrak{n}_+)^*$ has a basis dual to the basis of lexicographically ordered monomials in $U(\mathfrak{n}_+)$. We can choose this basis in such a way that each element is the product AB, where A is the product of $p_n, n \in I$ (a basis in $\mathfrak{a}_+$), and B is a product of elements of some basis in $\mathrm{Im}(\mathrm{ad}\, p_{-1}) \cap \mathfrak{n}_+$. Then the elements of $U(\mathfrak{n}_+)^*$ dual to the monomials of the second type form a basis of the space of $\mathfrak{a}_+$-invariants on $U(\mathfrak{n}_+)^*$. Hence the character of $\mathbb{C}[N_+/A_+]$ equals that of a free polynomial algebra with ℓ generators of each positive degree. Thus its character coincides with the character of $\mathbb{C}[\mathbf{U}]$, and this completes the proof. □

1.8 Definition of the mKdV hierarchy

Having identified $\mathbb{C}[\mathbf{U}]$ with $\mathbb{C}[N_+/A_+]$ in the theorem above, we can now view the derivation $p^R_{-n}, n \in I$ of the latter, as a derivation of $\mathbb{C}[\mathbf{U}]$, which we denote by ∂_n. Thus, we obtain an *infinite set of commuting derivations* of the ring of differential polynomials $\mathbb{C}[u_i^{(n)}]$, of degrees equal to the exponents of $\mathfrak{g}$ modulo the Coxeter number.

We call this set the *mKdV hierarchy* associated to the affine algebra $\mathfrak{g}$.

According to Theorem 1.1, to each map $\mathcal{D} \to \mathfrak{h}$ we can attach a map $\mathcal{D} \to N_+/A_+$ (here $\mathcal{D}$ is the formal disk). From analytic point of view, this means that we have a mapping that to a every smooth function $\mathbf{u}(z)$ from, say, $\mathbb{R}$ to $\mathfrak{h}$, assigns a smooth function $K(z) : \mathbb{R} \to N_+/A_+$. Moreover, this mapping is *local* in the sense that $K(z)$ depends on $\mathbf{u}$ only through the values of its derivatives at z (the jets of $\mathbf{u}(z)$ at z). Due to this property, $\mathbf{u}(z)$ can just as well be a smooth function on a circle, or an analytic function on a domain in $\mathbb{C}$.

Next we say that on N_+/A_+ we have a family of commuting vector fields p^R_{-n} that come from the right action of the Lie algebra $\mathfrak{a}_-$. The flows of the mKdV hierarchy are the corresponding flows on the space of smooth function $\mathbf{u}(z) : \mathbb{R} \to \mathfrak{h}$. Note that the function $\mathbf{u}(z)$ does not really have to be smooth everywhere; it may have singularities at certain points.

In principle, we have now answered the question posed in the Introduction as to where commuting derivations acting on rings of differential polynomials come from. But we would like to have explicit formulas for these derivations. From the construction itself we know that the first of them, ∂_1, is just ∂_z. But we don't know the rest. On the other hand, the mKdV hierarchies have been previously defined by other methods, and we want to make sure that our definition coincides with the earlier definitions.

These issues will be the subject of the next two lectures. We will first write the derivations ∂_n in the so-called zero-curvature form and then compare our definition of the mKdV hierarchy with the other definition to see that they are equivalent.

Lecture 2

2.1 Zero curvature representation

Let $\omega_i^\vee \in \mathfrak{h}$ be the fundamental coweights that satisfy: $(\alpha_i, \omega_j^\vee) = \delta_{i,j}$. Define $\mathbf{u}$ to be the element

$$\sum_{i=1}^{\ell} \omega_i^\vee \otimes u_i \in \mathfrak{h} \otimes \mathbb{C}[\mathbf{U}].$$

Denote by $\widehat{\mathfrak{g}}$ the completion of $\mathfrak{g}$, which is the inverse limit of $\mathfrak{g}/(\mathfrak{g}^0 \otimes t^m \mathbb{C}[t] \cap \mathfrak{g})$. Thus, if $\mathfrak{g} = \overline{\mathfrak{g}}[t, t^{-1}]$, then $\widehat{\mathfrak{g}} = \overline{\mathfrak{g}}((t))$, and in general $\widehat{\mathfrak{g}}$ is a Lie subalgebra of $\mathfrak{g}^0((t))$.

For each $K \in N_+/A_+$, $Kp_{-n}K^{-1}$ is a well-defined element of $\widehat{\mathfrak{g}}$. Thus we obtain a algebraic map $N_+/A_+ \to \widehat{\mathfrak{g}}$, or, equivalently, an element of $\widehat{\mathfrak{g}} \otimes \mathbb{C}[N_+/A_+]$.[1] By abuse of notation, we denote this element by $Kp_{-n}K^{-1}$, and its projection on $\mathfrak{b}_- \otimes \mathbb{C}[N_+/A_+]$ by $(Kp_{-n}K^{-1})_-$. Since $\mathbb{C}[N_+/A_+] \simeq \mathbb{C}[u_i^{(n)}]$, by Theorem 1.1, we can apply to it $1 \otimes \partial_m$, which we denote simply by ∂_m.

More explicitly, recall that the Lie algebra $\widehat{\mathfrak{g}}$ can be realized as a Lie subalgebra of the Lie algebra $\mathfrak{g}^0((t))$, where $\mathfrak{g}^0$ is a finite-dimensional simple Lie algebra; e.g., in the non-twisted case, $\mathfrak{g}^0 = \overline{\mathfrak{g}}$. If we choose a basis in $\mathfrak{g}^0$, we can consider an element of $\widehat{\mathfrak{g}}$ as a matrix, whose entries are Laurent power series. The entries of the matrix $Kp_{-n}K^{-1}$ are Laurent series in t whose coefficients are regular functions on N_+/A_+. Hence, under the isomorphism $\mathbb{C}[N_+/A_+] \simeq \mathbb{C}[u_i^{(n)}]$, each coefficient corresponds to a differential polynomial in u_i's. Applying ∂_m to $Kp_{-n}K^{-1}$ means applying ∂_n to each of these coefficients.

We are going to prove the following result.

Theorem 2.1 ([8], Theorem 2).

$$[\partial_z + p_{-1} + \mathbf{u}, \partial_n + (Kp_{-n}K^{-1})_-] = 0. \tag{2.1}$$

The equation (2.1) can be rewritten as

$$\partial_n \mathbf{u} = \partial_z (Kp_{-n}K^{-1})_- + [p_{-1} + \mathbf{u}, (Kp_{-n}K^{-1})_-]. \tag{2.2}$$

As explained above, this equation expresses $\partial_n u_i$ in terms of differential polynomials in u_i's. Since, by construction, ∂_n commutes with $\partial_1 \equiv \partial_z$, formula (2.2) uniquely determines ∂_n as an evolutionary derivation of $\mathbb{C}[u_i^{(n)}]$. Thus, we obtain an explicit formula for ∂_n.

Note that formula (2.1) looks like the zero curvature condition on a connection defined on a two-dimensional space – hence the name "zero curvature representation".

Remark 2.1. One should be careful in distinguishing the two variables: t and z, and the two factors, $\widehat{\mathfrak{g}}$ and $\mathbb{C}[N_+/A_+]$, in the formulas below. The variable z is

[1]More precisely, it lies in the completed tensor product, e.g., if $\widehat{\mathfrak{g}} = \overline{\mathfrak{g}}((t))$, then by $\widehat{\mathfrak{g}} \otimes \mathbb{C}[N_+/A_+]$ we mean $(\overline{\mathfrak{g}} \otimes \mathbb{C}[N_+/A_+])((t))$.

a "dynamical variable" connected with $\mathbb{C}[N_+/A_+]$ due to its isomorphism with $\mathbb{C}[u_i^{(n)}]$; in particular, ∂_z is the derivation of p_{-1}^R of $\mathbb{C}[N_+/A_+]$. On the other hand, t is just the formal variable entering the definition of the affine algebra $\widehat{\mathfrak{g}}$. We remark that in earlier works on soliton theory, t was denoted by λ^{-1}, and λ was called the *spectral parameter.* □

2.2 Proof of Theorem 2.1

We will actually prove a stronger result:

Proposition 2.1. *For $K \in N_+/A_+$,*

$$[\partial_m + (Kp_{-m}K^{-1})_-, \partial_n + (Kp_{-n}K^{-1})_-] = 0, \qquad \forall m, n \in I. \tag{2.3}$$

Let us first explain how to derive Theorem 2.1 from Proposition 2.1. We need to specialize (2.3) to $m = 1$ and to determine $(Kp_{-1}K^{-1})_-$ explicitly.

Lemma 2.1.

$$(Kp_{-1}K^{-1})_- = p_{-1} + \mathbf{u}. \tag{2.4}$$

Proof. It is clear that $(Kp_{-1}K^{-1})_- = p_{-1} + x$, where $x \in \mathfrak{h}$. Hence we need to show that $x = \mathbf{u}$, or, equivalently, that $(\alpha_i, x) = u_i, i = 1, \ldots, \ell$. We can rewrite the latter formula as $u_i = (\alpha_i, (Kp_{-1}K^{-1})_-)$, and hence as $u_i = (\alpha_i, Kp_{-1}K^{-1})$. But this is exactly the definition of the function $\widetilde{u}_i$ on N_+/A_+, which is the image of $u_i \in \mathbb{C}[\mathbf{U}]$ under the isomorphism $\mathbb{C}[\mathbf{U}] \to \mathbb{C}[N_+/A_+]$. □

Now specializing $m = 1$ in formula (2.3) and using Lemma 2.1 we obtain formula (2.1).

In order to prove formula (2.3), we need to find an explicit formula for the action of ∂_n on $Kp_{-m}K^{-1}$.

It follows from formula (1.3) that

$$a^R \cdot xvx^{-1} = [(xax^{-1})_+, xvx^{-1}], \qquad a, v \in \mathfrak{g}. \tag{2.5}$$

If a and v are both elements of $\mathfrak{a}$, then formula (2.5) does not change if we multiply x from the right by an element of A_+. Denote by K the coset of x in N_+/A_+. Then we can write:

$$\partial_n \cdot KvK^{-1} = [(Kp_{-n}K^{-1})_+, KvK^{-1}], \qquad v \in \mathfrak{a}. \tag{2.6}$$

Proof of Proposition 2.1. Substituting $v = p_{-m}$ into formula (2.6), we obtain:

$$\partial_n \cdot Kp_{-m}K^{-1} = [(Kp_{-n}K^{-1})_+, Kp_{-m}K^{-1}].$$

Hence

$$\begin{aligned}\partial_n \cdot (Kp_{-m}K^{-1})_- &= [(Kp_{-n}K^{-1})_+, Kp_{-m}K^{-1}]_- \\ &= [(Kp_{-n}K^{-1})_+, (Kp_{-m}K^{-1})_-]_-.\end{aligned}$$

Therefore we obtain:

$$\begin{aligned}
&[\partial_m + (Kp_{-m}K^{-1})_-, \partial_n + (Kp_{-n}K^{-1})_-] \\
&= \partial_m \cdot (Kp_{-n}K^{-1})_- - \partial_n \cdot (Kp_{-n}K^{-1})_- + [(Kp_{-m}K^{-1})_-, (Kp_{-n}K^{-1})_-] \\
&= [(Kp_{-m}K^{-1})_+, (Kp_{-n}K^{-1})_-]_- - [(Kp_{-n}K^{-1})_+, Kp_{-m}K^{-1}]_- \\
&\qquad + [(Kp_{-m}K^{-1})_-, (Kp_{-n}K^{-1})_-]_-.
\end{aligned}$$

Adding up the first and the last terms, we obtain

$$\begin{aligned}
&[Kp_{-m}K^{-1}, (Kp_{-n}K^{-1})_-]_- - [(Kp_{-n}K^{-1})_+, Kp_{-m}K^{-1}]_- \\
&\qquad = [Kp_{-m}K^{-1}, Kp_{-n}K^{-1}]_- = 0,
\end{aligned}$$

and Proposition 2.1 is proved. □

2.3 Recurrence relation

Now we have at our disposal explicit formulas for equations of the mKdV hierarchy. But from the practical point of view, it is still difficult to compute explicitly terms like $(Kp_{-n}K^{-1})_-$, since the coset K has been defined in Lecture 1 in a rather abstract way. In this section we will exhibit a simple property of K, which will enable us to compute everything via a straightforward recursive algorithm.

Given a vector space W, we will write $W[\mathbf{U}]$ for $W \otimes \mathbb{C}[\mathbf{U}]$.

Recall that for any $v \in \mathfrak{a}$, KvK^{-1} denotes the element of $\widehat{\mathfrak{g}}[\mathbf{U}] = \widehat{\mathfrak{g}} \otimes \mathbb{C}[\mathbf{U}]$, determined by the embedding $N_+/A_+ \to \widehat{\mathfrak{g}}$, which maps K to KvK^{-1}.

Lemma 2.2.

$$[\partial_z + p_{-1} + \mathbf{u}, KvK^{-1}] = 0, \qquad \forall v \in \mathfrak{a}. \tag{2.7}$$

Proof. Using formula (2.6) and Lemma 2.1 we obtain:

$$\begin{aligned}
\partial_z(KvK^{-1}) &= [(Kp_{-1}K^{-1})_+, KvK^{-1}] \\
&= -[(Kp_{-1}K^{-1})_-, KvK^{-1}] \\
&= -[p_{-1} + \mathbf{u}, KvK^{-1}].
\end{aligned}$$

□

Now suppose that $\mathcal{V} \in \widehat{\mathfrak{g}}[\mathbf{U}]$ satisfies

$$[\partial_z + p_{-1} + \mathbf{u}(z), \mathcal{V}] = 0. \tag{2.8}$$

Then we can decompose $\mathcal{V}$ with respect to the principal gradation on $\widehat{\mathfrak{g}}$ (the principal gradation on $\widehat{\mathfrak{g}}$ should not be confused with the gradation on $\mathbb{C}[\mathbf{U}] = \mathbb{C}[N_+/A_+]$!):

$$\mathcal{V} = \sum_{m \geq -n} \mathcal{V}_m, \qquad \mathcal{V}_{-n} \neq 0,$$

and write equation (2.8) in components. Recall that $\deg p_{-1} = -1, \deg \mathbf{u} = 0$, and $\deg \partial_z = 0$.

The first equation that we obtain, in degree $-n-1$, reads:

$$[p_{-1}, \mathcal{V}_{-n}] = 0.$$

By Proposition 1.2, it implies that $\mathcal{V}_{-n} \in \mathfrak{a}$. Hence $\mathcal{V}_{-n} = p_{-n}$ times an element of $\mathbb{C}[\mathbf{U}]$, where $n \in I \cup -I$.

Other equations have the form

$$\partial_z \mathcal{V}_m + [p_{-1}, \mathcal{V}_{m-1}] + [\mathbf{u}, \mathcal{V}_m] = 0, \qquad m \geq -n. \tag{2.9}$$

Recall from Proposition 1.2 that $\mathfrak{g} = \mathfrak{a} \oplus \mathrm{Im}(\mathrm{ad}\, p_{-1})$. So we can split each $x \in \widehat{\mathfrak{g}}$ into a sum $x = x^0 + x^\perp$, where $x^0 \in \mathfrak{a}$ and $x^\perp \in \mathrm{Im}(\mathrm{ad}\, p_{-1})$. Furhermore, we can split our equations (2.9) into two parts, lying in $\mathfrak{a}$ and $\mathrm{Im}(\mathrm{ad}\, p_{-1})$. Then we obtain two equations:

$$\begin{aligned} \partial_z \mathcal{V}^0_m + [p_{-1}, \mathcal{V}_{m+1}]^0 + [\mathbf{u}, \mathcal{V}_m]^0 &= 0, \\ \partial_z \mathcal{V}^\perp_m + [p_{-1}, \mathcal{V}_{m+1}]^\perp + [\mathbf{u}, \mathcal{V}_m]^\perp &= 0. \end{aligned}$$

But clearly, $[p_{-1}, \mathcal{V}_{m+1}]^0 = 0$, $[\mathbf{u}, \mathcal{V}^0_m]^0 = 0$, and $[p_{-1}, \mathcal{V}^0_{m+1}] = 0$. Also, $[p_{-1}, \mathcal{V}^\perp_{m+1}]^\perp = [p_{-1}, \mathcal{V}^\perp_{m+1}]$. Hence we obtain:

$$\partial_z \mathcal{V}^0_m + [\mathbf{u}, \mathcal{V}^\perp_m]^0 = 0, \tag{2.10}$$

$$\partial_z \mathcal{V}^\perp_m + [p_{-1}, \mathcal{V}^\perp_{m+1}] + [\mathbf{u}, \mathcal{V}_m]^\perp = 0. \tag{2.11}$$

We already know that $\mathcal{V}_{-n} \in \mathfrak{a}[\mathbf{U}]$, and so $\mathcal{V}^\perp_{-n} = 0$. Now we can find $\mathcal{V}^0_{-n}$ and $\mathcal{V}^\perp_{-n+1}$. First we obtain from equation (2.10):

$$\partial_z \mathcal{V}^0_{-n} = -[\mathbf{u}, \mathcal{V}^\perp_{-n}] = 0,$$

which implies that $\mathcal{V}_{-n} = p_{-n}$ times a *constant* factor. Without loss of generality we assume that $\mathcal{V}_{-n} = p_{-n}$. Next, we obtain from (2.11),

$$[p_{-1}, \mathcal{V}^\perp_{-n+1}] = -[\mathbf{u}, \mathcal{V}_{-n}]^\perp,$$

which uniquely determines $\mathcal{V}^\perp_{-n+1}$, since $\mathrm{ad}\, p_{-1}$ is invertible on $\mathrm{Im}(\mathrm{ad}\, p_{-1})$ (see Proposition 1.2).

Now assume by induction that we know $\mathcal{V}^0_k, -n \leq k \leq m-1$ and $\mathcal{V}^\perp_k, -n \leq k \leq m$. Then we obtain from formulas (2.10) and (2.11):

$$\partial_z \mathcal{V}^0_m = -[\mathbf{u}, \mathcal{V}^\perp_m]^0, \tag{2.12}$$

$$\mathcal{V}^\perp_{m+1} = (\mathrm{ad}\, p_{-1})^{-1} \left(-\partial_z \mathcal{V}^\perp_m - [\mathbf{u}, \mathcal{V}_m]^\perp \right). \tag{2.13}$$

Consider $[\mathbf{u}, \mathcal{V}^\perp_m]^0 \in \mathfrak{a}$. It can only be non-zero if $m \in \pm I$. In that case it is equal to $p_m \otimes P_m, P_m \in \mathbb{C}[\mathbf{U}]$, or $p^1_m \otimes P^1_m + p^2_m \otimes P^2_m$, if the multiplicity of the exponent is 2. If each P^i_m is a total derivative, $P^i_m = \partial_z Q^i_m$, then we find from equation (2.12):

$$\mathcal{V}^0_m = \sum_i p^i_m \otimes Q^i_m.$$

After that we can solve equation (2.13) for $\mathcal{V}^{\perp}_{m+1}$. But we know that $Kp_{-n}K^{-1} = p_{-n} + \dots$ does satisfy equation (2.9). Hence we know for sure that each P^i_m is a total derivative, and equation (2.12) can be resolved (i.e., the antiderivative can be taken whenever we need it). Therefore we can find all $\mathcal{V}_m$'s following this algorithm.

Note that the only ambiguity is introduced when we take the antiderivative, since we then have the freedom of adding $c_m \cdot p_m$ to $\mathcal{V}^0_m$, where $c_m \in \mathbb{C}$. This simply reflects the fact that adding $Kp_mK^{-1} = p_{-m} + \dots$ with $m > -n$ to $Kp_{-n}K^{-1}$, we do not violate (2.9). Thus, we have proved the following

Proposition 2.2. *Let $\mathcal{V}$ be an element of $\widehat{\mathfrak{g}}[\mathbf{U}]$ satisfying*

$$[\partial_z + p_{-1} + \mathbf{u}, \mathcal{V}] = 0. \tag{2.14}$$

Then $\mathcal{V} = KvK^{-1}$, where $v \in \mathfrak{a}$. Moreover, if $\mathcal{V} = p_{-n}+$ terms of degree higher than $-n$ with respect to the principal gradation on $\mathfrak{g}$ then $v = p_{-n}+$ terms of degree higher than $-n$.

Now recall that each $\mathcal{V}_n$ is an element of $\widehat{\mathfrak{g}}[\mathbf{U}]$, i.e., its matrix entries (in a particular representation of $\widehat{\mathfrak{g}}$) are Laurent power series in t, with coefficients from the ring $\mathbb{C}[\mathbf{U}]$. That ring has a $\mathbb{Z}$–gradation (which should not be confused with the principal gradation on $\mathfrak{g}$ itself!). The corresponding gradation on $\mathbb{C}[N_+/A_+]$ is defined by the vector field $-\rho^\vee$. Using this gradation, for each n we can distinguish a canonical solution $\mathcal{V}_n$ of the recurrence relations (2.9) that equals precisely $Kp_{-n}K^{-1}$.

Lemma 2.3. *Let $x \in \mathfrak{g}$ be homogeneous of degree k. Then the function $f \in \mathbb{C}[N_+/A_+]$ given by the formula $f(K) = (x, Kp_{-n}K^{-1})$ is a homogeneous element of $\mathbb{C}[N_+/A_+]$ of degree $n - m$.*

Proof. We have to show that $\rho^\vee \cdot f = (m-n)f$. But

$$\begin{aligned}(\rho^\vee \cdot f)(K) &= (x, [(K\rho^\vee K^{-1})_+, Kp_{-n}K^{-1}]) \\ &= (x, [K\rho^\vee K^{-1}, Kp_{-n}K^{-1}]) - (x, [(K\rho^\vee K^{-1})_-, Kp_{-n}K^{-1}]) \\ &= (x, K[\rho^\vee, p_{-n}]K^{-1}) + ([\rho^\vee, x], Kp_{-n}K^{-1}) \\ &= (m-n)(x, Kp_{-n}K^{-1}) = (m-n)f(K),\end{aligned}$$

and the lemma is proved. □

Our recurrence relations (2.10) and (2.11) imply that if x has degree m, then $(x, \mathcal{V}_n)$ has a term of degree $n - m$ in $\mathbb{C}[\mathbf{U}] \simeq \mathbb{C}[N_+/A_+]$ and possibly other terms of smaller degrees which result from addition of constants when taking anti-derivatives at the previous steps of the recursive procedure. It is clear that there is a unique solution $\mathcal{V}_n$ of equation (2.8), such that $(x, \mathcal{V}_n)$ is homogeneous of degree $n - m$, and this solution is $Kp_{-n}K^{-1}$.

Now we have an algorithm for finding $Kp_{-n}K^{-1}$. Hence we can construct explicitly the nth equations of the mKdV hierarchy by inserting $(Kp_{-n}K^{-1})_-$ into equation (2.1). In the next section we will do that in the case of $L\mathfrak{sl}_2$.

2.4 Example of $L\mathfrak{sl}_2$

We have

$$\mathbf{u} = \begin{pmatrix} \frac{1}{2}u & 0 \\ 0 & -\frac{1}{2}u \end{pmatrix}.$$

Hence

$$\partial_z + p_{-1} + \mathbf{u} = \partial_z + \begin{pmatrix} \frac{1}{2}u & t^{-1} \\ 1 & -\frac{1}{2}u \end{pmatrix}. \tag{2.15}$$

The first equation of the hierarchy is

$$[\partial_z + p_{-1} + \mathbf{u}, \partial_n + (Kp_{-n}K^{-1})_-] = 0.$$

By formula (2.4), $(Kp_{-1}K^{-1})_- = p_{-1} + \mathbf{u}$. Hence the first equation reads

$$[\partial_z + p_{-1} + \mathbf{u}, \partial_1 + p_{-1} + \mathbf{u}] = 0,$$

which is equivalent to

$$\partial_1 \mathbf{u} = \partial_z \mathbf{u},$$

as we already know (note that this is true for an arbitrary $\mathfrak{g}$).

Now we consider the next equation in the case of $L\mathfrak{sl}_2$, corresponding to $n = 3$. To write it down, we have to compute $(Kp_{-3}K^{-1})_-$. In order to do that we compute the first few terms of $Kp_{-3}K^{-1}$ following the algorithm of the previous section.

We introduce a basis of $L\mathfrak{sl}_2$ homogeneous with respect to the principal gradation:

$$p_{2j+1} = \begin{pmatrix} 0 & t^j \\ t^{j+1} & 0 \end{pmatrix}, \qquad q_{2j+1} = \begin{pmatrix} 0 & t^j \\ -t^{j+1} & 0 \end{pmatrix},$$

$$r_{2j} = \begin{pmatrix} t^j & 0 \\ 0 & -t^j \end{pmatrix}.$$

Then we write

$$Kp_{-3}K^{-1} = p_{-3} + R_{-2}r_{-2} + P_{-1}p_{-1} + Q_{-1}q_{-1} + R_0 r_0 + \dots .$$

The other terms are irrelevant for us now as we are only interested in $(Kp_{-3}K^{-1})_-$.

Equation (2.8) reads

$$[\partial_z + p_{-1} + \frac{1}{2} u r_0, \partial_3 + p_{-3} + R_{-2} r_{-2} + P_{-1} p_{-1} + Q_{-1} q_{-1} + R_0 r_0 + \ldots] = 0.$$

Rewriting it in components, we obtain the following equations:

$$2R_{-2} - u = 0, \qquad \partial_z R_{-2} - 2Q_{-1} = 0,$$

$$\partial_z P_{-1} + u Q_{-1} = 0, \qquad \partial_z Q_{-1} + u P_{-1} - 2R_0.$$

We find from these equations:

$$R_{-2} = \frac{1}{2} u, \quad Q_{-1} = \frac{1}{4} \partial_z u, \quad P_{-1} = -\frac{1}{8} u^2, \quad R_0 = -\frac{1}{16} u^3 + \frac{1}{8} \partial_z^2 u.$$

The equation of the mKdV hierarchy corresponding to $n = 3$ now reads:

$$\partial_3 \begin{pmatrix} \frac{1}{2} u & t^{-1} \\ 1 & -\frac{1}{2} u \end{pmatrix} =$$

$$\left[\begin{pmatrix} \frac{1}{2} u t^{-2} - \left(\frac{1}{16} u^3 - \frac{1}{8} \partial_z^2 u \right) & t^{-2} + \left(-\frac{1}{8} u^2 + \frac{1}{4} \partial_z u \right) t^{-1} \\ t^{-1} - \left(\frac{1}{8} u^2 + \frac{1}{4} \partial_z u \right) & -\frac{1}{2} u t^{-2} + \left(\frac{1}{16} u^3 - \frac{1}{8} \partial_z^2 u \right) \end{pmatrix}, \partial_z + \begin{pmatrix} \frac{1}{2} u & t^{-1} \\ 1 & -\frac{1}{2} u \end{pmatrix} \right].$$

This is equivalent to the equation

$$\partial_3 u = \frac{3}{8} u^2 \partial_z u - \frac{1}{4} \partial_z^3 u, \tag{2.16}$$

which is the *mKdV equation.* It is related to the KdV equation (0.1) by a change of variables (see Lecture 4).

Lecture 3

In the previous lecture we have written down explicitly the equations of the mKdV hierarchy in the zero curvature form. In this lecture we will discuss another approach to these equations, which is due to Drinfeld and Sokolov and goes back to Zakharov and Shabat. We will then establish the equivalence between the two approaches.

3.1 Generalities on zero curvature equations

Let us look at the general zero curvature equation

$$[\partial_z + p_{-1} + \mathbf{u}, \partial_\tau + V] = 0. \tag{3.1}$$

The element p_{-1} has degree -1 with respect to the principal gradation of $\mathfrak{g}$, while $\mathbf{u}$ has degree 0 (with respect to the gradation on $\mathfrak{g}$). This makes finding an element V that satisfies (3.1) a non-trivial problem. Indeed, equation (3.1) can be written as

$$\partial_\tau \mathbf{u} = [\partial_z + p_{-1} + \mathbf{u}, V]. \tag{3.2}$$

The left hand side of (3.2) has degree 0. Therefore V should be such that the expression in the right hand side of (3.2) is concentrated in degree 0.

Such elements can be constructed by the following trick (see [31, 7, 29]). Suppose we found some $\mathcal{V} \in \widehat{\mathfrak{g}}[\mathbf{U}]$ which satisfies

$$[\partial_z + p_{-1} + \mathbf{u}, \mathcal{V}] = 0. \tag{3.3}$$

We can split $\mathcal{V}$ into the sum $\mathcal{V}_+ + \mathcal{V}_-$ of its components of positive and non-positive degrees with respect to the principal gradation. Then $V = \mathcal{V}_-$ has the property that the right hand side of (3.2) has degree 0. Indeed, from (3.3) we find

$$[\partial_z + p_{-1} + \mathbf{u}(z), \mathcal{V}_-] = -[\partial_z + p_{-1} + \mathbf{u}(z), \mathcal{V}_+],$$

which means that both commutators have neither positive nor negative homogeneous components. Therefore equation (3.2) makes sense. Thus, the problem of finding elements V such that (3.2) makes sense, and hence constructing the mKdV hierarchy, has been reduced to the problem of finding solutions of equation (3.3).

In Proposition 2.2 we described the solutions of equation (3.3) in terms of the isomorphism $\mathbf{U} \simeq N_+/A_+$, which attaches $K \in N_+/A_+$ to $\mathbf{u}$.

There are other ways of solving this problem. In the next section we will discuss the approach of Drinfeld and Sokolov [7], closely related to the dressing method of Zakharov and Shabat [31]. Another approach, proposed by Wilson [29], will be mentioned briefly later.

3.2 Drinfeld-Sokolov approach

Consider the group $N_+[\mathbf{U}]$ of $\mathbf{U}$–points of N_+. Thus, in a representation of N_+ elements of $N_+[\mathbf{U}]$ are matrices, whose entries are differential polynomials in u_i's.

Proposition 3.1 ([7], Prop. 6.2). *There exists an element $M \in N_+[\mathbf{U}]$, such that*

$$M^{-1}\left(\partial_z + p_{-1} + \mathbf{u}(z)\right) M = \partial_z + p_{-1} + \sum_{i \in I} h_i p_i, \tag{3.4}$$

where $h_i \in \mathbb{C}[\mathbf{U}], \forall i \in I$. M is defined uniquely up to right multiplication by an element of $A_+[\mathbf{U}]$.

Proof. Since the exponential map $\widehat{\mathfrak{n}}_+ \to N_+$ is an isomorphism, we can write $M = \exp \mathbf{m}$, where $\mathbf{m} \in \widehat{\mathfrak{n}}_+[\mathbf{U}]$. For any $\alpha \in \mathfrak{g}$, we have:

$$M^{-1}\alpha M = \sum_{n\geq 0} \frac{(-1)^n}{n!}(\mathrm{ad}\,\mathbf{m})^n \cdot \alpha.$$

Decompose $\mathbf{m}$ with respect to the principal gradation:

$$\mathbf{m} = \sum_{j>0} \mathbf{m}_j.$$

Then for the nth component of equation (3.4) reads

$$[p_{-1}, \mathbf{m}_1] + \mathbf{u} = 0, \tag{3.5}$$

for $n = 0$, and

$$[p_{-1}, \mathbf{m}_{n+1}] + \text{terms involving } \mathbf{m}_i, i \leq n, \text{ and } \mathbf{u} = h_n p_n \tag{3.6}$$

for $n > 0$ (here we set $h_n = 0$, if $n \notin I$). Since $\mathbf{u} \in \mathrm{Im}(\mathrm{ad}\,p_{-1})$, we can solve equation (3.5) and start the inductive process. At the nth step of induction we would have found $\mathbf{m}_i, i \leq n$ and $h_i, i < n$. We can now split the nth equation (3.6) into two parts, lying in $\mathfrak{a}$ and $\mathrm{Im}(\mathrm{ad}\,p_{-1})$. We can then set $\mathbf{m}_{n+1} = (\mathrm{ad}\,p_{-1})^{-1}$ of the $\mathrm{Im}(\mathrm{ad}\,p_{-1})$ part, and $h_n p_n =$ the $\mathfrak{a}$ part of the equation (everything evidently works even if the multiplicity of n is 2). Clearly, both $\mathbf{m}_n$ and h_n is local. This inductive procedure gives us a unique solution M of equation (3.4), such that each $\mathbf{m}_{n+1}$ lies in $\mathrm{Im}(\mathrm{ad}\,p_{-1})$. □

Now observe that formula (3.4) implies:

$$[\partial_z + p_{-1} + \mathbf{u}, MvM^{-1}] = 0, \qquad \forall v \in \mathfrak{a}. \tag{3.7}$$

Hence the zero curvature equations

$$[\partial_z + p_{-1} + \mathbf{u}, \partial_{\tau_n} + (Mp_{-n}M^{-1})_-] = 0 \tag{3.8}$$

make sense. Drinfeld and Sokolov called these equations the mKdV hierarchy associated to $\mathfrak{g}$. We will now show that these equations coincide with our equations (2.1).

3.3 Equivalence of two constructions

In both [7] and [12], one assigns to a jet $(u_i^{(n)}) \in \mathrm{Spec}\,\mathbb{C}[\mathbf{U}]$, a coset in N_+/A_+. We now show that they coincide.

Theorem 3.1 ([8], Theorem 3). *The cosets M and K in N_+/A_+ assigned in [7] and [12], respectively, to a jet $(u_i^{(n)})$, coincide.*

Proof. According to formula (3.7),

$$[\partial_z + p_{-1} + \mathbf{u}, Mp_{-1}M^{-1}] = 0.$$

Since $Mp_{-1}M^{-1} = p_{-1}+$ terms of degree higher than -1 with respect to the principal gradation on $\mathfrak{g}$, we obtain from Lemma 2.2 that $Mp_{-1}M^{-1} = KvK^{-1}$, where $v \in \mathfrak{a}$. Let us show that this implies that $v = p_{-1}$ and that $K = M$ in N_+/A_+.

Indeed, from the equality

$$Mp_{-1}M^{-1} = KvK^{-1}$$

we obtain that $(K^{-1}M)p_{-1}(K^{-1}M)^{-1}$ lies in $\mathfrak{a}$. We can represent $K^{-1}M$ as $\exp y$ for some $y \in \widehat{\mathfrak{n}}_+$. Then $(K^{-1}M)p_{-1}(K^{-1}M)^{-1} = v$ can be expressed as a linear combination of multiple commutators of y and p_{-1}:

$$e^y p_{-1}(e^y)^{-1} = \sum_{n\geq 0} \frac{1}{n!}(\operatorname{ad} y)^n \cdot p_{-1}.$$

We can write $y = \sum_{j>0} y_j$, where y_j is the homogeneous component of y of principal degree j. It follows from Proposition 1.2 that $\widehat{\mathfrak{n}}_+ = \widehat{\mathfrak{a}}_+ \oplus \operatorname{Im}(\operatorname{ad} p_{-1})$. Therefore each y_j can be further split into a sum of $y_j^0 \in \widehat{\mathfrak{a}}_+$ and $y_j^1 \in \operatorname{Im}(\operatorname{ad} p_{-1})$.

Suppose that y does not lie in $\widehat{\mathfrak{a}}_+$. Let j_0 be the smallest number such that $y_{j_0}^1 \neq 0$. Then the term of smallest degree in $e^y p_{-1}(e^y)^{-1}$ is $[y_{j_0}^1, p_{-1}]$ which lies in $\operatorname{Im}(\operatorname{ad} p_{-1})$ and is non-zero, because $\operatorname{Ker}(\operatorname{ad} p_{-1}) = \widehat{\mathfrak{a}}_+$. Hence $e^y p_{-1}(e^y)^{-1}$ can not be an element of $\widehat{\mathfrak{a}}_+$.

Therefore $y \in \widehat{\mathfrak{a}}_+$ and so $K^{-1}M \in A_+$, which means that K and M represent the same coset in N_+/A_+, and that $v = p_{-1}$. □

Remark 3.1. In the proof of Theorem 3 of [8] one has to replace p_{-n} by p_{-1}. □

From the analytic point of view, Proposition 3.1 produces the same thing as Theorem 1.1: it associates to a smooth function $\mathbf{u}(z) : \mathbb{R} \to \mathfrak{h}$, a smooth function $M(z) : \mathbb{R} \to N_+/A_+$ (cf. Sect. 1.8). What we have just shown is that the two constructions are equivalent, and $K(z) = M(z)$.

Having established that $K = M$, we see that the two definitions of mKdV hierarchy: (2.1) and (3.8), coincide. Thus, now we have two equivalent ways of constructing these equations. One of them is based on Theorem 1.1 identifying $\operatorname{Spec} \mathbb{C}[u_i^{(n)}]$ with N_+/A_+, and the other uses the "dressing operator" M defined by formula (3.4). One can then use an element of N_+/A_+ (K or M) to solve equation (3.3), and produce the mKdV equation (3.2).

Wilson [29] proposed another approach, in which one solves (3.3) directly without constructing first the dressing operator. Recall from Sect. 2.3 that when one tries to solve equation (3.3), one has to be able to take the antiderivative at certain steps. If we know in advance the existence of solution (as we do in the two other approaches), then this is insured automatically. Wilson [29] gave an argument, which demonstrates directly that the antiderivatives can be taken at each step. Thus, he proved the existence of solution $\mathcal{V}$ of (3.3) of the form $\mathcal{V} = p_{-n}+$ terms of higher degree. He then constructed the equations of the mKdV hierarchy using these $\mathcal{V}$. We now see that all three approaches are equivalent.

In the next two sections we will give another, more illuminating, explanation of the fact that $K = M$.

Remark 3.2. There exist generalizations of the mKdV hierarchies which are associated to abelian subalgebras of $\mathfrak{g}$ other then $\mathfrak{a}$ (see Remark 1.4). It is known that the Drinfeld-Sokolov approach can be applied to these generalized hierarchies [5, 19]. On the other hand, our approach can also be applied; in the case of the non-linear Schrödinger hierarchy, which corresponds to the homogeneous abelian subalgebra of $\mathfrak{g}$, this has been done by Feigin and the author [13]. The results of this lecture can be extended to establish the equivalence between the two approaches in this general context. □

3.4 Realization of $\mathbb{C}[N_+]$ as a polynomial ring.

The approach to the mKdV and affine Toda equations used in [12] and here is based on Theorem 1.1 which identifies $\mathbb{C}[N_+/A_+]$ with the ring of differential polynomials $\mathbb{C}[u_i^{(n)}]_{i=1,\dots,\ell;n\geq 0}$. In this section we add to the latter ring new variables corresponding to A_+ and show that the larger ring thus obtained is isomorphic to $\mathbb{C}[N_+]$. In this and the following two sections we follow closely [8].

Consider $u_i^{(n)}, i = 1,\dots,\ell; n \geq 0$, as A_+–invariant regular functions on N_+. Recall that

$$u_i(x) = (\alpha_i, xp_{-1}x^{-1}), \qquad x \in N_+,$$

Now choose an element χ of $\widetilde{\mathfrak{h}}$, such that $(\chi, C) \neq 0$. Introduce the regular functions $\chi_n, n \in I$, on N_+ by the formula:

$$\chi_n(x) = (\chi, xp_{-n}x^{-1}), \qquad x \in N_+. \tag{3.9}$$

Theorem 3.2 ([8], Theorem 5). $\mathbb{C}[N_+] \simeq \mathbb{C}[u_i^{(n)}]_{i=1,\dots,\ell;n\geq 0} \otimes \mathbb{C}[\chi_n]_{n\in I}$.

Proof. Let us show that the functions $u_i^{(n)}$'s and χ_n's are algebraically independent. In order to do that, let us compute the values of the differentials of these functions at the origin. Those are elements of the cotangent space to the origin, which is isomorphic to the dual space $\widehat{\mathfrak{n}}_+^*$ of $\widehat{\mathfrak{n}}_+$.

Using the invariant inner product on $\widetilde{\mathfrak{g}}$ we identify $\widehat{\mathfrak{n}}_+^*$ with $\mathfrak{n}_-$. By Proposition 1.2 $\mathfrak{n}_- = \mathfrak{a}_- \oplus (\mathfrak{a}_+)^\perp$. Recall that with respect to the principal gradation, $(\mathfrak{a}_+)^\perp = \oplus_{j>0}(\mathfrak{a}_+)^\perp_{-j}$, where $\dim(\mathfrak{a}_+)^\perp_{-j} = \ell$, and $\operatorname{ad} p_{-1} : (\mathfrak{a}_+)^\perp_{-j} \to (\mathfrak{a}_+)^\perp_{-j-1}$ is an isomorphism for all $j > 0$.

By construction of u_i's given in the proof of Theorem 1.1, $du_i|_1, i = 1,\dots,\ell$, form a basis of $(\mathfrak{a}_+)^\perp_{-1}$, and hence $du_i^{(n)}|_1, i = 1,\dots,\ell$, form a basis of $(\mathfrak{a}_+)^\perp_{-n-1}$. Thus, the covectors $du_i^{(n)}|_1, i = 1,\dots,\ell; n \geq 0$, are linearly independent. Let us show now that the covectors $d\chi_n|_1$ are linearly independent from them and among themselves. For that it is sufficient to show that the pairing between $dF_m|_1$ and p_n is non-zero if and only if $n = m$. But we have:

$$p_n^R \cdot (\chi, xp_{-m}x^{-1}) = (\chi, x[p_n, p_{-m}]x^{-1}). \tag{3.10}$$

Since $p_n, n \in \pm I$, generate a Heisenberg subalgebra in $\widetilde{\mathfrak{g}}$ (see [22], Lemma 14.4),

$$[p_n, p_{-m}] = \alpha_n C \delta_{n,m}, \tag{3.11}$$

where $\alpha_n \neq 0, \forall n \in I$. Therefore the pairing between $dF_m|_1$ and p_n is equal to $\alpha_n(\chi, C)\delta_{n,-m}$, and it is non-zero if and only if $n = m$.

Thus, the functions $u_i^{(n)}$'s and χ_n's are algebraically independent. Hence we have an embedding $\mathbb{C}[u_i^{(n)}]_{i=1,\dots,\ell;n\geq 0} \otimes \mathbb{C}[\chi_n]_{n\in I} \to \mathbb{C}[N_+]$. But the characters of the two spaces with respect to the principal gradation are both equal to

$$\prod_{n\geq 0}(1-q^n)^{-\ell} \prod_{i\in I}(1-q^i)^{-1}.$$

Hence this embedding is an isomorphism. □

3.5 Another proof of Theorem 3.1.

Let $c_n = (\alpha_n(\chi, C))^{-1}$, where $\alpha_n, n \in I$, denote the non-zero numbers determined by formula (3.11).

Proposition 3.2. *Let K be an element of N_+. We associate to it another element of N_+,*

$$\overline{K} = K \exp\left(-\sum_{n\in I} c_n p_n \chi_n(K)\right)$$

In any finite-dimensional representation of N_+, $\overline{K}$ is represented by a matrix whose entries are Taylor series with coefficients in the ring of differential polynomials in $u_i, i = 1, \dots, \ell$

The map $N_+ \to N_+$ which sends K to $\overline{K}$ is constant on the right A_+-cosets, and hence defines a section $N_+/A_+ \to N_+$.

Proof. Each entry of

$$\overline{K} = K \exp\left(-\sum_{n\in I} c_n p_n \chi_n(K)\right)$$

is a function on N_+. According to Theorem 1.1 and Theorem 3.2, to prove the proposition it is sufficient to show that each entry of $\overline{K}$ is invariant under the right action of $\mathfrak{a}_+$. By formula (3.10) we obtain for each $m \in I$:

$$p_m^R \cdot \chi_n = c_n^{-1} \delta_{n,-m},$$

and hence

$$p^R_m \cdot \left(K \exp\Big(-\sum_{n\in I} c_n p_n \chi_n(K)\Big)\right) =$$
$$K p_m \exp\Big(-\sum_{n\in I} c_n p_n \chi_n\Big) + K \exp\Big(-\sum_{n\in I} c_n p_n \chi_n\Big)(-p_m) = 0.$$

Therefore $\overline{K}$ is right $\mathfrak{a}_+$–invariant.

To prove the second statement, let a be an element of A_+ and let us show that $\overline{Ka} = \overline{K}$. We can write: $a = \exp\left(\sum_{i\in I} \alpha_i p_i\right)$. Then according to formulas (3.9) and (3.10), $\chi_n(Ka) = (\chi, Kap_{-n}a^{-1}K^{-1}) = (\chi, Kp_{-n}K^{-1}) + c_n^{-1}\alpha_n = \chi_n(K) + c_n^{-1}\alpha_n$. Therefore

$$\overline{Ka} = Ka \exp\left(-\sum_{n\in I} \alpha_n p_n - \sum_{n\in I} c_n p_n \chi_n(K)\right) = \overline{K}.$$

□

Consider now the matrix $\overline{K}$. According to Proposition 3.2, the entries of $\overline{K}$ are Taylor series with coefficients in differential polynomials in u_i's. In other words, $\overline{K} \in N_+[\mathbf{U}]$. Hence we can apply to $\overline{K}$ any derivation of $\mathbb{C}[u_i^{(n)}]$, in particular, $\partial_n = p^R_{-n}$.

Lemma 3.1. *In any finite-dimensional representation of* N_+*, the matrix of* $\overline{K}$ *satisfies:*

$$\overline{K}^{-1}\Big(\partial_n + (\overline{K}p_{-n}\overline{K}^{-1})_-\Big)\overline{K} = \partial_n + p_{-n} - \sum_{i\in I} c_i (p^R_{-n}\cdot \chi_i) p_i. \tag{3.12}$$

Proof. Using formula (1.3), we obtain:

$$\begin{aligned}\overline{K}^{-1}\Big(\partial_n + (\overline{K}p_{-n}\overline{K}^{-1})_-\Big)\overline{K} &= \partial_n + \overline{K}^{-1}(p^R_{-n}\overline{K}) + \overline{K}^{-1}(\overline{K}p_{-n}\overline{K}^{-1})_-\overline{K} \\ &= \partial_n + \overline{K}^{-1}(\overline{K}p_{-n}\overline{K}^{-1})_+\overline{K} - \sum_{i\in I} c_i(p^R_{-n}\cdot\chi_i)p_i + \overline{K}^{-1}(\overline{K}p_{-n}\overline{K}^{-1})_-\overline{K} \\ &= \partial_n + p_{-n} - \sum_{i\in I} c_i(p^R_{-n}\cdot\chi_i)p_i,\end{aligned}$$

which coincides with (3.12). □

Let now K be the point of N_+/A_+ assigned to the jet $(u_i^{(n)})$ by Theorem 1.1. Then $\overline{K}$ is a well-defined element of N_+ corresponding to K under the map $N_+/A_+ \to N_+$ defined in Proposition 3.2. By construction, $\overline{K}$ lies in the A_+–coset K.

According to Lemma 2.1, $(\overline{K}p_{-1}\overline{K}^{-1})_{-} = p_{-1} + \mathbf{u}$. Letting $n = 1$ in formula (3.12), we obtain:

$$\overline{K}^{-1}(\partial_z + p_{-1} + \mathbf{u}(z))\overline{K} = \partial_z + p_{-1} - \sum_{i \in I} c_i(p_{-1}^R \cdot \chi_i)p_i.$$

This shows that $\overline{K}$ gives a solution to equation (3.4), and hence lies in the A_+-coset of the Drinfeld-Sokolov dressing operator M. Therefore the cosets of K and M coincide, and this completes our second proof of Theorem 3.2.

3.6 Baker-Akhiezer function

In this section we adopt the analytic point of view. Recall that in Lecture 1 we constructed a map, which assigns to every smooth function $\mathbf{u}(z) : \mathbb{R} \to \mathfrak{h}$, a smooth function $K(z) : \mathbb{R} \to N_+/A_+$, with the property that the action of nth mKdV flow on $\mathbf{u}(z)$ translates into the action of p_{-n}^R on N_+/A_+. One can ask whether it is possible to lift this map to the one that assigns to $\mathbf{u}$ a function $\mathbb{R} \to N_+$ with the same property. Our results from the previous section allow us to lift the function $K(z)$ to the function $\overline{K}(z) : \mathbb{R} \to N_+$, using the section $N_+/A_+ \to N_+$. But one can check easily that the action of the mKdV flows does not correspond to the action of $\mathfrak{a}_-^R$ on $\overline{K}(z)$. However, its modification

$$\widetilde{K}(z) = \overline{K}(z) \exp\left(\sum_{n \in I} c_n p_n \chi_n(\overline{K})\right)$$

does satisfy the desired property.

Unfortunately, $\chi_n \notin \mathbb{C}[u_i^{(n)}] \subset \mathbb{C}[N_+]$, so $\widetilde{K}(z)$ can not be written in terms of differential polynomials in u_i (i.e, the jets of $\mathbf{u}(z)$), so $\widetilde{K}(z)$ is "non-local". Nevertheless, one can show that $H_n = p_{-1}^R \cdot \chi_n$ does belong to $\mathbb{C}[u_i^{(n)}]$ (in fact, this H_n can be taken as the density of the hamiltonian of the nth mKdV equation, see Lecture 5). Therefore we can write:

$$\widetilde{K}(z) = \overline{K}(z) \exp\left(\sum_{n \in I} c_n p_n \int_{-\infty}^{z} H_n dz\right).$$

Now let $\mathbf{u}(\mathbf{t})$, where $\mathbf{t} = \{t_i\}_{i \in I}$ and t_i's are the times of the mKdV hierarchy, be a solution of the mKdV hierarchy. The Baker-Akhiezer function $\Psi(\mathbf{t})$ associated to $\mathbf{u}(\mathbf{t})$, is by definition a solution of the system of equations

$$(\partial_n + (Kp_{-n}K^{-1})_-)\Psi = 0, \qquad \forall n \in I. \tag{3.13}$$

In particular, for $n = 1$ we have:

$$(\partial_z + p_{-1} + \mathbf{u}(z))\Psi = 0. \tag{3.14}$$

From formula (3.12) we obtain the following explicit formula for the solution of the system (3.13) with the initial condition is $\Psi(\mathbf{0}) = \widetilde{K}(\mathbf{0})$:

$$\Psi(\mathbf{t}) = \widetilde{K}(\mathbf{t}) \exp\left(-\sum_{i \in I} p_{-i} t_i\right).$$

But by construction,

$$\widetilde{K}(\mathbf{t}) = \left(\widetilde{K}(\mathbf{0})\Gamma(\mathbf{t})\right)_+,$$

where

$$\Gamma(\mathbf{t}) = \exp\left(\sum_{i \in I} p_{-i} t_i\right)$$

and g_+ denotes the projection of $g \in B_- \cdot N_+ \subset G$ on N_+ (it is well-defined for almost all t_i's). Hence we obtain:

$$\Psi(\mathbf{t}) = \left(\widetilde{K}(\mathbf{0})\Gamma(\mathbf{t})\right)_+ \Gamma(\mathbf{t})^{-1}.$$

Similar formula for the Baker-Akhiezer function has been obtained by G. Segal and G. Wilson [28, 30]. Following the works of the Kyoto School [4], they showed that Baker-Akhiezer functions associated to solutions of the KdV equations naturally "live" in the Sato Grassmannian, and that the flows of the hierarchy become linear in these terms (see also [3]).

We have come to the same conclusion in a different way. We have identified the mKdV variables directly with coordinates on the big cell of $B_-\backslash G/A_+$, and constructed a map $\mathbf{u}(z) \to K(z)$. It is then straightforward to check, as we did above, that the Baker-Akhiezer function is simply a lift of this map to $B_-\backslash G$. In the case of KdV hierarchy, we obtain a map to $\overline{G}[t^{-1}]\backslash G$ (see the next lecture), which for $\overline{G} = SL_n$ is the formal version of the Grassmannian that Segal-Wilson had considered.

Lecture 4

In this lecture we define the generalized KdV hierarchy associated to an affine algebra $\mathfrak{g}$. We then show the equivalence between our definition and that of Drinfeld-Sokolov.

Throughout this lecture we restrict ourselves to non-twisted affine algebras.

4.1 The left action of $\overline{N}_+$

Let $\overline{\mathfrak{n}}_+$ be the Lie subalgebra of $\mathfrak{n}_+$ generated by $e_i, i = 1, \ldots, \ell$. Thus, $\overline{\mathfrak{n}}_+$ is the upper nilpotent subalgebra of the simple Lie algebra $\overline{\mathfrak{g}}$ – the "constant subalgebra" of $\widetilde{\mathfrak{g}}$, whose Dynkin diagram is obtained by removing the 0th nod of the Dynkin digram of $\widetilde{\mathfrak{g}}$. Let $\overline{N}_+$ be the corresponding subgroup of N_+. The group $\overline{N}_+$ acts on the main homogeneous space N_+/A_+ from the left.

Lemma 4.1. *The action of $\overline{N}_+$ on N_+/A_+ is free.*

Proof. We need to show that for each $K \in N_+/A_+$, the map τ from $\overline{\mathfrak{n}}_+$ to the tangent space of N_+/A_+ at each point K is injective. But the tangent space at K is naturally isomorphic to $\widehat{\mathfrak{n}}_+/(K\mathfrak{a}K^{-1})$, and τ is the compositon of the embedding $\overline{\mathfrak{n}}_+ \to \widehat{\mathfrak{n}}_+$ and the projection $\widehat{\mathfrak{n}}_+ \to \widehat{\mathfrak{n}}_+/(K\mathfrak{a}K^{-1})$. Thus, we need to show that $\overline{\mathfrak{n}}_+ \cap K\mathfrak{a}K^{-1} = 0$ for any $K \in N_+/A_+$. This is obvious because each element of $\overline{\mathfrak{n}}_+$ is a constant element of $\mathfrak{g}$, i.e. does not depend on t, while each element of $K\mathfrak{a}K^{-1}$ does have a t–dependence. □

4.2 The KdV jet space

Now we show that $\mathbb{C}[\overline{N}_+\backslash N_+/A_+]$ can be identified with a ring of differential polynomials.

We define the functions $v_i : N_+ \to \mathbb{C}, i = 1, \dots, \ell$, by the formula:

$$v_i(K) = (f_0, Kp_{-d_i}K^{-1}), \qquad K \in N_+. \tag{4.1}$$

It is clear that these function are right A_+–invariant. We also find:

$$(e_k^L \cdot v_i)(K) = (f_0, [e_k, Kp_{-n}K^{-1}]) = ([f_0, e_k], Kp_{-n}K^{-1}) = 0$$

for all $k = 1, \dots, \ell$. Therefore v_i is left $\overline{\mathfrak{n}}_+$–invariant, and hence left $\overline{N}_+$–invariant. Thus, each v_i gives rise to a regular function on $\overline{N}_+\backslash N_+/A_+$. Let us compute the degree of v_i. We obtain:

$$\begin{aligned}(-\rho^\vee \cdot v_i)(K) &= -(f_0, [(K\rho^\vee K^{-1})_+, Kp_{-d_i}K^{-1})] \\ &= -(f_0, [K\rho^\vee K^{-1}, Kp_{-d_i}K^{-1})] + (f_0, [(K\rho^\vee K^{-1})_+, Kp_{-d_i}K^{-1})] \\ &= -(f_0, K[\rho^\vee, p_{-1}]K^{-1}) + ([f_0, \rho^\vee], Kp_{-d_i}K^{-1})] \\ &= (d_i + 1)(f_0, Kp_{-d_i}K^{-1}) = (d_i + 1)v_i(K).\end{aligned}$$

Hence v_i is homogeneous of degree $d_i + 1$.

Now denote $v_i^{(n)} = (p_{-1}^R)^n \cdot v_i$.

Theorem 4.1.

$$\mathbb{C}[\overline{N}_+\backslash N_+/A_+] \simeq \mathbb{C}[v_i^{(n)}]_{i=1,\dots,\ell;n\geq 0}.$$

Proof. We follow the proof of Theorem 1.1. First we prove the algebraic independence of the functions $v_i^{(n)}$. To do that, we need to establish the linear independence of the values of their differentials $dv_i^{(n)}|_{\bar{1}}$ at the double coset $\bar{1}$ of the identity element. Note that $dv_i^{(n)}|_{\bar{1}} = (\operatorname{ad} p_{-1})^n \cdot dv_i|_{\bar{1}}$.

Using the invariant inner product on $\widetilde{\mathfrak{g}}$, we identify the cotangent space to $\bar{1}$ with $\mathcal{N} = (\mathfrak{a}_+ \oplus \overline{\mathfrak{n}}_+)^\perp \cap \mathfrak{n}_-$. It is a graded subspace of $\mathfrak{n}_-$ with respect to the principal gradation. Moreover, the action of $\operatorname{ad} p_{-1}$ on $\mathcal{N}$ is free. One can choose homogeneous elements $\beta_1, \dots, \beta_\ell \in \mathcal{N}$, such that $\{(\operatorname{ad} p_{-1})^n\beta_i\}_{i=1,\dots,\ell;n\geq 0}$ is a basis in $(\mathfrak{a}_+ \oplus \overline{\mathfrak{n}}_+)^\perp \cap \mathfrak{n}_+^*$. It is also easy to see that $\deg \beta_i = -(d_i + 1)$. This follows from Proposition 1.2 and the fact [23] that the set of degrees of a elements of a homogeneous basis in $\overline{\mathfrak{n}}_-$ is $\cup_{i=1}^{\ell} \{1, \dots, d_i\}$.

We claim that $dv_i|_{\bar{1}}$ can be taken as the elements β_i. Let us compute $dv_i|_{\bar{1}}$. From formula (4.1) we find, in the same way as in the proof of Theorem 1.1:

$$(x, dv_i|_{\bar{1}}) = -(f_0, [x, p_{-d_i}]) = ([f_0, p_{-d_i}], x) \qquad \forall x \in \mathfrak{n}_+.$$

Hence

$$dv_i|_{\bar{1}} = [f_0, p_{-d_i}] \tag{4.2}$$

as an element of $\mathcal{N} \subset \mathfrak{n}_-$. This formula shows that the degree of $dv_i|_{\bar{1}}$ equals $d_i + 1$ as we already know.

According to Proposition 1.2, $(\mathfrak{a}_+)^\perp \subset \mathfrak{n}_-$ has an ℓ–dimensional component $(\mathfrak{a}_+)^\perp_{-j}$ of each negative degree $-j$ (with respect to the principal gradation). Furthermore, since $\overline{\mathfrak{n}}_-$ has no homogeneous components of degrees less than or equal to $-h$, minus the Coxeter number, we obtain that $(\mathfrak{a}_+)^\perp_{-j} = \mathcal{N}_{-j}$ for all $j \geq h$. Now let

$$\gamma_i = (\operatorname{ad} p_{-1})^{h-d_i-1} \cdot dv_i|_{\bar{1}} = (\operatorname{ad} p_{-1})^{h-d_i-1} \cdot [f_0, p_{-d_i}] \in (\mathfrak{a}_+)^\perp_{-h} = \mathcal{N}_{-h} \tag{4.3}$$

for all $i = 1 \dots, \ell$. Recall that the operator $\operatorname{ad} p_{-1} : (\mathfrak{a}_+)^\perp_{-j} \to (\mathfrak{a}_+)^\perp_{-j-1}$ is an isomorphism. If the vectors γ_i are linearly independent, then the covectors

$$dv_i^{(j-d_i-1)}|_{\bar{1}} = (\operatorname{ad} p_{-1})^{j-d_i-1} \cdot [f_0, p_{-d_i}] \in \mathcal{N}_{-j},$$

are linearly independent for each $j > 1$ (here we ignore $dv_i^{(n)}|_{\bar{1}}$, if $n < 0$). We see that the linear independence of the covectors $\{dv_i^{(n)}|_{\bar{1}}\}_{i=1,\dots,\ell; n \geq 0}$ is equivalent to the following

Proposition 4.1. *The vectors $\{\gamma_i\}_{i=1,\dots,\ell}$ are linearly independent.*

The proof is given in the Appendix. This completes the proof of the algebraic independence of the functions $v_i^{(n)}$. Hence we obtain an injective homomorphism $\mathbb{C}[v_i^{(n)}] \to \mathbb{C}[\overline{N}_+ \backslash N_+/A_+]$. The fact that it is an isomorphism follows in the same way as in the proof of Theorem 1.1, from the computation of characters. Clearly,

$$\operatorname{ch} \mathbb{C}[v_i^{(n)}] = \prod_{i=1}^{\ell} \prod_{n_i > d_i} (1 - q^{n_i})^{-1}. \tag{4.4}$$

On the other hand, since $\overline{N}_+$ acts freely on N_+/A_+,

$$\operatorname{ch} \mathbb{C}[\overline{N}_+ \backslash N_+/A_+] = \operatorname{ch} \mathbb{C}[N_+/A_+](\operatorname{ch} \mathbb{C}[\overline{N}_+])^{-1},$$

and

$$\operatorname{ch} \mathbb{C}[\overline{N}_+] = \prod_{i=1}^{\ell} \prod_{n_i=1}^{d_i} (1 - q^{n_i})^{-1},$$

by [23] (here we use $-\rho^\vee$ as the gradation operator). Hence $\operatorname{ch} \mathbb{C}[\overline{N}_+ \backslash N_+ / A_+]$ is also given by formula (4.4), and the theorem is proved. □

Remark 4.1. One can prove that $\mathbb{C}[\overline{N}_+ \backslash N_+ / A_+]$ is isomorphic to a ring of differential polynomials without using formula (4.1) for the functions v_i. However, the proof given above is shorter and more explicit. □

4.3 KdV hierachy

Theorem 4.1 means that just like N_+/A_+, the double quotient $\overline{N}_+ \backslash N_+ / A_+$ can be identified with the space of ∞–jets of an ℓ–tuple of functions. We call this space the KdV jet space and denote it by $\mathbf{V}$. The vector fields p^R_{-n} still act on $\overline{N}_+ \backslash N_+ / A_+$ and hence give us an infinite set of commuting evolutionary derivations on $\mathbb{C}[\mathbf{V}]$. This is, by definition, the *KdV hierarchy* associated to $\overline{\mathfrak{g}}$.

The natural projection $N_+/A_+ \to \overline{N}_+ \backslash N_+ / A_+$ gives us a map $\mathbf{U} \to \mathbf{V}$, which amounts to expressing each v_i as a differential polynomial in u_j's. This map is called the generalized *Miura transformation.* It can be thought of as a change of variables transforming the mKdV hierarchy into the KdV hierarchy.

4.4 Drinfeld-Sokolov reduction

In this section we give another definition of the generalized KdV hierarchy following Drinfeld and Sokolov. We will show in the next section that the two definitions are equivalent.

Denote by $\mathbf{Q}$ the space of ∞–jets of functions $\mathbf{q} : \mathcal{D} :\to \overline{\mathfrak{b}}_+$, where $\overline{\mathfrak{b}}_+$ is the finite-dimensional Borel subalgebra $\mathfrak{h} \oplus \overline{\mathfrak{n}}_+$ (recall that $\overline{\mathfrak{n}}_+$ is generated by $e_i, i = 1, \ldots, \ell$). If we choose a basis $\{\omega_i^\vee, i = 1, \ldots, \ell\} \cup \{e_\alpha, \alpha \in \Delta_+\}$ of $\overline{\mathfrak{b}}_+$, then we obtain the corresponding coordinates $q_i^{(n)}, q_\alpha^{(n)}, n \geq 0$ on $\mathbf{Q}$. Thus, $\mathbb{C}[\mathbf{Q}]$ is a ring of differential polynomials equipped with an action of ∂_z.

Let $C(\overline{\mathfrak{g}})$ be vector the space of operators of the form

$$\partial_z + p_{-1} + \mathbf{q}, \tag{4.5}$$

where

$$\mathbf{q} = \sum_{i=1}^{\ell} \omega_i^\vee \otimes q_i + \sum_{\alpha \in \Delta_+} e_\alpha \otimes q_\alpha \in \overline{\mathfrak{b}}_+ \otimes \mathbb{C}[\mathbf{Q}].$$

The group $\overline{N}_+[\mathbf{Q}]$ acts naturally on $C(\overline{\mathfrak{g}})$:

$$x \cdot (\partial_z + p_{-1} + \mathbf{q}) = x(\partial_z + p_{-1} + \mathbf{q})x^{-1} = \partial_z + [x, p_{-1} + \mathbf{q}] - \partial_z x. \tag{4.6}$$

Note that

$$p_{-1} = \overline{p}_{-1} + e_{\max} \otimes t^{-1},$$

where

$$\overline{p}_{-1} = \sum_{j=1}^{\ell} \frac{(\alpha_i, \alpha_i)}{2} f_j \in \overline{\mathfrak{n}}_- \tag{4.7}$$

and $e_{\max}$ is a generator of the one-dimensional subspace of $\overline{\mathfrak{g}}$ corresponding to the maximal root.

We have a direct sum decomposition $\overline{\mathfrak{b}}_+ = \oplus_{i \geq 0} \overline{\mathfrak{b}}_{+,i}$ with respect to the principal gradation. The operator $\operatorname{ad} \overline{p}_{-1}$ acts from $\overline{\mathfrak{b}}_{+,i+1}$ to $\overline{\mathfrak{b}}_{+,i}$ injectively for all $i > 1$. Hence we can choose for each $j > 0$ a vector subspace $S_j \subset \overline{\mathfrak{b}}_{+,j}$, such that $\overline{\mathfrak{b}}_{+,j} = [\overline{p}_{-1}, \overline{\mathfrak{b}}_{+,j+1}] \oplus S_j$. Note that $S_j \neq 0$ if and only if j is an exponent of $\overline{\mathfrak{g}}$, and in that case $\dim S_j$ is the multiplicity of the exponent j. In particular, $S_0 = 0$. Let $S = \oplus_{j \in E} S_j \subset \overline{\mathfrak{n}}_+$, where E is the set of exponents of $\overline{\mathfrak{g}}$. Then, by construction, S is transversal to the image of the operator $\operatorname{ad} \overline{p}_{-1}$ in $\overline{\mathfrak{n}}_+$.

Proposition 4.2 **([7]).** *The action of $\overline{N}_+[\mathbf{Q}]$ on $C(\overline{\mathfrak{g}})$ is free. Furthermore, each $\overline{N}_+[\mathbf{Q}]$-orbit contains a unique operator* (4.5) *satisfying the condition that $\mathbf{q} \in S[\mathbf{Q}]$.*

Proof. Denote by $C'(\overline{\mathfrak{g}})$ the space of operators of the form

$$\partial_z + \overline{p}_{-1} + \mathbf{q}, \qquad \mathbf{q} \in \overline{\mathfrak{b}}_+ \otimes \mathbb{C}[\mathbf{Q}].$$

The group $\overline{N}_+[\mathbf{Q}]$ acts on $C'(\overline{\mathfrak{g}})$ by the formula analogous to (4.6). We have an isomorphism $C(\overline{\mathfrak{g}}) \to C'(\overline{\mathfrak{g}})$, which sends $\partial_z + p_{-1} + \mathbf{q}$ to $\partial_z + \overline{p}_{-1} + \mathbf{q}$. Since $[x, e_{\max}] = 0, \forall x \in \overline{\mathfrak{n}}_+$, this isomorphism commutes with the action of $\overline{N}_+[\mathbf{Q}]$. Thus, we can study the action of $\overline{N}_+[\mathbf{Q}]$ on $C'(\overline{\mathfrak{g}})$ instead of $C(\overline{\mathfrak{g}})$.

We claim that each element of $C(\overline{\mathfrak{g}})$ can be uniquely represented in the form

$$\partial_z + \overline{p}_{-1} + \mathbf{q} = \exp(\operatorname{ad} U) \cdot \left(\partial_z + \overline{p}_{-1} + \mathbf{q}^0\right), \tag{4.8}$$

where $U \in \overline{\mathfrak{n}}_+[\mathbf{Q}]$ and $\mathbf{q}^0 \in S[\mathbf{Q}]$. Decompose with respect to the principal gradation: $U = \sum_{j \geq 0} U_j$, $\mathbf{q} = \sum_{j \geq 0} \mathbf{q}_j$, $\mathbf{q}^0 = \sum_{j \geq 0} \mathbf{q}^0_j$. Equating the homogeneous components of degree j in both sides of (4.8), we obtain that $\mathbf{q}^0_i + [U_{i+1}, \overline{p}_{-1}]$ is expressed in terms of $\mathbf{q}_i, \mathbf{q}^0_1, \ldots, \mathbf{q}^0_{i-1}, U_1, \ldots, U_i$. The direct sum decomposition $\overline{\mathfrak{b}}_{+,i} = [\overline{p}_{-1}, \overline{\mathfrak{b}}_{+,i+1}] \oplus S_i$ then allows us to determine uniquely $\mathbf{q}^0_i$ and U_{i+1}. Hence U and $\mathbf{q}^0$ satisfying equation (4.8) can be found uniquely by induction, and lemma follows. □

From the analytic point of view, Proposition 4.2 means that every first order differential operator (4.5), where $\mathbf{q} : \mathbb{R} \to \overline{\mathfrak{b}}_+$ is a smooth function, can be brought to the form (4.5), where $\mathbf{q}$ takes values in $S \subset \overline{\mathfrak{n}}_+$, by gauge transformation with a function $x(z) : \mathbb{R} \to \overline{N}_+$. Moreover, $x(z)$ depends on the entries of $\mathbf{q}$ only through their derivatives at z, and on those it depends polynomially (i.e., it is local).

Remark 4.2. The statement of the lemma remains true if we replace the ring of differential polynomials $\mathbb{C}[\mathbf{Q}]$ by any differential ring R. For example, we can take $R = \mathbb{C}[[z]]$. Then the quotient of $C(\overline{\mathfrak{g}})$ by the action of $\overline{N}_+[[z]]$ is what Beilinson and Drinfeld call the space of *opers* on the formal disc. This space can be defined intrinsically without choosing a particular uniformizing parameter z (using the notion of connection on the formal disc). In this form it has been generalized by Beilinson and Drinfeld to the situation where the formal disc is replaced by any algebraic curve. For instance, in the case of $\overline{\mathfrak{g}} = \mathfrak{sl}_2$, the notion of oper coincides with that of projective connection. □

It is easy to see from the proof of Proposition 3.1 that its statement remains true if we replace the operator $\partial_z + p_{-1} + \mathbf{u}(z)$ by operator (4.5). Using this fact, Drinfeld and Sokolov construct in [7] the zero-curvature equations for the operator (4.5) in the same way as for the mKdV hierarchy using formulas (3.4). Drinfeld and Sokolov show that these equations preserve the corresponding $\overline{N}_+[\mathbf{Q}]$–orbits (see [7], Sect. 6.2). Thus, they obtain a system of compatible evolutionary equations on $\overline{N}_+[\mathbf{Q}]$–orbits in $C(\overline{\mathfrak{g}})$, which they call the generalized KdV hierarchy corresponding to $\mathfrak{g}$.

These equations give rise to evolutionary derivations acting on the ring of differential polynomials $\mathbb{C}[s_i^{(n)}]_{i=1,\dots,\ell;n\geq 0}$. Indeed, the space S is ℓ–dimensional. Let us choose homogeneous coordinates $s_1, \dots, s_\ell$ of S. Then according to Proposition 4.2, the KdV equations can be written as partial differential equations on s_i's. It is easy to see that the first of these derivations is just ∂_z itself. Hence others give rise to evolutionary derivations of $\mathbb{C}[s_i^{(n)}]_{i=1,\dots,\ell;n\geq 0}$.

4.5 Equivalence of two definitions

Recall that $\mathbf{Q}$ is the space of ∞–jets of $\mathbf{q}$ with coordinates $q_\alpha^{(n)}, q_i^{(n)}, n \geq 0$. Denote by $\mathbf{S}$ the space of ∞–jets of $\mathbf{q}^0$ with coordinates $s_i^{(n)}, i = 1, \dots, \ell, n \geq 0$.

Note that we have a natural embedding $\imath : \mathbf{U} \to \mathbf{Q}$, which sends $u_i^{(n)}$ to $q_i^{(n)}$. Let $\mu : \mathbf{U} \to \mathbf{S}$ be the composition of the embedding $\imath$ and the projection $\mathbf{Q} \to \mathbf{S}$.

Proposition 3.1, suitably modified for operators of the form (4.5), gives us a map $\nu : \mathbf{Q} \to N_+/A_+$ and hence a map $\widetilde{\nu} : \mathbf{S} \to \overline{N}_+\backslash N_+/A_+$. According to Theorem 3.1, the composition of the embedding $\imath$ with the map ν coincides with the isomorphism $\mathbf{U} \simeq N_+/A_+$ constructed in Theorem 1.1. Hence the maps ν and $\widetilde{\nu}$ are surjective.

Proposition 4.3. *The map $\widetilde{\nu} : \mathbf{S} \to \overline{N}_+\backslash N_+/A_+$ is an isomorphism.*

Proof. By construction, the homomorphism $\widetilde{\nu}$ is surjective and homogeneous with respect to the natural $\mathbb{Z}$–gradations. Hence, it suffices to show that the characters of the spaces $\mathbb{C}[\mathbf{S}]$ and $\mathbb{C}[\overline{N}_+\backslash N_+/A_+]$ coincide. Since, by construction, $\deg s_i = d_i + 1$,

$$\operatorname{ch} \mathbb{C}[\mathbf{S}] = \prod_{i=1}^{\ell} \prod_{n_i \geq d_i+1} (1 - q^{n_i})^{-1}.$$

But in the proof of Theorem 4.1 we showed that that $\operatorname{ch} \mathbb{C}[\overline{N}_+ \backslash N_+/A_+]$ is given by the same formula. □

Thus, we have shown that there is an isomorphism of rings

$$\mathbb{C}[v_i^{(n)}]_{i=1,\dots,\ell; n \geq 0} \to \mathbb{C}[s_i^{(n)}]_{i=1,\dots,\ell; n \geq 0},$$

which preserves the $\mathbb{Z}$–gradation and the action of ∂_z, and such that it sends the derivations of generalized KdV hierarchy defined in Sect. 4.3 to the derivations defined in Sect. 4.4 following Drinfeld-Sokolov [7]. It also follows that the map μ defined above as the composition of the embedding $\mathbf{U} \to \mathbf{Q}$ and the projection $\mathbf{Q} \to \overline{N}_+ \backslash \mathbf{Q} \simeq \mathbf{S}$ coincides with the projection $\mathbf{U} \simeq N_+/A_+ \to \overline{N}_+ \backslash N_+/A_+ \simeq \mathbf{S}$. This is the generalized Miura transformation.

4.6 Example of $\mathfrak{sl}_2$

The space $C'(\mathfrak{sl}_2)$ consists of operators of the form

$$\partial_z + \begin{pmatrix} a & b \\ 1 & -a \end{pmatrix}.$$

The group

$$\overline{N}_+ = \begin{pmatrix} 1 & x \\ 0 & 1 \end{pmatrix}$$

acts on $C'(\mathfrak{sl}_2)$ by gauge transformations (4.6). We write canonical representatives of $\overline{N}_+$–orbits in the form

$$\partial_z + \begin{pmatrix} 0 & s \\ 1 & 0 \end{pmatrix}.$$

We have:

$$\begin{pmatrix} 1 & -\frac{u}{2} \\ 0 & 1 \end{pmatrix} \left(\partial_z + \begin{pmatrix} \frac{u}{2} & 0 \\ 1 & -\frac{u}{2} \end{pmatrix} \right) \begin{pmatrix} 1 & -\frac{u}{2} \\ 0 & 1 \end{pmatrix}^{-1} = \begin{pmatrix} 0 & s \\ 1 & 0 \end{pmatrix},$$

where

$$s = \frac{1}{4} u^2 + \frac{1}{2} \partial_z u. \tag{4.9}$$

This formula defines the Miura transformation $\mathbb{C}[s^{(n)}] \to \mathbb{C}[u_i^{(n)}]$. If we apply this change of variables (4.9) to the mKdV equation (2.16) we obtain the equation

$$\partial_3 s = \frac{3}{2} s \partial_z s - \frac{1}{4} \partial_z^3 s, \tag{4.10}$$

which, as expected, closes on s and its derivatives. This equation becomes the KdV equation (0.1) after a slight redefinition of variables: $s \to -v$, $\tau_3 \to -4\tau$. Thus, the results of this lecture prove the existence of infinitely many higher KdV flows, i.e., evolutionary derivations on $\mathbb{C}[v^{(n)}]$, which commute with the KdV derivation defined by equation (0.1).

The fact that the KdV and mKdV equations are connected by a change of variables (4.9) was discovered by R. Miura, who also realized that it can be rewritten as

$$\partial_z^2 - s = \left(\partial_z - \frac{u}{2}\right)\left(\partial_z + \frac{u}{2}\right).$$

It was this observation that triggered the fascinating idea that the KdV equation should be considered as a flow on the space of Sturm-Liouville operators $\partial_z^2 - s(z)$ [15], which led to the concept of Inverse Scattering Method and the modern view of the theory of solitons (see [26]).

4.7 Explicit formulas for the action of $\mathfrak{n}_+$

In this section we obtain explicit formulas for the action of the generators $e_i, i = 0, \dots, \ell$, of $\mathfrak{n}_+$ on $\mathbb{C}[\mathbf{U}]$. This formulas can be used to find the KdV variables $v_i \in \mathbb{C}[\mathbf{U}]$, and we will also need them in the next lecture when we study Toda field theories.

In order to find these formulas, we need a geometric construction of modules contragradient to the Verma modules over $\mathfrak{g}$ and homomorphisms between them. This construction is an affine analogue of Kostant's construction [24] in the case of simple Lie algebras.

For $\lambda \in \mathfrak{h}^*$, denote by $\mathbb{C}_\lambda$ the one-dimensional representation of $\mathfrak{b}_+$, on which $\mathfrak{h} \subset \mathfrak{b}_+$ acts according to its character λ, and $\mathfrak{n}_+ \subset \mathfrak{b}_+$ acts trivially. Let M_λ be the Verma module over $\mathfrak{g}$ of highest weight λ:

$$M_\lambda = U(\mathfrak{g}) \otimes_{U(\mathfrak{b}_+)} \mathbb{C}_\lambda.$$

we denote by v_λ the highest weight vector of M_λ^*, $1 \otimes 1$.

Denote by $\langle \cdot, \cdot \rangle$ the pairing $M_\lambda^* \times M_\lambda \to \mathbb{C}$. Let ω be the Cartan anti-involution on $\mathfrak{g}$, which maps generators $e_0, \dots, e_\ell$ to $f_0, \dots, f_\ell$ and vice versa and preserves $\mathfrak{h}$ [22]. It extends to an anti-involution of $U(\mathfrak{g})$. Let M_λ^* be the module contragradient to M_λ. As a linear space, M_λ^* is the restricted dual of M_λ. The action of $x \in \mathfrak{g}$ on $y \in M_\lambda^*$ is defined as follows:

$$\langle x \cdot y, z \rangle = \langle y, \omega(x) \cdot z \rangle, \qquad z \in M_\lambda..$$

The module M_λ^* can be realized in the space of regular functions on N_+ (see [12], Sect. 4). Indeed, the $\mathfrak{n}_+$–module $\mathbb{C}[N_+]$ (with respect to the right action) is dual to a free module with one generator, and so is each M_λ^*. Hence we can identify M_λ^* and $\mathbb{C}[N_+]$ as $\mathfrak{n}_+$–modules for any λ. It is easy to see that for $\lambda = 0$, M_0^* is isomorphic to $\mathbb{C}[N_+]$ on which the action of $\mathfrak{g}$ is defined via the left infinitesimal action of $\mathfrak{g}$ on N_+ by vector fields, described in Lecture 1. For general λ, the action of $\mathfrak{g}$ is given by first order differential operators: for $a \in \mathfrak{g}$ this differential operator is equal to $a^R + f_\lambda(a)$, where $f_\lambda(a) \in \mathbb{C}[N_+]$. The function f_λ is the image of $a \cdot v_\lambda$ under the isomorphism $M_\lambda^* \simeq \mathbb{C}[N_+]$. Hence if a is homogeneous, then $f_\lambda(a)$ is also homogeneous of the same weight.

Here is another, homological, point of view on $f_\lambda(a)$. As a $\mathfrak{g}$–module, $\mathbb{C}[N_+] = M_0^*$ is coinduced from the trivial representation of $\mathfrak{b}_-$. By Shapiro's lemma (cf., e.g., [14], Sect. 1.5.4), $H^1(\mathfrak{g}, \mathbb{C}[N_+]) \simeq H^1(\mathfrak{b}_-, \mathbb{C}) \simeq (\mathfrak{b}_-/[\mathfrak{b}_-, \mathfrak{b}_-])^* = \mathfrak{h}^*$. We see that all elements of $H^1(\mathfrak{g}, \mathbb{C}[N_+])$ have weight 0. On the other hand, functions on N_+ can only have negative or 0 weights and the only functions, which have weight 0 are constants, which are invariant with respect to the action of $\mathfrak{g}$. Therefore the coboundary of any element of the 0th group of the complex, $\mathbb{C}[N_+]$, has a non-zero weight. Hence any cohomology class from $H^1(\mathfrak{g}, \mathbb{C}[N_+])$ canonically defines a one-cocycle f, i.e., a map $\mathfrak{g} \to \mathbb{C}[N_+]$. Thus, having identified $H^1(\mathfrak{g}, \mathbb{C}[N_+])$ with $\mathfrak{h}^*$, we can assign to each $\lambda \in \mathfrak{h}^*$ and each $a \in \mathfrak{g}$, a function on N_+ – this is our $f_\lambda(a)$.

The following two results will enable us to compute the action of e_i^L.

Proposition 4.4 ([24], Theorem 2.2). *Consider $\lambda \in \mathfrak{h}^*$ as an element of $\mathfrak{h}$, using the invariant inner product. Then*

$$f_\lambda(a)(x) = (\lambda, xax^{-1}). \tag{4.11}$$

Proposition 4.5 ([12], Prop. 3). *For any $a \in \mathfrak{g}$ we have:*

$$[e_i^L, a^R] = -f_{\alpha_i}(a) e_i^L, \qquad i = 0, \dots, \ell. \tag{4.12}$$

Substituting $\lambda = \alpha_i$ and $a = p_{-1}$ in formula (4.11), we obtain:

$$f_{\alpha_i}(p_{-1})(x) = (\alpha_i, x p_{-1} x^{-1}) = u_i.$$

Hence formula (4.12) gives:

$$[e_i^L, p_{-1}^R] = -u_i e_i^L. \tag{4.13}$$

Let us write

$$e_i^L = \sum_{1 \le j \le l, n \ge 0} C_{i,j}^{(m)} \frac{\partial}{\partial u_j^{(m)}},$$

where $C_{i,j}^{(m)} \in \mathbb{C}[u_i^{(n)}]_{1 \le i \le \ell; n \ge 0}$. Formula (4.13) now gives us recurrence relations for the coefficients of $\partial/\partial u_j^{(m-1)}$ in the vector field e_i:

$$C_{i,j}^{(m)} = -u_i C_{i,j}^{(m-1)} + \partial_z C_{i,j}^{(m-1)},$$

where we have identified p_{-1}^R with ∂_z and used the formula

$$\left[\frac{\partial}{\partial u_i^{(n)}}, \partial_z\right] = \frac{\partial}{\partial u_i^{(n-1)}},$$

if $n > 0$. We also have, according to formula (1.7),

$$e_i \cdot u_j = -(\alpha_i, \alpha_j).$$

This gives the initial condition for our recurrence relation. Combining them, we obtain the following formula:

$$e_i^L = -\sum_{n \geq 0} B_i^{(n)} \partial_i^{(n)}, \tag{4.14}$$

where

$$\partial_i^{(n)} = \sum_{1 \leq j \leq l} (\alpha_i, \alpha_j) \frac{\partial}{\partial u_j^{(n)}}, \tag{4.15}$$

and $B_i^{(n)}$'s are polynomials in $u_i^{(m)}$'s, which satisfy the recurrence relation:

$$B_i^{(n)} = -u_i B_i^{(n-1)} + \partial_z B_i^{(n-1)}, \tag{4.16}$$

with the initial condition $B_i^{(0)} = 1$.

Formula (4.14) is valid for all $i = 0, \ldots, \ell$. Using this formula, we can find explicitly the KdV variables $v_i, i = 1, \ldots, \ell$, following the proof of Theorem 4.1. Consider, for example, the case $\bar{\mathfrak{g}} = \mathfrak{sl}_2$. In this case $\mathbb{C}[\mathbf{U}] = \mathbb{C}[u^{(n)}]$, and $\mathbb{C}[\mathbf{V}] = \mathbb{C}[v^{(n)}]$ is the subspace of $\mathbb{C}[\mathbf{U}]$, which consists of differential polynomials annihilated by

$$e^L = -\sum_{n \geq 0} B^{(n)} \frac{\partial}{\partial u^{(n)}}.$$

We find from formula (4.16): $B^{(0)} = 1, B^{(1)} = -u$, etc.

The KdV variable v lies in the degree 2 component of $\mathbb{C}[\mathbf{U}]$, i.e., span$\{u^2, \partial_z u\}$, and we have: $e^L \cdot u^2 = -2u, e^L \cdot \partial_z u = u$. Hence

$$v = \frac{1}{4}u^2 + \frac{1}{2}\partial_z u \tag{4.17}$$

is annihilated by e^L and can be taken as a KdV variable. As expected, this agrees with formula (4.9) obtained by means of the Drinfeld-Sokolov reduction.

Lecture 5

In this lecture we consider the Toda equations and their local integrals of motion. The Toda equation associated to the simple Lie algebra $\bar{\mathfrak{g}}$ (respectively, affine algebra) reads

$$\partial_\tau \partial_z \phi_i(z,t) = \sum_{j \in J} (\alpha_i, \alpha_j) e^{-\phi_j(z,t)}, \quad i = 1, \dots, \ell, \tag{5.1}$$

where J is the set of vertices of the Dynkin diagram of $\overline{\mathfrak{g}}$ (respectively, $\mathfrak{g}$). Each $\phi_i(z,t), i = 1, \dots, \ell$, is a family of functions in z, depending on the time variable τ, and

$$\phi_0(z,t) = -\frac{1}{a_0} \sum_{i=1}^{\ell} a_i \phi_i(t,\tau).$$

If we set formally $u_i(t) = \partial_z \phi_i(t), i = 1, \dots, \ell$, then equations (5.1) can be rewritten as an equation on $\mathbf{u}(z)$:

$$\partial_\tau u_i = \sum_{j \in J} (\alpha_i, \alpha_j) e^{-\phi_j}, \quad i = 1, \dots, \ell, \tag{5.2}$$

where ϕ_j is now understood as the anti-derivative of u_j.

Certainly, equation (5.2) is not in the form (0.2), since the terms $e^{-\phi_j}$ on the right hand side are not differential polynomials in u_i's, so our differential algebra formalism can not be applied. However, we can modify our formalism to suit the Toda equations. Moreover, we will see that these equations are closely connected to the derivations e_i^L studied in the previous lecture. After that, we will introduce the notion of local integral of motion of the Toda equation. We will describe these local integrals of motion, and will show that the corresponding hamiltonian flows are the mKdV flows.

5.1 The differential algebra formalism of Toda equation

Let Λ be the $(\ell+1)$ dimensional lattice spanned by $\alpha_0, \dots, \alpha_\ell$. For any element $\lambda = \sum_{0 \leq i \leq l} \lambda_i \alpha_i$ of Λ, define the linear space $\pi_\lambda = \mathbb{C}[u_i^{(n)}] \otimes e^{\overline{\lambda}}$, where $\overline{\lambda} = \sum_{0 \leq i \leq l} \lambda_i \overline{\phi}_i$, equipped with an action of ∂_z by the formula

$$\partial_z \cdot (P \otimes e^{\overline{\lambda}}) = (\partial_z P) \otimes e^{\overline{\lambda}} + \left(\sum_{0 \leq i \leq l} \lambda_i u_i^{(0)} P \right) \otimes e^{\overline{\lambda}}, \tag{5.3}$$

where we put

$$u_0^{(n)} = -\frac{1}{a_0} \sum_{1 \leq i \leq l} a_i u_i^{(n)}.$$

This formula means that we set $\partial_z \phi_i = u_i$. In particular, $\pi_0 = \mathbb{C}[u_i^{(n)}]$.

We introduce a $\mathbb{Z}$-gradation on π_λ by setting $\deg P \otimes e^{\overline{\lambda}} = (\rho^\vee, \lambda) + \deg P$, where the gradation on $\mathbb{C}[u_i^{(n)}]$ is the one previously defined (recall that $\rho^\vee$ satisfies $(\rho^\vee, \alpha_i) = 1, i = 0, \dots, \ell)$.

An evolutionary operator $X : \pi_\lambda \to \pi_\mu$ is, by definition, a linear operator $\pi_\lambda \to \pi_\mu$, which commutes with the action of ∂_z. Defining an evolutionary operator X acting from π_0 to $\oplus_{j\in S}\pi_{\lambda_j}$ is the same as defining $X \cdot u_i$ as an element of $\oplus_{j\in S}\pi_{\lambda_j}$ for all $j \in S$. Hence given an equation of the form

$$\partial_\tau u_i = \sum_{j\in S} P_{ij},$$

where $P_j \in \pi_{\lambda_j}$, we obtain an evolutionary operator

$$X\{P_{ij}\} = \sum_{j\in S} X_j,$$

where

$$X_j = \sum_{n\geq 0}\sum_{i=1}^{\ell}(\partial_z^n \cdot P_{ij})\frac{\partial}{\partial u_i^{(n)}}.$$

Toda equation (5.2) gives rise to the evolutionary operator

$$\mathcal{H} : \pi_0 \to \oplus_{j\in J}\pi_{-\alpha_j}, \tag{5.4}$$

which is the sum of the terms

$$\widetilde{Q}_i = \sum_{n\geq 0}(\partial_z^n e^{-\phi_i})\partial_i^{(n)}, \tag{5.5}$$

where $\partial_i^{(n)}$ is given by formula (4.15). Here $\widetilde{Q}_i$ is an evolutionary operator acting from π_0 to $\pi_{-\alpha_i}$. Clearly,

$$\partial_z^n e^{-\phi_i} = b_i^{(n)} e^{-\phi_i},$$

where $b_i^{(n)}$'s are differential polynomials in $u_j^{(n)}$'s, satisfying the recurrence relation

$$b_i^{(n)} = -u_i b_i^{(n-1)} + \partial_z b_i^{(n-1)}.$$

This recurrence relation coincides with the recurrence relation (4.16) for the differential polynomials $B_i^{(n)}$ appearing in formula (4.14) for e_i^L. Hence we obtain the following result.

Proposition 5.1. *Let T_j be the operator of multiplication by $e^{-\phi_j}$ acting from π_0 to $\pi_{-\alpha_i}$. The evolutionary operator $\mathcal{H}$ associated to the Toda equation equals*

$$\mathcal{H} = -\sum_{j\in J} T_j e_j^L.$$

Thus, while the mKdV flows come from the right action of $\mathfrak{a}_+$ on N_+/A_+, the Toda flow comes from the left action of the generators of $\mathfrak{n}_+$ (or $\overline{\mathfrak{n}}_+$) on N_+/A_+.

Now we want to introduce the notion of local integral of motion of the Toda equation. In order to do that, we first review the general hamiltonian formalism of soliton equations.

5.2 Hamiltonian formalism

In this section we briefly discuss the hamiltonian structure on the space of local functionals. This is a special case of the formalism of generalized hamiltonian structures and of generalized hamiltonian operators developed by Gelfand, Dickey and Dorfman [16, 17, 18] (see [11], Sect. 2; [12], Sect. 2 for review).

First we introduce the concept of *local functional.* Consider the space $F(\mathfrak{h})$ of functions on the circle with values in the Cartan subalgebra $\mathfrak{h}$ of $\overline{\mathfrak{g}}$, $\mathbf{u}(z) = (u_1(z), \dots, u_\ell(z))$. Then each differential polynomial $P \in \mathbb{C}[u_i^{(n)}]$ gives rise to a functional on $F(\mathfrak{h})$ that sends $\mathbf{u}(z)$ to

$$\int P(u_i^{(n)})\, dz.$$

Such functional are called local functionals. It is easy to see that this functional is 0 if and only if P is a sum of a total derivative (i.e., $\partial_z Q$ for some $Q \in \mathbb{C}[u_i^{(n)}]$) and a constant. Hence we formally define the space of local functionals $\mathfrak{F}_0$ as the quotient of $\pi_0 = \mathbb{C}[u_i^{(n)}]$ by the subspace $\operatorname{Im}\partial_z \oplus \mathbb{C}$ spanned by total derivatives and constants. We introduce a $\mathbb{Z}$-gradation on $\mathfrak{F}_0$ by subtracting 1 from the gradation induced from π_0, and denote by $\int$ the projection $\pi_0 \to \mathfrak{F}_0$.

¿From now on, to simplify notation, we shall write ∂ for ∂_z.

Introduce variational derivatives:

$$\delta_i P = \sum_{n\geq 0} \sum_{1\leq j\leq l} (\alpha_i, \alpha_j)(-\partial)^n \frac{\partial P}{\partial u_j^{(n)}}, \qquad i = 1, \dots, \ell.$$

Now for each $P \in \pi_0$, define an evolutionary derivation ξ_P of π_0 by the formula.

$$\xi_P = \sum_{1\leq i\leq l, n\geq 0} (\partial^{n+1} \cdot \delta_i P) \frac{\partial}{\partial u_i^{(n)}}, \tag{5.6}$$

Since $\delta_i P = 0, i = 1, \dots, \ell, \forall P \in \operatorname{Im}\partial_z \oplus \mathbb{C}$, we see that ξ_P depends only on the image of P in $\mathfrak{F}_0$, $\int Pdz$. So sometimes we shall write $\xi_{\int Pdz}$ instead of ξ_P. We can view $\xi : P \to \xi_P$ as a linear map $\mathfrak{F}_0 \to \mathrm{Der}^{\partial_z}$, where $\mathrm{Der}^{\partial_z}$ is the space of evolutionary derivations on π_0 (note that it is a Lie algebra).

For instance, we find:

$$\partial = \sum_{1\leq i\leq l, n\geq 0} u_i^{(n+1)} \frac{\partial}{\partial u_i^{(n)}} = \xi_P, \qquad P = \frac{1}{2} \sum_{1\leq i\leq l} u_i u^i, \tag{5.7}$$

where $u^i, i = 1 \dots, \ell$, are dual to $u_i, i = 1 \dots, \ell$, with respect to the inner product defined by $(\cdot, \cdot)$.

Now we define a Poisson bracket on $\mathfrak{F}_0$ by the formula

$$\left\{ \int Pdz, \int Rdz \right\} = \int (\xi_P \cdot R)\, dz. \tag{5.8}$$

Proposition 5.2 ([16, 17, 18]). *Formula* (5.8) *defines a Lie algebra sructure on $\mathcal{F}_0$. Furthermore, the map $\xi : \mathcal{F}_0 \to \mathrm{Der}^{\partial}$ is a homomorphism of Lie algebras, so that*

$$\xi_{\{P,R\}} = [\xi_P, \xi_R], \qquad \forall P, R \in \pi_0. \tag{5.9}$$

Remark 5.1. Using the isomorphism $\pi_0 \simeq \mathbb{C}[N_+/A_+]$, B. Enriquez and the author have expressed the formula for Poisson bracket (5.8) in terms of the corresponding unipotent cosets [9]. □

Now we extend the map ξ to incorporate the spaces π_λ, following Kuperschmidt and Wilson [25, 29].

Let $\mathcal{F}_\lambda$ be the quotient of π_λ by the subspace of total derivatives and $\int$ be the projection $\pi_\lambda \to \mathcal{F}_\lambda, P \to \int P dz$. We define a $\mathbb{Z}$-gradation on $\mathcal{F}_\lambda$ by subtracting 1 from the gradation induced from π_λ.

For any $P \in \mathcal{F}_0$ the derivation $\xi_P : \pi_0 \to \pi_0$ can be extended to a linear operator on $\oplus_{\lambda \in \Lambda} \pi_\lambda$ by the formula

$$\xi_P = \sum_{1 \leq i \leq l, n \geq 0} (\partial^{n+1} \cdot \delta_i P) \frac{\partial}{\partial u_i^{(n)}} + \sum_{1 \leq i \leq l} \delta_i P \frac{\partial}{\partial \phi_i},$$

where $\partial/\partial\phi_i \cdot (Se^{\overline{\lambda}}) = \lambda_i S e^{\overline{\lambda}}$. This defines a structure of $\mathcal{F}_0$-module on π_λ.

For each $P \in \pi_0$ the operator ξ_P commutes with the action of derivative. Hence we obtain the structure of an $\mathcal{F}_0$-module on $\mathcal{F}_\lambda$, i.e., a map $\{\cdot,\cdot\} : \mathcal{F}_0 \times \mathcal{F}_\lambda \to \mathcal{F}_\lambda$:

$$\left\{ \int P dz, \int R dz \right\} = \int (\xi_P \cdot R)\, dz.$$

Similarly, any element $R \in \pi_\lambda$ defines a linear operator ξ_R, acting from π_0 to π_λ and commuting with ∂:

$$\xi_{Se^{\overline{\lambda}}} = \sum_{1 \leq i \leq l, n \geq 0} \partial^n \left(\partial(\delta_i S \cdot e^{\overline{\lambda}}) - S \frac{\partial e^{\overline{\lambda}}}{\partial \phi_i} \right) \frac{\partial}{\partial u_i^{(n)}}. \tag{5.10}$$

The operator ξ_R depends only on the image of R in $\mathcal{F}_\lambda$. Therefore it gives rise to a map $\{\cdot,\cdot\} : \mathcal{F}_\lambda \times \mathcal{F}_0 \to \mathcal{F}_\lambda$. We have for any $P \in \mathcal{F}_0, R \in \mathcal{F}_\lambda$:

$$\int (\xi_R \cdot P)\, dz = -\int (\xi_P \cdot R)\, dz.$$

Therefore our bracket $\{\cdot,\cdot\}$ is antisymmetric.

Note that, by construction, the operators ξ_R are evolutionary. Furthermore, formula (5.9) holds for any $P \in \mathcal{F}_0, R \in \oplus_{\lambda \in \Lambda} \mathcal{F}_\lambda$.

5.3 Local integrals of motion

We obtain from the definition of ξ:

$$\xi_{e^{-\phi_j}} = \sum_{n \geq 0} (\partial^n e^{-\phi_i}) \partial_i^{(n)},$$

which coincides with formula (5.5) for $\widetilde{Q}_i$. Hence

$$\widetilde{Q}_j = \left\{ \int e^{-\phi_j} dz, \cdot \right\},$$

is a hamiltonian operator, and we obtain:

Proposition 5.3. *The evolutionary operator $\mathcal{H}$ defined by the Toda equation is hamiltonian:*

$$\mathcal{H} = \sum_{j \in J} \left\{ \int e^{-\phi_j} dz, \cdot \right\}. \tag{5.11}$$

Consider the corresponding operator

$$\{H, \cdot\} : \mathcal{F}_0 \to \oplus_{0 \leq i \leq l} \mathcal{F}_{-\alpha_i},$$

where

$$H = \sum_{j \in J} \int e^{-\phi_j} dz$$

can be viewed as the hamiltonian of the Toda equation (5.2).

We define *local integrals of motion* of the (finite of affine) Toda field theory as local functionals, which lie in the kernel of the operator $\mathcal{H}$. Thus, local integrals of motion are quantities that are conserved with respect to the Toda equation.

Denote the space of all local integrals of motion associated to $\overline{\mathfrak{g}}$ by $I(\overline{\mathfrak{g}})$ and the space of local integrals of motion associated to $\mathfrak{g}$ by $I(\mathfrak{g})$. Denote by $\overline{Q}_i$ the operator $\mathcal{F}_0 \to \mathcal{F}_{-\alpha_i}$ induced by $\widetilde{Q}_i$. Then,

$$I(\overline{\mathfrak{g}}) = \bigcap_{i=1}^{\ell} \mathrm{Ker}_{\mathcal{F}_0} \overline{Q}_i, \qquad I(\mathfrak{g}) = \bigcap_{i=0}^{\ell} \mathrm{Ker}_{\mathcal{F}_0} \overline{Q}_i.$$

Thus, we have a sequence of embeddings: $I(\mathfrak{g}) \subset I(\overline{\mathfrak{g}}) \subset \mathcal{F}_0$. From the fact that $\widetilde{Q}_i$'s are hamiltonian operators, it follows that both $I(\mathfrak{g})$ and $I(\overline{\mathfrak{g}})$ are Poisson subalgebras of $\mathcal{F}_0$.

Recall that in the previous lecture we defined an embedding $\mathbb{C}[v_i^{(n)}] \to \mathbb{C}[u_i^{(n)}]$. Let $\mathcal{W}(\overline{\mathfrak{g}})$ be the quotient of $\mathbb{C}[v_i^{(n)}]$ by the total derivatives and constants. We have an embedding $\mathcal{W}(\overline{\mathfrak{g}}) \to \mathcal{F}_0$.

The following results describe the spaces $I(\overline{\mathfrak{g}})$ and $I(\mathfrak{g})$.

Theorem 5.1 ([11], (2.4.10)). *The space of local integrals of motion of the Toda equation associated to a simple Lie algebra $\overline{\mathfrak{g}}$ equals $\mathcal{W}(\overline{\mathfrak{g}})$.*

Theorem 5.2 ([12], Theorem 1). *The space of local integrals of motion of the Toda equation associated to an affine Kac-Moody algebra $\mathfrak{g}$ is linearly spanned by elements $I_m \in \mathcal{F}_0, m \in I$, where $\deg I_m = m$.*

The proof of these results is based on the following idea [11]. We need to compute the 0th cohomology of the two-step complex

$$\mathcal{F}_0 \to \oplus_{j\in J}\mathcal{F}_{-\alpha_j},$$

where the differential is given by the sum of $\overline{Q}_i, i \in S$. We will extend this complex further to the right (this certainly does not change its 0th cohomology), and then we will compute the cohomology of the resulting complex by using the Bernstein-Gelfand-Gelfand (BGG) resolution.

5.4 The complexes $F^\bullet(\mathfrak{g})$ and $F^\bullet(\overline{\mathfrak{g}})$

Recall [2, 27] that the dual of the BGG resolution of $\mathfrak{g}$ is a complex $B^\bullet(\mathfrak{g}) = \oplus_{j\geq 0}B^j(\mathfrak{g})$, where $B^j(\mathfrak{g}) = \oplus_{w\in W, l(w)=j}M^*_{w(\rho)-\rho}$. Here M^*_λ is the module contragradient to the Verma module of highest weight λ, W is the corresponding affine Weyl group, and $l : W \to \mathbb{Z}_+$ is the length function. The differentials of the complex are described in the Appendix. They commute with the action of $\mathfrak{g}$. The 0th cohomology of $B^\bullet(\mathfrak{g})$ is one-dimensional and all higher cohomologies of $B^\bullet(\mathfrak{g})$ vanish, so that $B^\bullet(\mathfrak{g})$ is an injective resolution of the trivial representation of $\mathfrak{n}_+$.

By construction, the right action of the Lie algebra $\mathfrak{g}$ on this complex commutes with the differentials. Therefore we can take the subcomplex of invariants $F^\bullet(\mathfrak{g}) = B^\bullet(\mathfrak{g})^{\mathfrak{a}_+^R}$ with respect to the action of the Lie algebra $\mathfrak{a}_+^R$. Since $M^*_\lambda \simeq \mathbb{C}[N_+]$ as an $\mathfrak{n}_+$-module (see Sect. 4.7), we can identify $(M^*_\lambda)^{\mathfrak{a}_+^R} \simeq \mathbb{C}[N_+/A_+]$ with $\pi_\lambda = \mathbb{C}[u_i^{(n)}] \otimes e^{\overline{\lambda}}$ using Theorem 1.1. Hence

$$F^0(\mathfrak{g}) = \pi_0, \qquad F^1(\mathfrak{g}) = \oplus_{i=0}^{\ell}\pi_{-\alpha_i}.$$

Moreover, according to Sect. A.2 of the Appendix, the differential $F^0(\mathfrak{g}) \to F^1(\mathfrak{g})$ equals $\mathcal{H}$ given by formula (5.11) with $J = \{0, \dots, \ell\}$.

By construction, the differential of the complex $F^\bullet(\mathfrak{g})$ commutes with the action of $\mathfrak{a}_-^R$, in particular, with the action of $\partial = p_{-1}^R$. It is also clear that the space of constants $\mathbb{C} \subset \pi_0 = F^0(\mathfrak{g})$ is a direct summand in $F^\bullet(\mathfrak{g})$. Hence the quotient $F^\bullet(\mathfrak{g})/(\mathbb{C} \oplus \operatorname{Im}\partial)$ is also a complex. We have, by definition,

$$\mathcal{F}^0(\mathfrak{g}) = \mathcal{F}_0, \qquad \mathcal{F}^1(\mathfrak{g}) = \oplus_{0\leq i\leq \ell}\mathcal{F}_{-\alpha_i},$$

and the differential $\bar{\delta}^0 : \mathcal{F}^0(\mathfrak{g}) \to \mathcal{F}^1(\mathfrak{g})$ is given by

$$\bar{\delta}^0 = \sum_{0\leq i\leq \ell} \widetilde{Q}_i.$$

Therefore the 0th cohomology of the complex $\mathcal{F}^\bullet(\mathfrak{g})$ is isomorphic to the $I(\mathfrak{g})$. This cohomology will be computed in the next section.

5.5 Proof of Theorem 5.2

First we compute the cohomology of the complex $F^\bullet(\mathfrak{g})$.

Lemma 5.1 ([12], Lemma 1). *The cohomology of the complex $F^\bullet(\mathfrak{g})$ is isomorphic to $H^\bullet(\mathfrak{n}_+, \pi_0)$. The action of $\mathfrak{a}_-^R$ on the cohomology of $F^\bullet(\mathfrak{g})$ is trivial.*

Here $H^\bullet(\mathfrak{n}_+, \pi_0)$ denotes the Lie algebra cohomology of $\mathfrak{n}_+$ with coefficients in $\pi_0 \simeq \mathbb{C}[N_+/A_+]$ (see [14] for the definition). The $\mathfrak{n}_+$–module $\mathbb{C}[N_+/A_+]$ is coinduced from the trivial representation of $\mathfrak{a}_+$. Hence, by Shapiro's lemma (see [14], Sect. 1.5.4), $H^\bullet(\mathfrak{n}_+, \pi_0) \simeq H^\bullet(\mathfrak{a}_+, \mathbb{C})$. Since $\mathfrak{a}_+$ is an abelian Lie algebra, $H^\bullet(\mathfrak{a}_+, \mathbb{C})$ is isomorphic to the exterior algebra $\bigwedge^\bullet(\mathfrak{a}_+^*) \simeq \bigwedge^\bullet(\mathfrak{a}_-)$.

Lemma 5.2. *The 0th cohomology of the complex $\mathcal{F}^\bullet(\mathfrak{g})$ is isomorphic to the 1st cohomology of the complex $F^\bullet(\mathfrak{g})$, and hence to $\mathfrak{a}_-$.*

This readily implies Theorem 5.2. Indeed, with respect to the $\mathbb{Z}$–gradation on $\mathcal{F}_\lambda$, the differentials of the complex are homogeneous of degree 0. Moreover, the corresponding $\mathbb{Z}$–gradation on cohomology coincides with the one induced by the principal gradation on $\mathfrak{a}_+$. Therefore the space $I(\mathfrak{g})$ is linearly spanned by elements $I_m, m \in I$, where $\deg I_m = m$.

The proof of Lemma 5.2 is given in the Appendix. Here we only explain how to construct an integral of motion in the affine Toda field theory starting from a class in the first cohomology of the complex $F^\bullet(\mathfrak{g})$.

Consider such a class $\widetilde{H} \in \oplus_{0 \leq i \leq \ell} \pi_{-\alpha_i}$. Since $\partial = p_{-1}^R \in \mathfrak{a}_-^R$ acts trivially on cohomologies of the complex $\mathcal{F}^\bullet(\mathfrak{g})$ (see, $\partial \widetilde{H}$ is a coboundary, i.e. there exists $H \in \pi_0$ that $\bar{\delta}^0 \cdot H = \partial \widetilde{H}$.

By construction, the element H has the property that $\widetilde{Q}_i \cdot H \in \pi_{-\alpha_i}$ is a total derivative for $i = 0, \ldots, \ell$. But it itself is not a total derivative, because otherwise $\widetilde{H}$ would also be a trivial cocycle. Therefore, $\int H dz \neq 0$. But then $\int H dz$ is an integral of motion of the mKdV hierarchy, because by construction $\bar{\delta}^0 \cdot \int H dz = \int (\bar{\delta}^0 \cdot H) \, dz = 0$ and hence $\widetilde{Q}_i \cdot \int H dz = 0$ for any $i = 0, \ldots, \ell$.

We denote by H_m the integral of motion corresponding to $p_{-m} \in \mathfrak{a}_-$ via Lemma 5.2. By Lemma 5.2, $I_m = \int H_m dz, m \in I$, span $I(\mathfrak{g})$. Explicit formulas for $\widetilde{H}_m$ and H_m can be found in [8], Sect. 5.

5.6 Proof of Theorem 5.1

Let $\overline{W}$ be the Weyl group of $\bar{\mathfrak{g}}$. It is a finite subgroup of the affine Weyl group W of $\mathfrak{g}$. Consider the subcomplex $F^\bullet(\bar{\mathfrak{g}})$ of $F^\bullet(\mathfrak{g})$, where

$$F^j(\bar{\mathfrak{g}}) = \oplus_{w \in \overline{W}} \pi_{w(\rho) - \rho}.$$

In particular,

$$F^0(\bar{\mathfrak{g}}) = \pi_0, \qquad F^1(\mathfrak{g}) = \oplus_{i=1}^{\ell} \pi_{-\alpha_i},$$

and the differential $F^0(\bar{\mathfrak{g}}) \to F^1(\bar{\mathfrak{g}})$ equals $\mathcal{H}$ given by formula (5.11) with $J = \{1, \ldots, \ell\}$.

Consider the quotient complex $F^\bullet(\overline{\mathfrak{g}})/(\mathbb{C} \oplus \operatorname{Im}\partial)$. We have, by definition,

$$\mathcal{F}^0(\overline{\mathfrak{g}}) = \mathcal{F}_0, \qquad \mathcal{F}^1(\overline{\mathfrak{g}}) = \oplus_{1 \le i \le \ell} \mathcal{F}_{-\alpha_i},$$

and the differential $\mathcal{F}^0(\overline{\mathfrak{g}}) \to \mathcal{F}^1(\overline{\mathfrak{g}})$ is given by $\sum_{0 \le i \le \ell} \widetilde{Q}_i$. Therefore the 0th cohomology of the complex $\mathcal{F}^\bullet(\overline{\mathfrak{g}})$ is isomorphic to the $I(\overline{\mathfrak{g}})$.

Lemma 5.3 ([11]). *The cohomology of the complex $F^\bullet(\overline{\mathfrak{g}})$ is isomorphic to $H^\bullet(\overline{\mathfrak{n}}_+, \pi_0)$. Hence the 0th cohomology of $F^\bullet(\overline{\mathfrak{g}})$ is isomorphic to $\mathbb{C}[v_i^{(n)}] \subset \mathbb{C}[u_i^{(n)}]$, and the higher cohomologies vanish.*

Theorem 5.1 immediately follows from this lemma. Thus, the space of integrals of motion of the Toda theory associated to $\overline{\mathfrak{g}}$ is the Poisson algebra $\mathcal{W}(\overline{\mathfrak{g}})$, which is a Poisson subalgebra of $\mathcal{F}_0$. It is called the *classical $\mathcal{W}$-algebra* associated to $\overline{\mathfrak{g}}$.

5.7 KdV hierarchy is hamiltonian

Denote by η_m the derivation ξ_{H_m}, where H_m is the density of the mth integral of motion of the affine Toda field theory associated to $\mathfrak{g}$ (it is defined up to adding a total derivative). In particular, we can choose as the one-cocycle $\widetilde{H}_1$, the vector $\sum_{0 \le i \le \ell} e^{-\phi_i}$. Then $\partial \widetilde{H}_1 = -\sum_{0 \le i \le \ell} u_i e^{-\phi_i}$ and $H_1 = \frac{1}{2} \sum_{1 \le i \le \ell} u_i u^i$, where u^i's are dual to u_i's with respect to the inner product on $\mathfrak{h}$. Hence $\eta_1 = \partial$, by formula (5.7).

Now we have for each $m \in I$, a hamiltonian vector field η_m on $\mathbf{U}$. On the other hand, we have a vector field p^R_{-m} coming from the right infinitesimal action of the Lie algebra $\mathfrak{a}_- \subset \mathfrak{g}$ on $N_+/A_+ \simeq \mathbf{U}$; these vector fields define the flows of the mKdV hierarchy.

Theorem 5.3 ([12], Theorem 3). *The vector field η_m coincides with the vector field p^R_{-m} up to a non-zero constant multiple for any $m \in I$.*

This, we can rescale H_m so as to make $\eta_m = p^R_{-m}$. The theorem means that the vector field p^R_{-m} is *hamiltonian*, with the hamiltonian being the mth integral of motion H_m of the affine Toda field theory. Note that we already know it when $m = 1$, since $\eta_1 = \partial = p^R_{-1}$.

This means that the mKdV equations are hamiltonian, that is the mth equation of the mKdV hierarchy can be written as

$$\partial_n u_i(z) = \left\{ u_i(z), \int H_m dz \right\}$$

(up to constant multiple). In this formula $u_i(z)$ stands for the delta-like functional whose value at $u_i : \mathbb{R} \to \mathfrak{h}$ is $u_i(z)$ (see [11]). Now recall that $\mathcal{W}(\overline{\mathfrak{g}})$ is a Poisson subalgebra of $\mathcal{F}_0$, by Theorem 5.1. It is clear from the definition that H_m can be chosen in such a way that it lies in $\mathbb{C}[v_i^{(n)}] \subset \mathbb{C}[u_i^{(n)}]$, so that

$\int H_m dz \in \mathcal{W}(\bar{\mathfrak{g}})$. Then we see that the KdV hierarchy is also hamiltonian, i.e., it can be written as

$$\partial_n v_i(z) = \left\{ v_i(z), \int H_m dz \right\}.$$

Proposition 5.4. *The integrals of motion of the affine Toda field theory (equivalently, the hamiltonians of the corresponding (m)KdV hierarchy) commute with each other:*

$$\left\{ \int H_n dz, \int H_m dz \right\} = 0$$

in $\mathcal{F}_0$ for any $n, m \in I$.

Proof. Since $p_{-m}, m \in I$, lie in a commutative Lie algebra, they commute with each other; so do the corresponding vector fields. By Theorem 5.3, the same holds for the vector fields $\eta_m, m \in I$: $[\eta_n, \eta_m] = 0$. By formula (5.9), injectivity of the map ξ on $\mathcal{F}_0$, and the definiton of the vector fields η_m, the corresponding integrals of motion also commute with each other. □

We conclude that the KdV and mKdV hierarchies are completely integrable hamiltonian systems.

5.8 Quantization

In conclusion of this lecture, I would like to describe the problem of quantization of Toda integrals of motion, which was the original motivation for developing the formalism of these lectures.

The map $\xi : \pi_0 \to \operatorname{End} \pi_0$ defined in Sect. 5.2 can be quantized in the following sense. Let $\pi_\lambda^\hbar = \pi_\lambda[[\hbar]]$. There exists a linear map $\xi^\hbar : \pi_0^\hbar \to \operatorname{End} \pi_0^\hbar$ such that

(1) $\xi^\hbar$ factors through $\mathcal{F}_0^\hbar = \mathcal{F}_0[[\hbar]] = \pi_0^\hbar / (\operatorname{Im} \partial \oplus \mathbb{C})[[\hbar]]$;

(2) the bracket $[\cdot, \cdot]_\hbar : \mathcal{F}_0^\hbar \times \mathcal{F}_0^\hbar \to \mathcal{F}_0^\hbar$, defined by the formula

$$\left[\int P dz, \int R dz \right]_\hbar = \int (\xi_P^\hbar \cdot R) \, dz,$$

where $\int$ denotes the projection $\pi_0^\hbar \to \mathcal{F}_0^\hbar$, is a Lie bracket;

(3) $\xi^\hbar = \hbar \xi^{(1)} + \hbar^2(\dots)$, and the map $\pi_0 \to \operatorname{End} \pi_0$ induced by $\xi^{(1)}$ coincides with ξ.

Such a map $\xi^\hbar$ can be defined using the vertex operator algebra structure on π_0. This is explained in detail in [11], Sect. 4 (where $\hbar$ is denoted by β^2). The key observation that enables us to construct this map is that π_0 can be viewed as a Fock representation of a Heisenberg algebra with generators $b_i(n), i = 1, \dots, \ell; n \in \mathbb{Z}$, and relations

$$[b_i(n), b_j(m)] = \hbar n (\alpha_i, \alpha_j) \delta_{n,-m}.$$

This module is endowed with a canonical vertex operator algebra structure $Y^\hbar : \pi_0^\hbar \to \operatorname{End}\pi_0^\hbar[[z, z^{-1}]], P \to Y^\hbar(P, z)$. Given $P \in \pi_0^\hbar$, we denote by $\xi_P^\hbar$ the linear endomorphism of $\pi_0^\hbar$ given by the residue, i.e. the (-1)st Fourier component, of $Y^\hbar(P, z)$. This gives us a map $\xi^\hbar : \pi_0^\hbar \to \operatorname{End}\pi_0^\hbar$, which satisfies the conditions above.

Thus, the Gelfand-Dickey-Dorfman structure on π_0 can be viewed as a classical limit of the structure of vertex operator algebra on $\pi_0^\hbar$.

In [11] we also defined quantum deformations of the maps $\xi : \pi_0 \to \operatorname{End}\pi_\lambda$ and $\pi_\lambda \to \operatorname{Hom}(\pi_0, \pi_\lambda)$. This enabled us to quantize the operators $\widetilde{Q}_i : \pi_0 \to \pi_{-\alpha_i}$ and $\overline{Q}_i : \mathcal{F}_0 \to \mathcal{F}_{-\alpha_i}$. Hence we can define the space of quantum integrals of motion as the intersection of the kernels of the quantum operators $\overline{Q}_0^\hbar, \ldots, \overline{Q}_\ell^\hbar$. This space could *a priori* be "smaller" than the space $I(\mathfrak{g})$ of classical integrals of motion, i.e., it could be that some (or even all) of them do not survive quantization. However, we proved in [11] that all integrals of motion of affine Toda field theory can be quantized.

Our proof was based on the fact that the quantum operators $\widetilde{Q}_i^\hbar$ in a certain sense generate the quantized universal enveloping algebra $U_q(\mathfrak{n}_+)$, where $q = \exp(\pi i\hbar)$ (recall that the operators $\widetilde{Q}_i$ generate $U(\mathfrak{n}_+)$). Using this fact, we were able to deform the whole complex $F^\bullet(\mathfrak{g})$ and derive the quantization property from a deformation theory argument (see [11]).

Appendix

A.1

Proof of Proposition 4.1. The proposition has been proved by B. Kostant (private communication). The proof given below is different, but it uses the ideas of Kostant's proof.

Recall that for the non-twisted $\mathfrak{g} = \overline{\mathfrak{g}} \otimes \mathbb{C}[t, t^{-1}]$,

$$p_{-1} = \overline{p}_{-1} + f_0,$$

where $\overline{p}_{-1}$ is given by formula (4.7) and $f_0 = e_{\max} \otimes t^{-1}$. Here $e_{\max}$ is a generator of the one-dimensional subspace of $\overline{\mathfrak{n}}_+$ corresponding to the maximal root. More generally, for $i = 1, \ldots, \ell$, we can also write

$$p_{-d_i} = \overline{p}_{-i} \otimes 1 + r_i \otimes t^{-1}, \tag{5.12}$$

where $\overline{p}_{-i} \in \overline{\mathfrak{n}}_-$, and $r_i \in \overline{\mathfrak{n}}_+$. It is clear that $[p_{-d_i}, p_{-d_j}] = 0$ implies

$$[\overline{p}_{-i}, \overline{p}_{-j}] = [r_i, r_j] = 0. \tag{5.13}$$

Since $[x, e_{\max}] = 0$ for all $x \in \overline{\mathfrak{n}}_+$,

$$\begin{aligned}[f_0, p_{-d_i}] &= -[\overline{p}_{-1}, p_{-d_i}] = -[\overline{p}_{-1}, \overline{p}_{-i} + r_i \otimes t^{-1}] \\ &= -[\overline{p}_{-1}, r_i] \otimes t^{-1}.\end{aligned}$$

Furthermore, we find from formula (4.3):

$$\gamma_i = -(\operatorname{ad} \overline{p}_{-1})^{h-d_i} \cdot r_i \otimes t^{-1} \in \mathfrak{h} \otimes t^{-1}.$$

Hence the linear independence of the vectors γ_i is equivalent to the linear independence of the vectors

$$\overline{\gamma}_i = (\operatorname{ad} \overline{p}_{-1})^{h-d_i} \cdot r_i, \qquad i = 1, \ldots, \ell.$$

Let $\{e, h, f\}$ be the principal $\mathfrak{sl}_2$ subalgebra of $\overline{\mathfrak{g}}$, such that $f = \overline{p}_{-1}$ and $h = 2\overline{\rho}^\vee$. Recall from [23] that as a principal $\mathfrak{sl}_2$–module, $\overline{\mathfrak{g}}$ splits into the direct sum of irreducible representations R_i of dimension $2d_i + 1$, where $i = 1, \ldots, \ell$. The multiplicity of R_i in the decomposition of $\overline{\mathfrak{g}}$ equals the multiplicity of the exponent d_i. Note that different components R_j are mutually orthogonal with respect to the invariant inner product $(\cdot, \cdot)$ on $\overline{\mathfrak{g}}$, and hence every element of $\overline{\mathfrak{g}}$ can be written canonically as a sum of its projections on various R_j's.

The linear independence of the vectors $\overline{\gamma}_i$ is equivalent to the statement that each r_i has a non-zero projection on R_i (and moreover, if d_i has multiplicity two, then the projections of the corresponding r_i^1, r_i^2 on R_i are linearly independent). Note that $\deg r_i = \ell - d_i$, with respect to the gradation operator $\overline{\rho}^\vee$. Thus, it suffices to show that $(r_i, \overline{p}_{-\ell+i}) \neq 0$ (resp., the pairing between $\operatorname{span}(r_i^1, r_i^2)$ and $\operatorname{span}(\overline{p}^1_{-\ell+i}, \overline{p}^2_{-\ell+i})$ induced by $(\cdot, \cdot)$ is non-degenerate).

Now recall that according to Kac [22], Lemma 14.4, the inverse image of $\mathfrak{a}$ in $\widetilde{\mathfrak{g}}$ is a (non-degenerate) Heisenberg Lie subalgebra $\mathfrak{a} \oplus \mathbb{C}C$. Thus, $[p_n, p_{-n}] \neq 0, \forall n \in I$ (and moreover, if n has multiplicity two, the pairing between $\operatorname{span}(p_n^1, p_n^2)$ and $\operatorname{span}(p_{-n}^1, p_{-n}^2)$ induced by the commutator is non-degenerate). But note that

$$p_{d_i} = \overline{p}_{-\ell+i} \otimes t + r_{\ell-i} \otimes 1. \tag{5.14}$$

We find from formulas (5.12) and (5.14) the following commutator in $\widetilde{\mathfrak{g}}$:

$$[p^k_{d_i}, p^l_{-d_i}] = (\overline{p}^k_{-\ell+i}, r^l_i)C.$$

The proposition follows.

A.2. The BGG resolution.

Vector y from the Verma module M_λ is called *singular vector* of weight $\mu \in \mathfrak{h}^*$, if $\mathfrak{n}_+ \cdot y = 0$ and $x \cdot y = \mu(x)y$ for any $x \in \mathfrak{h}$. We have $M_\lambda \simeq U(\mathfrak{n}_-) \cdot v_\lambda$, where v_λ, which is called the highest weight vector, is a generator of the space $\mathbb{C}_\lambda$. This vector is a singular vector of weight λ. Any singular vector of M_λ of weight μ can be uniquely represented as $P \cdot v_\lambda$ for some element $P \in U(\mathfrak{n}_-)$ of weight $\mu - \lambda$. This singular vector canonically defines a homomorphism of $\mathfrak{g}$-modules $i_P : M_\mu \to M_\lambda$, which sends $u \cdot v_\mu$ to $(uP) \cdot v_\lambda$ for any $u \in U(\mathfrak{n}_-)$. Denote by i_P^* the dual homomorphism $M_\lambda^* \to M_\mu^*$.

There is an isomorphism $U(\mathfrak{n}_-) \to U(\mathfrak{n}_+)$, which maps the generators $f_0, \dots, f_\ell$ to $-e_0, \dots, -e_\ell$. Denote by $\overline{P}$ the image of $P \in U(\mathfrak{n}_-)$ under this isomorphism.

The homomorphism sending $\alpha \in \mathfrak{n}_+$ to α^L, can be extended in a unique way to a homomorphism from $U(\mathfrak{n}_+)$ to the algebra of differential operators on N_+. Denote the image of $u \in U(\mathfrak{n}_+)$ under this homomorphism by u^L.

Proposition A.1 ([12], Prop. 2). *If $P \cdot v_\lambda$ is a singular vector in M_λ of weight μ, then the homomorphism $i_P^* : M_\lambda^* \to M_\mu^*$ is given by the differential operator $\overline{P}^L$.*

In the case of simple Lie algebras, these homomorphisms were first studied by B. Kostant [24]. Using Proposition A.1, one can explicitly construct the differentials of the dual BGG resolution $B^\bullet(\mathfrak{g})$ of Sect. 5.4.

It is known that for each pair of elements of the Weyl group, such that $w \prec w'$, there is a singular vector $P_{w,w'} \cdot v_{w(\rho)-\rho}$ in $M_{w(\rho)-\rho}$ of weight $w'(\rho) - \rho$. By Proposition A.1, this vector defines the homomorphism $\overline{P}^L_{w,w'} : M^*_{w(\rho)-\rho} \to M^*_{w'(\rho)-\rho}$.

It is possible to normalize all $P_{w,w'}$'s in such a way that $P_{w'_1,w''}P_{w,w'_1} = P_{w'_2,w''}P_{w,w'_2}$ for any quadruple of elements of the Weyl group, satisfying $w \prec w'_1, w'_2 \prec w''$. Then we obtain: $\overline{P}^L_{w,w'_1}\overline{P}^L_{w'_1,w''} = \overline{P}^L_{w,w'_2}\overline{P}^L_{w'_2,w''}$.

The differential $\delta^j : B^j(\mathfrak{g}) \to B^{j+1}(\mathfrak{g})$ of the BGG complex can be written as follows:

$$\delta^j = \sum_{l(w)=j, l(w')=j+1, w \prec w'} \epsilon_{w,w'} \, \overline{P}^L_{w,w'},$$

where $\epsilon_{w,w'} = \pm 1$ are chosen as in [2, 27].

A.3. Proof of Lemma 5.2.

Since $\partial = p^R_{-1}$ commutes with the differentials of the complex $F^\bullet(\mathfrak{g})$, we can consider the double complex

$$\mathbb{C} \longrightarrow F^\bullet(\mathfrak{g}) \xrightarrow{\pm p_{-1}} F^\bullet(\mathfrak{g}) \longrightarrow \mathbb{C}. \tag{A.3}$$

Here $\mathbb{C} \to \pi_0 \subset F^\bullet(\mathfrak{g})$ and $F^\bullet(\mathfrak{g}) \to \pi_0 \to \mathbb{C}$ are the embedding of constants and the projection on constants, respectively. We place $\mathbb{C}$ in dimensions -1 and 2 of our complex, and $F^\bullet(\mathfrak{g})$ in dimensions 0 and 1.

In the spectral sequence, in which $\pm p_{-1}$ is the 0th differential, the first term is the complex $\mathcal{F}^\bullet(\mathfrak{g})[-1]$, where

$$\mathcal{F}^j(\mathfrak{g}) \simeq \oplus_{l(w)=j} \mathcal{F}_{w(\rho)-\rho}.$$

Indeed, if $\lambda \neq 0$, then in the complex

$$\pi_\lambda \xrightarrow{p_{-1}} \pi_\lambda$$

the 0th cohomology is 0, and the first cohomology is, by definition, the space $\mathcal{F}_\lambda$. If $\lambda = 0$, then in the complex

$$\mathbb{C} \longrightarrow \pi_0 \xrightarrow{p_{-1}} \pi_0 \longrightarrow \mathbb{C}$$

the 0th cohomology is 0 and the first cohomology is, by definition, the space $\mathcal{F}_0$.

We need to find the 0th cohomology of the complex $\mathcal{F}^\bullet(\mathfrak{g})$, which is the same as the 1st cohomology of the double complex (A.3).

We can compute this cohomology, using the other spectral sequence associated to our double complex. Since $H^\bullet(F^\bullet(\mathfrak{g})) \simeq \bigwedge^\bullet(\mathfrak{a}_-)$, we obtain in the first term the following complex

$$\mathbb{C} \longrightarrow \wedge^\bullet(\mathfrak{a}_-) \xrightarrow{\pm p_{-1}} \wedge^\bullet(\mathfrak{a}_-) \longrightarrow \mathbb{C}.$$

By Lemma 5.1, the action of p_{-1} on $\bigwedge^\bullet(\mathfrak{a}_-)$ is trivial and hence the cohomology of the double complex (A.3) is isomorphic to $\bigwedge^\bullet(\mathfrak{a}_-)/\mathbb{C} \oplus \bigwedge^\bullet(\mathfrak{a}_-)/\mathbb{C}[-1]$. In particular, we see that the space of Toda integrals is isomorphic to $\mathfrak{a}_-$.

Bibliography

[1] V. Bazhanov, S. Lukyanov, A. Zamolodchikov, Commun. Math. Phys. **177** (1996) 381-398; Nucl.Phys. **B489** (1997) 487-531; Preprint hep-th/9604044.

[2] I.N. Bernstein, I.M. Gelfand, S.I. Gelfand, Differential operators on the basic affine space and a study of $\mathfrak{g}$–modules. In: I.M. Gelfand (ed.) Representations of Lie groups, Budapest 1971, pp. 21-64, London: Adam Hilder 1975.

[3] I. Cherednik, Funct. Anal. Appl. **17** (1983) 243-245; Russ. Math. Surv. **38**, N 6 (1983) 113-114.

[4] E. Date, M. Jimbo, M. Kashiwara, T. Miwa, in Non-linear Integrable Systems – Classical Theory and Quantum Theory, M. Jimbo, T. Miwa (eds.), pp. 39-120, Singapore, World Scientific, 1983.

[5] M.F. de Groot, T.J. Hollowood, J.L. Miramontes, Comm. Math. Phys. **145** (1992) 57-84.

[6] F. Delduc, L. Feher, J. Phys. **A 28** (1995) 5843-5882.

[7] V.G. Drinfeld, V.V. Sokolov, Sov. Math. Dokl. **23** (1981) 457-462; J. Sov. Math. **30** (1985) 1975-2035.

[8] B. Enriquez, E. Frenkel, Comm. Math. Phys. **185** (1997) 211-230 (q-alg/9606004).

[9] B. Enriquez, E. Frenkel, *Geometric interpretation of the Poisson structure in affine Toda field theories*, Preprint q-alg/9606023.

[10] B. Feigin, E. Frenkel, Phys. Lett. **B276**, 79-86 (1992)

[11] B. Feigin, E. Frenkel, in Lect. Notes in Math. **1620**, pp. 349-418, Springer Verlag, 1995 (hep-th/9310022).

[12] B. Feigin, E. Frenkel, Invent. Math. **120** (1995) 379-408 (hep-th/9311171).

[13] B. Feigin, E. Frenkel, *Non-linear Schrödinger equations and Wakimoto modules*, unpublished manuscript.

[14] D.B. Fuchs, Cohomology of Infinite-dimensional Lie Algebras, Plenum 1988.

[15] C.S. Gardner, J.M. Green, M.D. Kruskal, R.M. Miura, Phys. Rev. Lett. **19** (1967) 1095-1097.

[16] I.M. Gelfand, L.A. Dickey, Russ. Math. Surv. **30**(5) 77-113 (1975).

[17] I.M. Gelfand, L.A. Dickey, Funct. Anal. Appl., **10** 16-22 (1976).

[18] I.M. Gelfand, I.Ya. Dorfman, Funct. Anal. Appl. **15** 173-187 (1981).

[19] T. Hollowood, J.L. Miramontes, Comm. Math. Phys. **157** (1993) 99-117.

[20] V.G. Kac, Adv. Math. **30** (1978) 85-136.

[21] V.G. Kac, D. Peterson, in Anomalies, Geometry, Topology, Argonne, 1985, pp. 276-298. World Scientific, 1985.

[22] V.G. Kac, Infinite-dimensional Lie Algebras, 3rd Edition, Cambridge University Press, 1990.

[23] B. Kostant, Amer. J. Math. **81** (1959) 973-1032.

[24] B. Kostant, in Lect. Notes in Math. **466**, pp. 101-128, Springer Verlag, 1974.

[25] B.A. Kupershmidt, G. Wilson, Comm. Math. Phys. **81** (1981) 189-202.

[26] A.C. Newell, Solitons in Mathematics and Physics, SIAM, 1985.

[27] A. Rocha-Caridi, N. Wallach, Math. Z. **180** (1982) 151-177.

[28] G. Segal, G. Wilson, Loop groups and equations of KdV type, Publ. Math. IHES **63** (1985) 5-65.

[29] G. Wilson, Ergod. Th. and Dynam. Syst. **1** (1981) 361-380.

[30] G. Wilson, C. R. Acad. Sc. Paris **299**, Serie I (1984) 587-590; Phil. Trans. Royal Soc. London **A 315** (1985) 393-404.

[31] V.E. Zakharov, A.B. Shabat, Funct. Anal. Appl. **13** (1979) 166-174.

[32] A. Zamolodchikov, Adv. Stud. in Pure Math. **19**, 641-674 (1989).

Differential Geometry of the Space of Orbits of a Coxeter Group

Boris Dubrovin

Abstract

Differential-geometric structures on the space of orbits of a finite Coxeter group, determined by Grothéndieck residues, are calculated. This gives a construction of a 2D topological field theory for an arbitrary Coxeter group.

Introduction: Formulation of main results

Let W be a (finite) Coxeter group, i.e. a finite group of linear transformations of a n-dimensional Euclidean space V generated by reflections. The space of orbits

$$M = V/W$$

has a natural structure of affine variety: the coordinate ring of M coincides with the ring $R := S(V)^W$ of W-invariant polynomial functions on V. Due to Chevalley theorem this is a polynomial ring with the generators $x_1, \ldots, x_n$ being invariant homogeneous polynomials. The basic invariant polynomials are not specified uniquely. But their degrees $d_1, \ldots, d_n$ are invariants of the group (see below Sect. 2). The maximal degree h of the polynomials is called Coxeter number of the group W. (More details about Coxeter groups will be given in Sect. 2.)

A clue to understanding of a rich differential-geometric structure of the orbit spaces can be found in the singularity theory. According to this the complexified orbit space of an irreducible Coxeter group of A-D-E type is bi-holomorphic equivalent to the universal unfolding of a simple singularity [1, 10, 31, 41]. Under this identification the Coxeter group coincides with the monodromy group of vanishing cycles of the singularity. The discriminant of the Coxeter group (the set of irregular orbits) is identified with the bifurcation diagram of the singularity. The invariant Euclidean inner product on V coincides with the pairing on the cotangent bundle T_*M defined by the intersection form of vanishing cycles [45]. The bi-holomorphic equivalence is given by the period mapping.

Additional differential-geometric structures on a universal unfolding of an isolated hypersurface singularity are determined by the Grothéndieck residues (see [39]). Let me explain this for the simplest example of the group A_n where the formulae for the residues were rediscovered by R.Dijkgraaf, E. and H.Verlinde [15] (they also found new remarkable properties of these residues, see below). This is obtained from the group of all permutations of the coordinates $\xi_1, \ldots, \xi_{n+1}$ of $(n+1)$-dimensional space by restriction onto the subspace

$$V = \{\xi_1 + \ldots + \xi_{n+1} = 0\}.$$

The space of orbits of A_n can be identified with the universal unfolding of the simple singularity $f = z^{n+1}$,

$$M = \left\{ f(z; x_1, \ldots, x_n) = z^{n+1} + x_n z^{n-1} + \ldots + x_1 = \prod_{i=1}^{n+1} (z - \xi_i) \right\}.$$

The residue pairing defines a new metric on M: the inner product of two tangent vectors in a point $x = (x_1, \ldots, x_n)$ is defined by the formula

$$(\dot{f}(z; x(s_1)), \dot{f}(z; x(s_2)))_x := \operatorname{res}_{z=\infty} \frac{\dot{f}(z; x(s_1)) \dot{f}(z; x(s_2))}{f'(z; x)}. \tag{1}$$

Here the dots mean derivatives w.r.t. the parameters s_1, s_2 resp. on two curves through the point x, the prime means d/dz. This pairing does not degenerate on TM. We can define in a similar way a trilinear form on TM putting

$$c(\dot{f}(z; x(s_1)), \dot{f}(z; x(s_2)), \dot{f}(z; x(s_3)))_x := \operatorname{res}_{z=\infty} \frac{\dot{f}(z; x(s_1)) \dot{f}(z; x(s_2)) \dot{f}(z; x(s_3))}{f'(z; x)}. \tag{2}$$

This gives rise to an operation of multiplication of tangent vectors at any point $x \in M$

$$u,\ v \mapsto u \cdot v, \quad u, v \in T_x M$$

uniquely specified by the equation

$$c(u, v, w)_x = (u \cdot v, w)_x.$$

This is a commutative associative algebra with a unity for any x isomorphic to the algebra of truncated polynomials

$$\mathbf{C}[z]/(f'(z; x)).$$

At the origin $x = 0$ the algebra coincides with the local algebra of the A_n-singularity $\mathbf{C}[z]/(f'(z))$.

One can define in a similar way polylinear forms

$$c_k(u_1, \ldots, u_k)_x := \operatorname{res}_{z=\infty} \frac{\dot{f}(z; x(s_1)) \ldots \dot{f}(z; x(s_k))}{f'(z; x)}$$

where the tangent vectors in the point x have the form

$$u_i = \dot{f}(z; x(s_i)), i = 1, \ldots, k.$$

For $k > 3$ they can be expressed in a pure algebraic way via the multiplication of vectors and the pairing (,):

$$c_k(u_1, \ldots, u_k) = (u_1 \cdot u_2 \cdot \cdots \cdot u_{k-1}, u_k).$$

Note that this formula coincides with the factorization rule for the primary correlators in two-dimensional TFT [16, 50]. Further details about 2-dimensional topological field theory from the point of view of the theory of singularities can be found in [8].

Let us come back to orbit spaces of arbitrary Coxeter groups. As it was mentioned above the intersection form of the simple singularity corresponding to a Coxeter group (as a metric on the universal unfolding) on the space of orbits can be defined intrinsicaly being induced by the invariant Euclidean structure in V. V.I.Arnol'd in [3] formulated (for $A-D-E$-singularities; for other simple singularities see [28]) the problem of calculation of the local algebra structure in intrinsic terms, i.e. via the metric on the orbit space M (this metric was introduced by Arnol'd in [2]; it is called also *convolution of invariants*). In the same time K.Saito [36, 37] solved the problem of calculation in intrinsic terms the residue pairing metric. The ideas of the papers [3, 36, 37] (and of the paper [39] where the constructions of [36, 37] were developed for extended affine root systems) are very important for constructions of the present paper.

To develop the approaches of these works I am going to contribute to understanding of the problem of giving an intrinsic description of the differential-geometric structures on the space of orbits of a Coxeter group induced by the constructions of the theory of singularities. [This problem was spelled out by K.Saito in [39] (but the structures like (2) were not considered). He considered it as generalised Jacobi's inversion problem: to describe the image of the period mapping. An equivalent problem of axiomatization of the convolution of invariants was formulated by Arnol'd in [5, p.72].] I will give an intrinsic formula for calculation of the Grothéndieck residues for arbitrary Coxeter group without using the construction of the correspondent universal unfolding. My purpose is to obtain a complete differential-geometric characterization of the space of orbits in terms of these structures (see Conjecture at the end of this section).

I came to this problem from another side when I was trying to understand a geometrical foundation of two-dimensional topological field theories (TFT) [14–16, 42, 48–50]. The idea was to extend the Atiyah's axioms [7] of TFT (for the two-dimensional case) by the properties of the canonical moduli space of a TFT model proved in [15] (see also [16]). On this way I found a nice geometrical object that I called Frobenius manifold. Any model of two-dimensional TFT is encoded by a Frobenius manifold and I showed that many constructions of TFT (integrable hierarchies for the partition function, their bi-hamiltonian formalism and τ-functions, string equations, genus zero recursion relations for correlators) can be deduced from geometry of Frobenius manifolds [20, 22]. It looks like Frobenius manifolds play also an important rôle in the theory of singularities. Better understanding of the rôle could elucidate still misterious relations between the theory of singularities, theory of integrable systems, and intersection theories on moduli spaces of algebraic curves [12, 16, 29, 30, 48–52].

In the present paper I show that the orbit spaces of Coxeter groups carry a natural structure of Frobenius manifold. For the groups of $A-D-E$ series this gives an intrinsic description (i.e. only in terms of the Coxeter group) of the residue structures like (1), (2) (this coincides with the primary chiral algebra of the $A-D-E$-topological minimal models [15, 42]).

It's time to proceed to the definition of Frobenius manifold. This is a coordinate-free formulation of a differential equation arised in [15, 49] (I called it WDVV—Witten-Dijkgraaf-E.Verlinde-H.Verlinde equation). This is a system of equations for a function $F(t)$ of n variables $t = (t^1, \ldots, t^n)$ resulting from the following conditions:

1. The matrix
$$\eta_{\alpha\beta} := \frac{\partial^3 F(t)}{\partial t^1 \partial t^\alpha \partial t^\beta}$$
should be constant and not degenerate. Let us denote by $(\eta^{\alpha\beta})$ the inverse matrix.

2. The coefficients
$$c^\gamma_{\alpha\beta}(t) := \sum_\epsilon \eta^{\gamma\epsilon} \frac{\partial^3 F(t)}{\partial t^\epsilon \partial t^\alpha \partial t^\beta}$$
for any t should be structure constants of an associative algebra.

3. The function $F(t)$ should be quasihomogeneous of a degree $3-d$ where the degree of the variables are $1-q_\alpha := \deg t^\alpha$, $q_1 = 0$. (In physical literature d is called dimension of the TFT-model and q_α are called charges of the primary fields.)

To give a coordinate-free reformulation of the WDVV equations let me recall the notion of *Frobenius algebra.* This is a commutative* associative algebra with

*Also noncommutative Frobenius algebras are considered by algebraists.

a unity over a commutative ring R supplied with a symmetric nondegenerate R-bilinear inner product $(\ ,\ )$ being invariant† in the following sense

$$(ab, c) = (a, bc)$$

for any a, b, $c \in A$. We will consider the cases where $R = \mathbf{R}$, $\mathbf{C}$ or the ring of functions on a manifold. There is a natural operation of *rescaling* of a Frobenius algebra: for an invertible constant c we change the multiplication law, the unity e and the invariant inner product $(\ ,\ )$ putting

$$a \cdot b \mapsto ca \cdot b, \quad e \mapsto c^{-1}e, \quad (\ ,\ ) \mapsto \varphi(c)(\ ,\ )$$

for an arbitrary $\varphi(c)$.

Definition 1. A manifold M (real or complex) is called *Frobenius manifold* if the tangent planes T_xM are supplied with a structure of Frobenius algebra smoothly depending on the point x and satisfying the following properties.

1. The metric on M specified by the invariant inner product $(\ ,\ )$ is flat (i.e. the curvature of the metric vanishes).

2. The unity vector field e is covariantly constant

$$\nabla e = 0.$$

Here ∇ is the Levi-Civita connection for the metric $(\ ,\ )$.

3. Let c be the section of the bundle S^3T_*M (i.e. a symmetric trilinear form on TM) given by the formula

$$c(u, v, w) := (u \cdot v, w).$$

Then the tensor

$$(\nabla_z c)(u, v, w)$$

should be symmetric in u, v, w, z for any vector fields u, v, w, z.

4. A one-parameter group of diffeomorphisms should be defined on M acting as rescalings on the algebras T_xM.

We will denote by v the generator of the one-parameter group. It can be normalised by the condition on the commutator of the vector fields e, v

$$[e, v] = e.$$

We will call v *Euler vector field* on the Frobenius manifold. The eigenvalues of the linear operator $\nabla_i v^j$ are called *invariant degrees* of the Frobenius manifold.

† Invariant inner product on a Frobenius algebra is not unique: any linear functional $\omega \in A^*$ defines an invariant symmetric inner product on A by the formula $(a, b)_\omega := \omega(ab)$. This does not degenerate for generic ω. We consider a Frobenius algebra with a *marked* invariant inner product.

In the flat coordinates $t^1, \dots, t^n$ for $(\ ,\ )$ the metric is given by a constant matrix $(\eta_{\alpha\beta})$, the unity vector field also has constant coordinates (we can normalize the flat coordinates in such a way that $e = \partial/\partial t^1$) and the Euler vector field has the form

$$v = \sum (1 - q_\alpha) t^\alpha \frac{\partial}{\partial t^\alpha}.$$

The degrees $(1 - q_\alpha)$ of the coordinates t^α coincide with the invariant degrees of the Frobenius manifold. The tensor $c_{\alpha\beta\gamma}$ can be represented (at least localy) as the third derivatives of a function $F(t)$ satisfying WDVV equations.

The nondegenerate form $(\ ,\ )$ establishes an isomorphism

$$(\ ,\ ) : TM \to T_*M.$$

This provides also the cotangent planes with a structure of Frobenius algebra.

Note that the space of vector fields on a Frobenius manifold acquires a natural structure of a Frobenius algebra over the ring R of functions on M. This can be used for algebrization of the notion of Frobenius manifold for a suitable class of rings R as a Frobenius R-algebra structure on the R-module $\operatorname{Der} R$ of derivations of R satisfying the above properties [22]. Particularly, if $R = \mathbf{C}[x_1, \dots, x_n]$ is a polynomial ring then a Frobenius R-algebra structure on the polynomial vector fields $\operatorname{Der} R$ satisfying the conditions of Definition 1 will be called *polynomial Frobenius manifold* (see the algebraic reformulation of the notion of polynomial Frobenius manifold in Appendix to this paper). In this case $M = \operatorname{Spec} R$ is an affine space and the correspondent solution of WDVV is a quasihomogeneous polynomial. Polynomial solutions of WDVV with integer and rational coefficients are of special interest due to their probable relation to intersection theories on moduli spaces of algebraic curves and their holomorphic maps [6, 52].

Let us come back to the orbit space M of a Coxeter group. We denote by $\langle\ ,\ \rangle^*$ the metric on the cotangent bundle T_*M induced by the W-invariant Euclidean structure on V. There are two marked vector fields on M: the Euler vector field

$$E := \sum d_i x_i \frac{\partial}{\partial x_i}$$

and the vector field

$$e := \frac{\partial}{\partial x_1}$$

corresponding to the polynomial of the maximal degree $\deg x_1 = h$. The vector field e is defined uniquely up to a constant factor. The *Saito metric* on M (inner product on T_*M) is defined as

$$(\ ,\ )^* := \mathcal{L}_e \langle\ ,\ \rangle^*$$

(the Lie derivative along e). This is a flat globaly defined metric on M [36, 37] (for convenience of the reader I reprove this statement in Sect. 2). Our main technical observation inspired by the differential-geometric theory of S.P. Novikov and the author of Poisson brackets of hydrodynamic type [24, 25] and by bi-hamiltonian formalism [33] is that any linear combination $a\langle\ ,\ \rangle^* + b(\ ,\ )^*$ of the flat metrics is again flat.

Theorem 1. *There exists a unique (up to rescaling) polynomial Frobenius structure on the space of orbits of a finite irreducible Coxeter group with the charges and dimension*

$$q_\alpha = 1 - \frac{d_\alpha}{h}, \quad d = 1 - \frac{2}{h}, \tag{3}$$

the unity e, the Euler vector field $\frac{1}{h}E$, and the Saito invariant metric such that for any two invariant polynomials f, g the following formula holds

$$i_v(df \cdot dg) = \langle df, dg\rangle^*. \tag{4}$$

Here i_v is the operator of inner derivative (contraction) along the vector field v.

The formula (4) gives an effective method [23] for calculation of the structure constants of the Frobenius manifold in the flat coordinates for the Saito metric (see formula (2.25) below). If the Saito flat coordinates are chosen to be invariant polynomials with rational coefficients then the polynomial $F(t)$ also has rational coefficients (it follows from (2.25)).

In the origin $t = 0$ the structure constants $c_{ij}^k(0)$ of the Frobenius algebra on T_0M coincide with the structure constants of the local algebra of the correspondent simple singularity $F(z) = 0$

$$\phi_i(z)\phi_j(z) = c_{ij}^k(0)\phi_k(z)\,(\operatorname{mod} F'(z)).$$

Here $\phi_i(z) := [\partial F(z; x_1, \ldots, x_n)/\partial x_i]_{x=0}$, $F(z; x_1, \ldots, x_n)$ is the versal deformation of the singularity $F(z) \equiv F(z; 0) = 0$. In the origin the formula (4) thus coincides with the formula of Arnol'd [3, 28] related the local algebra with the linearized convolution of invariants (i.e., with the linear part of the Euclidean metric) where the identification of T_0M with the cotangent plane $T_{*0}M$ should be given by the Saito metric. But the formula (4) gives more providing a possibility to calculate the local algebra via the convolution of invariants.

Let $R = \mathbf{C}[x_1, \ldots, x_n]$ be the coordinate ring of the orbit space M. The Frobenius algebra structure on the tangent planes T_xM for any $x \in M$ provides the R-module $\operatorname{Der} R$ of invariant vector fields with a structure of Frobenius algebra over R. To describe this structure let us consider such a basis of invariant polynomials $x_1, \ldots, x_n$ of the Coxeter group that $\deg x_1 = h$. Let $D(x_1, \ldots, x_n)$ be the discriminant of the group. We introduce a polynomial of degree n in an auxiliary variable u putting

$$P(u; x_1, \ldots, x_n) := D(x_1 - u, x_2, \ldots, x_n). \tag{5}$$

Let $D_0(x_1, \ldots, x_n)$ be the discriminant of this polynomial in u. It does not vanish identicaly on the space of orbits.

Theorem 2. *The map*

$$1 \mapsto e, \quad u \mapsto v \tag{6a}$$

can be extended uniquely to an isomorphism of R-algebras

$$\mathbf{C}[u, x_1, \dots, x_n]/(P(u; x)) \to \operatorname{Der} R. \tag{6b}$$

Corollary. *The algebra on T_xM has no nilpotents outside the zeroes of the polynomial $D_0(x_1, \dots, x_n)$.*

A non-nilpotent Frobenius algebra (over $\mathbf{C}$) can be decomposed into a direct sum of one-dimensional Frobenius algebras. Warning: by no means this implies even local decomposability of a massive (see below) Frobenius manifold into a direct sum of one-dimensional Frobenius manifolds.

Definition 2. A Frobenius manifold M is called *massive* if the algebra on the tangent planes T_xM is non-nilpotent for a generic $x \in M$.

In physical language massive Frobenius manifolds correspond to massive TFT models. Examples of massless TFT models where the algebra structure on the tangent planes is identicaly nilpotent are given by topological sigma-models with a Calabi-Yau target space [52].

Conjecture. Any massive polynomial Frobenius manifold with positive invariant degrees is isomorphic to the orbit space of a finite Coxeter group.

In other words, the constructions of Theorem 1 (and their direct products) give all massive polynomial solutions of WDVV with

$$0 \le q_\alpha \le d < 1.$$

This could give a simple approach to classification of 2-dimensional topological field theories with $d < 1$. An alternative approach was developed recently by S. Cecotti and C. Vafa [12]. It is based on studying of Hermitean metrics on a Frobenius manifold obeying certain system of differential equations (the so-called equation of topological-antitopological fusion [11], see also [21]). The approach of [12] also gives rise to Coxeter groups (and their generalizations) in classification of topological field theories.

The conjecture can be "improved" a little: instead of polynomiality it is sufficiently to assume that the function $F(t)$ is analytic in the origin. For positive invariant degrees analyticity in the origin implies polynomiality.

The Conjecture can be verified easily for 2- and 3-dimensional manifolds. There are other strong evidences in support of the conjecture. I am going to discuss them in a separate publication.

Historical Remark. I started to think about polynomial solutions of WDVV trying to answer a question of Vafa [43]: what are the 2-dimensional topological field theories (in the approach based on WDVV equations) for which the partition function is a power series in the coupling constants with rational coefficients? The question was motivated by the interpretation, due to E.Witten [42–44], of the logarythm of the partition function as a generating function of intersection numbers of cycles on some orbifolds. On this way I found the solutions (2.46)–(2.48); the sense of the solutions (2.47) and (2.48) from the point of view of known topological field theories was not clear. In December 1992 during my talk at I.Newton institute Arnol'd immediately recognized in the degrees of the polynomials (2.46)–(2.48) the Coxeter numbers (plus one) of the three Coxeter groups in the three-dimensional space. This became the starting point of the present work.

1 Differential-geometric preliminaries

The name *contravariant metric* (or, briefly, *metric*) will mean a symmetric nondegenerate bilinear form $\langle\ ,\ \rangle^*$ on the cotangent bundle T_*M to a manifold M. In a local coordinate system $x^1, \dots, x^n$ the metric is given by its components

$$g^{ij}(x) := \langle dx^i, dx^j\rangle^* \tag{1.1}$$

where (g^{ij}) is an invertible symmetric matrix. The inverse matrix $(g_{ij}) := (g^{ij})^{-1}$ specifies a *covariant metric* $\langle\ ,\ \rangle$ on the manifold M (usually it is also called metric on the manifold) i.e. a nondegenerate inner product on the tangent bundle TM

$$\langle \partial_i, \partial_j\rangle := g_{ij}(x) \tag{1.2}$$

$$\partial_i := \frac{\partial}{\partial x^i}.$$

The *Levi-Civita connection* ∇_k for the metric is uniquely specified by the conditions

$$\nabla_k g_{ij} := \partial_k g_{ij} - \Gamma^s_{ki} g_{sj} - \Gamma^s_{kj} g_{is} = 0 \tag{1.3a}$$

or, equivalently,

$$\nabla_k g^{ij} := \partial_k g^{ij} + \Gamma^i_{ks} g^{sj} + \Gamma^j_{ks} g^{is} = 0 \tag{1.3b}$$

and

$$\Gamma^k_{ij} = \Gamma^k_{ji}. \tag{1.4}$$

(Summation over twice repeated indices here and below is assumed. We will keep the symbol of summation over more than twice repeated indices.) Here the coefficients Γ^k_{ij} of the connection (the Christoffel symbols) can be expressed via the metric and its derivatives as

$$\Gamma^k_{ij} = \frac{1}{2} g^{ks}\left(\partial_i g_{sj} + \partial_j g_{is} - \partial_s g_{ij}\right). \tag{1.5}$$

For us it will be more convenient to work with the *contravariant components* of the connection

$$\Gamma^{ij}_k := \langle dx^i, \nabla_k dx^j \rangle^* = -g^{is}\Gamma^j_{sk}. \tag{1.6}$$

The equations (1.3) and (1.4) for the contravariant components read

$$\partial_k g^{ij} = \Gamma^{ij}_k + \Gamma^{ji}_k \tag{1.7}$$

$$g^{is}\Gamma^{jk}_s = g^{js}\Gamma^{ik}_s. \tag{1.8}$$

It is also convenient to introduce operators

$$\nabla^i = g^{is}\nabla_s \tag{1.9a}$$

$$\nabla^i \xi_k = g^{is}\partial_s \xi_k + \Gamma^{is}_k \xi_s. \tag{1.9b}$$

For brevity we will call the operators ∇^i and the correspondent coefficients Γ^{ij}_k *contravariant connection.*

The *curvature tensor* R^k_{slt} of the metric measures noncommutativity of the operators ∇_i or, equivalently ∇^i

$$(\nabla_s \nabla_l - \nabla_l \nabla_s)\xi_t = -R^k_{slt}\xi_k \tag{1.10a}$$

where

$$R^k_{slt} = \partial_s \Gamma^k_{lt} - \partial_l \Gamma^k_{st} + \Gamma^k_{sr}\Gamma^r_{lt} - \Gamma^k_{lr}\Gamma^r_{st}. \tag{1.10b}$$

We say that the metric is *flat* if the curvature of it vanishes. For a flat metric local *flat coordinates* $p^1, \ldots, p^n$ exist such that in these coordinates the metric is constant and the components of the Levi-Civita connection vanish. Conversely, if a system of flat coordinates for a metric exists then the metric is flat. The flat coordinates are determined uniquely up to an affine transformation with constant coefficients. They can be found from the following system

$$\nabla^i \xi_j = g^{is}\partial_s \partial_j p + \Gamma^{is}_j \partial_s p = 0, \; i,j = 1, \ldots, n. \tag{1.11}$$

If we choose the flat coordinates orthonormalized

$$\langle dp^a, dp^b \rangle^* = \delta^{ab} \tag{1.12}$$

then for the components of the metric and of the Levi-Civita connection the following formulae hold

$$g^{ij} = \frac{\partial x^i}{\partial p^a}\frac{\partial x^j}{\partial p^a} \tag{1.13a}$$

$$\Gamma^{ij}_k dx^k = \frac{\partial x^i}{\partial p^a}\frac{\partial^2 x^j}{\partial p^a \partial p^b} dp^b. \tag{1.13b}$$

All these facts are standard in geometry (see, e.g., [26]). We need to represent the formula (1.10b) for the curvature tensor in a slightly modified form (cf. [25, formula (2.18)]).

Lemma 1.1. *For the curvature of a metric the following formula holds*

$$R_l^{ijk} := g^{is} g^{jt} R_{slt}^k = g^{is}\left(\partial_s \Gamma_l^{jk} - \partial_l \Gamma_s^{jk}\right) + \Gamma_s^{ij}\Gamma_l^{sk} - \Gamma_s^{ik}\Gamma_l^{sj}. \qquad (1.14)$$

Proof. Multiplying the formula (1.10b) by $g^{is}g^{jt}$ and using (1.6) and (1.7) we obtain (1.14). The lemma is proved. □

Let us consider now a manifold supplied with two nonproportional metrics $\langle\ ,\ \rangle_1^*$ and $\langle\ ,\ \rangle_2^*$. In a coordinate system they are given by their components g_1^{ij} and g_2^{ij} resp. I will denote by Γ_{1k}^{ij} and Γ_{2k}^{ij} the correspondent Levi-Civita connections ∇_1^i and ∇_2^i. Note that the difference

$$\Delta^{ijk} = g_2^{is}\Gamma_{1s}^{jk} - g_1^{js}\Gamma_{2s}^{ik} \qquad (1.15)$$

is a tensor on the manifold.

Definition 1.1. We say that the two metrics form a *flat pencil* if:

1. The metric
$$g^{ij} = g_1^{ij} + \lambda g_2^{ij} \qquad (1.16a)$$
is flat for arbitrary λ and

2. The Levi-Civita connection for the metric (1.16a) has the form
$$\Gamma_k^{ij} = \Gamma_{1k}^{ij} + \lambda \Gamma_{2k}^{ij}. \qquad (1.16b)$$

Proposition 1.1. *For a flat pencil of metrics a vector field $f = f^i \partial_i$ exists such that the difference tensor (1.15) and the metric g_1^{ij} have the form*

$$\Delta^{ijk} = \nabla_2^i \nabla_2^j f^k \qquad (1.17a)$$

$$g_1^{ij} = \nabla_2^i f^j + \nabla_2^j f^i + c g_2^{ij} \qquad (1.17b)$$

for a constant c. The vector field should satisfy the equations

$$\Delta_s^{ij}\Delta_l^{sk} = \Delta_s^{ik}\Delta_l^{sj} \qquad (1.18)$$

where

$$\begin{aligned} \Delta_k^{ij} &:= g_{2ks}\Delta^{sij} = \nabla_{2k}\nabla_2^i f^j, \\ (g_1^{is}g_2^{jt} &- g_2^{is}g_1^{jt})\nabla_{2s}\nabla_{2t} f^k = 0. \end{aligned} \qquad (1.19)$$

Conversely, for a flat metric g_2^{ij} and for a solution f of the system (1.18), (1.19) the metrics g_2^{ij} and (1.17b) form a flat pencil.

Proof. Let us assume that $x^1, \dots, x^n$ is the flat coordinate system for the metric g_2^{ij}. In these coordinates we have

$$\Gamma_{2k}^{ij} = 0, \ \Delta_k^{ij} := g_{2ks}\Delta^{sij} = \Gamma_{1k}^{ij}. \tag{1.20}$$

The equation $R_l^{ijk} = 0$ in these coordinates reads

$$\left(g_1^{is} + \lambda g_2^{is}\right)\left(\partial_s \Delta_l^{jk} - \partial_l \Delta_s^{jk}\right) + \Delta_s^{ij}\Delta_l^{sk} - \Delta_s^{ik}\Delta_l^{sj} = 0. \tag{1.21}$$

Vanishing of the linear in λ term provides existence of a tensor f^{ij} such that

$$\Delta_k^{ij} = \partial_k f^{ij}.$$

The rest part of (1.21) gives (1.18). Let us use now the condition of symmetry (1.8) of the connection (1.16b). In the coordinate system this reads

$$\left(g_1^{is} + \lambda g_2^{is}\right)\partial_s f^{jk} = \left(g_1^{js} + \lambda g_2^{js}\right)\partial_s f^{ik}. \tag{1.22}$$

Vanishing of the terms in (1.22) linear in λ provides existence of a vector field f such that

$$f^{ij} = g_2^{is}\partial_s f^j.$$

This implies (1.17a). The rest part of the equation (1.22) gives (1.19). The last equation (1.7) gives (1.17b). The first part of the proposition is proved. The converse statement follows from the same equations. □

Remark. The theory of S.P. Novikov and the author establishes a one-to-one correspondence between flat contravariant metrics on a manifold M and Poisson brackets of hydrodynamic type on the loop space

$$L(M) := \{\text{smooth maps } S^1 \to M\}$$

with certain nondegeneracy conditions [24, 25]. For a flat metric $g^{ij}(x)$ and the correspondent contravariant connection ∇^i the Poisson bracket of two functionals of the form

$$I = I[x] = \frac{1}{2\pi}\int_0^{2\pi} P(s, x(s))\, ds, \quad J = J[x] = \frac{1}{2\pi}\int_0^{2\pi} Q(s, x(s))\, ds,$$

$x = (x^i(s))$, $x(s + 2\pi) = x(s)$ is defined by the formula

$$\{I, J\} := \frac{1}{2\pi}\int_0^{2\pi} \frac{\delta I}{\delta x^i(s)}\nabla^i \frac{\delta J}{\delta x^j(s)}\, dx^j(s) + \frac{1}{2\pi}\int_0^{2\pi} \frac{\delta I}{\delta x^i(s)} g^{ij}(x) d_s \frac{\delta J}{\delta x^j(s)}.$$

Here the variational derivative $\delta I/\delta x^i(s) \in T_* M|_{x=x(s)}$ is defined by the equality

$$I[x + \delta x] - I[x] = \frac{1}{2\pi}\int_0^{2\pi} \frac{\delta I}{\delta x^i(s)}\delta x^i(s)\, ds + o(|\delta x|);$$

$\delta J/\delta x^j(s)$ is defined by the same formula, $d_s := ds\frac{\partial}{\partial s}$. The Poisson bracket can be uniquely extended to all "good" functionals on the loop space by Leibnitz rule [24, 25]. Flat pencils of metrics correspond to compatible pairs of Poisson brackets of hydrodynamic type. By the definition, Poisson brackets $\{\ ,\ \}_1$ and $\{\ ,\ \}_2$ are called compatible if an arbitrary linear combination

$$a\{\ ,\ \}_1 + b\{\ ,\ \}_2$$

again is a Poisson bracket. Compatible pairs of Poisson brackets are important in the theory of integrable systems [33].

The main source of flat pencils is provided by the following statement.

Lemma 1.2. *If for a flat metric in some coordinate system $x^1, \dots, x^n$ both the components $g^{ij}(x)$ of the metric and $\Gamma^{ij}_k(x)$ of the correspondent Levi-Civita connection depend linearly on the coordinate x^1 then the metrics*

$$g_1^{ij} := g^{ij} \text{ and } g_2^{ij} := \partial_1 g^{ij} \tag{1.23}$$

form a flat pencil if $\det(g_2^{ij}) \neq 0$. The correspondent Levi-Civita connections have the form

$$\Gamma^{ij}_{1k} := \Gamma^{ij}_k,\ \Gamma^{ij}_{2k} := \partial_1 \Gamma^{ij}_k. \tag{1.24}$$

Proof. The equations (1.7), (1.8) and the equation of vanishing of the curvature have constant coefficients. Hence the transformation

$$g^{ij}(x^1, \dots, x^n) \mapsto g^{ij}(x^1 + \lambda, \dots, x^n),\ \Gamma^{ij}_k(x^1, \dots, x^n) \mapsto \Gamma^{ij}_k(x^1 + \lambda, \dots, x^n)$$

for an arbitrary λ maps the solutions of these equations to the solutions. By the assumption we have

$$g^{ij}(x^1 + \lambda, \dots, x^n) = g_1^{ij}(x) + \lambda g_2^{ij}(x),\ \Gamma^{ij}_k(x^1 + \lambda, \dots, x^n) = \Gamma^{ij}_{1k}(x) + \lambda \Gamma^{ij}_{2k}(x).$$

The lemma is proved. □

All the above considerations can be applied also to complex (analytic) manifolds where the metrics are quadratic forms analyticaly depending on the point of M.

2 Frobenius structure on the space of orbits of a Coxeter group

Let W be a *Coxeter group*, i.e. a finite group of linear transformations of real n-dimensional space V generated by reflections. We always can assume the transformations of the group to be orthogonal w.r.t. a Euclidean structure on V. The complete classification of irreducible Coxeter groups was obtained in [13]; see also [9]. The complete list consists of the groups (dimension of the space V equals the subscript in the name of the group) A_n, B_n, D_n, E_6, E_7, E_8, F_4, G_2 (the Weyl groups of the correspondent simple Lie algebras), the groups H_3 and H_4 of symmetries of the regular icosahedron and of the regular 600-cell in the 4-dimensional space resp. and the groups $I_2(k)$ of symmetries of the regular k-gon on the plane. The group W also acts on the symmetric algebra $S(V)$ (polynomials of the coordinates of $V = V^*$) and on the $S(V)$-module $\Omega(V)$ of differential forms on V with polynomial coefficients. The subring $R = S(V)^W$ of W-invariant polynomials is generated by n algebraicaly independent homogeneous polynomials $x^1, \ldots, x^n$ [9]. The submodule $\Omega(V)^W$ of the W-invariant differential forms with polynomial coefficients is a free R-module with the basis $dx^{i_1} \wedge \ldots \wedge dx^{i_k}$ [9]. Degrees of the basic invariant polynomials are uniquely defined by the Coxeter group. They can be expressed via the *exponents* $m_1, \ldots, m_n$ of the group, i.e. via the eigenvalues of a Coxeter element C in W [9]

$$d_i := \deg x^i = m_{n-i+1} + 1, \tag{2.1a}$$

$$\{\mathrm{eigen}\, C\} = \left\{ \exp \frac{2\pi i(d_1 - 1)}{h}, \ldots, \exp \frac{2\pi i(d_n - 1)}{h} \right\}. \tag{2.1b}$$

The maximal degree h is called *Coxeter number* of W. I will use the reversed ordering of the invariant polynomials

$$d_1 = h > d_2 \geq \ldots \geq d_{n-1} > d_2 = 2. \tag{2.2}$$

The degrees satisfy the *duality condition*

$$d_i + d_{n-i+1} = h + 2, \; i = 1, \ldots, n. \tag{2.3}$$

The list of the degrees for all the Coxeter groups is given in Table 1.

W	$d_1, \; \ldots, \; d_n$
A_n	$d_i = n + 2 - i$
B_n	$d_i = 2(n - i + 1)$
$D_n, \; n = 2k$	$d_i = 2(n - i), \; i \leq k,$
	$d_i = 2(n - i + 1), \; k + 1 \leq i$
$D_n, \; n = 2k + 1$	$d_i = 2(n - i), \; i \leq k,$
	$d_{k+1} = 2k + 1,$
	$d_i = 2(n - i + 1), \; k + 2 \leq i$
E_6	12, 9, 8, 6, 5, 2
E_7	18, 14, 12, 10, 8, 6, 2
E_8	30, 24, 20, 18, 14, 12, 8, 2
F_4	12, 8, 6, 2
G_2	6, 2
H_3	10, 6, 2
H_4	30, 20, 12, 2
$I_2(k)$	k, 2

Table 1.

I will extend the action of the group W to the complexified space $V \otimes \mathbf{C}$. The space of orbits

$$M = V \otimes \mathbf{C}/W$$

has a natural structure of an affine algebraic variety: the coordinate ring of M is the (complexified) algebra R of invariant polynomials of the group W. The coordinates $x^1, \ldots, x^n$ on M are defined up to an invertible transformation

$$x^i \mapsto x^{i'}(x^1, \ldots, x^n), \tag{2.4}$$

where $x^{i'}(x^1, \ldots, x^n)$ is a graded homogeneous polynomial of the same degree d_i in the variables $x^1, \ldots, x^n$, $\deg x^k = d_k$. Note that the Jacobian $\det(\partial x^{i'}/\partial x^j)$ is a constant (it should not be zero). The transformations (2.4) leave invariant the vector field $\partial_1 := \partial/\partial x^1$ (up to a constant factor) due to the strict inequality $d_1 > d_2$. The coordinate x^n is determined uniquely within a factor. Also the vector field

$$E = d_1 x^1 \partial_1 + \ldots + d_n x^n \partial_n = p^a \frac{\partial}{\partial p^a} \tag{2.5}$$

(the generator of scaling transformations) is well-defined on M.

Let $\langle \ , \ \rangle$ denotes the W-invariant Euclidean metric in the space V. I will use the orthonormal coordinates $p^1, \ldots, p^n$ in V with respect to this metric. The invariant x^n can be chosen as

$$x^n = \frac{1}{2}((p^1)^2 + \ldots + (p^n)^2). \tag{2.6}$$

We extend $\langle \ , \ \rangle$ onto $V \otimes \mathbf{C}$ as a complex quadratic form.

The factorization map $V \otimes \mathbf{C} \to M$ is a local diffeomorphism on an open subset of $V \otimes \mathbf{C}$. The image of this subset in M consists of *regular orbits* (i.e. the number of points of the orbit equals $\# W$). The complement is the *discriminant* $\mathrm{Discr}\, W$. By the definition it consists of all irregular orbits. Note that the linear coordinates in V can serve also as local coordinates in small domains in $M \setminus \mathrm{Discr}\, W$. This defines a metric $\langle \ , \ \rangle$ (and $\langle \ , \ \rangle^*$) on $M \setminus \mathrm{Discr}\, W$. The contravariant metric can be extended onto M according to the following statement (cf. [39, Sections 5 and 6]).

Lemma 2.1. *The Euclidean metric of V induces polynomial contravariant metric $\langle \ , \ \rangle^*$ on the space of orbits*

$$g^{ij}(x) = \langle dx^i, dx^j \rangle^* := \frac{\partial x^i}{\partial p^a}\frac{\partial x^j}{\partial p^a} \tag{2.7}$$

and the correspondent contravariant Levi-Civita connection

$$\Gamma^{ij}_k(x)dx^k = \frac{\partial x^i}{\partial p^a}\frac{\partial^2 x^j}{\partial p^a \partial p^b} dp^b \tag{2.8}$$

also is a polynomial one.

Proof. The right-hand sides in (2.7)/(2.8) are W-invariant polynomials/differential forms with polynomial coefficients. Hence $g^{ij}(x)/\Gamma_k^{ij}(x)$ are polynomials in $x^1, \dots, x^n$. Lemma is proved. □

Remark. The matrix $g^{ij}(x)$ does not degenerate on $M \setminus \operatorname{Discr} W$ where the factorization $V \otimes \mathbf{C} \to M$ is a local diffeomorphism. So the polynomial (also called *discriminant* of W)

$$D(x) := \det(g^{ij}(x)) \tag{2.9}$$

vanishes precisely on the discriminant $\operatorname{Discr} W$ where the variables $p^1, \dots, p^n$ fail to be local coordinates. Due to this fact the matrix $g^{ij}(x)$ often is called *discriminant matrix* of W. The operation $x^i, x^j \mapsto g^{ij}(x)$ is also called *convolution of invariants* (see [2]). Note that the image of V in the real part of M is specified by the condition of positive semidefiniteness of the matrix $(g^{ij}(x))$ (cf. [34]). The Euclidean connection (2.8) on the space of orbits is called *Gauss-Manin connection.*

Corollary 2.1. *The functions $g^{ij}(x)$ and $\Gamma_k^{ij}(x)$ depend linearly on x^1.*

Proof. From the definition one has that $g^{ij}(x)$ and $\Gamma_k^{ij}(x)$ are graded homogeneous polynomials of the degrees

$$\deg g^{ij}(x) = d_i + d_j - 2 \tag{2.10}$$

$$\deg \Gamma_k^{ij}(x) = d_i + d_j - d_k - 2. \tag{2.11}$$

Since $d_i + d_j \leq 2h = 2d_1$ these polynomials can be at most linear in x^1. Corollary is proved. □

Corollary 2.2 (K. Saito). *The matrix*

$$\eta^{ij}(x) := \partial_1 g^{ij}(x) \tag{2.12}$$

has a triangular form

$$\eta^{ij}(x) = 0 \text{ for } i + j > n + 1, \tag{2.13}$$

and the antidiagonal elements

$$\eta^{i(n-i+1)} =: c_i \tag{2.14}$$

are nonzero constants. Particularly,

$$c := \det(\eta^{ij}) = (-1)^{\frac{n(n-1)}{2}} c_1 \dots c_n \neq 0. \tag{2.15}$$

Proof. One has

$$\deg \eta^{ij}(x) = d_i + d_j - 2 - h.$$

Hence $\deg \eta^{i(n-i+1)} = 0$ (see (2.3)) and $\deg \eta^{ij} < 0$ for $i + j > n + 1$. This proves triangularity of the matrix and constancy of the antidiagonal entries c_i. To prove nondegenerateness of $(\eta^{ij}(x))$ we consider, following Saito, the discriminant (2.9) as a polynomial in x^1

$$D(x) = c(x^1)^n + a_1(x^1)^{n-1} + \ldots + a_n$$

where the coefficients $a_1, \ldots, a_n$ are quasihomogeneous polynomials in $x^2, \ldots, x^n$ of the degrees $h, \ldots, nh$ resp. and the leading coefficient c is given in (2.15). Let γ be the eigenvector of a Coxeter transformation C with the eigenvalue $\exp(2\pi i/h)$. Then

$$x^k(\gamma) = x^k(C\gamma) = x^k(\exp(2\pi i/h)\gamma) = \exp(2\pi i d_k/h)x^k(\gamma).$$

For $k > 1$ we obtain

$$x^k(\gamma) = 0, \ k = 2, \ldots, n.$$

But $D(\gamma) \neq 0$ [9]. Hence the leading coefficient $c \neq 0$. Corollary is proved. □

Corollary 2.3. *The space M of orbits of a finite Coxeter group carries a flat pencil of metrics $g^{ij}(x)$ (2.7) and $\eta^{ij}(x)$ (2.12) where the matrix $\eta^{ij}(x)$ is polynomialy invertible globaly on M.*

We will call (2.12) *Saito metric* on the space of orbits. This metric will be denoted by $(\, , \,)^*$ (and by $(\, , \,)$ if considered on the tangent bundle TM). Let us denote by

$$\gamma_k^{ij}(x) := \partial_1 \Gamma_k^{ij}(x) \tag{2.16}$$

the components of the Levi-Civita connection for the metric $\eta^{ij}(x)$. These are quasihomogeneous polynomials of the degrees

$$\deg \gamma_k^{ij}(x) = d_i + d_j - d_k - h - 2. \tag{2.17}$$

Corollary 2.4 (K. Saito). *There exist homogeneous polynomials $t^1(p), \ldots, t^n(p)$ of degrees $d_1, \ldots, d_n$ resp. such that the matrix*

$$\eta^{\alpha\beta} := \partial_1 \langle dt^\alpha, dt^\beta \rangle^* \tag{2.18}$$

is constant.

The coordinates $t^1, \ldots, t^n$ on the orbit space will be called *Saito flat coordinates.* They can be chosen in such a way that the matrix (2.18) is antidiagonal

$$\eta^{\alpha\beta} = \delta^{\alpha+\beta, n+1}.$$

Then the Saito flat coordinates are defined uniquely up to an η-orthogonal transformation

$$t^\alpha \mapsto a_\beta^\alpha t^\beta,$$

$$\sum_{\lambda+\mu=n+1} a_\lambda^\alpha a_\mu^\beta = \delta^{\alpha+\beta, n+1}.$$

Proof. From flatness of the metric $\eta^{ij}(x)$ it follows that the flat coordinates $t^\alpha(x)$, $\alpha = 1, \dots, n$ exist at least localy. They are to be determined from the following system

$$\eta^{is}\partial_s\partial_j t + \gamma_j^{is}\partial_s t = 0 \tag{2.19}$$

(see (1.11)). The inverse matrix $(\eta_{ij}(x)) = (\eta^{ij}(x))^{-1}$ also is polynomial in $x^1, \dots, x^n$. So rewriting the system (2.19) in the form

$$\partial_k\partial_l t + \eta_{il}\gamma_k^{is}\partial_s t = 0 \tag{2.20}$$

we again obtain a system with polynomial coefficients. It can be written as a first-order system for the entries $\xi_l = \partial_l t$,

$$\partial_k\xi_l + \eta_{il}\gamma_k^{is}\xi_s = 0, \; k, l = 1, \dots, n \tag{2.21}$$

(the integrability condition $\partial_k\xi_l = \partial_l\xi_k$ follows from (1.4)). This is an overdetermined completely integrable system. So the space of solutions has dimension n. We can choose a fundamental system of solutions $\xi_l^\alpha(x)$ such that $\xi_l^\alpha(0) = \delta_l^\alpha$. These functions are analytic in x for sufficiently small x. We put $\xi_l^\alpha(x) =: \partial_l t^\alpha(x)$, $t^\alpha(0) = 0$. The system of solutions is invariant w.r.t. the scaling transformations

$$x^i \mapsto c^{d_i}x^i, \; i = 1, \dots, n.$$

So the functions $t^\alpha(x)$ are quasihomogeneous in x of the same degrees $d_1, \dots, d_n$. Since all the degrees are positive the power series $t^\alpha(x)$ should be polynomials in $x^1, \dots, x^n$. Because of the invertibility of the transformation $x^i \mapsto t^\alpha$ we conclude that $t^\alpha(x(p))$ are polynomials in $p^1, \dots, p^n$. Corollary is proved. □

We need to calculate particular components of the metric $g^{\alpha\beta}$ and of the correspondent Levi-Civita connection in the coordinates $t^1, \dots, t^n$ (in fact, in arbitrary homogeneous coordinates $x^1, \dots, x^n$).

Lemma 2.2. *Let the coordinate t^n be normalized as in (2.6). Then the following formulae hold:*

$$g^{n\alpha} = d_\alpha t^\alpha \tag{2.22}$$

$$\Gamma_\beta^{n\alpha} = (d_\alpha - 1)\delta_\beta^\alpha. \tag{2.23}$$

(In the formulae there is no summation over the repeated Greek indices!)

Proof. We have

$$g^{n\alpha} = \frac{\partial t^n}{\partial p^a}\frac{\partial t^\alpha}{\partial p^a} = p^a\frac{\partial t^\alpha}{\partial p^a} = d_\alpha t^\alpha$$

due to the Euler identity for the homogeneous functions $t^\alpha(p)$. Furthermore,

$$\Gamma_\beta^{n\alpha}dt^\beta = \frac{\partial t^n}{\partial p^a}\frac{\partial^2 t^\alpha}{\partial p^a\partial p^b}dp^b = p^a\frac{\partial^2 t^\alpha}{\partial p^a\partial p^b}dp^b = p^a d\left(\frac{\partial t^\alpha}{\partial p^a}\right) =$$

$$d\left(p^a\frac{\partial t^\alpha}{p^a}\right) - \frac{\partial t^\alpha}{\partial p^a}dp^a = (d_\alpha - 1)dt^\alpha.$$

Lemma is proved. □

We can formulate now the main result of this section.

Main lemma. *Let $t^1, \ldots, t^n$ be the Saito flat coordinates on the space of orbits of a finite Coxeter group and*

$$\eta^{\alpha\beta} = \partial_1 \langle dt^\alpha, dt^\beta \rangle^* \tag{2.24}$$

be the correspondent constant Saito metric. Then there exists a quasihomogeneous polynomial $F(t)$ of the degree $2h+2$ such that

$$\langle dt^\alpha, dt^\beta \rangle^* = \frac{(d_\alpha + d_\beta - 2)}{h} \eta^{\alpha\lambda} \eta^{\beta\mu} \partial_\lambda \partial \mu F(t). \tag{2.25}$$

The polynomial $F(t)$ determines on the space of orbits a polynomial Frobenius structure with the structure constants

$$c_{\alpha\beta}^\gamma(t) = \eta^{\gamma\epsilon} \partial_\alpha \partial_\beta \partial_\epsilon F(t) \tag{2.26a}$$

the unity

$$e = \partial_1 \tag{2.26b}$$

and the invariant inner product η.

Proof. Because of Corollary 2.3 in the flat coordinates the tensor $\Delta_\gamma^{\alpha\beta} = \Gamma_\gamma^{\alpha\beta}$ should satisfy the equations (1.17)–(1.19) where $g_1^{\alpha\beta} = g^{\alpha\beta}(t)$, $g_2^{\alpha\beta} = \eta^{\alpha\beta}$. First of all according to (1.17a) we can represent the tensor $\Gamma_\gamma^{\alpha\beta}(t)$ in the form

$$\Gamma_\gamma^{\alpha\beta}(t) = \eta^{\alpha\epsilon} \partial_\epsilon \partial_\gamma f^\beta(t) \tag{2.27}$$

for a vector field $f^\beta(t)$. The equation (1.8) (or, equivalently, (1.19)) for the metric $g^{\alpha\beta}(t)$ and the connection (2.27) reads

$$g^{\alpha\sigma} \Gamma_\sigma^{\beta\gamma} = g^{\beta\sigma} \Gamma_\sigma^{\alpha\gamma}.$$

For $\alpha = n$ because of Lemma 2.2 this gives

$$\sum_\sigma d_\sigma t^\sigma \eta^{\beta\epsilon} \partial_\sigma \partial_\epsilon f^\gamma = (d_\gamma - 1) g^{\beta\gamma}.$$

Applying to the l.h.s. the Euler identity (here $\deg \partial_\epsilon f^\gamma = d_\gamma - d_\epsilon + h$) we obtain

$$(d_\gamma - 1) g^{\beta\gamma} = \sum_\epsilon \eta^{\beta\epsilon} (d_\gamma - d_\epsilon + h) \partial_\epsilon f^\gamma = (d_\gamma + d_\beta - 2) \eta^{\beta\epsilon} \partial_\epsilon f^\gamma. \tag{2.28a}$$

From this one obtains the symmetry

$$\frac{\eta^{\beta\epsilon} \partial_\epsilon f^\gamma}{d_\gamma - 1} = \frac{\eta^{\gamma\epsilon} \partial_\epsilon f^\beta}{d_\beta - 1}.$$

Let us denote

$$\frac{f^\gamma}{d_\gamma - 1} =: \frac{F^\gamma}{h}. \tag{2.28b}$$

We obtain

$$\eta^{\beta\epsilon}\partial_\epsilon F^\gamma = \eta^{\gamma\epsilon}\partial_\epsilon F^\beta.$$

Hence a function $F(t)$ exists such that

$$F^\alpha = \eta^{\alpha\epsilon}\partial_\epsilon F. \tag{2.28c}$$

It is clear that $F(t)$ is a quasihomogeneous polynomial of the degree $2h+2$. From the formula (2.28) one immediately obtains (2.25).

Let us prove now that the coefficients (2.26a) satisfy the associativity condition. It is more convenient to work with the dual structure constants

$$c_\gamma^{\alpha\beta}(t) = \eta^{\alpha\lambda}\eta^{\beta\mu}\partial_\lambda\partial_\mu\partial_\gamma F.$$

Because of (2.27), (2.28) one has

$$\Gamma_\gamma^{\alpha\beta} = \frac{d_\beta - 1}{h} c_\gamma^{\alpha\beta}.$$

Substituting this in (1.18) we obtain associativity. Finaly, for $\alpha = n$ the formulae (2.22), (2.23) imply

$$c_\beta^{n\alpha} = h\delta_\beta^\alpha.$$

Since $\eta^{1n} = h$, the vector (2.26b) is the unity of the algebra. Lemma is proved. □

Proof of Theorem 1. Existence of a Frobenius structure on the space of orbits satisfying the conditions of Theorem 1 follows from Main lemma. We are now to prove uniqueness. Let us consider a polynomial Frobenius structure on M with the charges and dimension (3) and with the Saito invariant metric. In the Saito flat coordinates we have

$$dt^\alpha \cdot dt^\beta = \eta^{\alpha\lambda}\eta^{\beta\mu}\partial_\lambda\partial_\mu\partial_\gamma F(t)dt^\gamma.$$

The l.h.s. of (4) reads

$$i_v(dt^\alpha \cdot dt^\beta) = \sum_\gamma d_\gamma t^\gamma \eta_{\alpha\lambda}\eta^{\beta\mu}\partial_\lambda\partial_\mu\partial_\gamma F(t) = (d_\alpha + d_\beta - 2)\eta_{\alpha\lambda}\eta^{\beta\mu}\partial_\lambda\partial_\mu F(t).$$

This should be equal to $h\langle dt^\alpha, dt^\beta\rangle^*$. So the function $F(t)$ should satisfy (2.25). It is determined uniquely by this equation up to terms quadratic in t^α. Such an ambiguity does not affect the Frobenius structure. Theorem is proved. □

An algebraic remark: let T be a n-dimensional space and $U : T \to T$ an endomorphism (linear operator). Let

$$P_U(u) := \det(U - u \cdot 1)$$

be the characteristic polynomial of U. We say that the endomorphism U is semisimple if all the n roots of the characteristic polynomial are simple. For a semisimple endomorphism there exists a cyclic vector $e \in T$ such that

$$T = \mathrm{span}(e, Ue, \dots, U^{n-1}e).$$

The map

$$\mathbf{C}[u]/(P_U(u)) \to T, \quad u^k \mapsto U^k e, \; k = 0, 1, \dots, n-1 \tag{2.29}$$

is an isomorphism of linear spaces.

Let us fix a point $x \in M$. We define a linear operator

$$U = (U^i_j(x)) : T_x M \to T_x M \tag{2.30}$$

(being also an operator on the cotangent bundle) taking the ratio of the quadratic forms g^{ij} and η^{ij}

$$(U\omega_1, \omega_2)^* = \langle \omega_1, \omega_2 \rangle^* \tag{2.31}$$

or, equivalently,

$$U^i_j(x) := \eta_{js}(x) g^{si}(x). \tag{2.32}$$

Lemma 2.3. *The characteristic polynomial of the operator $U(x)$ is given up to a nonzero factor c^{-1} (2.15) by the formula (5).*

Proof. We have

$$\begin{aligned} P(u; x^1, \dots, x^n) &:= \det(U - u \cdot 1) = \det(\eta_{js}) \det(g^{si} - u\eta^{si}) = \\ & c^{-1} \det(g^{si}(x^1 - u, x^2, \dots, x^n) = c^{-1} D(x^1 - u, x^2, \dots, x^n). \end{aligned}$$

Lemma is proved. □

Corollary 2.5. *The operator $U(x)$ is semisimple at a generic point $x \in M$.*

Proof. Let us prove that the discriminant $D_0(x^1, \dots, x^n)$ of the characteristic polynomial $P(u; x^1, \dots, x^n)$ does not vanish identicaly on M. Let us fix a Weyl chamber $V_0 \subset V$ of the group W. On the inner part of V_0 the factorization map

$$V_0 \to M_{Re}$$

is a diffeomorphism. On the image of V_0 the discriminant $D(x)$ is positive. It vanishes on the images of the n walls of the Weyl chamber:

$$D(x)_{i\text{-th wall}} = 0, \quad i = 1, \dots, n. \tag{2.33}$$

On the inner part of the i-th wall (where the surface (2.33) is regular) the equation (2.33) can be solved for x^1:

$$x^1 = x^1_i(x^2, \dots, x^n). \tag{2.34}$$

Indeed, on the inner part

$$(\partial_1 D(x))_{i\text{-th wall}} \neq 0.$$

This holds since the polynomial $D(x)$ has simple zeroes at the generic point of the discriminant of W (see, e.g., [2]) .

Note that the functions (2.34) are the roots of the equation $D(x) = 0$ as the equation in the unknown x^1. It follows from above that this equation has simple roots for generic $x^2, \dots, x^n$. The roots of the characteristic equation

$$D(x^1 - u, x^2, \dots, x^n) = 0$$

are therefore

$$u_i = x^1 - x_i^1(x^2, \dots, x^n), \quad i = 1, \dots, n. \tag{2.35}$$

Generically these are distinct. Lemma is proved. □

Lemma 2.4. *The operator U on the tangent planes $T_x M$ coincides with the operator of multiplication by the Euler vector field $v = \frac{1}{h}E$.*

Proof. We check the statement of the lemma in the Saito flat coordinates:

$$\sum_\sigma \frac{d_\sigma}{h} t^\sigma c^\alpha_{\sigma\beta} = \frac{h - d_\beta + d_\alpha}{h} \eta^{\alpha\epsilon} \partial_\epsilon \partial_\beta F =$$

$$\sum_\lambda \frac{d_\lambda + d_\alpha - 2}{h} \eta_{\beta\lambda} \eta^{\alpha\epsilon} \eta^{\lambda\mu} \partial_\epsilon \partial_\mu F = \eta_{\beta\lambda} g^{\alpha\lambda} = U^\alpha_\beta.$$

Lemma is proved. □

Proof of Theorem 2. Because of Lemmas 2.3, 2.4 the vector fields

$$e, \ v, \ v^2, \dots, v^{n-1} \tag{2.36}$$

genericaly are linear independent on M. It is easy to see that these are polynomial vector fields on M. Hence e is a cyclic vector for the endomorphism U acting on $\operatorname{Der} R$. So in generic point $x \in M$ the map (6a) is an isomorphism of Frobenius algebras

$$\mathbf{C}[u]/(P(u; x)) \to T_x M.$$

This proves Theorem 2. □

Remark 1. The Euclidean metric (2.7) also defines an invariant inner product for the Frobenius algebras (on the cotangent planes $T_* M$). It can be shown also that the trilinear form

$$\langle \omega_1 \cdot \omega_2, \omega_3 \rangle^*$$

can be represented (localy, outside the discriminant $\operatorname{Discr} W$) in the form

$$(\hat\nabla^i \hat\nabla^j \hat\nabla^k \hat F(x)) \partial_i \otimes \partial_j \otimes \partial_k$$

for some function $\hat F(x)$. Here $\hat\nabla$ is the Gauss-Manin connection (i.e. the Levi-Civita connection for the metric (2.7)). The unity dt^n/h of the Frobenius algebra on $T_* M$ is not covariantly constant w.r.t. the Gauss-Manin connection.

Remark 2. The vector fields

$$l^i := g^{is}(x)\partial_s, \quad i = 1, \dots, n \tag{2.37}$$

form a basis of the R-module $\mathrm{Der}_R(-\log(D(x))$ of the vector fields on M tangent to the discriminant [2]. By the definition, a vector field $u \in \mathrm{Der}_R(-\log(D(x))$ iff

$$uD(x) = p(x)D(x)$$

for a polynomial $p(x) \in R$. The basis (2.37) of $\mathrm{Der}_R(-\log(D(x))$ depends on the choice of coordinates on M. In the Saito flat coordinates commutators of the basic vector fields can be calculated via the structure constants of the Frobenius algebra on T_*M. The following formula holds:

$$[l^\alpha, l^\beta] = \frac{d_\beta - d_\alpha}{h} c_\epsilon^{\alpha\beta} l^\epsilon. \tag{2.38}$$

This can be proved using (2.25).

Remark 3. The eigenvalues $u_1(x), \dots, u_n(x)$ of the endomorphism $U(x)$ can be chosen as new local coordinates near a generic point $x \in M$ (such that $D_0(x) \neq 0$). As it follows from [20, 22] these are *canonical coordinates* on the Frobenius manifold M: by the definition, this means that the law of multiplication of the coordinate vector fields has the form

$$\partial_i \cdot \partial_j = \delta_{ij}\partial_i \tag{2.39}$$

$$\partial_i = \frac{\partial}{\partial u_i}.$$

In these coordinates the Saito metric (,) is given by a diagonal Egoroff metric (see [20] for the definition)

$$\eta_{ij}(u) = \eta_{\alpha\beta}\frac{\partial t^\alpha}{\partial u_i}\frac{\partial t^\beta}{\partial u_i}\delta_{ij}. \tag{2.40}$$

The Euclidean metric $\langle\ ,\ \rangle$ outside of the discriminant $u_1 \dots u_n = 0$ in these coordinates is written as another diagonal Egoroff metric with the diagonal entries $\eta_{ii}(u)/u_i$. The unity vector field has the form

$$e = \sum_{i=1}^n \partial_i \tag{2.41}$$

and the Euler vector field

$$\frac{1}{h}E = v = \sum_{i=1}^n u_i\partial_i. \tag{2.42}$$

I recall that, according to the theory of [20] the metric (2.40) satisfies the Darboux-Egoroff equations

$$\partial_k\gamma_{ij} = \gamma_{ik}\gamma_{kj}, \quad i, j, k \text{ are distinct}, \tag{2.43a}$$

$$\sum_{k=1}^{n} \partial_k \gamma_{ij} = 0 \tag{2.43b}$$

$$\sum_{k=1}^{n} u_k \partial_k \gamma_{ij} = -\gamma_{ij} \tag{2.43c}$$

where the rotation coefficients $\gamma_{ij}(u) = \gamma_{ji}(u)$ are defined by the formula

$$\gamma_{ij}(u) := \frac{\partial \sqrt{\eta_{jj}(u)}}{\sqrt{\eta_{ii}(u)}}, \quad i \neq j. \tag{2.44}$$

The system (2.43) is empty for $n = 1$; it is linear for $n = 2$. For the first nontrivial case $n = 3$ it can be reduced to a particular case of the Painlevé-VI equation [27] using the first integral

$$(u_1 - u_2)^2 \gamma_{12}^2 + (u_1 - u_3)^2 \gamma_{13}^2 + (u_2 - u_3)^2 \gamma_{23}^2 = R^2. \tag{2.45}$$

For any $n \geq 3$ the system (2.43) can be reduced to a system of ordinary differential equations. It coincides with the equations of isomonodromy deformations of a certain linear differential operator with rational coefficients [20, 22]. Thus the eqs. (2.43) can be called a high-order analogue of the Painlevé-VI. The constructions of the present paper for the groups A_3, B_3, H_3 specify three distinguished solutions of the correspondent Painlevé-VI eqs.. The function $F(t)$ for these groups has the form

$$F_{A_3} = \frac{t_1^2 t_3 + t_1 t_2^2}{2} + \frac{t_2^2 t_3^2}{4} + \frac{t_3^5}{60} \tag{2.46}$$

$$F_{B_3} = \frac{t_1^2 t_3 + t_1 t_2^2}{2} + \frac{t_2^3 t_3}{6} + \frac{t_2^2 t_3^3}{6} + \frac{t_3^7}{210} \tag{2.47}$$

$$F_{H_3} = \frac{t_1^2 t_3 + t_1 t_2^2}{2} + \frac{t_2^3 t_3^2}{6} + \frac{t_2^2 t_3^5}{20} + \frac{t_3^{11}}{3960}. \tag{2.48}$$

The correspondent constants R in (2.45) equal 1/4, 1/3 and 2/5 resp.

Concluding remarks

1. The results of this paper can be generalised for the case where W is the Weyl group of an extended affine root system of codimension 1 (see the definition in [39]). In this case the Frobenius structure will be polynomial in all the coordinates but one and it will be a modular form in this exceptional coordinate. The solutions of WDVV of [32, 46] are just of this type. We consider the orbit spaces of these groups in a subsequent publication.

2. The two metrics on the space of orbits of the group A_n are closely related to the two hamiltonian structures of the nKdV hierarchy (see [18–20, 22]).

The Saito metric is obtained by the semiclassical limit of [24, 25] from the first Gelfand-Dickey Poisson bracket of nKdV, and the Euclidean metric is obtained by the same semiclassical limit from the second Gelfand-Dickey Poisson bracket. The Saito and the Euclidean coordinates on the orbit space are the Casimirs for the corresponding Poisson brackets. The factorization map $V \to M = V/W$ is the semiclassical limit of the Miura transformation. Probably, the semiclassical limit of the bi-hamiltonian structure of the $D - E$ Drinfeld-Sokolov hierarchies [17] give the two flat metrics on the orbit spaces of the groups D_n and E_6, E_7, E_8 resp. But this should be checked.

It is still an open question if it is possible to relate integrable hierarchies to the Coxeter groups not of $A - D - E$ series. A partial answer to this question is given in [20, 22]: the unknown integrable hierarchies for any Frobenius manifold are constructed in a semiclassical (i.e., in the dispersionless) approximation.

3. A closely related question: what is the algebraic-geometrical description of the TFT models related to the polynomial solutions of WDVV constructed in this paper? For $A - D - E$ groups the correspondent TFT models are the topological minimal models of [15]. For other Coxeter groups the TFT can be constructed as equivariant topological Landau-Ginsburg models using the results of [44, 47] for $W \neq H_4$ (the singularity theory related to H_4 was partialy developed in [35, 40]). For the group A_n a nice algebraic-geometrical reformulation of the correspondent TFT as the intersection theory on a certain covering over the moduli space of stable algebraic curves, was proposed in [50, 51] (for the topological gravity $W = A_1$ this conjecture was proved by M.Kontsevich [29, 30]). What are the moduli spaces whose intersection theories are encoded by the orbit spaces of other Coxeter groups? Note that a part of these intersection numbers should coincide with the coefficients of the polynomials $F(t)$ (these are rational but not integer numbers since the moduli spaces are not manifolds but orbifolds).

Acknowledgements

I am grateful to V.I.Arnol'd for fruitful discussions.

Appendix: Algebraic version of the definition of polynomial Frobenius manifold

Let k be a field of the characteristic $\neq 2$ and

$$R := k[x^1, \ldots, x^n] \tag{A.1}$$

be the ring of polynomials with the coefficients in k. By $\operatorname{Der} R$ we denote the R-module of k-derivations of R. This is a free R-module with the basis

$$\partial_i := \frac{\partial}{\partial x^i}, \quad i = 1, \dots, n.$$

A map

$$\operatorname{Der} R \times R \to R$$

is defined by the formula

$$(u = u^i \partial_i, \ p) \mapsto up := u^i \partial_i p. \tag{A.2}$$

A R-bilinear symmetric inner product

$$\begin{gathered} \operatorname{Der} R \times \operatorname{Der} R \to R \\ u, \ v \mapsto (u, v) \in R \end{gathered} \tag{A.3}$$

is called nondegenerate if from the equations

$$(u, v) = 0 \ \text{ for any } v \in \operatorname{Der} R$$

it follows that $u = 0$.

As it was mentioned in the introduction, a polynomial Frobenius manifold is a structure of Frobenius R-algebra on $\operatorname{Der} R$ satisfying certain conditions. We obtain here these conditions by reformulating the Definition 1 in a pure algebraic way.

The first standard step is to reformulate the notion of the Levi-Civita connection. By the definition, this is a map

$$\begin{gathered} \operatorname{Der} R \times \operatorname{Der} R \to \operatorname{Der} R \\ u, \ v \mapsto \nabla_u v \end{gathered} \tag{A.4}$$

R-linear in the first argument and satisfying the Leibnitz rule in the second one

$$\nabla_u (pv) = p \nabla_u v + (up) v \tag{A.5}$$

uniquely specified by the equations

$$u(v, w) = (\nabla_u v, w) + (v, \nabla_u w) \tag{A.6a}$$

$$\nabla_u v - \nabla_v u = [u, v] \tag{A.6b}$$

(the commutator of the derivations). Equivalently, it can be determined from the equation

$$\begin{aligned} \langle \nabla_u v, w \rangle = & \frac{1}{2} [u \langle v, w \rangle + v \langle w, u \rangle - w \langle u, v \rangle \\ & + \langle [u, v], w \rangle + \langle [w, u], v \rangle + \langle [w, v], u \rangle] \end{aligned} \tag{A.6c}$$

for arbitrary $u, \ v, \ w \in \operatorname{Der} R$.

Now the assumptions 1–3 of Definition 1 for the Frobenius R-algebra $\operatorname{Der} R$ can be reformulated as follows:

1. For any u, v, w the following identity holds
$$(\nabla_u \nabla_v - \nabla_v \nabla_u - \nabla_{[u,v]})w = 0.$$
2. For the unity $e \in \operatorname{Der} R$ and for arbitrary $u \in \operatorname{Der} R$
$$\nabla_u e = 0.$$
3. The identity
$$\nabla_u(v \cdot w) - \nabla_v(u \cdot w) + u \cdot \nabla_v w - v \cdot \nabla_u w = [u, v] \cdot w \tag{A.7}$$
holds for any three derivations fields u, v, w.

To reformulate the assumption 4 of Definition 1 let us assume that $\operatorname{Der} R$ is a graded algebra over a graded ring R with a graded invariant inner product $(\ ,\ )$. That means that two gradings deg and deg′ are defined on R and on $\operatorname{Der} R$ resp., i.e. real numbers
$$P_i := \deg x^i, \quad Q_i := \deg' \partial_i \tag{A.8}$$
are assigned to the generators $x^1, \ldots, x^n$ and to the basic derivations $\partial_1, \ldots, \partial_n$ resp. By the definition, the degree of a monomial
$$p = (x^1)^{m_1} \ldots (x^n)^{m_n}$$
equals
$$\deg p := m_1 P_1 + \cdots + m_n P_n.$$
Homogeneous elements of $\operatorname{Der} R$ are defined by the assumption that the operators $p \mapsto up$ shifts the grading in R to $\deg' u - Q_0$ for a constant Q_0, i.e.
$$\deg(up) = \deg' u + \deg p - Q_0. \tag{A.9}$$
The R-algebra structure on $\operatorname{Der} R$ should be consistent with the grading, i.e. for any homogeneous elements p, q of R and u, v of $\operatorname{Der} R$ the following formulae hold:
$$\deg'(pu) = \deg' u + \deg p \tag{A.10}$$
$$\deg(pq) = \deg p + \deg q \tag{A.11}$$
$$\deg'(u \cdot v) = \deg' u + \deg' v. \tag{A.12}$$
The invariant inner product $(\ ,\ )$ should be graded of a degree D, i.e.
$$(u, v) = 0 \ \text{ if } \ \deg' u + \deg' v \neq D \tag{A.13}$$
for arbitrary homogeneous u, $v \in \operatorname{Der} R$. Note that the Euler vector field is homogeneous of the degree Q_0. We consider only the case $Q_0 \neq 0$.

The numbers P_i, Q_i, Q_0, D are defined up to rescaling. One can normalise these in such a way that $Q_0 = 1$. Then we have
$$Q_i := q_i, \ P_i = 1 - q_i, \ D = d$$
in the notations of Introduction.

The constructions of this paper give such an algebraic structure for $k = \mathbf{Q}$.

Bibliography

[1] Arnol'd V.I., Normal forms of functions close to degenerate critical points. The Weyl groups A_k, D_k, E_k, and Lagrangian singularities, *Functional Anal.* **6** (1972) 3–25.

[2] Arnol'd V.I., Wave front evolution and equivariant Morse lemma, *Comm. Pure Appl. Math.* **29** (1976) 557–582.

[3] Arnol'd V.I., Indices of singular points of 1-forms on a manifold with boundary, convolution of invariants of reflection groups, and singular projections of smooth surfaces, *Russ. Math. Surv.* **34** (1979) 1–42.

[4] Arnol'd V.I., Gusein-Zade S.M., and Varchenko A.N., Singularities of Differentiable Maps, volumes I, II, Birkhäuser, Boston-Basel-Berlin, 1988.

[5] Arnol'd V.I., Singularities of Caustics and Wave Fronts, Kluwer Acad. Publ., Dordrecht - Boston - London, 1990.

[6] Aspinwall P.S., Morrison D.R., Topological field theory and rational curves, *Comm. Math. Phys.* **151** (1993) 245–262.

[7] Atiyah M.F., Topological quantum field theories, *Publ. Math. I.H.E.S.* **68** (1988) 175.

[8] Blok B. and Varchenko A., Topological conformal field theories and the flat coordinates, *Int. J. Mod. Phys.* **A7** (1992) 1467.

[9] Bourbaki N., Groupes et Algèbres de Lie, Chapitres 4, 5 et 6, Masson, Paris-New York-Barcelone-Milan-Mexico-Rio de Janeiro, 1981.

[10] Brieskorn E. Singular elements of semisimple algebraic groups, In: Actes Congres Int. Math., **2**, Nice (1970), 279–284.

[11] Cecotti S. and Vafa C., *Nucl. Phys.* **B367** (1991) 359.

[12] Cecotti S. and Vafa C., On classification of $N = 2$ supersymmetric theories, Preprint HUTP-92/A064 and SISSA-203/92/EP, December 1992.

[13] Coxeter H.S.M., Discrete groups generated by reflections, *Ann. Math.* **35** (1934) 588–621.

[14] Dijkgraaf R. and Witten E., *Nucl. Phys.* **B 342** (1990) 486

[15] Dijkgraaf R., E.Verlinde, and H.Verlinde, *Nucl. Phys.* **B 352** (1991) 59; Notes on topological string theory and 2D quantum gravity, Preprint PUPT-1217, IASSNS-HEP-90/80, November 1990.

[16] Dijkgraaf R., Intersection theory, integrable hierarchies and topological field theory, Preprint IASSNS-HEP-91/91, December 1991.

[17] Drinfel'd V.G. and Sokolov V.V., *J. Sov. Math.* **30** (1985) 1975.

[18] Dubrovin B., Differential geometry of moduli spaces and its application to soliton equations and to topological field theory, Preprint No.117, Scuola Normale Superiore, Pisa (1991).

[19] Dubrovin B., Hamiltonian formalism of Whitham-type hierarchies and topological Landau-Ginsburg models, *Comm. Math. Phys.* **145** (1992) 195–207.

[20] Dubrovin B., Integrable systems in topological field theory, *Nucl. Phys.* **B 379** (1992) 627–689.

[21] Dubrovin B., Geometry and integrability of topological-antitopological fusion, Pre print INFN-8/92-DSF, to appear in *Comm. Math. Phys.*

[22] Dubrovin B., Integrable systems and classification of 2-dimensional topological field theories, Preprint SISSA 162/92/FM, September 1992, to appear in "Integrable Systems", Proceedings of Luminy 1991 conference dedicated to the memory of J.-L. Verdier.

[23] Dubrovin B., Topological conformal field theory from the point of view of integrable systems, Preprint SISSA 12/93/FM, January 1993, to appear in Proceedings of 1992 Como workshop "Quantum Integrable Systems".

[24] Dubrovin B. and Novikov S.P., The Hamiltonian formalism of one-dimensional systems of the hydrodynamic type and the Bogoliubov - Whitham averaging method,*Sov. Math. Doklady* **27** (1983) 665–669.

[25] Dubrovin B. and Novikov S.P., Hydrodynamics of weakly deformed soliton lattices. Differential geometry and Hamiltonian theory, *Russ. Math. Surv.* **44:6** (1989) 35–124.

[26] Dubrovin B., Novikov S.P., and Fomenko A.T., Modern Geometry, Parts 1–3, Springer Verlag.

[27] Fokas A.S., Leo R.A., Martina L., and Soliani G., *Phys. Lett.* **A115** (1986) 329.

[28] Givental A.B., Convolution of invariants of groups generated by reflections, and connections with simple singularities of functions, *Funct. Anal.* **14** (1980) 81–89.

[29] Kontsevich M., *Funct. Anal.* **25** (1991) 50.

[30] Kontsevich M., *Comm. Math. Phys.* **147** (1992) 1.

[31] Looijenga E., A period mapping for certain semiuniversal deformations, *Compos. Math.* **30** (1975) 299–316.

[32] Maassarani Z., *Phys. Lett.* **273B** (1992) 457.

[33] Magri F., *J. Math. Phys.* **19** (1978) 1156.

[34] Procesi C. and Schwarz G., Inequalities defining orbit spaces, *Invent. Math.* **81** (1985) 539–554.

[35] Roberts R.M. and Zakalyukin V.M., Symmetric wavefronts, caustic and Coxeter groups, to appear in Proceedings of Workshop in the Theory of Singularities, Trieste 1991.

[36] Saito K., On a linear structure of a quotient variety by a finite reflection group, Preprint RIMS-288 (1979).

[37] Saito K., Yano T., and Sekeguchi J., On a certain generator system of the ring of invariants of a finite reflection group, *Comm. in Algebra* **8(4)** (1980) 373–408.

[38] Saito K., Period mapping associated to a primitive form, *Publ. RIMS* **19** (1983) 1231–1264.

[39] Saito K., Extended affine root systems II (flat invariants), *Publ. RIMS* **26** (1990) 15–78.

[40] Shcherbak O.P., Wavefronts and reflection groups, *Russ. Math. Surv.* **43:3** (1988) 149–194.

[41] Slodowy P., Einfache Singularitaten und Einfache Algebraische Gruppen, Preprint, Regensburger Mathematische Schriften **2**, Univ. Regensburg (1978).

[42] Vafa C., *Mod. Phys. Let.* **A4** (1989) 1169.

[43] Vafa C., Private communication, September 1992.

[44] Varchenko A.N. and Chmutov S.V., Finite irreducible groups, generated by reflections, are monodromy groups of suitable singularities, *Func. Anal.* **18** (1984) 171–183.

[45] Varchenko A.N. and Givental A.B., Mapping of periods and intersection form, *Funct. Anal.* **16** (1982) 83–93.

[46] Verlinde E. and Warner N., *Phys. Lett.* **269B** (1991) 96.

[47] Wall C.T.C., A note on symmetry of singularities, *Bull. London Math. Soc.* **12** (1980) 169–175;
A second note on symmetry of singularities, *ibid.*, 347–354.

[48] Witten E., *Comm. Math. Phys.* **117** (1988) 353;
ibid., **118** (1988) 411.

[49] Witten E., *Nucl. Phys.* **B 340** (1990) 281.

[50] Witten E., *Surv. Diff. Geom.* **1** (1991) 243.

[51] Witten E., Algebraic geometry associated with matrix models of two-dimensional gravity, Preprint IASSNS-HEP-91/74.

[52] Witten E., Lectures on mirror symmetry, In: Proceedings MSRI Conference on mirror symmetry, March 1991, Berkeley.

Boris Dubrovin
International School for Advanced Studies (SISSA)
Via Beirut, 2–4
I-34013 TRIESTE, Italy
E-mail: dubrovin@sissa.it

Differential Geometry of Moduli Spaces and its Applications to Soliton Equations and to Topological Conformal Field Theory

Boris Dubrovin

Abstract

We construct flat Riemann metrics on moduli spaces of algebraic curves with marked meromorphic function. This gives a new class of exact algebraic-geometry solutions of some non-linear equations in terms of functions on the A_n moduli spaces. We show that the Riemann metrics on moduli spaces coincide with two-point correlators in topological conformal field theory and calculate the analogue of the partition function for A_n-model for arbirary genus. A universal method for constructing complete families of conservation laws for Whitham-type hierarchies of PDE also is proposed.

Introduction

The recent progress in the study of matrix models [1] of QFT revealed a remarkable connection with hierarchies of integrable equations of the KdV-type. It was shown also [2]–[4] that so-called topological conformal field theories (TCFT) are very important in the study of the low-dimensional string theories and of the matrix models (the general notion of topological field theory was introduced by E. Witten [5]).

The Landau-Ginsburg superpotentials machinery [6], [7] (see below Section 4) in TCFT was analyzed from different points of view. The relation of it with the singularity theory was investigated in refs. [6], [7], [8] (see also ref. [9]). Very recently Krichever [10] has observed the relation of this machinery with the so-called averaged KdV-type hierarchy [11]–[15] (or *Whitham-type hierarchy*). He showed that the target space for this Whitham-type hierarchy coincides with the coupling space of zero genus TCFT and the dependence of the Landau-Ginsburg potential on the coupling constants is determined via solving the equation of this hierarchy (in fact, a very particular solution proved to be involved.)

Our main observation is that the flat metric on the target space of Whitham-type hierarchy being involved in the Hamiltonian description of it (see refs. [11], [12], [13], [15]) coincides with the two-point correlator of the corresponding TCFT. Starting from this point we have found a very general construction of flat Riemann metrics on moduli spaces M of algebraic curves of given genus with marked meromorphic function. This function in TCFT plays the role of Landau-Ginsburg superpotential (we consider only the A_{n-1}-theories) and the relevant moduli space M being the coupling space. It turns out that the equations of flatness of these Riemann metrics coincide with well-known in the soliton theory N-wave interaction system. We obtain therefore a new class of exact solutions of the N-wave system in terms of some special functions on moduli spaces M (the simplest solution of this class has been found in ref. [16]). Some global properties of moduli spaces of the type being described above also follow from our considerations. We construct also the general class of Whitham-type hierarchies of dynamical systems in the loop spaces $\mathcal{L}M$. We describe the bi-Hamiltonian structure and recurrence operator for this hierarchy and construct explicitly the complete family of conservation laws. As a result of these considerations the explicit formula for the non-zero genus TCFT partition function is obtained. In the appendix we discuss the relation of TCFT to the theory of Frobenius algebras.

1 Orthogonal systems of curvilinear coordinates, integrable equations and Hamiltonian formalism

We start with some information on the geometry of curvilinear orthogonal coordinate systems.

Let

$$ds^2 = \sum_{i=1}^{N} g_{ii}(u)\,(du^i)^2 \tag{1.1}$$

be a diagonal metric on some manifold $M = M^N$ (we give all the formulae for positive definite metrics; indefinite metrics can be considered in a similar way). The variables $u^1, \dots, u^N$ determine a curvilinear coordinate system in Euclidean space iff the curvature of (1.1) vanishes:

$$R_{ijkl}(ds^2) = 0 \tag{1.2}$$

This is a very complicated system of nonlinear PDE. But there is a special subclass [16] of metrics for which the system (1.2) is an integrable one.

Definition 2. The diagonal metric ds is called Egoroff metric (it was proposed by Darboux [18]) iff the rotation coefficients

$$\gamma_{ij} = \frac{\partial_j \sqrt{g_{ii}}}{\sqrt{g_{jj}}}, \quad \partial_j = \partial/\partial u^j \tag{1.3}$$

satisfy the symmetry

$$\gamma_{ji} = \gamma_{ij} \tag{1.4}$$

Equivalently, there exists a potential $V(u)$ for the metric g_{ii}:

$$g_{ii}(u) = \partial_i V(u), \quad i = 1, \dots, N \tag{1.5}$$

Proposition 1. *(see ref. [16]). The equations of zero curvature for Egoroff metric have the form*

$$\partial_k \gamma_{ij} = \gamma_{ik}\gamma_{kj}, \quad i, j, k \text{ are distinct,} \tag{1.6}$$

$$\partial \gamma_{ij} = 0, \quad i \neq j, \quad \partial = \sum_{i=1}^{N} \partial_i \tag{1.6$'$}$$

The corresponding linear problem has the form

$$\partial_j \Psi_i = \gamma_{ij} \Psi_j, \quad i \neq j \tag{1.7}$$

$$\partial \Psi_i = \sigma \Psi_i, \quad \sigma \text{ is the spectral parameter.} \tag{1.7$'$}$$

Remark 1. The linear system (1.7), (1.7′) essentially is equivalent to a system of ODE of N-th order. It has N-dimensional space of solutions for given σ. For example, if $\Psi_i^{(\alpha)}, \alpha = 1, \dots, N$, form the basis of solutions of the system (1.7), (1.7′) for $\sigma = 0$ then the flat coordinates $v^1, \dots, v^N$ for the metric ds^2 can be found from the system

$$\partial_i v^\alpha = \sqrt{g_{ii}}\, \Psi_i^{(\alpha)}, \quad \alpha = 1, \dots, N, \tag{1.8}$$

$$\eta^{\alpha\beta} = \sum_{i=1}^{N} g_{ii}^{-1} \partial_i v^\alpha \partial_i v^\beta = \text{const}, \quad \alpha, \beta = 1, \dots, N. \tag{1.9}$$

Remark 2. It was shown in ref. [16] that the system (1.6), (1.6′) with the symmetry (1.4) is equivalent to the pure imaginary reduction of the N-wave interaction problem (see e.g. ref. [19]). The system (1.6), (1.6′) is invariant under the scaling transformations

$$u^i \to cu^i, \quad \gamma_{ij} \to c^{-1}\gamma_{ij} \tag{1.10}$$

The corresponding similarity reduction of the system is equivalent to some nonlinear ODE. In the first nontrivial case $N = 3$ this reduction has the form

$$\Gamma'_{23} = \Gamma_{21}\Gamma_{13}, \quad (z\Gamma_{13})' = -\Gamma_{12}\Gamma_{23}, \quad [(z-1)\Gamma_{13}]' = \Gamma_{12}\Gamma_{23} \tag{1.11}$$

Here

$$z = \frac{u^1 - u^3}{u^2 - u^3}, \quad \gamma_{ij}(u) = \frac{1}{u^2 - u^3}\Gamma_{ij}(z). \tag{1.11′}$$

The system (1.11) can be reduced [20] to a system of the second order equivalent to the Painlevé-VI equation using the first integral

$$\Gamma_{23}^2 + (z\Gamma_{13})^2 + [(z-1)\Gamma_{13}]^2 = \text{ const.} \tag{1.11″}$$

Remark 3. If $\vec{r} = \vec{r}(u), \vec{r} = (v^1, \dots, v^N)$ is the realization of the curvilinear orthogonal coordinate system in Euclidean space then the law of transport along the u^j-axis of the corresponding orthonormal frame

$$\vec{\eta_i} = \partial_i\vec{r}/\sqrt{g_{ii}} \tag{1.12}$$

has the form

$$\begin{aligned} \partial_j\vec{\eta_i} &= \gamma_{ij}\vec{\eta_j}, \quad j \neq i \\ \partial_i\vec{\eta_i} &= -\sum_{j\neq i}\gamma_{ji}\vec{\eta_j} \end{aligned} \tag{1.13}$$

This explains the name "rotation coefficients" for γ_{ij}. It follows from (1.13) that this transport of the frame is invariant for Egoroff metric under the diagonal translations of the coordinates:

$$u^i \to u^i + \Delta u, \quad i = 1, \dots, N. \tag{1.14}$$

In general the Egoroff metric is not invariant under these translations. If it is invariant, i.e. $\partial g_{ii} = 0$, then we shall call it ∂-invariant.

Using the flat metric (1.4) we introduce the following Poisson structure $\{\,,\} = \{\,,\}_{ds^2}$ on the loop space

$$\mathcal{L}M \equiv M^{S^1}$$

of functions of $x \in S^1$ having their values in M (Poisson brackets of hydrodynamic type [11]–[13], [15]) via the formula

$$\{I_1, I_2\}_{ds^2} = \int \frac{\delta I_1}{\delta u^i(x)} g^{ij}(u) \nabla_k \left(\frac{\delta I_2}{\delta u^j(x)} \right) du^k(x). \tag{1.15}$$

Here

$$g^{ij} = \delta^{ij} \cdot [g_{ii}(u)]^{-1}$$

and ∇_k is the Levi-Cività connection for the metric ds^2. The corresponding Hamiltonian systems for Hamiltonians of the form

$$H = \int h(u)\, dx \tag{1.16}$$

have the form of the first order evolutionary systems of PDE linear in derivatives

$$u^i_t = \{u^i(x), H\}_{ds^2} = \nabla^i \nabla_j h(u) u^j_x. \tag{1.17}$$

In the flat coordinates $v^\alpha = v^\alpha(u)$ (see (1.9)) the P.B. (1.15) has a constant form

$$\{v^\alpha(x), v^\beta(y)\}_{ds^2} = \eta^{\alpha\beta} \delta'(x-y) \tag{1.18}$$

The P.B. $\{\, ,\}_{ds^2}$ is degenerate: the functionals

$$\int v^1\, dx, \dots, \int v^N\, dx \tag{1.19}$$

are the Casimirs of it.

Definition 3. (cfr. ref. [21]). The family $\mathcal{H}$ of functionals H on the loop space $\mathcal{L}M$ is called a Lagrangian family if all of them commute pairwise and if it is complete. This means that the skew-gradients of these functionals span the tangent space to their common level surface.

All the Casimirs (1.19) are to belong to $\mathcal{H}$.

It follows from the results of Tsarev [21] that for the P.B. (1.18) Lagrangian families $\mathcal{H}$ of functionals of the form (1.16) are in one-to-one correspondence to systems of curvilinear orthogonal coordinates in the flat space with the metric $\eta^{\alpha\beta}$. The explicit construction of $\mathcal{H}$ is as follows. For P.B. of the form (1.15) for any flat diagonal metric (1.1) the Lagrangian family of functionals of the form (1.16) can be constructed as the family of solutions of the system

$$\partial_i \partial_j h = \Gamma^j_{ji} \partial_j h + \Gamma^i_{ij} \partial_i h, \quad i \neq j. \tag{1.20}$$

The corresponding commuting flows (1.17) have a diagonal form

$$u^i_t = w^i(u) u^i_x, \quad i = 1, \dots, N. \tag{1.21}$$

All of them are completely integrable [21]. The system (1.20) for finding the commuting Hamiltonians of the Lagrangian family $\mathcal{H}$ can be rewritten in the form (1.7) via the substitution

$$\partial_i h = \sqrt{g_{ii}}\,\Psi_i, \quad i = 1, \dots, N. \tag{1.22}$$

The coefficients $w^i(u)$ of the commuting flows (1.21) also can be found from the same system (1.7) (for Egoroff metric) via the substitution

$$\Psi_i = \sqrt{g_{ii}}\,w^i, \quad i = 1, \dots, N. \tag{1.23}$$

Therefore we obtain a mapping

$$\text{(commuting Hamiltonians)} \to \text{(commuting flows)} \tag{1.24}$$

of the form

$$\left(H = \int h(u)\,dx\right) \to \left(u^i_t = g_{ii}^{-1}\partial_i h(u) u^i_x, \quad i = 1, \dots, N\right). \tag{1.25}$$

Warning: this is not the skew-gradient mapping (but in some cases — see below section 3 — it is related to the second Hamiltonian structure of the system (1.21)). For ∂-invariant metric (i.e. $\partial g_{ii} = 0$) the skew-gradient mapping has the form

$$\left(h(u)\right) \to \left(u^i_t = g_{ii}^{-1}\partial_i \partial h(u) u^i_x, \quad i = 1, \dots, N\right). \tag{1.26}$$

For ∂-invariant metric the operator ∂ plays the role of "recursion operator": if $h(u)$ is one of the Hamiltonians in the Lagrangian family $\mathcal{H}$ then ∂h also belongs to $\mathcal{H}$; also the operator ∂^{-1} can be defined on $\mathcal{H}$ with the same property. It is possible to construct a dense subset [21] in the Lagrangian family $\mathcal{H}$ using the operator ∂^{-1} starting from the Casimirs (1.19). The densities of the functionals of this subset have the form

$$h^{(\alpha,m)}(u) = \partial^{-m} v^\alpha, \quad \alpha = 1, \dots, N, \quad m = 0, 1, \dots \tag{1.27}$$

2 Flat metrics on moduli spaces

Let us consider for given integers (g, m, n), $g \geq 0, m > 0, n \geq m$, a moduli space $M = M^N, N = 2g + n + m - 2$ of sets $(C, Q_1, \dots, Q_m, \lambda)$, where C is a smooth algebraic curve of genus g with m marked points $Q_1, \dots, Q_m$ and with a meromorphic function λ of degree n such that $\lambda^{-1}(\infty) = Q_1 \cup \dots \cup Q_m$. To specify a component of M one has to fix also the local degrees $n_1, \dots, n_m$ of λ in the points $Q_1, \dots, Q_m$. These are arbitrary positive integers such that $n_1 + \dots + n_m = n$. We need that the λ-projections $u^1, \dots, u^N$ of the branch points $P_1, \dots, P_N$

$$d\lambda|_{P_j} = 0, \quad u^j = \lambda(P_j), \quad j = 1, \dots, N \tag{2.1}$$

(i.e the critical points of λ) are good local coordinates in an open domain in M. Another assumption is that the one-dimensional affine group acts on M as

$$(C, Q_1, \ldots, Q_m, \lambda) \to (C, Q_1, \ldots, Q_m, a\lambda + b). \tag{2.2}$$

In the coordinates $u^1, \ldots, u^N$ it acts as

$$u^i \to au^i + b, \quad i = 1, \ldots, N \tag{2.3}$$

The tautological fiber bundle is defined

$$\underset{M}{\downarrow} C \tag{2.4}$$

such that the fiber over $u \in M$ is the curve $C(u)$. The canonical connection is defined on (2.4): the operators ∂_i are lifted on (2.4) in such a way that

$$\partial_i \lambda \equiv 0 \tag{2.5}$$

Example 1. Here $g = 0, m = 1$. The space M is the set of all polynomials of the form

$$\lambda(p) = p^n + q_{n-2}p^{n-2} + \cdots + q_0, \quad q_0, q_1, \ldots, q_{n-2} \in \mathbb{C}. \tag{2.6}$$

The branch points $p_1, \ldots, p_{n-1}$ can be determined from the equation

$$\lambda'(p) = 0 \tag{2.7}$$

The affine transformations $\lambda \to a\lambda + b$ have the form

$$p \to a^{1/n}p, \quad q_i \to q_i a^{\frac{i}{n}-1}, \quad i > 0, \quad q_0 \to aq_0 + b \tag{2.8}$$

Example 2. Here $g = 0, m = n$ (let us redenote $m = n \to n + 1$). The space M consists of all rational functions of the form

$$\lambda(p) = p + \sum_{i=1}^{n} \frac{\eta_i}{p + q_i}, \quad \eta_i, q_i \in \mathbb{C} \tag{2.9}$$

Here $Q_i = \{p = -q_i\}$, $i = 1, \ldots, n$, $Q_{n+1} = \{p = \infty\}$.

Example 3. $g > 0, m = 1, n = 2$. Here M is the set of all hyperelliptic curves

$$\mu^2 = \prod_{j=1}^{2g+1} (\lambda - u^j), \tag{2.10}$$

the pairwise distinct parameters $u^1, \ldots, u^{g+1}$ are the local coordinates on M.

Example 4. $g > 0, m = 1, n > g$. Here M is the set of all curves of genus g with marked point Q_1 and with marked meromorphic function $\lambda(P)$ having a pole of n-th order in Q_1 only.

Let $\tilde{M}$ be the covering of M being obtained by fixing a canonical basis $a_1, \dots, a_g$, $b_1, \dots, b_g$ in $H_1(C, \mathbb{Z})$ (for $g = 0, \tilde{M} = M$). We add small cycles $\gamma_1, \dots, \gamma_{m-1}$ around the points $Q_1, \dots, Q_{m-1}$ (for $m > 1$) to obtain a basis in $H_1(C \backslash (Q_1 \cup \dots \cup Q_m), \mathbb{Z})$. Let us define multivalued Abelian differential on C as Abelian differentials on the universal covering of the punctured curve $C \backslash (Q_1 \cup \dots \cup Q_m)$ such that

$$\Omega(P + \gamma) = \Omega(P) + \sum_k c_k(\gamma) \lambda^k \, d\lambda \tag{2.11}$$

for any cycle $\gamma \in H_1(C \backslash (Q_1 \cup \dots \cup Q_m), \mathbb{Z})$. Such a multivalued differential is said to be holomorphic in the point $P \in C$ iff some branch of it is holomorphic in P. It is called normalized iff

$$\oint_{a_\alpha} \Omega = 0, \quad \alpha = 1, \dots, g \tag{2.12}$$

Definition 4. A family $\Omega = \Omega(P, u)$ of multivalued Abelian differentials on the curve $C = C(u)$ smoothly depending on the parameter $u \in \tilde{M}$ is called horizontal if:

1. It is holomorphic for any u on $C \backslash (Q_1 \cup \dots \cup Q_m)$.
2. Its covariant derivatives $\partial_j \Omega$ are Abelian differentials of the second kind on C (i.e. with zero residues) with double poles only in the branch points $P_1, \dots, P_N$ and with zero a-periods.

Let $\mathcal{D}(\tilde{M})$ be the quotient of the space of all horizontal differentials over the subspace of differentials of the form

$$\sum c_k \lambda^k \, d\lambda, \quad \partial_j c_k \equiv 0. \tag{2.13}$$

Proposition 1. *The basis of the space $\mathcal{D}(\tilde{M})$ can be constructed as follows:*

1. *Normalized Abelian differentials $\Omega_a^{(k)}$ of the second kind with a single pole in the point Q_a and with the principal part*

$$\Omega_a^{(k)}(P) = \frac{dz_a}{z_a^{k+1}} + \textit{regular terms}, \quad P \to Q_a \tag{2.14}$$

$$z_a = \lambda^{-1/n_a}, \tag{2.15}$$

$a = 1, \dots, m, k = 1, 2, \dots$. *The following linear constraints in $\mathcal{D}(\tilde{M})$ hold for these differentials*

$$\sum_{a=1}^{m} n_a \Omega_a^{(kn_a)} = 0, \quad k = 1, 2, \dots. \tag{2.16}$$

2. *Normalized Abelian differentials* $\Psi_a \equiv \Psi_a^{(0)}$ *(for* $m > 1, a \neq m$*) of the third kind with simple poles in* Q_a, Q_m *and with residues* ± 1 *resp.*

3. *Holomorphic differentials* $\omega_\alpha \equiv \omega_\alpha^{(0)}, \alpha = 1, \ldots, g$ *normalized as follows:*

$$\oint_{a_\beta} \omega_\alpha = 2\pi i \delta_{\alpha\beta}. \tag{2.17}$$

4. *Multivalued normalized holomorphic on* C *differentials* $\sigma_a^{(k)}, k = 1, 2, \ldots, g$ *with increments of the form*

$$\sigma_a^{(k)}(P + b_\alpha) - \sigma_a^{(k)}(P) = -k\lambda^{k-1}\, d\lambda, \tag{2.18}$$

other increments vanish.

5. *Multivalued normalized holomorphic on* C *differentials* $\omega_\alpha^{(k)}, \alpha = 1, \ldots, g$, $k = 1, 2, \ldots$ *with increments of the form*

$$\omega_\alpha^{(k)}(P + a_\alpha) - \omega_\alpha^{(k)}(P) = -k\lambda^{k-1}\, d\lambda, \tag{2.19}$$

6. *Multivalued normalized differentials* $\Psi_a^{(k)}, a = 1, \ldots, m-1, k = 1, 2, \ldots$, *holomorphic on* $C \backslash (Q_a \cup Q_m)$ *with singularities of the form*

$$\begin{aligned} \Psi_\alpha^{(k)} &= -\frac{1}{n_a}\, d\Psi_k(\lambda) + \text{ regular terms}, \quad P \to Q_a \\ &= -\frac{1}{n_m}\, d\Psi_k(\lambda) + \text{ regular terms}, \quad P \to Q_m \end{aligned} \tag{2.20}$$

$$\Psi_k(\lambda) = \frac{\lambda^k}{k!}\left[\log\lambda - \left(1 + \frac{1}{2} + \ldots + \frac{1}{k}\right)\right], \quad k > 0. \tag{2.21}$$

The proof is straightforward.

Lemma 1. *Let* $\Omega^{(i)}, i = 1, 2$ *be any two horizontal differentials such that*

$$\Omega^{(i)} = \sum_k c_{ka}^{(i)} z_a^k\, dz_\alpha + d\sum_{k>0} r_{ka}^{(i)} \frac{\lambda^k \log\lambda}{n_a}, \quad P \to Q_a, \tag{2.22}$$

$$\oint_{a_\alpha} \Omega^{(i)} = 2\pi i A_\alpha^{(i)} \tag{2.23}$$

$$\Omega^{(i)}(P + a_\alpha) - \Omega^{(i)}(P) = dp_\alpha^{(i)}, \quad p_\alpha^{(i)} = \sum_{s>0} p_{s\alpha}^{(i)} \lambda^s, \tag{2.24}$$

$$\Omega^{(i)}(P+b_\alpha)-\Omega^{(i)}(P)=dq_\alpha^{(i)},\quad q_\alpha^{(i)}=\sum_{s>0}q_{s\alpha}^{(i)}\lambda^s, \tag{2.25}$$

Then

$$res_{P_j}\frac{\Omega^{(1)}\Omega^{(2)}}{d\lambda}=\partial_j V_{\Omega^{(1)}\Omega^{(2)}}(u) \tag{2.26}$$

where

$$\begin{aligned}V_{\Omega^{(1)}\Omega^{(2)}}(u)=&-\sum_{a=1}^{m}\Bigg[\sum_{k\geq 0}\frac{C^{(1)}_{-k-2,a}}{k+1}C^{(2)}_{k,a}+\\&+C^{(1)}_{-1,a}\ v.p.\int_{Q_0}^{Q_a}\Omega^{(2)}-\\&-\ v.p.\int_{Q_0}^{Q_a}r^{(1)}_{k,a}\lambda^k\Omega^{(2)}+\pi i\sum_{k>0}r^{(1)}_{k,a}C^{(2)}_{n_ak-1}\Bigg]+\\&+\sum_{\alpha=1}^{g}\Bigg[-\frac{1}{2\pi i}\oint_{a_\alpha}q^{(1)}_\alpha(\lambda)\Omega^{(2)}+\\&+\frac{1}{2\pi i}\oint_{b_\alpha}p^{(1)}_\alpha(\lambda)\Omega^{(2)}+A^{(1)}_\alpha\oint_{b_\alpha}\Omega^{(2)}\Bigg].\end{aligned} \tag{2.27}$$

The regularized integrals are defined with respect to the local parameters (2.15). *The sum over a in* (2.27) *does not depend on $Q_0\in C$. We recall that all the numbers $A^{(i)},p^{(i)}_{s\alpha},q^{(i)}_{s\alpha},r^{(i)}_{s\alpha}$ and $c^{(i)}_{ka}$ for negative k are constants.*

Proof. Let $\tilde{C}$ be the polygon with $4g$ edges obtained by cutting C along the cycles $a_1,\dots,a_g,b_1,\dots,b_g$ passing through a point $Q_0\in C$. Let us choose also some curves in C from Q_0 to $Q_1,\dots,Q_m$ and cut $\tilde{C}$ along these curves to obtain a domain $\tilde{C}_0$. We assume that the λ-images of all of these cuttings do not depend on u — at least in some neighborhood of the point $u\in\tilde{M}$, and that $\lambda(Q_0)\equiv 0$. Then we have an identity

$$\frac{1}{2\pi i}\oint_{\partial C_0}\left[\Omega^{(1)}(P)\int_{Q_0}^{P}\partial_j\Omega^{(2)}\right]=\mathrm{res}_{P_j}\frac{\Omega^{(1)}\Omega^{(2)}}{d\lambda}. \tag{2.28}$$

After calculation of all the residues and of all the contour integrals we obtain (2.26). □

Let $\Omega\in\mathcal{D}(\tilde{M})$ be any non-zero horizontal differential. It defines a metric ds^2_Ω on $(\tilde{M}_\Omega)$ being diagonal in the coordinates $u^1,\dots,u^N$:

$$ds^2_\Omega=\sum_{i=1}^{N}g^\Omega_{ii}(u)\,(du^i)^2 \tag{2.29}$$

via the formula

$$g_{ii}^{\Omega} = \mathrm{res}_{P_i} \frac{\Omega^2}{d\lambda}, \quad i = 1, \dots, N. \tag{2.30}$$

Here the subspace $\tilde{M}_\Omega$ consists of all pairs (C, λ) such that Ω does not vanish at $P_1 \dots P_N$ (we recall that $P_1, \dots, P_N$ are the critical points of λ). (In fact we consider the complex analogue of metric. So the coordinates are complex. We need only non-degeneracy of the metric (2.29)).

Theorem 1. The metric (2.29), (2.30) is a flat Egoroff metric on $\tilde{M}_\Omega$. Its rotation coefficients $\gamma_{ij} = \gamma_{ij}(u)$ (1.3) do not depend on Ω. They are invariant under scaling transformations (1.10).

Corollary. For any given $g, m, n_1, \dots, n_m (n_1 + \dots + n_m = n)$ the rotation coefficients $\gamma_{ij}(u), u \in (\tilde{M})^N, N = 2g + n + m - 2$, of the metric (2.29), give a self-similar solution of the system (1.6), (1.6$'$).

Proof. From the Lemma 1 the potential (see (1.5)) of the metric (2.29) has the form

$$g_{ii}^{\Omega} = \partial_i V_{\Omega\Omega}(u), \quad 1 \leq i \leq N. \tag{2.31}$$

Hence the rotation coefficients $\gamma_{ij}^{\Omega}(u)$ of the metric (2.29),(2.30) are symmetric in i, j. To prove the identity (1.6) for γ_{ij}^{Ω} let us consider the differential

$$\partial_i \partial_j \Omega \int \partial_k \Omega$$

for distinct i, j, k. It has poles only in the branch points P_i, P_j, P_k. The contour integral of the differential along $\partial \tilde{C}_0$ equals zero. Hence the sum of the residues vanishes. This reads as

$$\partial_j \sqrt{g_{ii}^{\Omega}} \partial_k \sqrt{g_{ii}^{\Omega}} + \partial_i \sqrt{g_{jj}^{\Omega}} \partial_k \sqrt{g_{jj}^{\Omega}} = \sqrt{g_{kk}^{\Omega}} \partial_i \partial_j \sqrt{g_{kk}^{\Omega}}.$$

This can be written in the form (1.6) due to the symmetry (1.4).

Let us prove now that the rotation coefficients γ_{ij}^{Ω} do not depend on Ω. Let us consider the differential

$$\partial_i \Omega^{(1)} \int \partial_j \Omega^{(2)}, \quad i \neq j,$$

for any two horizontal differentials $\Omega^{(1)}, \Omega^{(2)}$. From vanishing of the sum of its residues we obtain

$$\sqrt{g_{jj}^{\Omega^{(2)}}} \partial_i \sqrt{g_{jj}^{\Omega^{(1)}}} = \sqrt{g_{ii}^{\Omega^{(1)}}} \partial_j \sqrt{g_{ii}^{\Omega^{(2)}}}.$$

Using the symmetry (1.4) we immediately obtain

$$\gamma_{ij}^{\Omega^{(1)}} = \gamma_{ij}^{\Omega^{(2)}}.$$

Now we are to prove the last identity (1.6′). It is sufficient to prove it for a holomorphic normalized differential Ω. Let us define an operator D on functions $f = f(P, u)$ by the formula

$$Df = \frac{\partial f}{\partial \lambda} + \partial f, \quad \partial = \sum_{j=1}^{N} \partial_j. \tag{2.32}$$

The operator D is extended to differentials Ω in such a way that $Dd = dD$. For any normalized holomorphic differential we have

$$D\Omega = 0$$

(cfr. Lemma 2 below). Hence

$$\partial g_{jj}^{\Omega} = 0 \Rightarrow \partial \gamma_{ij} = 0.$$

The flatness of the metric (2.29) is proved. The scaling invariance can be verified easily again for holomorphic Ω (in this case $g_{ii}^{\Omega}(cu) = c^{-1} g_{ii}^{\Omega}(u)$). The theorem is proved. □

A horizontal differential such that $D\Omega = 0$ we shall call *primary* differential.

Lemma 2. *The subspace in $\mathcal{D}(\tilde{M})$ of all primary differentials is N-dimensional. It is spanned by the differentials (see* (2.14)–(2.21) *for the notations)*

$$\begin{aligned} &\Omega_a^{(k)}, \quad k = 1, \dots, n_a, \quad a = 1, \dots, m, \quad \sum_{a=1}^{m} n_a \Omega_a^{(n_a)} = 0 \\ &\Psi_a^{(0)}, \quad a = 1, \dots, m-1, \\ &\omega_\alpha, \sigma_\alpha \equiv \sigma_\alpha^{(1)}, \quad \alpha = 1, \dots, g. \end{aligned} \tag{2.33}$$

Proof. It is clear that for any differential Ω of the form (2.33) $D\Omega$ is a holomorphic single-valued differential on C. It is easy to see that is has zero a-periods. Hence $D\Omega = 0$. Conversely, we have

$$\begin{aligned} D\Omega_a^{(k)} &= \frac{n_a - k}{n_a} \Omega_a^{(k-n_a)}, \quad k > n_a \\ D\sigma_\alpha^{(k)} &= k\sigma_\alpha^{(k-1)}, \quad k > 1 \\ D\omega_\alpha^{(k)} &= k\omega_\alpha^{(k-1)}, \quad k > 1 \\ D\Psi_a^{(k)} &= \Psi_a^{(k-1)}, \quad k \geq 1. \end{aligned} \tag{2.34}$$

Hence any primary Ω is a linear combination of the differentials (2.33). Lemma is proved. □

Let us denote by $\mathcal{D}_0(\tilde{M}) \subset \mathcal{D}(\tilde{M})$ the subspace of all primary differentials.

Let us fix any primary differential Ω. It defines the mapping

$$\mathcal{D}(\tilde{M}) \mapsto \text{Functions}(\tilde{M}) \tag{2.35}$$

via the formula

$$\Omega' \mapsto V_{\Omega\Omega'} \tag{2.36}$$

for any horizontal differential Ω'. We call the function $V_{\Omega\Omega'}$ *conjugate* to Ω'. The image of the mapping (2.36) will be denoted as $\mathcal{A}_\Omega(\tilde{M})$. Vice versa, for any function $f \in \mathcal{A}_\Omega(\tilde{M})$ the unique *conjugate* differential $\Omega_f \in \mathcal{D}(\tilde{M})$ is determined such that

$$V_{\Omega\Omega_f} = f. \tag{2.37}$$

The basis in the space $\mathcal{A}_\Omega(\tilde{M})$ is given as follows:

$$\operatorname{res}_{Q_a} z_a^{-k}\Omega, \quad k = 1, 2, \dots ,$$

with linear constraints

$$\sum_{a=1}^{m} \operatorname{res}_{Q_a} z_a^{-kn_a}\Omega = 0, \quad k = 1, 2, \dots$$

$$\oint_{a_\alpha} \lambda^k\Omega, \quad k \geq 1, \qquad \oint_{b_\alpha} \lambda^k\Omega, \quad k \geq 0, \quad \alpha = 1, \dots, g \tag{2.38}$$

$$\text{v.p.} \int_{Q_m}^{Q_a} \lambda^k\Omega, \quad k \geq 1, \quad a \neq m$$

Note that these functions are well-defined globally on $\tilde{M}$.

Lemma 3. *For any two functions $f(u)$, $h(u) \in \mathcal{A}_\Omega(\tilde{M})$ the following identities hold:*

$$\langle df, dh\rangle_\Omega = \sum_{j=1}^{N} res_{P_j} \frac{\Omega_f\Omega_h}{d\lambda} = \partial V_{\Omega_f\Omega_h}. \tag{2.39}$$

Here $\langle\, ,\rangle_\Omega$ is the scalar product of gradients of the functions f, h with respect to the metric ds^2_Ω.

Proof. This immediately follows from the definition of the conjugate differentials (2.37) and from the Lemma 1. □

This Lemma gives us a bridge between Riemannian geometry of the moduli spaces and TCFT (see Section 4 below).

We want the explicit formulae for acting of the translation generator ∂ on the basis (2.38).

Lemma 4. *For any primary differential $\Omega \in \mathcal{D}_0(\tilde{M})$ the following identities hold:*

$$\partial\ res_{Q_a}(z_a^{-k}\Omega) = \frac{k}{n_a} res_{Q_a}(z_a^{-k+n_a}\Omega)$$

$$\partial \oint_{a_\alpha} \lambda^k \Omega = k \oint_{a_\alpha} \lambda^{k-1}\Omega, \quad \partial \oint_{b_\alpha} \lambda^k \Omega = k \oint_{b_\alpha} \lambda^{k-1}\Omega \tag{2.40}$$

$$\partial\ v.p. \int_{Q_0}^{Q_a} \Omega = \frac{\Omega}{d\lambda}\bigg|_{Q_0}, \qquad \partial\ v.p. \int_{Q_0}^{Q_a} \lambda^k \Omega = 0, \quad k > 0.$$

The proof is straightforward.

Theorem 2. For any primary differential Ω the flat coordinates for the metric ds^2_Ω have the form

$$t_{k,a} = -\frac{1}{k} \mathrm{res}_{Q_a} z_a^{-k}\Omega, \quad 1, \ldots, m, \quad k = 1, \ldots, n_a,$$

$$t_{0,a} = \ \mathrm{v.p.} \int_{Q_0}^{Q_a} \Omega + t_0, \tag{2.41}$$

$$t'_\alpha = -\frac{1}{2\pi i} \oint_{a_\alpha} \lambda\Omega, \quad t''_\alpha = \oint_{b_\alpha} \Omega, \quad \alpha = 1, \ldots, g$$

with two constraints

$$\sum_{a=1}^{m} t_{0,a} = \sum_{a=1}^{m} t_{n_a,a} = 0. \tag{2.42}$$

The matrix η of the metric ds^2_Ω in the coordinates (2.41) can be obtained from the matrix

$$\begin{aligned} \eta^{(a,k),(b,l)} &= \delta^{a,b}\delta^{k+1,n_a} \\ \eta^{\alpha',\beta''} &= \delta^{\alpha,\beta} \end{aligned} \tag{2.43}$$

other components vanish, via the restriction on the subspace (2.42). The conjugate differentials have the form

$$\Omega_{t_{(k,a)}} = \Omega_a^{(k)}, \quad 1 \le k \le n_a$$

$$\Omega_{[t_{(0,a)} - t_{(0,m)}]} = \Psi_a^{(0)}, \quad a \neq m \tag{2.44}$$

$$\Omega_{t'_\alpha} = \omega_\alpha, \quad \Omega_{t''_\alpha} = \sigma_\alpha$$

The proof immediately follows from the formula (2.27) and from the Lemmas 3, 4.

Corollary 6. *For any primary differential Ω the metric ds^2_Ω is well-defined and non-degenerate globally on $\tilde{M}$.*

Corollary 7. *For any primary differential Ω the mapping $\tilde{M}_\Omega \mapsto \mathbb{C}^N$ being given by the flat coordinates (2.41) is regular everywhere and therefore is a covering.*

It is interesting to find the degree of this covering. For $g = 0$ it equals one.

Remark. For any horizontal differential Ω it is possible to construct another flat metric on $\tilde{M}'_\Omega \equiv \{(C, \lambda) \in \tilde{M}_\Omega \,|\, \lambda(P_i) \neq 0, i = 1, \dots, N\}$,

$$d\tilde{s}^2_\Omega = \sum_{i=1}^{N} \tilde{g}^\Omega_{ii} (du^i)^2 \tag{2.45}$$

where

$$\tilde{g}^\Omega_{ii} = \operatorname{res}_{P_i} \frac{\Omega^2}{\lambda d\lambda} = \frac{g^\Omega_{ii}}{u^i}. \tag{2.46}$$

It is an Egoroff metric in the coordinates

$$z^i = \log u^i, \quad i = 1, \dots, N \tag{2.47}$$

with the rotation coefficients

$$\tilde{\gamma}_{ij}(z^1, \dots, z^N) = \exp\left(\frac{z^i + z^j}{2}\right) \gamma_{ij}(e^{z^1}, \dots, e^{z^N}) \tag{2.48}$$

Hence the functions $\tilde{\gamma}_{ij}(z)$ also enjoy the system (1.6), (1.6′) (but they are not scaling invariant!) The flat coordinates for (2.45) also can be calculated explicitly for scaling invariant Ω.

3 Poisson structures on the loop space $\mathcal{L}\tilde{M}$.

We recall that the flat metric ds^2_Ω on $\tilde{M}$ determines a Poisson structure of the form (1.15) on the loop space $\mathcal{L}\tilde{M}$. Let us denote it by $\{\,,\}_\Omega$. Let Ω be a primary differential.

Theorem 3. 1. For any horizontal differential Ω' the t-flow on the loop space $\mathcal{L}\tilde{M}$ of the form

$$\partial_t \Omega = \partial_x \Omega' \tag{3.1}$$

is a Hamiltonian flow with respect to P.B. $\{\,,\}_\Omega$ with the Hamiltonian $H = \int h(u)\, dx$ such that

$$\partial h = V_{\Omega\Omega'} \tag{3.2}$$

2. The functionals $H = \int h(u)\,dx$, $h(u) \in \mathcal{A}_\Omega(\tilde{M})$, form a Lagrangian family for the P.B. $\{\,,\}_\Omega$.

3. For any horizontal Ω' the flow (3.1) is completely integrable. The (locally) general solution of (3.1) can be written in the form

$$\{x\Omega + t\Omega' + \Omega_0\}_{P_j} = 0, \quad j = 1, \dots, N \tag{3.3}$$

for any horizontal differential Ω_0.

Proof. The equation (3.2) can be obtained from the definition (1.26) of the P.B. $\{\,,\}_\Omega$ (note that the metric ds^2_Ω is ∂-invariant). The completeness of the functionals with densities in $\mathcal{A}_\Omega(\tilde{M})$ follows from the Lemma 4. Indeed, these functionals can be constructed starting from the Casimirs (2.41) using the recursion procedure (1.27). The formula (3.3) for general solution can be proved as in ref. [14]. □

Remark 1. The flow (3.1) can be considered also as x-flow on the space of functions on t (cf. ref. [22]). Its Hamiltonian structure is defined by the bracket P.B. $\{\,,\}_{\Omega'}$.

Remark 2. Let the primary differential Ω be scaling invariant:

$$\lambda \to c\lambda, \quad u^i \to cu^i, \quad \Omega \to c^q\Omega. \tag{3.4}$$

Then all the flows (3.1) are Hamiltonian flows also for the P.B. $\{\,,\}^{\tilde{}}_\Omega$ being determined by the flat metric (2.45). The corresponding recursion operator coincides with ∂ (up to some constant). If $f \in \mathcal{A}_\Omega(\tilde{M})$ is a homogeneous function then the corresponding flow

$$u^i_t = \left\{u^i(x), \int f\,dx\right\}^{\tilde{}}_\Omega \tag{3.5}$$

can be written in the form

$$\partial_t\Omega = \partial_x\Omega_f. \tag{3.6}$$

Here the differential Ω_f is defined by the formula (2.37). The definition of the conjugate differential Ω_f can be written therefore in the form

$$\Omega_f = \partial_x^{-1}\left\{\Omega(P, u(x)), \int f\,dx\right\}^{\tilde{}}_\Omega \tag{3.7}$$

The system of equations of the form (3.1) where Ω' is any of the basic differentials (2.14)–(2.20) we shall call *Whitham-type hierarchy* (or W-hierarchy) for given primary differential Ω. It is put in order by action of the recursion operator D^{-1} on the differentials Ω'.

4 Main examples. Application to TCFT.

Example 1. For the family of polynomials $M = \{\lambda = p^n + q_{n-2}p^{n-1} + \ldots + q_0\}$ the equations of the W-hierarchy for $\Omega = dp$ have the form

$$\partial_{t_i} dp(\lambda) = \partial_x dr^{(i)}(\lambda), \quad i \neq kn \tag{4.1}$$

where

$$r^{(i)} = \frac{[\lambda^{i/n}(p)]_+}{i} \tag{4.2}$$

(the polynomial part in p). Note that $t_1 = x$. They can be rewritten in the form [10]

$$\partial_{t_i} \lambda = \partial_p r^{(i)} \partial_x \lambda - \partial_x r^{(i)} \partial_p \lambda. \tag{4.3}$$

Here $\lambda = \lambda(p)$, $r^{(i)} = r^{(i)}(p)$ are polynomials. Equation (4.3) can be obtained by averaging of the Kdv-type hierarchy (or the Gel'fand-Dikii hierarchy)

$$\partial_i L = \frac{1}{i}[L, [L^{i/n}]_+], \quad i \neq kn, \tag{4.4}$$

$$L = \partial^n + q_{n-2}\partial^{n-2} + \ldots + q_0 \tag{4.4$'$}$$

over the family of the constant solutions $q_j = \text{const}$, $j = 0, \ldots, n-2$. The metric ds^2_Ω has the form

$$ds^2_\Omega = \sum_{i=1}^{n-1} \frac{(du^i)^2}{\lambda''(p_i)} \tag{4.5}$$

Here $p_1, \ldots, p_{n-1}$ are the critical points of $\lambda(p)$,

$$\lambda'(p_i) = 0$$

and

$$u^i = \lambda(p_i).$$

For $n = 4$ the rotation coefficients of the metric (4.5) give a nontrivial algebraic solution of the system (1.6), (1.6$'$) (and, therefore, of the system (1.11)). The flat coordinates $t_1, \ldots, t_{n-1}$ have the form

$$t_{n-i} = \text{res}_{\lambda=\infty} \frac{\lambda^{i/n}}{i} dp, \quad i = 1, \ldots, n-1, \tag{4.6}$$

$$\langle dt^i, dt^j \rangle_{dp} = \frac{\delta_{i+j,n}}{n}. \tag{4.6$'$}$$

Remark. We also can take any other differential $\Omega = dr^{(i)}$, $i = 2, \dots, n-1$ to determine a flat metric on M. All these flat geometries are inequivalent one to another.

To explain the relation [10] of (4.2), (4.3) to TCFT we recall here the Landau-Ginsburg superpotentials approach [6], [7]. In TCFT all the correlation functions do not depend on coordinates (but do depend on coupling parameters $t_1, \dots, t_N$) and can be expressed in terms of the two-point and the three-point correlation functions

$$\eta_{\alpha\beta} = \langle\phi_\alpha\phi_\beta\rangle = \text{const}, \quad \det(\eta_{\alpha\beta}) \neq 0 \tag{4.7}$$

$$c_{\alpha\beta\gamma}(t) = \langle\phi_\alpha\phi_\beta\phi_\gamma\rangle \tag{4.8}$$

by the factorization formulae

$$\begin{aligned} \langle\phi_\alpha\phi_\beta\phi_\gamma\phi_\delta\rangle &= c_{\alpha\beta}^\epsilon c_{\epsilon\gamma\delta} \\ \langle\phi_\alpha\phi_\beta\phi_\gamma \int \phi_\delta\rangle &= \partial_\delta c_{\alpha\beta\gamma}, \quad \partial_\delta = \frac{\partial}{\partial t_\delta} \end{aligned} \tag{4.9}$$

etc. (the raising of indices using the metric $\eta_{\alpha\beta}$). For the primary free energy $F = F(t_1, \dots, t_N)$ of the model the following identities hold:

$$\partial_1\partial_\alpha\partial_\beta F = \eta_{\alpha\beta}, \tag{4.10}$$

$$\partial_\alpha\partial_\beta\partial_\gamma F = c_{\alpha\beta\gamma}. \tag{4.11}$$

The function $F(t)$ is quasihomogeneous in $t_1, \dots, t_N$. To find these correlation functions for genus zero let us consider the set of polynomials $\lambda(p)$ of the form given above (Landau-Ginsburg superpotentials, A_{n-1}-model) depending on the coupling parameters $t_1, \dots, t_N$ $(N = n-1)$ in such a way that

$$\partial_\alpha\lambda = -\phi_\alpha, \quad \alpha = 1, \dots, N. \tag{4.12}$$

Here $\phi_1, \dots, \phi_N$ is the basis of polynomials of degrees $0, 1, \dots, N-1$ orthogonal with respect to the scalar product

$$\langle\phi, \varphi\rangle = \operatorname{res}_{\lambda=\infty} \frac{\phi\varphi}{d\lambda/dp}, \tag{4.13}$$

$$\langle\phi_\alpha, \varphi_\beta\rangle = \delta_{\alpha+\beta,n} \tag{4.14}$$

Proposition 1. *The family* $\lambda = \lambda(p, t_1, \dots, t_{n-1})$ *is a particular solution of the system* (4.1) *where* $i = 1, \dots, n-1$ *with the initial data*

$$\lambda|_{t_1=\dots=t_{n-1}=0} = p^n. \tag{4.15}$$

The crucial point in the proof is in the observation [4] that the orthogonal polynomials (4.13), (4.14) have the form

$$\phi_\alpha = dr^{(\alpha)}/dp, \quad \alpha = 1, \dots, n-1. \tag{4.16}$$

The dependence of the coefficients $q_0, q_1, \dots, q_{n-2}$ of the polynomial $\lambda(p)$ on $t_1, \dots, t_{n-1}$ can be found from the equations (4.6). It can be represented in the form [10]

$$\begin{aligned} &f(p_k) = 0, \quad k = 1, \dots, n-1, \\ &f(p) = \frac{d}{dp}\left[r^{(n+1)}(p) + \sum_{i=1}^{n-1} t_i r^{(i)}(p)\right] \end{aligned} \tag{4.16'}$$

The triple correlators have the form [6], [7]

$$c_{\alpha\beta\gamma}(t) = \operatorname{res}_{p=\infty} \frac{\phi_\alpha \phi_\beta \phi_\gamma}{d\lambda/dp}$$

where the polynomials $\phi_\alpha = \phi_\alpha(p;t)$ are determined by (4.16). The coefficients $c^\gamma_{\alpha\beta}(t) = c_{\alpha\beta\epsilon}(t)\eta^{\epsilon\gamma}$ for any $t = t_1, \dots, t_{n-1}$ are the structure constants of an associative commutative algebra A with a unit ϕ_1 and with a constant (in t) invariant scalar product $\eta_{\alpha\beta}$ (chiral algebra of primary fields). We recall that those algebras are called Frobenius algebras [25]. In this example it is isomorphic to the truncated polynomials

$$A \cong C[p]/(\lambda'(p)).$$

The free energy $F = F(t)$ was found by Krichever. It has the form

$$F(t) = \frac{1}{2}\operatorname{res}_{p=\infty}\left\{\left[\int f(p)\,dp\right]_+ f(p)\,dp\right\}.$$

The function $f(p)$ is determined by (4.16′); for any function $g(p)$ the symbol $[g(p)]_+$ means the polynomial part of $g(p)$ with respect to the parameter $z = \lambda^{1/n}$. Krichever also argued that the function $F(t)$ should be considered as a logarithm of τ-function of the hierarchy (4.1) for the particular solution (4.16′).

Example 2. For the family of rational functions (2.9) the W-hierarchy has the form

$$\partial_{t_{i,a}} dp = \partial_x d[\lambda^i]_a, \qquad a = 1, \dots, n, \quad i = 0, 1, \dots \tag{4.17}$$

$$\partial_{s_{i,a}} dp = \partial_x d[\varphi_i(\lambda)]_a, \qquad a = 1, \dots, n, \quad i = 0, 1, \dots \tag{4.18}$$

Here the operation $[\,]_a$ means that one should kill singularities in all the points $Q_1, \dots, Q_n$ but Q_a; the functions $\varphi_i(\lambda)$ are defined in (2.21). The flat coordinates for the primary differential $\Omega = dp$ are η_i, q_i. The hierarchy (4.17), (4.18) is the hierarchy of the "highest Benney equations" — it can be deduced from Zakharov's paper [23].

Example 3. For the moduli space of hyperelliptic curves (2.10) the differential $\Omega = \Omega^{(1)}$ is a primary one. The part of the corresponding W-hierarchy of the form

$$\partial_{t_i}\Omega = \partial_x \Omega^{(2i+1)}, \quad i = 0, 1, \dots \tag{4.19}$$

coincides with the KdV hierarchy averaged over the family of g-gap solutions. This was proved by Flaschka, Forest and McLaughlin [24]. One should add the equations

$$\begin{aligned} \partial_{t'_{\alpha,k}}\Omega &= \partial_x \omega_\alpha^{(k)}, \quad k = 0, 1, \dots \\ \partial_{t''_{\alpha,k}}\Omega &= \partial_x \sigma_\alpha^{(k)}, \quad k = 1, 2, \dots \end{aligned} \tag{4.20}$$

to obtain the complete hierarchy ($\alpha = 1, \dots, g$). The functionals $H = \int h(u)\,dx$ with the densities of the form

$$\mathrm{res}_{\lambda=\infty} \lambda^{(2i+1)/2}\Omega, i = 0, 1, \dots, \qquad \oint_{a_k} \lambda^k \Omega, k > 0, \qquad \oint_{b_k} \lambda^k \Omega, k \geq 0, \tag{4.21}$$

give the complete family of conservation laws for the hierarchy (4.19), (4.20) (the Lagrangian family $\mathcal{A}_\Omega(\tilde{M})$). The P.B. $\{\,,\}_\Omega$ coincides [16] with the averaged [11]–[13], [15] Gardner-Zakharov-Faddeev (GZF) P.B. of the KdV-hierarchy. The flat coordinates (2.41) for the metric ds^2_Ω are:

$t_1 =$ annihilator of the GZF P.B.,
$t'_1, \dots, t'_g$ action variables for the GZF P.B.
$t''_1, \dots, t''_g$ components of the wave-number vector

(see ref. [13] for details). It can be proved also that the second P.B. $\{\,,\}\tilde{}_\Omega$ coincides with the averaged Magri bracket [13].

Example 4. For the moduli space M of all curves of genus g with a marked point Q_1 and a meromorphic function λ with pole in Q_1 of order n the primary differentials are

$$\Omega^{(i)} = \frac{dz}{z^{i+1}} + \text{regular terms}, \quad z = \lambda^{-1/n}, \quad i = 1, \dots, n-1$$

$$\oint_{a_\alpha} \Omega^{(i)} = 0, \quad \alpha = 1, \dots, g, \tag{4.22}$$

$$\omega_\alpha, \sigma_\alpha, \quad \alpha = 1, \dots, g,$$

(for the definitions of ω_α, $\sigma_\alpha \equiv \sigma_\alpha^{(1)}$ see (2.17), (2.18)). Let us use these differentials for calculations of correlation functions of genus g minimal model of TCFT. Here the space $\tilde{M}_\Omega$ plays the role of the coupling space (we recall that $\dim M = N = 2g + n - 1$). Let us redenote the primary differentials (4.22) as $\Phi_1, \ldots, \Phi_N$ in such a way that

$$\begin{aligned} \Phi_\alpha &= -n\Omega^{(\alpha)}, \quad \alpha = 1, \ldots, n-1, \\ \left.\begin{aligned} \Phi_{n-1+\alpha} &= \omega_\alpha \\ \Phi_{g+n-1+\alpha} &= \sigma_\alpha \end{aligned}\right\} & \quad \alpha = 1, \ldots, g. \end{aligned} \tag{4.23}$$

Let $\wedge = (\wedge_{\alpha\beta}(u))$ be the symmetric $N \times N$-matrix of the form

$$\wedge_{\alpha\beta}(u) = V_{\Phi_\alpha \Phi_\beta}. \tag{4.24}$$

Under scaling transformations

$$\lambda \to c^n \lambda \tag{4.25}$$

the matrix transforms as follows:

$$\wedge \to S \wedge S \tag{4.26}$$

where $S = \mathrm{diag}(c^{-1}, \ldots, c^{-n+1}, 1, \ldots, 1, c^n, \ldots, c^n)$. For any fixed β the variables $t_1 = \wedge_{\beta\beta}$, $t_2 = \wedge_{\beta\beta+1}, \ldots$ provide the flat coordinates for the metric $ds^2_{\Phi_\beta}$.

Let us choose the differential $\Omega = \Omega^{(1)}$ for defining the flat metric ds^2_Ω (this choice seems to be natural since it does not depend on n and g). Let us denote it as $\Omega = dp$. The flat coordinates on $\tilde{M}$ are

$$\begin{aligned} t_{n-\alpha} &= \mathrm{res}_{\lambda=\infty} \frac{\lambda^{\alpha/n}}{\alpha} dp, \quad \alpha = 1, \ldots, n-1, \\ \left.\begin{aligned} t_{n-1+\alpha} &= \tfrac{1}{2\pi i} \oint_{a_\alpha} \lambda\, dp \\ t_{g+n-1+\alpha} &= \oint_{b_\alpha} dp \end{aligned}\right\} & \quad \alpha = 1, \ldots, g \end{aligned} \tag{4.27}$$

The point of $\tilde{M}$ with coordinates $t_1, \ldots, t_N$ is determined by the following system (cf. (3.3))

$$(n\Omega^{(n+1)} + \hat{\Omega})|_{P_j} = 0, \ j = 1, \ldots, N \tag{4.28}$$

where

$$\hat{\Omega} = \sum_{i=1}^{N} t_i \Phi_i \tag{4.29}$$

Indeed, if $u \in \tilde{M}$ is the point with the coordinates $t_1, \ldots, t_N$ then the following identity holds on the corresponding curve $C = C(u)$:

$$p\, d\lambda = n\Omega^{(n+1)} + \hat{\Omega} \tag{4.30}$$

The l.h.s. of (4.30) vanishes in all the branch points $P_1, \dots, P_N$. This proves (4.28).

The correlation functions have the form

$$\langle \Phi_\alpha \Phi_\beta \rangle = \sum \operatorname{res}_{d\lambda=0} \frac{\Phi_\alpha \Phi_\beta}{d\lambda} = \eta_{\alpha\beta} \tag{4.31}$$

$$\langle \Phi_\alpha \Phi_\beta \Phi_\gamma \rangle = \sum \operatorname{res}_{d\lambda=0} \frac{\Phi_\alpha \Phi_\beta \Phi_\gamma}{d\lambda dp} = c_{\alpha\beta\gamma} \tag{4.32}$$

etc. The correlation functions (4.31) do not depend on t. This follows from Lemma 3 and from Theorem 2. They have the form

$$\begin{aligned} \langle \Phi_\alpha \Phi_\beta \rangle &= n\delta_{\alpha+\beta,n}, \quad 1 \le \alpha, \beta \le n-1 \\ \langle \Phi_{n-1+\alpha} \Phi_{g+n-1+\beta} \rangle &= \delta_{\alpha\beta}, \quad 1 \le \alpha, \beta \le g \end{aligned} \tag{4.33}$$

otherwise zero.

The corresponding Landau-Ginsburg superpotential is $\lambda = \lambda(p)$, $p = \int dp$. In other words

$$\partial_{t_\alpha} (\lambda\, dp)_{p=const} = -\Phi_\alpha, \quad \alpha = 1, \dots, N. \tag{4.34}$$

This follows from identities

$$\partial_{t_\alpha} (p\, d\lambda) = \Phi_\alpha, \quad \alpha = 1, \dots, N \tag{4.35}$$

and

$$\partial_{t_\alpha} (p\, d\lambda)_{\lambda=const} = -\partial_{t_\alpha} (\lambda\, dp)_{p=const}. \tag{4.36}$$

Theorem 4. The free energy $F = F(t_1, \dots, t_N)$ satisfying (4.10), (4.11) for (4.31), (4.32) has the form

$$-2F = V_{p\,d\lambda, p\,d\lambda} \equiv n^2 V_{\Omega^{(n+1)}, \Omega^{(n+1)}} + 2n \sum_{i=1}^N t_i V_{\Omega^{(n+1)}, \Phi_i} + \sum_{i,j=1}^N t_i t_j V_{\Phi_i, \Phi_j}. \tag{4.37}$$

Proof. From (4.28), (4.29) we obtain

$$\sum_{i=1}^N t_i \partial_{t_\alpha} V_{\Phi_i, \Phi_j} + n \partial_{t_\alpha} V_{\Omega^{(n+1)}, \Phi_j} = 0, \quad j = 1, \dots, N,$$

$$\sum_{i=1}^N t_i \partial_{t_\alpha} V_{\Omega^{(n+1)}, \Phi_j} + n \partial_{t_\alpha} V_{\Omega^{(n+1)}, \Omega^{(n+1)}} = 0.$$

Hence

$$\partial_{t_\alpha}\partial_{t_\beta}F = -V_{\Phi_\alpha\Phi_\beta} \tag{4.38}$$

Let us prove now that

$$\partial_{t_\gamma}V_{\Phi_\alpha\Phi_\beta} = -c_{\alpha\beta\gamma}. \tag{4.39}$$

Indeed,

$$\partial_{t_\gamma}V_{\Phi_\alpha,\Phi_\beta} = \sum_{i=1}^{N}\frac{\partial}{\partial u^\delta}V_{\Phi_\alpha,\Phi_\beta}\frac{\partial u^\delta}{\partial t_\gamma} = -\sum_{\delta=1}^{N}\operatorname{res}_{P_\delta}\frac{\Phi_\alpha\Phi_\beta}{d\lambda}\frac{\Phi_\gamma}{dp}\bigg|_{P_\delta} = -c_{\alpha\beta\gamma}.$$

Here we use (2.26) and (4.34). This completes the proof. □

The corresponding Frobenius algebra of primary fields (see above) has the form

$$\Phi_i\Phi_j = c_{ij}^k\Phi_k\,dp\,(\operatorname{mod}\,\mathcal{D}_0(\tilde{M})\cdot d\lambda) \tag{4.40}$$

where

$$c_{ij}^k = \eta^{kl}c_{ijl}. \tag{4.40$'$}$$

Remark 1. It follows from (4.38) that the Hessian

$$\tau_{\alpha\beta} = -(\partial_{t_{n-1+\alpha}}\partial_{t_{n-1+\beta}}F), \quad 1 \le \alpha, \beta \le g \tag{4.41}$$

coincides with the period matrix of the curve C. We shall consider the linear Virasoro-type constraints for F in the next publication.

Remark 2. Probably the exactness of the differential $d\lambda$ is not necessary. Almost all the constructions of this section seem to be realizable also for any normalized Abelian differential $d\lambda$ with poles in $Q_1, \ldots, Q_m$. This possibility also is to be investigated.

The quasihomogeneous property of the partition function (4.37) has the form

$$F(c^n t_1, \ldots, c^2 t_{n-1}, \ldots, c^{n+1}t_{n+g-1}, ct_{n+g}, \ldots, ct_N) = c^{2(n+1)}F(t). \tag{4.42}$$

We shall also analyze the problem of glueing all the Riemann manifolds $\tilde{M}$ with different genera g in the next publication.

Appendix. Deformation of Frobenius algebras and partition functions of TCFT.

The logarithm of partition function $F = F(t_1, \ldots, t_N)$ of TCFT satisfies the following system of nonlinear equations: its third derivatives (after raising of an index) for any t form a set of structure constants of a commutative associative N-dimensional algebra with a unity with invariant nondegenerate scalar product (in fact, only the equation of associativity is nontrivial). Such algebras are well known as Frobenius algebras. We see that the free energy $F(t)$ determines some deformation of Frobenius algebra

$$c^i_{jk}(t) = \eta^{is} c_{sjk}(t), \quad c_{ijk}(t) = \partial_{t_i}\partial_{t_j}\partial_{t_k} F(t) \tag{A.1}$$

such that the corresponding invariant scalar product η^{ij} does not depend on t (let us call (A.1) F-deformations). Here we shall construct some class of F-deformations of any Frobenius algebra using the results of ref. [25].

Let A be any N-dimensional Frobenius algebra and $M = A^*$ (the dual space). A multiplication is defined on T^*M: if $u^1, \ldots, u^N$ is a basis in A (providing the coordinate system in M) then

$$du^i \cdot du^k = c^{ij}_k du^k, \tag{A.2}$$

c^{ij}_k being the structure constants of A. The non-degenerate scalar product on T^*M (and, therefore, a metric on M) is defined by the formula

$$\langle df, dg \rangle = 2 i_\partial (df \cdot dg), \tag{A.3}$$

$\partial = u^i \frac{\partial}{\partial u^i}$ is the dilation generator. It was observed [25] that the metric (A.3) is flat and the corresponding Levi-Civita connection has the form

$$\nabla^i T^j = \partial^i T^j - c^{ij}_s T^s \tag{A.4}$$

(raising of indices using the metric (A.3)). The flat coordinates $t_1, \ldots, t_N$ can be introduced via appropriate quadratic substitution of the form [25]

$$u^i = \frac{1}{2} a^i_{\alpha\beta} t^\alpha t^\beta \tag{A.5}$$

$$\langle dt^\alpha, dt^\beta \rangle = \eta^{\alpha\beta} = \text{const.} \tag{A.5$'$}$$

Let us introduce the coefficients

$$c^{i_1, \ldots, i_n} = 2 c^{i_1 i_2}_{s_1} c^{s_1 i_3}_{s_2} \ldots c^{s_{n-2} i_n}_{s_{n-1}} u^{s_{n-1}} \tag{A.6}$$

and the functions

$$\tilde{c}^{\alpha\beta}_\gamma(t) = \frac{\partial t^\alpha}{\partial u^i} \frac{\partial t^\beta}{\partial u^j} \frac{\partial u^k}{\partial t^\alpha} c^{ij}_k. \tag{A.7}$$

Proposition. The function (A.7) defines a F-deformation of the Frobenius algebra A with constant scalar product (A.5$'$) and with the "free energy"

$$F(t) = \sum_{n=3}^{\infty} \frac{2^{n-1}}{n} c^{i_1 \ldots i_n} u_{i_1} \ldots u_{i_n}. \tag{A.8}$$

Proof. It is sufficient to prove that in the curvilinear coordinates $u^1, \dots, u^N$ the function (A.8) satisfies the equation

$$\nabla^i \nabla^j \nabla^k F = c^{ijk}. \tag{A.9}$$

The proof of (A.9) is straightforward using the identities

$$\nabla^l c_k^{ij} = -c_s^{ij} c_k^{sl}, \quad \nabla^k u_i = \frac{1}{2}\delta_i^k. \tag{A.10}$$

□

Acknowledgments. This work was made in June, 1991 during the International semester *Infinite dimensional algebras and algebraic geometry* being organized by V. G. Kac and C. De Concini in Scuola Normale Superiore, Pisa. I am grateful to SNS for hospitality. I wish to thank I. Krichever and M. Kontsevich for many useful discussions.

June 28, 1991

Bibliography

[1] E. Brezin and V. Kazakov, Phys. Lett. B236 (1990) 144; M. Douglas and S. Shenker, Nucl. Phys. B335 (1990) 635; D. J. Gross and A. Migdal, Phys. Rev. Lett. 64 (1990) 127.

[2] T. Eguchi and S.-K. Yang, *$N = 2$ Superconformal Models as Topological Field Theories*, Tokyo preprint UT-564.

[3] K. Li, *Topological Gravity with Minimal Matter*, Caltech preprint CALT-68-1662; *Recursion Relations in Topological Gravity with Minimal Matter*, Caltech preprint CALT-68-1670.

[4] E. Verlinde and H. Verlinde, *A Solution of Two-Dimensional Topological Gravity*, preprint IASSNS-HEP-90/45 (1990).

[5] E. Witten, Comm. Math. Phys. 117 (1988) 353; Comm. Math. Phys. 118 (1988) 411.

[6] R. Dijkgraaf, E. Verlinde and H. Verlinde, *Topological strings in $d < 1$*, preprint PUPT-1024

[7] R. Dijkgraaf, E. Verlinde and H. Verlinde, *Notes on Topological Strings Theory and Two-Dimensional Gravity*, PUPT-1217, IASSNS-HEP-90/80

[8] C. Vafa, *Topological Landau-Ginsburg Models*, preprint HUTP-90/A064.

[9] B. Blok and A. Varchenko, *Topological Conformal Field Theories and the Flat Coordinates*, preprint IASSNS-HEP-91/5.

[10] I. Krichever, May 1991, private communication.

[11] B. Dubrovin and S. Novikov, Sov. Math. Doklady 27 (1983), 665.

[12] S. Novikov, Russian Math. Surveys 40:4 (1985), 79.

[13] B. Dubrovin and S. Novikov, Russian Math. Surveys 44:6 (1989), 35.

[14] I. Krichever, Funct. Anal. Appl. 22 (1988), 206.

[15] B. Dubrovin, Nucl. Phys. B (Proc. Suppl.) 18A (1990), 23.

[16] B. Dubrovin, Funct. Anal. Appl. 24 (1990).

[17] I. M. Gelfand and L. A. Dikii, Russian Math. Surveys 30:5 (1975), 77.

[18] G. Darboux, *Leçons sur les systemes ortogonaux et les coordonnees curvilignes.* Paris, 1897.

[19] S. P. Novikov (Ed.), *Theory of Solitons. The Inverse Scattering Method*, Plenum, New York, 1984.

[20] A. S. Fokas, R. A. Leo, L. Martina and G. Soliani, Phys. Lett. A115 (1986), 329.

[21] S. Tsarev, Math. USSR Izvestiya (1990).

[22] S. Tsarev, Math. Notes 45 (1989).

[23] V. Zakharov, Funct. Anal. Appl. 14 (1980), 89.

[24] H. Flaschka, M. G. Forest and D. W. McLaughlin, Comm. Pure Appl. Math. 33 (1980), 739.

[25] S. Novikov and A. Balinkski, Sov. Math. Doklady 32 (1985), 228.

Boris Dubrovin
International School for Advanced Studies (SISSA)
Via Beirut, 2–4
I-34013 TRIESTE, Italy
E-mail: dubrovin@sissa.it

Symplectic Forms in the Theory of Solitons*

I.M. Krichever† and D.H. Phong‡

† Department of Mathematics
Columbia University
New York, NY 10027
and
Landau Institute for Theoretical Physics
Moscow 117940, Russia
e-mail: krichev@math.columbia.edu

‡ Department of Mathematics
Columbia University
New York, NY 10027
‡ e-mail: phong@math.columbia.edu

Abstract

We develop a Hamiltonian theory for 2D soliton equations. In particular, we identify the spaces of doubly periodic operators on which a full hierarchy of commuting flows can be introduced, and show that these flows are Hamiltonian with respect to a universal symplectic form $\omega = \mathrm{Res}_{\infty}\langle \Psi_0^* \delta L \wedge \delta \Psi_0 \rangle dk$. We also construct other higher order symplectic forms and compare our formalism with the case of 1D solitons. Restricted to spaces of finite-gap solitons, the universal symplectic form agrees with the symplectic forms which have recently appeared in non-linear WKB theory, topological field theory, and Seiberg-Witten theories. We take the opportunity to survey some developments in these areas where symplectic forms have played a major role.

*Research supported in part by the National Science Foundation under grant DMS-95-05399.

I. Introduction

There is increasing evidence that symplectic structures for solitons may provide a unifying thread to many seemingly unrelated developments in geometry and physics. In soliton theory, the space of finite-gap solutions to the equation $[\partial_y - L, \partial_t - A] = 0$ is a space $\mathcal{M}_g(n,m)$ of punctured Riemann surfaces Γ and pair of Abelian integrals E and Q with poles of order less than n and m at the punctures. The fibration over $\mathcal{M}_g(n,m)$ with Γ as fiber carries a natural meromorphic one-form, namely $d\lambda = Q\,dE$. It is a remarkable and still mysterious fact that the form $d\lambda$ is actually central to several theories with very distinct goals and origins. These include the non-linear WKB (or Whitham) theory [22][23][26][36][38], two-dimensional topological models [11][12][13], and Seiberg-Witten exact solutions of N=2 supersymmetric gauge theories [18][40][52][53]. The form $d\lambda$ can be viewed as a precursor of a symplectic structure. Indeed, it can be extended as a 1-form $\sum_{i=1}^{k} d\lambda(z_i)$ on the fibration over $\mathcal{M}_g(n,m)$ with fiber a symmetric gth-power of Γ. Its differential ω becomes single-valued when restricted to a suitable g-dimensional leaf of a canonical foliation on $\mathcal{M}_g(n,m)$, and defines a symplectic form [39]. Earlier special cases of this type of construction were pioneered by Novikov and Veselov [50] in the context of hyperelliptic surfaces and 1D solitons, and by Seiberg, Witten, and Donagi [18][53] in the context of N=2 SUSY gauge theories.

The goal of this paper is twofold. Our first and primary objective is to construct the foundations of a Hamiltonian theory of 2D solitons.

- For this, we provide an improved formulation of 2D hierarchies, since the classical formulations (e.g. Sato [54]) are less pliable than in the 1D case, and inadequate for our purposes. In particular, the new formulation allows us to identify suitable spaces $\mathcal{L}(b)$ of doubly periodic operators on which a full hierarchy of commuting flows $\partial_m L = \partial_y A_m + [A_m, L]$ can be introduced;

- We can then define a universal symplectic form ω on these spaces $\mathcal{L}(b)$ by

$$\omega = \mathrm{Res}_\infty \langle \Psi_0^* \delta L \wedge \delta \Psi_0 \rangle dk, \tag{1.1}$$

where Ψ_0 and Ψ_0^* are the formal Bloch and dual Bloch functions for L. This form had been shown in [39] to restrict to the geometric symplectic form $\delta \sum_{i=1}^{g} d\lambda(z_i)$ when finite-gap solitons are imbedded in the space of doubly periodic operators. Here we show that it is a symplectic form in its own right on $\mathcal{L}(b)$, and that with respect to this form, the hierarchy of 2D flows is Hamiltonian. Their Hamiltonians are shown to be $2nH_{n+m}$, where H_s are the coefficients of the expansion of the quasi-momentum in terms of the quasi-energy.

- Our formalism is powerful enough to encompass many diverse symplectic structures for 1D solitons. For example, ω reduces to the Gardner-Faddeev-Zakharov symplectic structure for KdV, while its natural modifications for y-independent equations (see (2.71) and (2.73) below), reproduce the infinite set of Gelfand-Dickey as well as Adler-Magri symplectic structures.

• The symplectic form (1.1) is algebraic in nature. However, it suggests new higher symplectic forms,

$$\omega_{m_0} = \mathrm{Res}_\infty \langle \Psi_0^* (A_{m_0}^{(1)} \delta L - L^{(1)} \delta A_{m_0}) \wedge \delta \Psi_0 \rangle_0 \, dk \tag{1.2}$$

which are well-defined only on certain spaces of operators with suitable growth or ergodicity conditions. For Lax equations $\partial_{t_m} L = [A_m, L]$, these higher symplectic forms have a remarkable interpretation: they are forms with respect to which the eigenvalues of A_{m_0}, suitably averaged, can serve as Hamiltonians, just as the eigenvalues of L are Hamiltonians with respect to the basic symplectic structure (1.1). It would be very interesting to understand these new forms in an *analytic* theory of solitons.

Our second objective is to take this opportunity to provide a unified survey of some developments where the form $d\lambda$ (or its associated symplectic form ω) played a central role. Thus $d\lambda$ emerges as the generating function for the Whitham hierarchy, and its coefficients and periods are Whitham times (Section IV). The same coefficients are deformation parameters of topological Landau-Ginzburg models in two dimensions (Section V), while for N=2 SUSY four-dimensional gauge theories, the periods of $d\lambda$ generate the lattice of Bogomolny-Prasad-Sommerfeld states (Section VI). Together with $d\lambda$, another notion, that of a prepotential $\mathcal{F}$, emerges repeatedly, albeit under different guises. In non-linear WKB methods, $\mathcal{F}$ is the exponential of the τ-function of the Whitham hierarchy. In topological Landau-Ginzburg models, it is the free energy. In N=2 supersymmetric gauge theories, it is the prepotential of the Wilson effective action. It is an unsolved, but clearly very important problem, to determine whether these coincidences can be explained from first principles.

II. Hamiltonian Theory of 2D Soliton Equations

Solitons arose originally in the study of shallow water waves. Since then, the notion of soliton equations has widened considerably. It embraces now a wide class of non-linear partial differential equations, which all share the characteristic feature of being expressible as a compatibility condition for an auxiliary pair of linear differential equations. This is the viewpoint we also adopt in this paper. Thus the equations of interest to us are of the form

$$[\partial_y - L, \partial_t - A] = 0, \tag{2.1}$$

where the unknown functions $\{u_i(x,y,t)\}_{i=0}^n$, $\{v_j(x,y,t)\}_{j=0}^m$ are the $N \times N$ matrix coefficients of the ordinary differential operators

$$L = \sum_{i=0}^n u_i(x,y,t)\partial_x^i, \ A = \sum_{j=0}^m v_j(x,y,t)\partial_x^j \,. \tag{2.2}$$

A preliminary classification of equations of the form (2.1) is by the orders n, m of the operators L and A, and by the dimension N of the square matrices

$u_i(x,y,t)$, $v_j(x,y,t)$. In what follows, we assume that the leading coefficients of L and A are constant diagonal matrices $u_n^{\alpha\beta} = u_n^\alpha \delta_{\alpha\beta}$, $v_m^{\alpha\beta} = v_m^\alpha \delta_{\alpha\beta}$. Under this assumption, the equation (2.1) is invariant under the gauge transformations $L, A \to L' = g(x)^{-1} L g(x)$, $A' = g(x) A g(x)^{-1}$ where $g(x)$ is a diagonal matrix. We fix the gauge by the condition $u_n^{\alpha\beta} = \delta^{\alpha\beta} u_\alpha$, $u_{n-1}^{\alpha\alpha} = 0$. We shall refer to (2.1) as a *zero curvature* or *2D soliton equation.* The 1D soliton equation corresponds to the special case of y-independent operators L and A. In this case the equation (2.1) reduces to a *Lax equation* $L_t = [A, L]$.

A. Difficulties in a Hamiltonian Theory of 2D Solitons

The Hamiltonian theory of 1D solitons is a rich subject which has been developed extensively over the years [10][25]. However, much less is known about the 2D case. We illustrate the differences between 1D and 2D equations in the basic example of the hierarchies for the Korteweg-deVries (KdV)

$$u_t - \frac{3}{2} u\, \partial_x u - \frac{1}{4} \partial_x^3 u = 0 \tag{2.3}$$

and the Kadomtsev-Petviashvili (KP) equations

$$\frac{3}{4} u_{yy} = \partial_x \left(u_t - \frac{3}{2} u\, \partial_x u - \frac{1}{4} \partial_x^3 u \right). \tag{2.4}$$

The KP equation arises from the choice $N = 1$, $n = 2$, $m = 3$, and $L = \partial_x^2 + u$, $A = \partial_x^3 + \frac{3}{2} u\, \partial_x + v_0$ in (2.1). We obtain in this way the system

$$\partial_x v_0 = \frac{3}{4} \partial_x^2 u + \frac{3}{4} u_y, \quad v_{0,y} = u_t - \frac{1}{4} \partial_x^3 u + \frac{3}{4} \partial_x u_y - \frac{3}{2} u\, \partial_x u \tag{2.5}$$

which is equivalent to (2.4) (up to an (x,y)-independent additive term in v_0, which does not affect the commutator $[\partial_y - L, \partial_t - A]$). Taking L and A independent of y gives the KdV equation.

The basic mechanism behind this construction is that the zero curvature equation actually determines A in terms of L. This remains the case for the 1D Lax equation $L_t = [A, L]$ even when A is taken to be of arbitrarily high order m, but not for the 2D zero curvature equation $L_t - A_y = [A, L]$. The point is that $[A, L]$ is a differential operator of order $m + 1$. The Lax equation requires that it be in fact of order 0, while the 2D zero curvature equation requires only that it be of order $\leq m - 1$. The order 0 constraint is quite powerful. Expressed as differential constraints on the coefficients of A, it implies readily that the space of such A's for fixed L is of dimension m. An explicit basis can be obtained by the Gelfand-Dickey construction [10][27], which we present for a general operator L of order n. Let a pseudo-differential operator of order n be a formal Laurent series $\sum_{i=-\infty}^{n} w_i \partial_x^i$ in ∂_x, with ∂_x and ∂_x^{-1} satisfying the identities

$$\partial_x u = u \partial_x + u', \ \partial_x^{-1} u = \sum_{i=0}^{\infty} (-)^i u^{(i)} \partial_x^{-i-1}.$$

Then there exists a unique pseudo-differential operator $L^{1/n}$ of order 1 satisfying $(L^{1/n})^n = L$. Evidently, the coefficients of $L^{1/n}$ are differential polynomials in the coefficients of L. For example, for $L = \partial_x^2 + u$, we find $L^{1/2} = \partial_x + \frac{1}{2}u\,\partial_x^{-1} - \frac{1}{4}u'\partial_x^{-2} + \ldots$. We set

$$L^{i/n} = L_+^{i/n} + L_-^{i/n},$$

where the first term on the right hand side is the differential part of the pseudo-differential operator $L^{i/n}$, and the second term on the right hand side is of order ≤ -1. Then $[L, L_+^{i/n}] = [L, L^{i/n}] - [L, L_-^{i/n}] = -[L, L_-^{i/n}]$. Since the commutator $[L, L_-^{i/n}]$ is of order at most $n-2$, this shows that the differential operators $L_+^{i/n}$ provide the desired basis.

Associated to L are then an infinite hierarchy of flows, obtained by introducing "times" $t_1, \ldots, t_m, \ldots$, and considering the evolutions of $L = \sum_{i=0}^{n} u(x; t_1, \ldots, t_n)\partial_x^i$ defined by

$$\partial_m L = [L, L_+^{m/n}]. \tag{2.6}$$

where we have denoted by ∂_m the partial derivative with respect to the time t_m. A key property of these flows is their commutativity, i.e.

$$[\partial_i - L_+^{i/n}, \partial_j - L_+^{j/n}] = 0. \tag{2.7}$$

To see this, we note first that if L evolves according to a flow $\partial_t L = [L, A]$, then L^α evolves according to $\partial_t L^\alpha = [L^\alpha, A]$. Thus we have $\partial_i L^{j/n} = [L_+^{i/n}, L^{j/n}]$, $\partial_j L^{i/n} = [L_+^{j/n}, L^{i/n}]$, and the left hand side of (2.7) can be rewritten as $-[L_+^{i/n}, L^{j/n}] + [L_+^{j/n}, L^{i/n}] + [L_+^{i/n}, L_+^{j/n}]$. If we replace $L_+^{i/n}$, $L_+^{j/n}$ by $L^{i/n} - L_-^{i/n}$ and $L^{j/n} - L_-^{j/n}$, all terms cancel, except for $[L_-^{i/n}, L_-^{j/n}]$. This term is however pseudo-differential, of order ≤ -3, and cannot occur in the left hand side of (2.7), which is manifestly a differential operator.

The flows (2.6) are known to be Hamiltonian with respect to an infinite number of symplectic structures with different Hamiltonians. For example, the KdV equation itself can be rewritten in two Hamiltonian forms

$$u_t = \partial_x \frac{\delta H}{\delta u}, \quad u_t = (\partial_x^3 + 2(u\,\partial_x + \partial_x u))\frac{\delta H'}{\delta u}$$

where the skew-symmetric operators $K = \partial_x$, $K' = \partial_x^3 + 2(u\,\partial_x + \partial_x u)$ correspond to two different symplectic structures (called respectively the Gardner-Faddeev-Zakharov [10][25] and Adler-Magri structures [1][41]), and $H = \frac{1}{4}u^3 - \frac{1}{8}u_x^2$, $H' = \frac{1}{8}u^2$ are the corresponding Hamiltonians.

The situation for the 2D zero curvature equation is much less simple, since the arguments narrowing A to an m-dimensional space of operators break down. Although formally, we may still introduce the KP hierarchy as $\partial_m L = \partial_y A_m + [A_m, L]$, with A_m an operator of order m which should also be viewed as an unknown, this is not a closed system of equations for the coefficients of L, as it was in the case of the Lax equation. Another way, due to Sato [54], is to introduce the KP hierarchy as a system of commuting flows

$$\partial_m \mathrm{L} = [\mathrm{L}_+^m, \mathrm{L}] \tag{2.8}$$

on the coefficients $(v_i(x,t))_{i=1}^{i=\infty}$ of a *pseudo-differential* operator L

$$\mathrm{L} = \partial + \sum_{i=1}^{\infty} v_i(x,t)\partial^{-i}.$$

In this form, the KP hierarchy can be viewed as a completely integrable Hamiltonian system. However, it now involves an infinite number of functions v_i, and its relation to the original KP equation (which is an equation for a single function of two variables (x,y)) requires additional assumptions.

A. Quasi-Energy and Quasi-Momentum

Our first main task is then to identify the space of differential operators with periodic coefficients on which the KP equation and its higher order analogues can be considered as completely integrable Hamiltonian systems. Our approach actually applies systematically to general 2D soliton equations. We present these results at the end of this section, and concentrate for the moment on the simplest case of a differential operator L of order n. We begin with the construction of the formal Bloch eigenfunction for two-dimensional linear operators with periodic coefficients.

Theorem 4. *Let L be an arbitrary linear differential operator of order n with doubly periodic coefficients*

$$L = \partial_x^n + \sum_{i=0}^{n-2} u_i(x,y)\partial_x^i, \qquad u_i(x+1,y) = u_i(x,y+1) = u_i(x,y) \tag{2.9}$$

Then there exists a unique formal solution $\Psi_0(x,y;k)$ of the equation

$$(\partial_y - L)\Psi_0(x,y;k) = 0, \tag{2.10}$$

which satisfies the following properties

(i) *$\Psi_0(x,y;k)$ has the form*

$$\Psi_0(x,y;k) = \left(1 + \sum_{s=1}^{\infty} \xi_s(x,y)k^{-s}\right) e^{\left(kx + k^n y + \sum_{i=0}^{n-2} B_i(y)k^i\right)}$$

(ii) $\Psi_0(0,0;k) = 1$

(iii) $\Psi_0(x, y; k)$ *is a Bloch function with respect to the variable* x, *i.e.,*

$$\Psi_0(x+1, y, k) = w_1(k)\Psi_0(x, y, k), \quad w_1(k) = e^k \ . \tag{2.11}$$

The formal solution $\Psi_0(x, y, k)$ *is then also a Bloch function with respect to the variable* y *with a Bloch multiplier* $w_2(k)$

$$\begin{aligned}\Psi_0(x, y+1; k) &= w_2(k)\Psi_0(x, y; k), \\ w_2(k) &= (1 + \sum_{s=1}^{\infty} J_s k^{-s}) e^{\left(k^n + \sum_{i=0}^{n-2} B_i(1))k^i\right)} \ .\end{aligned} \tag{2.12}$$

Proof. To simplify the notation, we begin with the proof in the case of $n = 2$, with $L = \partial_x^2 + u(x, y)$. The formal solution has then the form

$$\Psi_0(x, y; k) = \left(1 + \sum_{s=1}^{\infty} \xi_s(x, y) k^{-s}\right) e^{kx + k^2 y + B_0(y)}$$

Substituting this formal expansion in the equation $(\partial_y - L)\Psi_0(x, y; k) = 0$ gives the following equations for the coefficients ξ_s

$$\partial_y \xi_s + \partial_y B_0 \xi_s = 2\xi'_{s+1} + \xi''_s + u\xi_s. \tag{2.13}$$

(Here and henceforth, we also denote derivatives in x by primes.) These equations are solved recursively by the formula:

$$\begin{aligned}\xi_{s+1}(x, y) &= c_{s+1}(y) + \xi^0_{s+1}(x, y), \\ \xi^0_{s+1}(x, y) &= \frac{1}{2}\int_0^x (\partial_y \xi_s(\tilde{x}, y) - \xi''_s(\tilde{x}, y) + (\partial_y B_0 - u(\tilde{x}, y))\xi_s(\tilde{x}, y)) d\tilde{x}\end{aligned} \tag{2.14}$$

where $c_s(y)$ are *arbitrary* functions of the variable y with the only requirement that $c_s(0) = 0$, which is dictated by (ii).

Our next step is to show by induction that the Bloch property (ii), which is equivalent to the periodicity condition

$$\xi_s(x+1, y) = \xi_s(x, y), \tag{2.15}$$

uniquely defines the functions $c_s(y)$. Assume then that $\xi_{s-1}(y)$ is known and satisfies the condition that the corresponding function $\xi^0_s(x, y)$ is periodic. The first step of the induction, namely the periodicity in x of $\xi_1(x, y)$, requires $B_0(y)$ to be

$$B_0(y) = \int_0^y \int_0^1 u(x', y') \, dx' \, dy' \ . \tag{2.16}$$

The choice of the function $c_s(y)$ does not affect the periodicity property of $\xi_s(x, y)$, but it does affect the periodicity in x of the function $\xi^0_{s+1}(x, y)$. In order to make $\xi^0_{s+1}(x, y)$ periodic, the function $c_s(y)$ should satisfy the equation

$$\partial_y c_s = -\int_0^1 (\partial_y \xi^0_s(x, y) - (\xi^0_s)''(x, y) + (\partial_y B_0 - u(x, y))\xi^0_s(x, y)) dx \ . \tag{2.17}$$

Together with the initial condition $c_s(0) = 0$, this defines c_s uniquely

$$c_s = -\int_0^y dy' \int_0^1 (\partial_y \xi_s^0(x,y') - (\xi_s^0)''(x,y') + (\partial_y B_0 - u(x,y'))\xi_s^0(x,y'))dx, \tag{2.18}$$

and completes the induction step.

We can now establish (2.12), also by induction. Assume that ξ_{s-1} satisfies the relation

$$\xi_{s-1}(x,y+1) - \xi_{s-1}(x,y) = \sum_{i=1}^{s-1} J_i \xi_{s-i-1}(x,y) \ , \tag{2.19}$$

where $J_1, \dots, J_{s-1}$ are constants. Then (2.14) implies that

$$\xi_s(x,y+1) - \xi_s(x,y) = \sum_{i=1}^{s-1} J_i \xi_{s-i}(x,y) + J_s = \sum_{i=1}^{s} J_i \xi_{s-i}(x,y), \tag{2.20}$$

with

$$J_s = c_s(y+1) - c_s(y) - \sum_{i=1}^{s-1} J_i c_{s-i}(y).$$

We claim that J_s is actually constant. In fact, it follows from (2.14) and (2.19) that

$$\xi_s^0(x,y+1) - \xi_s^0(x,y) = \sum_{i=1}^{s-1} J_i \xi_{s-i}^0(x,y) = \sum_{i=1}^{s-1} J_i(\xi_{s-i}(x,y) - c_{s-i}(y)).$$

Thus (2.18) implies that

$$\begin{aligned}\partial_y c_s(y+1) - \partial_y c_s(y) &= -\sum_{i=1}^{s-1} J_i \int_0^1 (\partial_y \xi_{s-i}^0 - (\xi_s^0)'' + (\partial_y B_0 - u)\xi_s^0)dx \\ &= \sum_{i=1}^{s-1} J_i(\partial_y c_{s-i} - \xi_{s+1}^0(1,y) - \xi_{s+1}^0(0,y)) = \sum_{i=1}^{s-1} J_i \partial_y c_{s-i}.\end{aligned}$$

In particular the derivative of J_s vanishes. This proves Theorem 1 when $n = 2$.

The proof can be easily adapted to the case of general n. Let $\xi_s^0(x,y)$ be the coefficients of the formal series

$$\Psi^{(0)}(x,y,k) = \frac{\Psi_0(x,y,k)}{\Psi_0(0,y,k)} = (1 + \sum_{s=1}^{\infty} \xi_s^0(x,y,k)k^{-s})e^{kx}.$$

Then $(\partial_y - L)\Psi_0 = 0$ is equivalent to the system of equations

$$\sum_{l=0}^{n-1} C_n^l \left(\partial_x^{n-l} \xi_{s+l}^0\right) + \sum_{i=0}^{n-2} u_i \sum_{l=0}^{i} C_i^l \left(\partial_x^{i-l} \xi_{s+l}^0\right) = \partial_y \xi_s^0 + \sum_{j=-s}^{n-2} b_j \xi_{j+s}^0, \quad s \geq -n+2, \tag{2.21}$$

where we have assumed that $\xi_s^0 = 0$ for $s < 0$, and set $b_j = b_j(y)$ to be the coefficients of the series

$$\frac{\partial_y \Psi_0(0,y,k)}{\Psi_0(0,y,k)} = k^n + \sum_{j=-\infty}^{n-2} b_j(y) k^j. \tag{2.22}$$

These equations are of the form

$$n\partial_x \xi^0_{s+n-1} = b_{-s} + F_s(\xi^0_{s'}, b_{s''}) \; . \tag{2.23}$$

Here F is a linear combination of the $\xi^0_{s'}$, $s' \leq s+n-2$ and their derivatives, with coefficients which are themselves linear in u_i and $b_{s''}$, $s'' < s$ (c.f. (2.13)). The equations define recursively $b_{-s}(y)$ and ξ^0_{s+n-1}. The coefficient b_{-s} follows from the periodicity in x of (2.23)

$$b_{-s}(y) = -\int_0^1 F_s(\xi^0_{s'}, b_{s''})dx, \tag{2.24}$$

and the coefficient ξ^0_{s+n-1} follows in turn

$$n\xi^0_{s+n-1} = b_{-s}(y)x + \int_0^x F_s(\xi^0_{s'}, b_{s''})dx'. \tag{2.25}$$

We can now integrate (2.22) and find that Ψ_0 has the form (i) with

$$B_i(y) = \int_0^y b_i(y), \quad i = 0, \ldots, n-2. \tag{2.26}$$

We obtain at the same time the Bloch property for Ψ_0 with respect to y with the Bloch multiplier

$$w_2(k) = \int_0^1 \left(k^n + \sum_{i=-\infty}^{n-2} b_i(y)k^i\right)dy. \tag{2.27}$$

The proof of Theorem 1 is complete. □

In Theorem 1, we have chosen the simple form $w_1(k) = e^k$ for the Bloch multiplier in x. If we view x as a "space" variable, this identifies the spectral parameter k with the "quasi-momentum" (up to a factor of $i = \sqrt{-1}$). The variable y can then be interpreted as a time variable in the Schrödinger-like equation $(\partial_y - L)\Psi_0 = 0$. This identifies (again up to a factor of i) the logarithm of the second Bloch multiplier $w_2(k)$ with the quasi-energy $E(k)$. Alternatively, we may change spectral parameters, and introduce the spectral parameter K as well as the coefficients $(\mathcal{E}_i)_{i=-n+2}^{\infty}$ of the expansion in k of the quasi-energy by

$$\begin{aligned} K^n &= E(k) = \log w_2(k) = k^n + \sum_{i=-n+2}^{\infty} \mathcal{E}_i k^{-i} \; , \\ K &= k + O(k^{-1}) \; . \end{aligned} \tag{2.28}$$

We observe that a change of spectral parameter of the form $k \to K = k + O(k^{-1})$ transforms a formal series in k of the form (i) in Theorem 1 into another formal series of the same form in K. With K as spectral parameter, the second Bloch multiplier $w_2(K)$ reduces to $w_2(K) = \exp(K^n)$, but the first Bloch multiplier $w_1(K)$ becomes non-trivial. The coefficients $(H_s)_{s=1}^{\infty}$ of the expansion in K of its logarithm

$$w_1(K) = e^{k(K)}, \quad k(K) = K + \sum_{s=1}^{\infty} H_s K^{-s} \tag{2.29}$$

will play an important role in the sequel.

B. Basic Constraints

The coefficients $\mathcal{E}_i$ and H_s of the expansions (2.28) and (2.29) of the spectral parameters are uniquely defined by the coefficients $(u_i)_{i=0}^{n-2}$ of the operator L, and hence can be considered as functionals on the space of periodic operators. We now restrict ourselves to the subspace $\mathcal{L}(b)$ of operators L satisfying the constraints

$$b_i(y) = b_i, \ 0 \leq i \leq n-2, \tag{2.30}$$

where $b_i(y) = dB_i(y)/dy$, with the functions $B_i(y)$ defined by (i) in Theorem 1, and $b = (b_0, \ldots, b_{n-2})$ are $(n-1)$ fixed constants. On the space $\mathcal{L}(b)$, the essential singularity in the second Bloch multiplier $w_2(k)$ simplifies to $k^n + \sum_{i=0}^{n-2} b_i k^i$ (c.f. (2.12)). Comparing with (2.28), we see that the constraints (2.30) fix the values of the first $(n-1)$ functionals $\mathcal{E}_{-i} = b_i$. This is in turn clearly equivalent to fixing the values of the first $(n-1)$ functionals $H_1, \ldots, H_{n-1}$ in (2.29). We claim that these constraints can also be expressed under the form

$$\int_0^1 h_i(u_j)dx = H_s = \text{const}, \quad i = 1, \ldots, n-1, \tag{2.31}$$

where $h_s(u_j)$, $j \geq (n-i-1)$, are universal differential polynomials depending only on n. In fact, the constraints (2.30) imply that the essential singularity in the Bloch function $\Psi_0(x, y; k)$ is of the form

$$kx + \left(k^n + \sum_{i=0}^{n-2} b_i k^i\right) y = Kx + K^n y + O(K^{-1}) \ .$$

Since the expression $\exp\left(O(K^{-1})\right)$ contains no essential singularity and can be expanded as a formal series in K^{-1}, we have shown that the formal Bloch solution $\Psi_0(x, y; k)$ can be rewritten in terms of the spectral parameter K under the form

$$\Psi_0(x, y, K) = \left(1 + \sum_{s=1}^{\infty} \zeta_s(x, y) K^{-s}\right) e^{Kx + K^n y} \ . \tag{2.32}$$

Substituting (2.32) in $(\partial_y - L)\Psi_0(x, y; K) = 0$ gives for the first $(n-1)$ coefficients $\zeta_1, \ldots, \zeta_{n-1}$ a system of ordinary differential equations

$$\sum_{l=0}^{n-1} C_n^l \left(\partial_x^{n-l}\zeta_{s+l}^0\right) + \sum_{i=0}^{n-2} u_i \sum_{l=0}^{i} C_i^l \left(\partial_x^{i-l}\zeta_{s+l}^0\right) = 0, \quad s = -n+2 \ldots, 0, \tag{2.33}$$

which just coincides with the first equations defining formal eigenfunctions for ordinary differential operators (see [33]). Let $\zeta_i(x, y; x_0)$, be the solution of (2.33) with the normalization $\zeta_0 = 1$ and $\zeta_i(x_0, y; x_0) = 0$. Then the equations (2.33) define recursively differential polynomials $h_i(u_j(x, y))$ such that

$$\partial_x \zeta_i(x, y; x_0)|_{x=x_0} = h_i(u_j(x, y))|_{x=x_0}. \tag{2.34}$$

The above left hand side is equal to the first coefficients of the logarithmic derivative of Ψ_0 at $x = x_0$, i.e.

$$K + \sum_{s=1}^{n-1} \left(\partial_x \zeta_s(x, y; x_0)|_{x_0=x}\right) K^{-s} = \frac{\partial_x \Psi_0(x, y, K)}{\Psi_0(x, y, K)} + O(K^{-n}). \tag{2.35}$$

Integrating gives the first (constant) coefficients of (2.29) and establishes our claim.

Henceforth we will always assume that L is in the space $\mathcal{L}(b)$ of operators with periodic coefficients $(u_i)_{i=0}^{n-2}$ satisfying either one of the equivalent constraints (2.30) or (2.31).

C. Commuting Flows

On the space $\mathcal{L}(b)$ we can now define an infinite set of mutually commuting flows as follows. First, we observe that for any formal series of the form (i) in Theorem 1, there exists for each integer $m \geq 1$ a unique differential operator A_m of the form

$$A_m = \partial_x^m + \sum_{i=0}^{m-2} u_{i,m}(x, y)\partial_x^i, \tag{2.36}$$

which satisfies the condition

$$(A_m - K^m(k))\Psi_0(x, y; k) = O(k^{-1})\Psi_0(x, y; k), \tag{2.37}$$

where $K(k)$ is defined by (2.28). Indeed, this condition is equivalent to the following finite system of equations for the coefficients $(u_{i,m})$ of A_m

$$\sum_{i=0}^{m} u_{i,m} \sum_{l=0}^{i} C_i^l (\partial_x^{i-l}\zeta_{s+l}) = \zeta_{s+m}, \quad s = -m+2, \ldots, 0.$$

This system is triangular, and identifies uniquely the coefficients $u_{i,m}$ as differential polynomials in the first $m-1$ coefficients ξ_s of $\Psi_0(x, y; k)$. For example, we find

$$u_{m-2,m} = -m\zeta_1', \quad u_{m-3,m} = -m\zeta_2' - C_m^2\zeta_1''.$$

Let L be now an operator in $\mathcal{L}(b)$, $\Psi_0(x,y;k)$ its Bloch function, and A_m be the operators obtained by the preceding construction. Then

$$[\partial_y - L, A_m]\Psi_0(x,y,k) = (\partial_y - L)O(k^{-1})\Psi_0(x,y,k) = O(k^{n-2})e^{Kx+K^n y}, \tag{2.38}$$

which implies that the operator $[\partial_y - L, A_m]$ has order less or equal to $n-2$. We set

$$[\partial_y - L, A_m] = \partial_y A_m - [L, A_m] = \sum_{i=0}^{n-2} F_{i,m}(x,y)\partial_x^i \; . \tag{2.39}$$

The functions $F_{i,m}(x,y)$ are uniquely defined by $u_j(x,y)$ and can by expressed in terms of multiple integrals of differential polynomials in u_j. Thus they can also be expressed as

$$F_{i,m}(x,y) = \hat{F}_{i,m}(u(x,y)), \quad u = (u_0, \dots, u_{n-2}). \tag{2.40}$$

where $\hat{F}_{i,m}(.)$ is *a nonlocal* but exact functional of $u_j(x,y)$.

Theorem 5. *The system of equations*

$$\partial_m u_i = \hat{F}_{i,m}(u) \longleftrightarrow \partial_m L = \partial_y A_m + [A_m, L] \tag{2.41}$$

defines an infinite set of commuting flows on the subspace $\mathcal{L}(b)$ *of doubly periodic operators.*

Since $\hat{F}_{i,m}$ and A_m are well-defined functions of L, we need only check the commutativity of the flows. For this, we need the following version of the uniqueness of Bloch solutions, namely, that if $\Psi(x,y;k)$ is a Bloch solution ($\Psi(x+1,y;k) = w_1(k)\Psi(x,y;k)$) of the equation $(\partial_y - L)\Psi = 0$ having the form

$$\Psi_0(x,y;k) = \left(\sum_{s=-N}^{\infty} \xi_s(x,y)k^{-s}\right) e^{kx+(k^n+\sum_{i=0}^{n-2} b_i k^i)y}, \tag{2.42}$$

then $\Psi(x,y;k)$ must be given by

$$\Psi(x,y;k) = a(k)\Psi_0(x,y;k), \tag{2.43}$$

where $a(k)$ is a constant Laurent series $a(k) = \sum_{s=N}^{\infty} a_s k^{-s}$. This is because the leading coefficient ξ_N in any formal solution Ψ is a constant. As a consequence, if $\Psi(x,y;k)$ is a Bloch solution, then the ratio $\Psi(x,y;k)\Psi^{-1}(0,0;k)$ is also a solution. This expression has all the properties of, and can be identified with Ψ_0. Since $\Psi(0,0;k)$ is a Laurent series, our assertion follows.

Returning to the proof of Theorem 2, we observe that if Ψ_0 is the Bloch solution to (2.22) then $(\partial_m - A_m)\Psi_0$ is also a Bloch solution to the same equation. The preceding uniqueness property implies

$$(\partial_m - A_m)\Psi_0 = -\tilde{\Omega}_m(k)\Psi_0, \quad \tilde{\Omega}_m(k) = K^m + O(k^{-1}) \tag{2.44}$$

for a suitable (x, y) independent Laurent series $\Omega_m(k)$. In particular, $[\partial_m - A_m, \partial_n - A_n]\Psi_0 = O(k^{-1})\Psi_0$. The last equality implies that $[\partial_m - A_m, \partial_n - A_n]$ is an ordinary differential operator in x of order less than zero. Therefore, it must vanish identically

$$[\partial_m - A_m, \partial_n - A_n] = 0 \; . \tag{2.45}$$

This establishes the commutativity of the flows. The proof of Theorem 2 is complete.

Conversely, let $u(x, y, t, k)$, $t = (t_1, t_2, \ldots)$ be a solution to the hierarchy (2.41). Then there exists a unique formal Bloch solution $\Psi(x, y, t; K)$ of the equations

$$(\partial_y - L)\Psi = 0, \quad (\partial_m - A_m)\Psi = 0 \tag{2.46}$$

having the form

$$\Psi(x, y, t, K) = \left(1 + \sum_{s=1}^{\infty} \zeta_s(x, y, t) K^{-s}\right) e^{Kx + K^n y + \sum_m K^m t_m}. \tag{2.47}$$

The above expression identifies the original variables x and y with the first and the n-th *times* of the hierarchy,

$$x = t_1, \quad y = t_n, \tag{2.48}$$

respectively. More generally, the preceding results show that for periodic operatos L, an equation of the zero curvature form $[\partial_y - L, \partial_t - A] = 0$ must be equivalent to a pencil of equations for the coefficients of L only. In other words, there must exist constants c_i such that

$$A = \sum_{i=1}^{m} c_i A_i \tag{2.49}$$

and the flow is along the basic times $t_i = c_i t$, of the hierarchy.

Finally, we point out that for all n, the equations of the corresponding hierarchy for $L = L_n$ have the same form $[\partial_n - L_n, \partial_m - L_m] = 0$, and can be considered as reductions of this system. Our approach to these reductions is to select two particular times which we treat as spatial variables, and to impose periodicity conditions in these variables.

D. Dual Formal Bloch Solutions

A key ingredient in our construction of symplectic structures on spaces of periodic operators L is the notion of dual Bloch functions $\Psi_0^*(x, y; k)$. In the one-dimensional case, dual Bloch functions were introduced in [7]. In our setup, its main properties are as follows:

• Let $\Psi_0(x,y,k)$ be a formal series of the form (i) in Theorem 1. Then there exists a unique formal series $\Psi_0^*(x,y;k)$ of the form

$$\Psi_0^*(x,y;k) = e^{-kx-(k^n+\sum_{i=0}^{n-2} b_i k^i)y}\left(1+\sum_{s=1}^{\infty}\xi_s^*(x,y)k^{-s}\right), \tag{2.50}$$

such that for all non-negative integers m the equalities

$$\mathrm{Res}_\infty\left(\Psi_0^*(x,y;k)\partial_x^m\Psi_0(x,y;k)dk\right)=0, \quad m=0,1,\ldots \tag{2.51}$$

are fulfilled;

• If Ψ_0 is a Bloch function with Bloch multipliers $w_i(k)$, then $\Psi_0^*(x,y;k)$ is a Bloch function as well with inverse Bloch multipliers

$$\Psi_0^*(x+1,y;k)=w_1^{-1}(k)\Psi_0^*(x,y;k), \quad \Psi_0^*(x,y+1;k)=w_2^{-1}(k)\Psi_0^*(x,y;k), \tag{2.52}$$

• If $\Psi_0(x,y;k)$ is a solution to the equation $(\partial_y - L)\Psi_0 = 0$, then the series $\Psi_0^*(x,y;k)$ is a solution to the adjoint equation

$$\Psi_0^*(\partial_y - L) = 0, \tag{2.53}$$

where the action on the left of a differential operator is defined as a formal adjoint action, i.e. for any function f^*

$$(f^*\partial_x^i) = (-\partial_x)^i f^*. \tag{2.54}$$

To see this, we begin by noting that, although each of the factors in (2.51) has an essential singularity, their product is a meromorphic differential and the residue is well-defined. It has the form

$$\mathrm{Res}_\infty\left(\Psi_0^*\partial_x^m\Psi_0\,dk\right)=\xi_m+\xi_m^*+g_m(\xi_1,\ldots,\xi_{m-1};\xi_1^*,\ldots,\xi_{m-1}^*),$$

where g_m is linear in ξ_s^*, in ξ_s and their derivatives, $s<m$. The condition (2.51) defines then ξ_m recursively as differential polynomials in ξ_s, $s=1,\ldots,m$. For example, we have

$$\xi_1^*=-\xi_1, \quad \xi_2^*=-\xi_2+\xi_1^2-\xi_1'.$$

This shows the existence and uniqueness of Ψ_0^*. Since the second statement is a direct corollary of the uniqueness of Ψ_0^*, we turn to the proof of the last statement. First, we show that if $\Psi^*(x,y;k)$ is a formal series

$$\Psi^*(x,y;k) = e^{-kx-(k^n+\sum_{i=0}^{n-2} b_i k^i)y}\left(\sum_{s=-N}^{\infty}\xi_s^*(x,y)k^{-s}\right), \tag{2.55}$$

satisfying the equations (2.51), then there exists a unique degree N ordinary linear differential operator D such that

$$\Psi^*(x,y;k)=\Psi_0^*(x,y;k)D.$$

Since $\partial_x^i \Psi_0^*$ satisfies the equations (2.51), we can find D satisfying the condition

$$\Psi^*(x,y;k) - \Psi_0^*(x,y;k)D = O(k^{-1})\Psi_0^*(x,y;k).$$

The above right hand side has the form (2.55) with $N < 0$ and satisfies (2.51). Evaluating the leading term, we find that it must vanish identically.

Let Ψ_0 be a solution of $(\partial_y - L)\Psi_0 = 0$. Then

$$\mathrm{Res}_\infty\,(\partial_y \Psi_0^* \partial_x^m \Psi_0\, dk) = \partial_y\, \mathrm{Res}_\infty\,(\Psi_0^* \partial_x^m \Psi_0\, dk) - \mathrm{Res}_\infty\,(\Psi_0^* \partial_x^m L \Psi_0\, dk) = 0$$

In particular, there exists a differential operator $\tilde{L}$ such that

$$\partial_y \Psi_0^* = \Psi_0^* \tilde{L}\ .$$

Let $f(x)$ be an arbitrary periodic function on one variable. We have

$$\partial_y \langle f(x) \Psi_0^* \Psi_0 \rangle_x = \langle f(x) \Psi_0^* (\tilde{L} + L) \Psi_0) \rangle_x,$$

where we have denoted as usual the average value in x of any periodic function $g(x)$ by $\langle g \rangle_x$. The above left hand side is of order -1 in k. On the other hand, if $L + \tilde{L}$ is not equal to zero and $g_i(x,y)$, $0 \le i \le n-2$ are its leading coefficients, then the right hand side is of the form $\langle f(x) g_i(x,y) \rangle_x k^i + O(k^{i-1})$. This implies that $\langle f g_i \rangle_x = 0$. Since f was arbitrary, we conclude that $g_i = 0$, establishing the last desired property of dual Bloch functions.

We conclude our discussion of dual Bloch functions with several useful remarks. The first is that the identity

$$\langle \Psi_0^*(x,y;k) \Psi_0(x,y;k) \rangle_x = 1. \tag{2.56}$$

holds for any formal series $\Psi_0(x,y;k)$ of the form (i) in Theorem 1 and its dual Bloch series $\Psi_0^*(x,y;k)$. Indeed, just as in Section II.C, we can show the existence of a unique pseudo-differential operator $\Phi = 1 + \sum_{s=0}^\infty w_s(x,y)\partial_x^{-s}$ so that

$$\Psi_0(x,y;k) = \Phi e^{kx + (k^n + \sum_{i=0}^n b_i k^i)y}\ . \tag{2.57}$$

As in [10], this implies

$$\Psi_0^*(x,y;k) = \left(e^{-kx - (k^n - \sum_{i=0}^n b_i k^i)y}\right) \Phi^{-1}. \tag{2.58}$$

More precisely, let $Q = \sum_{s=N}^\infty q_s(x)\partial_x^{-s}$ be a pseudo-differential operator. Then we may define its residue $\mathrm{res}_\partial\, Q$ by

$$\mathrm{res}_\partial\, Q = \mathrm{Res}_\infty (e^{-kx}(Qe^{kx}))dk = -q_1.$$

The point is that, while the ring of pseudo-differential operators is not commutative, the residue is, after averaging

$$\langle \mathrm{res}_\partial (Q_1 Q_2) \rangle_x = \langle \mathrm{res}_\partial (Q_2 Q_1) \rangle_x.$$

This shows that the series defined by the right hand side of (2.58) satisfies (2.51), and hence must coincide with Ψ_0^*. The desired identity (2.56) is now a direct consequence of the two preceding identities and of the associativity of the left and right actions under averaging.

Secondly, we would like to stress that, although Ψ_0^* is a Bloch solution of the adjoint equation $\Psi_0^*(\partial_y - L) = 0$, its normalization is different from that used for Ψ_0. This symmetry may be restored if we introduce

$$\Psi_0^+(x, y; k) = \frac{\Psi_0^*(x, y; k)}{\Psi_0^*(0, 0; k)}. \tag{2.59}$$

The inverse relation is then

$$\Psi_0^*(x, y; k) = \frac{\Psi_0^+(x, y; k)}{\langle \Psi_0^+ \Psi_0 \rangle_x}.$$

Finally, the definition of the action on the left of a differential operator adopted earlier implies that for any degree N differential operator

$$D = \sum_{i=0}^{N} w_i(x) \partial_x^i$$

there exist degree $(N - i)$ differential operators $D^{(i)}$ such that for any pair of functions f^+ and g the equality

$$(f^* D) g = f^*(Dg) + \sum_{i=0}^{n} \partial_x^i (f^*(D^{(i)} g))$$

holds. The set of operators $D^{(i)}$ was introduced in [36]. Of particular interest is of course $D^{(0)} = D$, and the "first descendant" of D, namely

$$D^{(1)} = \sum_{i=0}^{n} i w_i(x) \partial_x^{i-1}. \tag{2.60}$$

E. The Basic Symplectic Structure

We are now in position to introduce a symplectic structure on the space $\mathcal{L}(b)$ of periodic operators L subject to the constraints (2.30), and to show that the infinite set of commuting flows constructed in Theorem 2 are Hamiltonian.

The main ingredients are the one-forms δL and $\delta \Psi_0$. The one-form δL is given by

$$\delta L = \sum_{i=0}^{n-2} \delta u_i \partial_x^i,$$

and can be viewed as an operator-valued one-form on the space of operators $L = \partial_x^n + \sum_{i=0}^{n-2} u_i \partial_x^i$. Similarly, the coefficients of the series Ψ_0 are explicit integro-differential polynomials in u_i. Thus $\delta\Psi_0$ can be viewed as a one-form on the space of operators with values in the space of formal series. More concretely, we can write

$$\delta\Psi_0 = \left(\sum_{s=1}^{\infty} \delta\xi_s k^{-s}\right) e^{kx+(k^n+\sum_{i=0}^{n-2} b_i k^i)y} = \left(\sum_{s=1}^{\infty} \delta\zeta_s K^{-s}\right) e^{Kx+K^n y}.$$

The coefficients $\delta\xi_s$ (or $\delta\zeta_s$) can be found from the variations of the formulae (2.24), (2.25) for ξ_s, or recursively from the equation

$$(\partial_y - L)\delta\Psi_0 = (\delta L)\Psi_0. \tag{2.61}$$

Let $f(x, y)$ be a function of the variables x and y. We denote its mean value by

$$\langle f \rangle = \int_0^1 \int_0^1 f(x, y) dx\, dy.$$

Theorem 6. (a) *The formula*

$$\omega = \text{Res}_\infty \langle \Psi_0^* \delta L \wedge \delta\Psi_0 \rangle dk \tag{2.62}$$

defines a symplectic form, i.e., a closed non-degenerate two-form on the space $\mathcal{L}(b)$ *of operators* L *with doubly periodic coefficients.* (b) *The form* ω *is actually independent of the normalization point* $(x_0 = 0, y_0 = 0)$ *for the formal Bloch solution* $\Psi_0(x, y; k)$. (c) *The flows (2.41) are Hamiltonian with respect to this form, with the Hamiltonians* $2nH_{m+n}(u)$ *defined by (2.29).*

Proof. We require the following formula, which is a generalization of the well-known expression for the variation of energy for one-dimensional operators. Let $E(k)$ be the quasi-energy which is defined by (2.28). Its coefficients are nonlocal functionals on the space $\mathcal{L}(b)$ of periodic functions $u_i(x, y)$ subject to the constraints (2.30). Then we have

$$\delta E(k) = \langle \Psi_0^* \delta L \Psi_0 \rangle\ . \tag{2.63}$$

Indeed, from the equation $(\partial_y - L)\Psi_0 = 0$ and (2.53), it follows that

$$\partial_y \langle \Psi_0^* \delta\Psi_0 \rangle_x = \langle \Psi_0^* \delta L \Psi_0 \rangle_x.$$

Taking the integral over y and using the following monodromy property of $\delta\Psi_0$

$$\delta\Psi_0(x, y+1; k) = w_2(k)(\delta\Psi_0(x, y; k) + \delta E(k)\Psi_0).$$

we obtain (2.63).

We begin by checking that the form $\langle \Psi_0^* \delta L \wedge \delta\Psi_0) \rangle_x$ is periodic in y. The shift of the argument $y \to y + 1$ gives

$$\langle \Psi_0^* \delta L \wedge \delta\Psi_0 \rangle_x \longrightarrow \langle \Psi_0^* \delta L \wedge \delta\Psi_0 \rangle_x + \langle \Psi_0^* \delta L \Psi_0 \rangle_x \wedge \delta E\ .$$

The second term on the right hand side can be rewritten as $\delta E \wedge \delta E$ and hence vanishes, due to the skew-symmetry of the wedge product.

Next, we show that (b) is a consequence of the basic constraints defining the space $\mathcal{L}(b)$. Let Ψ_1 be the formal Bloch solution with the normalization $\Psi_1(x_1, y_1, k) = 1$. Then

$$\Psi_1(x, y; k) = \Psi_0(x, y; k)\Psi_0^{-1}(x_1, y_1; k)$$

and

$$\mathrm{Res}_\infty \langle \Psi_1^* \delta L \wedge \delta \Psi_1 \rangle dk = \omega - \mathrm{Res}_\infty \left(\delta E \wedge \frac{\delta \Psi_0(x_1, y_1; k)}{\Psi_0(x_1, y_1; k)} \right) dk.$$

In view of the constraints (2.30), we have $\delta E = O(k^{-1})$. On the other hand, the second factor in the last term of the above right hand side also has order $O(k^{-1})$. The product has therefore order $O(k^{-2})$ and its residue equals zero.

To see that ω is a closed form, we express the operator L as

$$L = \Phi D \Phi^{-1} + O(\partial_x^{-1}), \ D = \partial_x^n + \sum_{i=0}^{n-2} b_i \partial_x^i, \tag{2.64}$$

which can be done in view of (2.57) and (2.58). Therefore

$$\omega = \langle \mathrm{res}_\partial (D\Phi^{-1}\delta\Phi\Phi^{-1} \wedge \delta\Phi) \rangle = -\delta \langle \mathrm{res}_\partial (D\Phi^{-1}\delta\Phi) \rangle \tag{2.65}$$

and ω is closed.

We turn now to the non-degeneracy of ω on $\mathcal{L}(b)$. Let V be a vector field such that $\omega(V_1, V) = 0$ for all vector fields V_1. Let $\Psi_1 = \delta\Psi(V)$ be the evaluation of the one-form $\delta\Psi$ on V. Then the equality

$$\omega(V_1, V) = \mathrm{Res}_\infty \langle \Psi_0^* L_1 \Psi_1 \rangle dk = 0, \tag{2.66}$$

holds for all degree $n-2$ operators $L_1 = \delta L(V_1)$. Since L_1 is arbitrary, it follows that $\Psi_1 = O(k^{-n})\Psi_0$. In view of (2.61) we have then

$$\delta L(V)\Psi_0 = (\partial_y - L)\Psi_1 = O(k^{-1})\Psi_0.$$

Hence $\delta L(V) = 0$. This means that $V = 0$, and the non-degeneracy of ω is established.

It remains to exhibit the flows (2.41) as Hamiltonian flows. We recall the classical definition of the Hamiltonian vector field ∂_t corresponding to a Hamiltonian H and a two-form ω. The contraction $i(\partial_t)\omega$ of ω with the vector field ∂_t should be the one-form given by the differential of the Hamiltonian, i.e. the equality

$$i(\partial_t)\omega(X) = \omega(X, \partial_t) = dH(X), \tag{2.67}$$

should be fulfilled for all vector-fields X.

The contraction of the form ω defined by (2.62) with the vector-field ∂_m (2.41) is equal to

$$i(\partial_m)\omega = \mathrm{Res}_\infty \langle \Psi_0^* \delta L \partial_m \Psi_0 \rangle dk - \mathrm{Res}_\infty \langle \Psi_0^* (\partial_m L) \delta \Psi_0 \rangle dk. \tag{2.68}$$

Here we use the fact that the evaluations of the forms δL and $\delta\Psi_0$ on the vector field ∂_m are equal by definition to

$$\delta L(\partial_m) = \partial_m L, \quad \delta\Psi_0(\partial_m) = \partial_m \Psi_0.$$

From (2.44) it follows that

$$\mathrm{Res}_\infty \langle \Psi_0^* \delta L \partial_m \Psi_0 \rangle dk = \mathrm{Res}_\infty \Big(\langle \Psi_0^* \delta L A_m \Psi_0 \rangle + \tilde{\Omega}_m(k) \langle \Psi_0^* \delta L \Psi_0 \rangle \Big)\, dk.$$

The first term in the right hand side is zero due to the definition of Ψ_0^*. The usual formula for the implicit derivative

$$\delta E(k)\, dk = -\delta k(K)\, dE, \tag{2.69}$$

implies that the second term is equal to

$$\begin{aligned} -\mathrm{Res}_\infty \tilde{\Omega}_m(K) \delta k(K)\, dE &= -\mathrm{Res}_\infty \left(K^m + O(K^{-1})\right) \left(\sum_{s=n}^{\infty} \delta H_s k^{-s} \right) dK^n \\ &= n\, \delta H_{n+m}. \end{aligned} \tag{2.70}$$

(Recall that $\delta H_s = 0,\ s < n$ due to the constraints.) Consider now the second term in the right hand side of (2.68). The equation (2.41) for $\partial_m L$ and the defining equations for Ψ_0 and Ψ_0^* imply

$$\mathrm{Res}_\infty \langle \Psi_0^* (\partial_m L) \delta \Psi_0 \rangle_x dk = \mathrm{Res}_\infty \partial_y \left(\langle \Psi_0^* A_m \delta \Psi_0 \rangle_x \right) dk.$$

Therefore,

$$\int_0^1 \partial_y \langle \Psi_0^* A_m \delta \Psi_0 \rangle_x dy = \delta E(k) \langle \Psi_0^* A_m \Psi_0 \rangle_x |_{y=0}.$$

The equality (2.37) implies

$$\langle \Psi_0^* A_m \Psi_0 \rangle_x = K^m + O(K^{-1}).$$

Hence, the second term in (2.68) is equal to

$$\mathrm{Res}_\infty \langle \Psi_0^* (\partial_m L) \delta \Psi_0 \rangle dk = -n \delta H_{m+n}$$

and Theorem 3 is proved. □

Example 1. For $n = 2$, the operator L is the second order differential operator of the form $L = \partial_x^2 + u(x, y)$. The space $\mathcal{L}(b_0)$ is the space of periodic functions with fixed mean value in x

$$\langle u \rangle_x = H_1 \to \langle \delta u \rangle_x = 0.$$

and the symplectic form ω becomes

$$\omega = -\left\langle \delta u \wedge \int_{x_0}^{x} \delta u \, dx \right\rangle .$$

This symplectic form reduces to the Gardner-Faddeev-Zakharov symplectic form when $u(x, y) = u(x)$ is a function of a single variable x. In this case the KP equation reduces to the KdV equation.

Example 2. For $n = 3$, the operator L is the third order differential operator $L = \partial_x^3 + u\partial_x + v$. The space $\mathcal{L}(b_0, b_1)$ is the space of doubly periodic functions $u = u(x, y)$, $v = v(x, y)$ satisfying the constraints

$$\langle u \rangle_x = \text{const}, \ \langle v \rangle_x = \text{const}.$$

The symplectic form ω works out to be

$$\omega = -\frac{2}{3}\left(\delta u \wedge \int_{x_0}^{x} \delta v \, dx + \delta v \wedge \int_{x_0}^{x} \delta u \, dx \right) .$$

In the case where u and v are functions of a single variable x, this form gives a symplectic structure for the Boussinesq equation hierarchy

$$u_t = 2v_x - u_{xx}, \ v_t = v_{xx} - \frac{2}{3}u_{xxx} - \frac{2}{3}uu_x.$$

Note that the usual form of the Boussinesq equation, $u_{tt} + \left(\frac{4}{3}uu_x + \frac{1}{3}u_{xxx}\right)_x = 0$, as an equation in one unknown function u, is the result of eliminating v from the above system.

F. Lax Equations

In this section, we compare the results obtained in our formalism with the one-dimensional case, where the zero curvature equation reduces to the Lax equation, and where there is a rich theory of Hamiltonian structures. It turns out that the symplectic structure constructed above reduces then to the so-called first (or generalized) Gardner-Faddeev-Zakharov symplectic structure. Thus our approach gives a new representation for this structure, as well as a new proof of its well-known properties (c.f. [10][25]). As we shall see below, the second (Adler-Magri) symplectic structure requires a slight modification, which explains why it is special to the one dimensional case and has no analogue in the proposed Hamiltonian theory of two-dimensional systems.

Our construction of the basic symplectic form ω easily extends to the construction of an infinite sequence of symplectic structures:

Theorem 7. *Let l be any integer ≥ 0. Then the formula*

$$\omega^{(l)} = \text{Res}_{\infty}\ E^{l}(k)\langle \Psi_0^* \delta L \wedge \delta \Psi_0 \rangle_x dk, \tag{2.71}$$

defines a closed two-form on the space $\mathcal{L}(H_1, \ldots, H_{nl-1})$ of doubly periodic operators L, subject to the constraints $H_s = const$, $s = 1, \ldots, nl-1$. The equations (2.8) are Hamiltonian with respect to this form, with the Hamiltonians $2nH_{m+n(l+1)}(u)$ defined by (2.29).

The proof of the theorem is identical to the proof for the basic structure ω. Specializing to the subspace of periodic L with coefficients depending only on x, we can easily verify that the symplectic forms $\omega = \omega^{(0)}$ coincide with the generalized Gardner-Faddeev-Zakharov forms.

The construction of the Adler-Magri symplectic structure is less obvious, although formally it has the form (2.71) with $l = -1$ and the residue at infinity is replaced by the residue at $E = 0$. Let L be an ordinary linear differential operator of order n with periodic coefficients. Then for generic values of the complex number E, there exist n linearly independent Bloch solutions $\Psi_i(x, E)$ of the equation

$$(L - E)\Psi_i = 0, \tag{2.72}$$

with different Bloch multipliers $w_i(E)$,

$$\Psi_i(x+1, E) = w_i(E)\Psi_i(x, E).$$

The value $p_i(E) = \log w_i(E)$ is called the quasi-momentum. Its differential $dp_i(E)$ is well-defined. (In our previous formal theory of Bloch solutions, there are also n different solutions corresponding to the same E, due to the relation $E = k^n + O(k^{n-2}) = K^n$ which defines k and K only up to a root of unity.) We fix the Bloch solutions Ψ_j^* of the adjoint equation

$$\Psi_i^*(L - E) = 0$$

by the condition

$$\langle \Psi_i^*(x, E)\Psi_j(x, E)\rangle_x = \delta_{ij}\ .$$

Theorem 8. *The formula*

$$\omega^{(-1)} = \sum_{i=1}^{n} R_i^{-1}\langle \Psi_i^*(x, 0)\delta L \wedge \delta \Psi_i(x, 0)\rangle, \tag{2.73}$$

where the constants R_i are given by

$$R_i = \langle \Psi_i^* L^{(1)} \Psi_i \rangle,$$

defines a closed two-form on the space of operators L with coefficients depending only on the variable x, and obeying the constraints $H_s = const$, $s = 1, \ldots, n-1$. The equations (2.8) are Hamiltonian with respect to this symplectic form, with the Hamiltonians $nH_m(u)$ defined by (2.29).

We observe that for $n = 2$, the operator $L^{(1)}$ reduces to $L^{(1)} = 2\partial_x$. The expression $\langle\Psi_i^* L^{(1)}\Psi_i\rangle_x$ is then just the Wronskian of two solutions of the Schrödinger equation.

The proof of Theorem 5 is analogous to the proof of Theorem 3. The formula (2.73) can be rewritten in the form

$$\omega^{(-1)} = -\operatorname{Res}_0 \frac{dE}{E} \sum_{i=1}^{n} \frac{\langle \Psi_i^*(x,E)\delta L \wedge \delta\Psi_i(x,E)\rangle}{\langle \Psi_i^*(x,E)L^{(1)}\Psi_i(x,E)\rangle}. \tag{2.74}$$

Due to the summation over i, this expression is independent of the labeling of the Bloch functions. Thus on the right hand side, we have the residue of a well-defined function of E. The formula we need for the differential of the branch of the quasi-momentum corresponding to the Bloch solution $\Psi(x,E)$ of (2.72) is the following

$$i\, dp\langle \Psi_i^*(x,E)L^{(1)}\Psi_i(x,E)\rangle = dE. \tag{2.75}$$

Its proof is identical to the proof in the finite gap theory (see [36]). Consider the differential $d\Psi$ in the variable E of the Bloch function. Then

$$(L-E)d\Psi = -\Psi\, dE.$$

Integrating from x_0 to x_0+1 the identity

$$0 = (\Psi^*(L-E))d\Psi = -(\Psi^*\Psi)dE + \sum_{j=1}^{n} \partial_x^j(\Psi^*(L^j\, d\Psi)).$$

we obtain

$$dE = i\, dp\big(\Psi_i^*(x_0,E)L^{(1)}\Psi_i(x_0,E) + \sum_{j=2}^{n} \partial_x^{j-1}(\Psi^*(x_0,E)(L^j\Psi(x_0,E))\big).$$

The desired formula follows after averaging in x_0 this last identity.

With the formula (2.75) for dE and the analyticity in E for $E \neq 0$ of all relevant expressions, we can, in the computation of the contracted form $i(\partial_m)\omega^{(-1)}$, reduce the residues at $E = 0$ to the residue at $E = \infty$ and get the desired result. For example, we have

$$-\operatorname{Res}_0 \sum_{i=1}^{n} \langle \Psi_i^*\delta L\partial_m\Psi_i\rangle \frac{dp_i(E)}{E} = \operatorname{Res}_\infty \langle \Psi^*(x,K)\delta L\partial_m\Psi(x,K)\rangle \frac{dK}{E} = n\delta H_m.$$

G. Higher Symplectic Structures in the Two-Dimensional Case

In this section, we introduce higher Hamiltonian structures which exist in both one and two dimensions. We would like to emphasize that, in contrast with the previous results which have basically an *algebraic nature*, the following results require in general some additional assumptions on the long-time behavior of the solutions to the hierarchy of flows $\partial_m L = \partial_t A_m + [A_m, L]$. In these higher Hamiltonian structures, the role of the quasi-momentum $k(K)$ for the basic structure is assumed by the quasi-momentums $\Omega_m(K)$ corresponding to the higher times of the hierarchy.

Our first step is to study in greater detail the quasi-momentums $\Omega_m(K)$, corresponding to higher times of the hierarchy. Their "densities" $\tilde{\Omega}_m(K)$ made their first appearance in (2.44). They can be re-expressed as

$$\tilde{\Omega}_m(K) = A_m \Psi_0(x, y; k)|_{x=y=0} = K^m + \sum_{s=1}^{\infty} \tilde{\Omega}_{s,m} K^{-s} . \tag{2.76}$$

The coefficients $\tilde{\Omega}_{s,m}$ of $\tilde{\Omega}_m(k)$ are integro-differential polynomials in the coefficients of the operator L. As stated above, they do not depend on (x, y), but they do depend on the times t if the operator L evolves according to the equation (2.41), i.e. $\tilde{\Omega} = \tilde{\Omega}(K, t)$. From (2.44) and (2.45) it follows that

$$\partial_i \tilde{\Omega}_m = \partial_m \tilde{\Omega}_i. \tag{2.77}$$

Since the coefficients $\tilde{\Omega}_{s,m}$ are independent of the choice of normalization point, they can be considered as functionals on the space of periodic operators L.

The subsequent arguments are based on the variational formulas for these functionals, which were found originally in the case of finite-gap solutions in [36]. Following [36], we use the identity

$$\begin{aligned} &\sum_{j>1} \partial_x^{j-1} \Big(\partial_m (\Psi_0^* (L^{(j)} \delta \Psi_0) - \partial_y (\Psi_0^* (A_m^{(j)} \delta \Psi_0) \Big) \\ &= \sum_{j \geq 1} \partial_x^{j-1} \Big((\Psi_0^* (L^{(j)} (\delta A_m + \delta \tilde{\Omega}_m) \Psi_0) - (\Psi_0^* (A^{(j)} \delta L \Psi_0) \Big) \\ &+ \sum_{k,j \geq 1} \partial_x^{j+k-1} \Big(\Psi_0^* (A_m^{(k)} L^{(j)} - L^{(k)} A_m^{(j)}) \delta \Psi_0 \Big) , \end{aligned} \tag{2.78}$$

where $L^{(j)}$ and $A^{(j)}$ are the descendants of the operators L and A defined by (2.60). Note that if L and A_m satisfy (2.44), then the equality

$$\partial_m L^{(j)} - \partial_y A_m^{(j)} + \sum_k [L^{(k)}, A_m^{(j-k)}] = 0 \tag{2.79}$$

holds. We now average (2.78) first in x and y, and then in the normalization point x_0 (the last averaging eliminates all terms with $j > 1$). The outcome is the equality

$$\delta \tilde{\Omega}_m (K, t) \langle \Psi_0^* L^{(1)} \Psi_0 \rangle = \langle \Psi_0^* (L^{(1)} \delta A_m - A_m^{(1)} \delta L) \Psi_0 \rangle + \partial_m \langle \Psi_0^* L^{(1)} \delta \Psi_0 \rangle. \tag{2.80}$$

Let $\mathcal{D}_{m_0}$, be the space of all periodic operators L which are stationary under the first m_0-th flow, i.e. which satisfy the condition

$$\partial_y A_{m_0} = [L, A_{m_0}]. \tag{2.81}$$

It should be emphasized that due to (2.77) onto this space the corresponding density of the corresponding quasi-momentum is a constant, i.e. does not depend on the times

$$\Omega_{m_0}(K) = \tilde{\Omega}_{m_0}(K).$$

The space carrying a higher symplectic structure is a subspace $\mathcal{D}_{m_0}^{(I)}$ of $\mathcal{D}_{m_0}$, consisting of the stationary operators L satisfying in addition the following higher constraints

$$\Omega_{m_0} = K^{m_0} + \sum_{s=1}^{\infty} \Omega_{s,m_0} K^{-i}, \qquad \Omega_{s,m_0} = I_s, \tag{2.82}$$

for a set $I = (I_1, \ldots, I_{n-1})$ of $(n-1)$ fixed constants. These constraints replace the constraints (2.30) of our previous considerations. The subspace $\mathcal{D}_{m_0}^{(I)}$ is invariant with respect to all the other flows corresponding to times t_i.

Theorem 9. *The formula*

$$\omega_{m_0} = \mathrm{Res}_\infty \langle \Psi_0^* (A_{m_0}^{(1)} \delta L - L^{(1)} \delta A_{m_0}) \wedge \delta \Psi_0 \rangle dk, \tag{2.83}$$

defines a closed two-form on $\mathcal{D}_{m_0}^{(I)}$. *The restrictions of the equations (2.41) to this space are Hamiltonian with respect to this form, with Hamiltonians* $2n\Omega_{m_0,n+m}$.

Remark 1. This statement has an obvious generalization if we replace the stationary condition (2.81) by the condition that L be stationary with respect to a linear combination of the first m_0 flows, i.e.

$$\partial_y A = [L, A], \quad A = \sum_{i=0}^{m_0} c_i A_i.$$

Remark 2. For $m_0 = 1$, we have $A_1 = \partial$, $A_1^{(1)} = 1$, and the formula (2.83) becomes identical to (2.62), i.e. $\omega_1 = \omega$.

The proof of Theorem 6 is identical to that of Theorem 3, after replacing of (2.63) by the formula

$$\delta \Omega_{m_0}(K) dE = \langle \Psi_0^* (L^{(1)} \delta A_{m_0} - A_{m_0}^{(1)} \delta L) \Psi_0 \rangle dk, \tag{2.84}$$

which is valid on $\mathcal{D}_{m_0}$. This formula is itself a direct corollary of (2.80) and (2.75).

As an example, we consider the case $n = 2, m_0 = 3$. For $n = 2$, the equations (2.41) define the KP hierarchy on the space of periodic functions $u(x, y)$ of two variables. For $A = L_3$, the condition (2.81) describes the stationary solutions of the original KP equation, i.e. the space of functions described by the equation

$$3u_{yy} + (6uu_x + u_{xxx})_x = 0. \tag{2.85}$$

Theorem 6 asserts that, besides of the basic Hamiltonian structure, the restriction of any flow of the KP-hierarchy to the space of functions $u(x, y)$ subject to (2.85) is Hamiltonian with respect to the structure given by (2.83).

In the one-dimensional case, the constraint (2.81) is equivalent to the restriction of the Lax hierarchy on the space of finite-gap solutions. This space is described by the following commutativity condition for the ordinary differential operators L and A of respective degrees n and m_0

$$[L, A_{m_0}] = 0. \tag{2.86}$$

This condition is equivalent to a system of ordinary differential equations for the coefficients $u_i(x)$ of L. Theorem 6 asserts that the restriction of the Lax hierarchy to the space of solutions to (2.86) is Hamiltonian with respect to the symplectic form (2.83). In particular, the first flow (which is just a shift in x) is Hamiltonian. For the KdV case the corresponding symplectic structure coincides with the *stationary* Hamiltonian structure found in [5].

Example 3. We return to the case $n = 3$ of Example 2, and consider this time operators A of order $m = 2$. Thus the operators L and A are given by $L = \partial_x^3 + u\,\partial_x + v$, $A = \partial_x^2 + \frac{2}{3}u$. The space $\mathcal{D}_c$ is the space of two quasi-periodic functions $u(x, y)$ and $v(x, y)$ satisfying the constraints

$$\langle u \rangle_x = \text{const}, \ \langle v \rangle_x = \text{const}, \ \langle u^2 \rangle_x = \text{const}.$$

The operators $L^{(1)}$ and $A^{(1)}$ are given by

$$L^{(1)} = 3\partial_x^2 + u, \quad A^{(1)} = 2\partial_x$$

and the symplectic form ω_2 of (2.83) becomes

$$\omega = 2\left\langle \frac{3}{4} u\delta u \wedge \int_{x_0}^{x} \delta u\, dx + 2\delta v \wedge \int_{x_0}^{x} \delta v\, dx + 2\delta u \wedge \delta v - \frac{3}{2}\delta u_x \wedge \delta u \right\rangle.$$

H. Symplectic Structures under Ergodicity Assumptions

It may be worthwhile to point out that the existence of the higher Hamiltonian structures obtained in the previous section requires less that the stationary condition (2.81). The only item which was necessary to the argument was the possibility of dropping the last term (which was a full derivative in t_{m_0}) in the formula (2.80).

This suggests considering the space $\mathcal{D}_{m_0}^{erg}$ of all operators L with smooth periodic coefficients, for which the corresponding solution $L(t_{m_0})$, $L(0) = L$

of the equation (2.41) for $m = m_0$ exists for all t_{m_0} with uniformly bounded coefficients. In this case we may introduce the quasi-momentum

$$\Omega_{m_0}(K) = \lim_{T\to\infty} \frac{1}{T} \int_0^T \tilde{\Omega}_{m_0}(K,t)dt_{m_0}.$$

Although in this definition, only the dependence on t_{m_0} is eliminated through averaging, the quasi-momentum Ω_{m_0} is actually also independent of all the other times t_i, in view of (2.77).

For $L \in \mathfrak{D}_{m_0}^{erg}$ the formula

$$\delta\Omega_{m_0}(K)dE = \langle \Psi_0^*(L^{(1)}\delta A_{m_0} - A_{m_0}^{(1)}\delta L)\Psi_0\rangle_0 \, dk, \tag{2.87}$$

where $\langle f(x,y,t)\rangle_0$ stands for

$$\langle f(x,y,t)\rangle_0 = \lim_{T\to\infty} \frac{1}{T} \int_0^T \langle f(x,y,t)\rangle dt,$$

holds. Here we make use of the fact that if δL is variation of the initial Cauchy data $L(0)$ for (1.47) then the variation $\delta L(t_{m_0})$ is defined by the linearized equation

$$\partial_{m_0}\delta L - \partial_y \delta A_{m_0} + [\delta L, A_{m_0}] + [L, \delta A_{m_0}].$$

With (2.87), it is now easy to establish the following theorem

Theorem 10. *The formula*

$$\omega_{m_0} = \mathrm{Res}_\infty \langle \Psi_0^*(A_{m_0}^{(1)}\delta L - L^{(1)}\delta A_{m_0}) \wedge \delta\Psi_0\rangle_0 \, dk, \tag{2.88}$$

defines a closed two-form on subspaces of $\mathfrak{D}_{m_0}^{erg}$ subject to the constraints (2.82). The restrictions of the equations (2.41) to this space are Hamiltonian with respect to this form, with the Hamiltonians $2n\Omega_{m_0,n+m}$.

The space $\mathfrak{D}_{m_0}^{erg}$ appears to be a complicated space, and we do not have at this moment an easier description for it. As noticed in [36], it contains (for an arbitrary m_0) all the finite-gap solutions. There exist a few other cases where we can justify the ergodicity assumption. For example, for the KdV hierarchy, the ergodicity assumption is fulfilled for smooth periodic functions with sufficiently rapidly decreasing Fourier coefficients. Indeed, if $u(x)$ can be extended as an analytic function in a complex neighborhood of real values for x and y,

$$|u(x)| < U, \quad |\mathrm{Im}\, x| < q \tag{2.89}$$

then $u(x,t)$ is bounded by the same constant for all t due to trace formulae. Using the approximation theorem [37] for all periodic solutions to (2.5) (also called the KP-2 equation, by contrast with the KP-1 equation given by (3.19) below) by finite-gap solutions, we can prove the ergodicity assumption in the case when the Fourier coefficients u_{ij} of u satisfy the condition $|u_{ij}| < Uq^{|i|+|j|}$.

Important Remark. In order to clarify the meaning of the higher symplectic forms and the higher Hamiltonians, it is instructive to explain its analogue for the usual Lax equations. The Lax equations $\partial_m L = [M_m, L]$ obviously imply that the eigenvalues of L are integrals of motion, and usually they serve as Hamiltonians for the basic symplectic structure. The higher Hamiltonians correspond to the eigenvalues of the operator M_{m_0} instead of L. Of course, they are time dependent, but after averaging with respect to one of the times, namely t_{m_0}, they become nonlocal integrals of motion and can serve as Hamiltonians for the corresponding symplectic structures.

I. The Matrix Case and the 2D Toda Lattice

Our formalism extends without difficulty to a variety of more general settings. We shall discuss briefly the specific cases of matrix equations and of the Toda lattice, which correspond respectively to the cases where L is matrix-valued, and where the differential operator ∂_x is replaced by a difference operator.

Let $L = \sum_{i=0}^{n} u_i(x,y)\partial_x^i$ be then an operator with matrix coefficients $u_i = (u_i^{\alpha\beta})$ which are smooth and periodic functions of x and y, whose leading term $u_n^{\alpha\beta} = u_n^\alpha \delta_{\alpha\beta}$ is diagonal with distinct diagonal elements $u_n^\alpha \neq u_n^\beta$ for $\alpha \neq \beta$, and which satisfies $u_{n-1}^{\alpha\alpha} = 0$. Then, arguing as in Section II.B, we can show that there exists a unique matrix formal solution $\Psi_0 = (\Psi_0^{\alpha\beta}(x,y;k))$ of the equation $(\partial_y - L)\Psi_0 = 0$, which has the form

$$\Psi_0(x,y;k) = \left(I + \sum_{s=1}^{\infty} \xi_s(x,y)k^{-s}\right) \exp\left(kx + u_n k^n y + \sum_{i=0}^{n-2} B_i(y)k^i\right) \tag{2.90}$$

(where I is the identity matrix, $\xi_s = (\xi_s^{\alpha\beta})$ are matrix functions, and $B_i(y) = (B_i^{\alpha\beta}(y)) = (B_i^\alpha(y)\delta_{\alpha\beta})$ are diagonal matrices), and which has the Bloch property

$$\Psi_0(x+1,y;k) = \Psi_0(x,y;k)w_1(k), \;\; w_1(k) = e^k.$$

The formal solution $\Psi_0(x,y;k)$ has the Bloch property with respect to y as well,

$$\Psi_0(x,y+1;k) = \Psi_0(x,y;k)w_2(k)$$

with the Bloch multiplier $w_2(k)$ of the form

$$w_2(k) = \left(1 + \sum_{s=1}^{\infty} J_s k^{-s}\right) \exp\left(u_n k^n + \sum_{i=0}^{n-2} B_i k^i\right),$$

where J_s and B_i are diagonal matrices.

As noted at the end of Section II.B, the second Bloch multiplier defines the quasi-energy $E(k)$. This defines in turn the functionals $\mathcal{E}_i$ just as in (2.28), with the only difference the fact that they are now diagonal matrices. If we introduce the diagonal matrix K by the equality

$$u_n K^n = E(k) = \log w_2(k) = u_n k^n + \sum_{i=-n+2}^{\infty} \mathcal{E}_i k^{-i} \tag{2.91}$$

then we may define diagonal matrices $H_s = (H_s^\alpha \delta_{\alpha\beta})$ by

$$kI = K + \sum_{i=1}^{\infty} H_s K^{-s} \ . \tag{2.92}$$

The definition of the commuting flows in the matrix case is then just the same as in the scalar case. The only difference is that the number of these flows is now N times larger. The corresponding times are denoted by $t = (t_{\alpha,m})$, and the flows are given by

$$[\partial_y - L, \partial_{\alpha,m} - L_{\alpha,m}] = 0 \ , \tag{2.93}$$

where $L_{\alpha,m}$ is the unique operator of the form

$$L_{\alpha,m} = v_\alpha \partial_x^m + \sum_{i=1}^{m-1} u_{i,(\alpha,m)}(t)\partial_x^i, \ v_\alpha^{\beta\gamma} = \delta_{\alpha\beta}\delta^{\beta\gamma} \ ,$$

which satisfies the condition

$$L_{\alpha,m}\Psi_0(x,y;k) = \Psi_0(x,y;k)(v_\alpha K^m + O(K^{-1})). \tag{2.94}$$

As before, the dual Bloch formal series $\Psi_0^*(x,y;k)$ is defined as being of the form (2.58), and satisfying the equation

$$\mathrm{Res}_\infty \mathrm{Tr}(\Psi_0^* v \partial_x^m \Psi_0)dk = 0, \ m \geq 0 \tag{2.95}$$

for arbitrary matrices v. We have then

Theorem 11. *The formula*

$$\omega = \mathrm{Res}_\infty Tr\langle \Psi_0^* \delta L \wedge \delta \Psi_0 \rangle dk \tag{2.96}$$

defines a closed non-degenerate two-form on the space of periodic operators L subject to the constraints $H_s^\beta = constant$, $\beta = 1,\ldots,l$, $s = 1,\ldots,n-1$. The equations (2.93) are Hamiltonian with respect to this form, with Hamiltonians $2nH_{m+n}^\alpha$ defined by (2.92).

Example 4. Consider the case $n = 1$, where the operator L is of the form $L = A\partial_x + u(x,y)$, with A the $N \times N$ matrix $A^{\alpha,\beta} = a_\alpha \delta_{\alpha,\beta}$ and $u(x,y)$ is an $N \times N$ matrix with zero diagonal entries $u^{\alpha,\alpha} = 0$. In this case the symplectic form (2.96) becomes

$$\omega = \sum_{\alpha \neq \beta} \frac{1}{a_\alpha - a_\beta} \delta u^{\beta,\alpha} \wedge \delta u^{\alpha,\beta}.$$

Finally, as a basic example of a system corresponding to an auxiliary linear equation where the differential operator ∂_x is replaced by a difference operator acting on spaces of infinite sequences, we consider the $2D$ Toda lattice.

The 2D Toda lattice is the system of equations for the unknown functions $\varphi_n = \varphi_n(t_+, t_-)$

$$\frac{\partial^2}{\partial t_+ \partial t_-}\varphi_n = e^{\varphi_n - \varphi_{n-1}} - e^{\varphi_{n+1} - \varphi_n}.$$

It is equivalent to the compatibility conditions for the following auxiliary linear problem

$$\partial_+ \psi_n = \psi_{n+1} + v_n \psi_n, \; v_n = \partial_+ \varphi_n, \quad \partial_- \psi_n = c_n \psi_{n-1}, \; c_n = e^{\varphi_n - \varphi_{n-1}}.$$

We consider solutions of this system which are periodic in the variables n and $y = (t_+ + t_-)$. The relevant linear operator is the difference operator

$$L\Psi_n = \Psi_{n+1} + v_n \Psi_n + c_n \Psi_{n-1}$$

with periodic coefficients $v_n(y) = v_{n+N}(y) = v_n(y+l)$ and $c_n(y) = c_{n+N}(y) = c_n(y+1)$. Then, arguing as in Section II.B we can show that there exist unique formal solutions $\Psi^{(\pm)} = \Psi_n^{(\pm)}(y;k)$ of the equation

$$(\partial_y - L)\Psi^{(\pm)} = 0,$$

which have the form

$$\Psi_n^{(\pm)}(y;k) = k^{\pm n}\left(\sum_{s=0}^{\infty} \xi_s^{(\pm)}(n,y)k^{-s}\right)e^{ky+B(y)}, \quad \xi_0^{(+)} = 1, \tag{2.97}$$

the Bloch property

$$\Psi_{n+N}^{(\pm)}(y;k) = \Psi_n^{(\pm)}(y;k)w_1^{(\pm)}(k), \quad w_1^{(\pm)}(k) = k^{\pm N},$$

and which are normalized by the condition

$$\Psi_0^{(\pm)}(0;k) = 1.$$

The coefficients $\xi_s^{(\pm)}$ can be found recursively. The initial value $\xi_0^{(+)} = 1$ and the condition that $\xi_1^{(+)}$ is periodic in n define the function $B(y)$ in (2.97)

$$B(y) = N^{-1}\int_0^y \left(\sum_{n=1}^{N} v_n\right) dy.$$

The only difference with the previous differential case is the definition of the leading term $\xi_0^{(-)}(n,y)$. Let us introduce $\varphi_n(y) = \log \xi_0^{(-)}(n,y)$. Then we have

$$c_n(y) = e^{\varphi_n(y) - \varphi_{n-1}(y)}.$$

The periodicity condition for $\xi_1^{(-)}$ requires the equality

$$\sum_{n=0}^{N-1} \partial_y \varphi_n = 0,$$

which allows us to define φ_n uniquely through c_n:

$$\varphi_n = \sum_{n=1}^{n-1} \log c_n - \int_0^y \left(N^{-1} \sum_{n=0}^{N-1} \log c_n \right) dy \ .$$

The formal solutions $\Psi_n^{(\pm)}(y,k)$ have the Bloch property with respect to y as well,

$$\Psi_n^{(\pm)}(y+1;k) = \Psi_n^{(\pm)}(y;k) w_2^{(\pm)}(k)$$

with the Bloch multipliers $w_2^{(\pm)}(k)$ of the form

$$w_2^{(\pm)}(k) = e^{k+B} \left(\sum_{s=0}^{\infty} J_s^{(\pm)} k^{-s} \right) .$$

As noted at the end of Section II.B, the second Bloch multiplier defines the quasi-energy $E^{(\pm)}(k)$ and the functionals $\mathcal{E}_i^{(\pm)}$ just as in (2.28). If we introduce the variable

$$K = E^{(\pm)}(k) = \log w_2^{(\pm)}(k) = k + \sum_{i=0}^{\infty} \mathcal{E}_i^{(\pm)} k^{-i}$$

then we may define the functionals $H_s^{(\pm)}$ by

$$\log k = \log K \pm \sum_{s=0} H_s^{(\pm)} K^{-s}. \tag{2.98}$$

The definition of the commuting flows in the discrete case is then the same as in the scalar case. The basic constraints that specify the space of periodic functions v_n and c_n have the form

$$I_0 = \sum_{n=1}^{n} \log c_n(y) = \text{const}, \quad I_1 = \sum_{n=1}^{N} v_n(y) = \text{const}. \tag{2.99}$$

The corresponding times are denoted by $t = (t_{\pm,m})$ and the flows are given by

$$[\partial_y - L, \partial_{\pm,m} - L_{\pm,m}] = 0, \tag{2.100}$$

where $L_{\pm,m}$ is the unique operator of the form

$$L_{\pm,m}\Psi_n = \sum_{i=0}^{m} u_{i,(\pm,m)}(n,t)\Psi_{n\pm i}, \ u_{i,(+,m)} = 1, \ u_{i,(-,m)}(n,t) = e^{\varphi_n - \varphi_{n-i}}$$

which satisfies the condition

$$L_{\pm,m}\Psi_n^{(\pm)} = \Psi_n^{(\pm)}(K^m + O(K^{-1})).$$

The dual formal series $\Psi_n^{*,(\pm)}$ are defined as the formal series of the form

$$\Psi_n^{*,(\pm)}(y,k) = k^{\mp n}\left(\sum_{s=0}^{\infty} \xi_s^{*,(\pm)}(n,y)k^{-s}\right) e^{-(ky+B(y))}, \quad \xi_0^{*,(+)} = 1,$$

which satisfy the equations

$$\mathrm{Res}_\infty \left(\Psi_n^{*,(+)}\Psi_m^{(+)}) - \Psi_n^{*,(-)}\Psi_m^{(-)}\right)\frac{dk}{k} = 0, \; m \in \mathbf{Z},$$

for all integers m.

Theorem 12. *The formula*

$$\omega = \mathrm{Res}_\infty\langle \Psi^{*,(+)}\delta L \wedge \delta\Psi^{(+)} - \Psi^{*,(-)}\delta L \wedge \delta\Psi^{(-)}\rangle \frac{dk}{k} = \langle \delta(2v_n - \partial_y\varphi_n) \wedge \delta\varphi_n\rangle \tag{2.101}$$

defines a closed non-degenerate two-form on the space of periodic operators L subject to the constraints (2.99). The equations (2.100) are Hamiltonian with respect to this form, with Hamiltonians $H_m^{\pm}$ defined by (2.98).

Finally, we would like to conclude this section by calling the reader's attention to [59][60], where symplectic structures are discussed from a group theoretic viewpoint, and where applications of soliton theory to harmonic maps are given.

III. Geometric Theory of 2D Solitons

In Section I, we had developed a general Hamiltonian theory of 2D solitons. The central notion was the symplectic form (1.1), which was defined on the infinite-dimensional space $\mathcal{L}(b)$ of doubly periodic operators obeying suitable constraints. Our main goal in this section is to present and extend the results of [39]. In this work, as described in the Introduction, a natural symplectic form $\omega_{\mathcal{M}}$ was constructed on Jacobian fibrations over the leaves of moduli spaces $\mathcal{M}_g(n,m)$ of finite-gap solutions to soliton equations. Imbedded in the space of doubly periodic functions, the form $\omega_{\mathcal{M}}$ was shown to coincide with ω (this was in fact our motivation for constructing a general Hamiltonian theory based on ω in this paper). Although the infinite-dimensional symplectic form ω and its variants in Section II can be expected to play an important role in an analytic theory of solitons, it is the geometric and finite-dimensional form $\omega_{\mathcal{M}}$ which has provided a unifying theme with topological and supersymmetric field theories.

In Section II, we have seen how a differential operator L determined a Bloch function Ψ_0, which was a formal series in a spectral parameter k or K. The key to the construction of finite-gap solutions of soliton equations is the reverse process, namely the association of an operator L to a series of the form of Ψ_0. To allow for evolutions in an arbirary time t_m, it is convenient for us to incorporate a factor $e^{t_m k^m}$ in Ψ_0 for each t_m, and consider series $\Psi_0(t;k)$ of the form

$$\Psi_0(t;k) = \left(1 + \sum_{s=1}^{\infty} \xi_s(t) k^{-s}\right) \exp\left(\sum_{i=1}^{\infty} t_i k^i\right). \tag{3.1}$$

As usual, all the times t_i except for a finite number have been set to 0. Then the operators L_m are uniquely defined by the requirement that

$$(\partial_m - L_m)\Psi_0(t;k) = O(k^{-1}) \exp\left(\sum_{i=1}^{\infty} t_i k^i\right)$$

(this is equivalent to the earlier requirement that $(L_m - k^m)\Psi_0 = O(k^{-1})\Psi_0$ in the case of $\xi_s(t)$ independent of t_m). In particular, we have the following identity between formal power series

$$[\partial_n - L_n, \partial_m - L_m]\Psi_0 = O(k^{-1}) \exp\left(\sum_{i=1}^{\infty} t_i k^i\right). \tag{3.2}$$

This identity assumes its full value when the formal series $\Psi_0(t;k)$ is a genuine convergent function of k and has an analytic continuation as a meromorphic function with g poles on a Riemann surface of genus g. In this case, the equation (3.2) with zero right hand side becomes exact. The null space of $[\partial_n - L_n, \partial_m - L_m]$ is parametrized then by k and is infinite-dimensional. Since $[\partial_n - L_n, \partial_m - L_m]$ is an ordinary differential operator, it must vanish. Thus a convergent $\Psi_0(t;k)$ gives rise to a solution of the zero curvature equation $[\partial_n - L_n, \partial_m - L_m] = 0$. The algebraic-geometric theory of solitons provides precisely the geometric data which leads to convergent Bloch functions. These functions are now known as Baker-Akhiezer functions.

A. Geometric Data and Baker-Akhiezer Functions

In a Baker-Akhiezer function, the spectral parameter k is interpreted as the inverse $k = z^{-1}$ of a local coordinate z on a Riemann surface. Thus let a "geometric data" (Γ, P, z) consist of a Riemann surface Γ of some fixed genus g, a puncture P on Γ, and a local coordinate k^{-1} near P. Let $\gamma_1, \ldots, \gamma_g$ be g points of Γ in general position. Then for any $t = (t_i)_{i=1}^{\infty}$, only a finite number of which are non-zero, there exists a unique function $\Psi(t;z)$ satisfying

(i) Ψ is a meromorphic on $\Gamma \setminus P$, with at most simple poles at $\gamma_1, \ldots, \gamma_g$;

(ii) in a neighborhood of P, Ψ can be expressed as a convergent series in k of the form appearing on the right hand side of (3.1).

The exponential factor in (3.1) describes the essential singularity of $\Psi(t;z)$ near P. Alternatively, we can view it as a transition function (on the overlap between $\Gamma \setminus P$ and a neighborhood of P) for a line bundle $\mathcal{L}(t)$ on Γ. The Baker-Akhiezer function Ψ is then a section of $\mathcal{L}(t)$, meromorphic on the whole of Γ.

The form of the essential singularity of Ψ implies that Ψ has as many zeroes as it has poles (equivalently, the line bundle $\mathcal{L}(t)$ has vanishing Chern class). Indeed, $d\Psi/\Psi = d(\sum_{i=1}^{\infty} t_i k^{-i}) +$ regular, and thus has no residue at P. From this, the uniqueness of the Baker-Akhiezer function follows, since the ratio $\Psi/\tilde{\Psi}$ of two Baker-Akhiezer functions would be a meromorphic function on the whole of Γ, with at most g poles (corresponding to the zeroes of $\tilde{\Psi}$). By the Riemann-Roch Theorem, it must be constant. Finally, the existence of Ψ can be deduced most readily from an explicit formula. Let $A_1, \ldots, A_g$ $B_1, \ldots, B_g$ be a canonical homology basis for Γ

$$A_j \cap A_k = B_j \cap B_k = 0, \ A_j \cap B_k = \delta_{jk}, \tag{3.3}$$

and let

(a) $d\omega_j$, τ_{jk} be respectively the dual basis of holomorphic abelian differentials and the period matrix

$$\oint_{A_j} d\omega_k = \delta_{jk}, \ \oint_{B_j} d\omega_k = \tau_{jk} \ ;$$

(b) $\theta(z|\tau)$ the Riemann θ-function;

(c) $d\Omega_i^0$ the Abelian differential of the second kind with unique pole of the form

$$d\Omega_i^0 = (1 + O(k^{-i-1}))dk^i \tag{3.4}$$

normalized to have vanishing A-periods

$$\oint_{A_I} d\Omega_i^0 = 0, \ 1 \leq I \leq g; \tag{3.5}$$

(d) P_0 a fixed reference point, with which we can define the Abel map $A : z \in \Gamma \to A(z) \in \mathbf{C}^g$ and Abelian integrals Ω_i^0 by

$$A_j(z) = \int_{P_0}^{z} d\omega_j, \ \Omega_i^0(z) = \int_{P_0}^{z} d\Omega_i^0 \tag{3.6}$$

Here the Abel map as well as the Abelian integrals are path dependent, and we need to keep track of the path, which is taken to be the same in both cases.

(e) $Z = K - \sum_{s=1}^{g} A(\gamma_s)$, where K is the vector of Riemann constants (c.f. [48]).

Using the transformation laws for the θ-function, it is then easily verified that the following expression is well defined, and must coincide with $\Psi(t; z)$

$$\Psi(t;z) = \frac{\theta(A(z) + \frac{1}{2\pi i}\sum_{i=1}^{\infty} t_i \oint_B d\Omega_i^0 + Z|\tau)\theta(A(P) + Z)}{\theta(A(z) + Z)\theta(A(P) + \frac{1}{2\pi i}\sum_{i=1}^{\infty} t_i \oint_B d\Omega_i^0 + Z|\tau)} \exp\Big(\sum_{i=1}^{\infty} t_i \Omega_i^0(z)\Big). \tag{3.7}$$

In this formalism, the role of the quasi-momentum p is now assumed by the Abelian integral Ω_1 with unique pole at P of the first order normalized to have pure imaginary periods on Γ. With the Baker-Akhiezer function assuming now the former role of formal Bloch functions, we can construct a hierarchy of operators L_n as in (2.37). The requirement that $(\partial_m - L_m)\Psi(t;z) = O(z)\Psi(t;z)$ determines recursively the coefficients of L_m as differential polynomials in the ξ_s. The crucial improvement over formal Bloch functions is that here, this requirement actually implies that

$$(\partial_n - L_n)\Psi(t;z) = 0 \tag{3.8}$$

identically. In fact, $(\partial_n - L_n)\Psi(t;z)$ satisfies all the conditions for a Baker-Akhiezer function, except for the fact that the Taylor expansion of its coefficient in front of the essential singularity $\exp(\sum_{i=1}^{\infty} t_i k^i)$ starts with k^{-1}. If this function is not identically zero, it can be used to generate distinct Baker-Akhiezer functions from any given one, contradicting the uniqueness of Baker-Akhiezer functions. As noted earlier, (3.8) implies that $[\partial_n - L_n, \partial_m - L_m]\Psi(t;z) = 0$, and hence that the zero curvature equation $[\partial_n - L_n, \partial_m - L_m] = 0$ holds. In summary, we have defined in this way a "geometric map' $\mathcal{G}$ which sends the geometric data $(\Gamma, P, z; \gamma_1, \ldots, \gamma_g)$ to an infinite hierarchy of operators [33][34]

$$\mathcal{G} : (\Gamma, P, z; \gamma_1, \ldots, \gamma_g) \to (L_n)_{n=2}^{\infty} \tag{3.9}$$

The expression (3.7) for $\Psi(t;z)$ leads immediately to explicit solutions for a whole hierarchy of soliton equations. Let $t_1 = x$, $t_2 = y$, $t_3 = t$, and $n = 2$. We find then solutions $u(x,y,t)$ of the KP equation (2.4) expressed as [34]

$$u(x,y,t) = 2\partial_x^2 \log\theta\left(x \oint_B d\Omega_1^0 + y \oint_B d\Omega_2^0 + t \oint_B d\Omega_3^0 + Z|\tau\right) + \text{const.} \tag{3.10}$$

This formula is at the origin of a remarkable application of the theory of non-linear integrable models, namely to a solution of the famous Riemann-Schottky problem. According to the Torelli theorem, the period matrix defines uniquely the algebraic curve. The Riemann-Schottky problem is to describe the symmetric matrices with positive imaginary part which are period matrices of algebraic curves. Novikov conjectured that the function $u(x,y,t) = 2\partial_x^2 \log\theta(Ux + Vy + Wt|\tau)$ is a solution of the KP equation if and only if the matrix τ is the period matrix of an algebraic curve, and U, V, W are the B-periods of the corresponding normalized meromorphic differentials with poles only at a fixed point of the curve. This conjecture was proved in [55].

The dual Baker-Akhiezer function

For later use, we also recall here the main properties of the dual Baker-Akhiezer function $\Psi^+(t;z)$ which coincides with the formal dual series defined in Section II.D. To define $\Psi^+(t;z)$, we note that, given g points $\gamma_1, \ldots, \gamma_g$ in general position, the unique meromorphic differential $d\Omega = d(z^{-1} + \sum_{s=2}^{\infty} a_s z^s)$ with double pole at P and zeroes at $\gamma_1, \ldots, \gamma_g$, must also have g other zeroes, by the Riemann-Roch theorem. Let these additional zeroes be denoted by $\gamma_1^+, \ldots, \gamma_g^+$. Then the dual Baker-Akhiezer function $\Psi^+(t;z)$ is the unique function $\Psi^+(t;z)$ which is meromorphic everywhere except at P, has at most simple poles at $\gamma_1^+, \ldots, \gamma_g^+$, and admits the following expansion near P

$$\Psi^+(t;z) = \exp\left(-\sum_{n=1}^{\infty} t_n z^{-n}\right)\left(1 + \sum_{s=1}^{\infty} \xi_s^+(t) z^s\right)$$

To compare the dual Baker-Akhiezer function with the formal dual Bloch function Ψ_0^+ of Section II.E, it suffices to observe that

$$\mathrm{Res}_P\, \Psi^+(t;z)\big(\partial_x^m \Psi(t;z)\big) d\Omega = 0,$$

since the differential on the left hand side is meromorphic everywhere, and holomorphic away from P. Together with the normalization $\Psi^+(0;z) = 1$, this implies that Ψ^+ indeed coincides with the formal dual function Ψ_0^+. An exact formula for $\Psi^+(t;z)$ can be obtained from (3.7) by changing signs for t and by replacing the vector Z by Z^+. From the definition of the dual set of zeroes $\gamma_1^+, \ldots, \gamma_g^+$, this vector satisfies the equation $Z + Z^+ = 2P + K$, where K is the canonical class. Recalling that the quasi-momentum p was defined to be $p = \Omega_1$, we also obtain the following formula for the differential $d\Omega$ we introduced earlier

$$d\Omega = \frac{dp}{\langle \Psi^+ \Psi \rangle}\ . \tag{3.11}$$

The Multi-Puncture Case

The above formalism extends easily to the case of N punctures P_α (with one marked puncture P_1). The Baker-Akhiezer function Ψ is required then to have the essential singularity

$$\Psi(t;z) = \left(\sum_{s=0}^{\infty} \xi_{s\alpha}(t) k_\alpha^{-s}\right) \exp\left(\sum_{i=1}^{\infty} t_{i\alpha} k_\alpha^i\right)\ , \tag{3.12}$$

where k_α^{-1} are local coordinates near each puncture P_α, $t_{i\alpha}$ are given "times", only a finite number of which are non-zero, and the coefficient ξ_{s1} at P_1 is normalized to be 1 for $s = 0$. We can introduce as before $d\Omega_{i\alpha}^0$ associated now to each puncture P_α and their Abelian integrals $\Omega_{i\alpha}^0$. Then the Baker-Akhiezer function $\Psi(t;z)$ becomes

$$\Psi(t;z) = \frac{\theta(A(z) + \frac{1}{2\pi i}\sum_{\alpha=1}^{N}\sum_{i=1}^{\infty} t_{i\alpha} \oint_B d\Omega_{i\alpha}^0 + Z|\tau)\theta(A(P_1) + Z)}{\theta(A(z) + Z)\theta(A(P_1) + \frac{1}{2\pi i}\sum_{\alpha=1}^{N}\sum_{i=1}^{\infty} t_{i\alpha} \oint_B d\Omega_{i\alpha}^0 + Z|\tau)} \times \exp\left(\sum_{\alpha=1}^{N}\sum_{i=1}^{\infty} t_{i\alpha}\Omega_{i\alpha}^0(z)\right)\ . \tag{3.13}$$

For each pair (α, n) there is now a unique operator $L_{\alpha n}$ of the form (2.36) so that $(\partial_{\alpha n} - L_{\alpha n})\psi(t, z) = 0$, with $\partial_{\alpha n} = \partial/\partial t_{\alpha n}$. The operators $L_{\alpha n}$ satisfy the compatibility condition $[\partial_{\alpha n} - L_{\alpha n}, \partial_{\beta m} - L_{\beta m}] = 0$.

Periodic Solutions

In general, the finite-gap solutions of soliton equations obtained by the above construction are meromorphic, quasi-periodic functions in each of the variables $t_{\alpha i}$ (a quasi-periodic function of one variable is the restriction to a line of a periodic function of several variables). We would like to single out the geometric data which leads to periodic solutions. For this we need the following slightly different formula for the Baker-Akhiezer function.

Let $d\Omega_{i\alpha}$ be the unique differential with pole of the form (3.4) near P_α, but normalized so that all its periods be purely imaginary, and define the function $\Phi(\zeta_1, \ldots, \zeta_{2g}; z)$ by the formula

$$\Phi(\zeta; z) = \frac{\theta(A(z) + \zeta_k e_k + \zeta_{k+g}\tau_k + Z|\tau)\theta(A(P_1) + Z)}{\theta(A(z) + Z)\theta(A(P_1) + \zeta_k e_k + \zeta_{k+g}\tau_k + Z|\tau)} \exp\left(2\pi i \sum_{k=1}^{g} A_k(z)\zeta_{k+g}\right), \tag{3.14}$$

where $e_k = (0, \ldots, 0, 1, 0, \ldots, 0)$ are the basis vectors in $\mathbf{C}^g$, and τ_k are the vectors with components τ_{jk}. We observe that Φ is periodic of period 1 in each of the variables $\zeta_1, \ldots, \zeta_{2g}$. Then the Baker-Akhiezer function can be expressed as

$$\Psi(t; z) = \Phi\left(\sum_{i\alpha} t_{i\alpha} U_{i\alpha}; z\right) \exp\left(\sum_{i\alpha} t_{i\alpha}\Omega_{i\alpha}\right), \tag{3.15}$$

where we have denoted by $U_{i\alpha}$ the real, $2g$-vector of periods of $d\Omega_{i\alpha}$

$$U_{i\alpha}^k = \frac{1}{2\pi i}\oint_{A_k} d\Omega_{i\alpha}, \quad U_{i\alpha}^{k+g} = -\frac{1}{2\pi i}\oint_{B_k} d\Omega_{i\alpha}.$$

In particular, for geometric data $\{\Gamma, P_\alpha, z_\alpha\}$ satisfying the condition

$$\frac{1}{2\pi i}\sum_{i\alpha} a_{i\alpha} U_{i\alpha} \in \mathbf{Z}, \tag{3.16}$$

the Baker-Akhiezer function is a Bloch function with respect to the variable $\tilde{t}$ if we set $t_{i\alpha} = a_{i\alpha}\tilde{t}$, with Bloch multiplier $w = \exp(\sum_{i\alpha} a_{i\alpha}\Omega_{i\alpha}(z))$. The coefficients of the operators $L_{\alpha n}$ are then periodic functions of $\tilde{t}$. As an example, we consider the one-puncture case. If we express the data under the form $U_1^k = 2\pi m_k/l_1$, $U_2^k = 2\pi n_k/l_2$, with $m_k, n_k \in \mathbf{Z}$, then the corresponding solution of the KP hierarchy is periodic in the variables $x = t_1$, $y = t_2$, with periods l_1 and l_2 respectively.

Real and Smooth Solutions

There are two types of conditions which guarantee that the solutions obtained by the above geometric construction are real and smooth for real values of $t_{i\alpha}$. We present them in the case of the KP hierarchy.

Assume that the geometric data defining the Baker-Akhiezer function is real, in the sense that

(a) the algebraic curve Γ admits an anti-holomorphic involution $\iota : \Gamma \to \Gamma$;

(b) the puncture P_1 is a fixed point of ι;

(c) the local coordinate k^{-1} in a neighborhood of P_1 satisfies the condition $k(\iota(z)) = \overline{k(z)}$;

(d) the divisor $(\gamma_1, \dots, \gamma_g)$ is invariant under ι, i.e., $\iota(\gamma_s) = \gamma_{\sigma(s)}$, where σ is a permutation.

Then the Baker-Akhiezer function satisfies the reality condition

$$\Psi(t; \iota(z)) = \overline{\Psi(t; z)}. \tag{3.17}$$

This is an immediate consequence of the uniqueness of the Baker-Akhiezer function and the fact that both sides of the equation have the same analytic properties. In particular, the coefficients of L_n and the corresponding solutions of the KP hierarchy are real.

In order to have real and *smooth* solutions, it is necessary to restrict further the geometric data. In general, the set of fixed points of any anti-holomorphic involution on a smooth Riemann surface is a union of disjoint cycles. The number of these cycles is less or equal to $g + 1$. The algebraic curves which admit an anti-involution with exactly $g+1$ fixed cycles are called *M-curves*. We claim that the coefficients of L_n are real and smooth functions of all variables t_i when Γ is an M-curve with fixed cycles $A_0, A_1, \dots, A_g$, and $P \in A_0$, $\gamma_s \in A_s$, $s = 1, \dots, g$. To see this, we note that, from the explicit expression for the Baker-Akhiezer function, the coefficients of L_n have poles at some value of t_i if and only if

$$\theta\Big(A(P_1) + \sum_i U_i t_i + Z\Big) = 0. \tag{3.18}$$

The monodromy properties of the θ-function imply that the zeros of the function $\theta(A(z) + \sum_i U_i t_i + Z)$ are well-defined on Γ, even though the function itself is multi-valued. The number of these zeroes is g. They coincide with the zeroes of $\Psi(t, z)$. In view of (3.17), the Baker-Akhiezer function is real on the cycles A_s. On each of the cycles $A_1, \dots, A_g$, there is one pole of Ψ. There must then be at least one zero on the same cycle. Hence all zeroes of $\theta(A(z) + \sum_i U_i t_i + Z)$ are located on cycles A_s. Since $P_1 \in A_0$, the equation (3.18) cannot be fulfilled for real values of t_i.

We observe that the real and smooth solutions of the KP hierarchy corresponding to M-curves with a fixed puncture, are parametrized by the points of a real g-dimensional torus which is the product of the g cycles A_s. If we choose these cycles (as our notation suggests) as half of a canonical basis of cycles, then this torus is the real part of the Jacobian.

In the theory of real and smooth solutions the equation (2.4) is called the KP-2 equation. The other, so-called, KP-1 equation is the other real form of the same equation. It can be obtained from (2.4) by changing y to iy. Thus the KP-1 equation is given explicitly by

$$-\frac{3}{4}u_{yy} = \left(u_t - \frac{3}{2}uu_x - \frac{1}{4}u_{xxx}\right)_x . \tag{3.19}$$

As complex equations (2.4) and (3.19) are equivalent. But the conditions which single out real and smooth solutions are different. These conditions for the KP-1 equation may be found in [37]. Briefly they are :

Assume that the geometric data defining the Baker-Akhiezer function is real, in the sense that

(a) the algebraic curve Γ admits an anti-holomorphic involution $\iota : \Gamma \to \Gamma$;

(b) the puncture P_1 is a fixed point of ι;

(c) the local coordinate k^{-1} in a neighborhood of P_1 satisfies the condition $k(\iota(z)) = -\overline{k(z)}$;

(d) the divisor $(\gamma_1, \ldots, \gamma_g)$ under ι becomes the dual divisor $\gamma_1^+, \ldots, \gamma_g^+$ i.e., $\iota(\gamma_s) = \gamma_{\sigma(s)}^+$, where σ is a permutation.

Then the Baker-Akhiezer function satisfies the reality condition

$$\Psi^+(t'; \iota(z)) = \overline{\Psi(t'; z)}, \tag{3.20}$$

where the new variables $t' = (t_1', \ldots)$ are equal to $t_{2m+1}' = t_{2m+1}$, $t_{2m} = it_{2m}$. As before, this is an immediate consequence of the uniqueness of the Baker-Akhiezer function and the fact that both sides of the equation have the same analytic properties. In particular, the coefficients of L_n and the corresponding solutions of the KP-1 hierarchy are real for real values of t'.

The further restriction of geometric data corresponding to real and *smooth* solutions of the KP-1 hierarchy is as follows. The fixed cycles $a_1, \ldots, a_l$ of ι should divide Γ into two disconnected domaines $\Gamma^\pm$. The complex domain Γ^+ defines the orientation on the cycles considered as its boundary. The differential $d\Omega$ of (3.11) should be positive on a_s with respect to this orientation.

B. Moduli Spaces of Surfaces and Abelian Integrals

The space $\{\Gamma, P, z, \gamma_1, \ldots, \gamma_g\}$ provides geometric data for solutions of a complete hierarchy of soliton equations, and is infinite-dimensional. In the remaining part of this paper, we concentrate rather on a single equation of zero curvature form $[\partial_y - L, \partial_t - A] = 0$. The geometric data associated with the pair (L, A) corresponds to the Jacobian bundle over a finite-dimensional moduli space $\mathcal{M}_g(n,m)$ of Riemann surfaces with a pair of Abelian integrals (E,Q) with poles of order n and m respectively at the puncture P. The associated operators (L, A) are then operators of order n and m, and are obtained by the basic construction (3.9), after imbedding $\mathcal{M}_g(n,m)$ in the space (Γ, P, z) of geometric data. Alternatively, we may choose to represent the equation $[\partial_y - L, \partial_t - A] = 0$ as a dynamical system on a space of operators L, with t as time variable. In this case, a finite-dimensional and geometric space of operators L is obtained by the same construction as just outlined, starting instead from the Jacobian bundle over the moduli space $\mathcal{M}_g(n)$ of Riemann surfaces Γ with just one Abelian integral with pole of order n at the puncture.

More precisely, given (Γ, E), a geometric data (Γ, P, z) is obtained by setting the local coordinate $z \equiv K^{-1}$ near the puncture P to be

$$E = z^{-n} + R^E \log z \; , \tag{3.21}$$

where $n \geq 1$ and R^E are respectively the order of the pole of E and its residue at P. When $n = 0$, we set instead

$$E = R^E \log z. \tag{3.22}$$

This gives immediately a map

$$\begin{gathered}(\Gamma, E) \to (\Gamma, P, z), \\ (\Gamma, E, \gamma_1, \ldots, \gamma_g) \to (\Gamma, P, z, \gamma_1, \ldots, \gamma_g) \to \; L,\end{gathered} \tag{3.23}$$

where the operator L is characterized by the condition $(\partial_y - L)\Psi = 0$, with $\Psi(x, y; k)$ the Baker-Akhiezer function having the essential singularity $\exp(kx + k^n y)$, $k = z^{-1}$. In presence of a second Abelian integral Q, we can select a second time t, by writing the singular part $Q_+(k)$ of Q as a polynomial in k and setting

$$\begin{gathered}Q_+(k) = a_1 k + \ldots + a_m k^m, \\ t_i = a_i t, \;\; 1 \leq i \leq m.\end{gathered} \tag{3.24}$$

This means that we consider the Baker-Akhiezer function $\Psi(x, y, t; k)$ with the essential singularity $\exp(kx + k^n y + Q_+(k)t)$, and construct the operators L and A by requiring that $(\partial_y - L)\Psi = (\partial_t - A)\Psi = 0$. The pair (L, A) provides then a solution of the zero-curvature equation. By rescaling t, we can assume that A is monic. Altogether, we have restricted the geometric map $\mathcal{G}$ of (3.9) to a map on finite-dimensional spaces, which we still denote by $\mathcal{G}$

$$\begin{gathered}\mathcal{G} : (\Gamma, E; \gamma_1, \ldots \gamma_g) \to (\Gamma, P, z) \to \; (L), \\ \mathcal{G} : (\Gamma, E, Q; \gamma_1, \ldots \gamma_g) \to (\Gamma, P, z, t) \to \; (L, A).\end{gathered} \tag{3.25}$$

Here we have indicated explicitly the choice of time in the geometric data. The proper interpretation of the full geometric data $(\Gamma, E, Q; \gamma_1, \ldots \gamma_g)$ is as a point in the bundle $\mathcal{N}_g^g(n,m)$ over $\mathcal{M}_g(n,m)$, whose fiber is the g-th symmetric power $S^g(\Gamma)$ of the curve. The g-th symmetric power can be identified with the Jacobian of Γ via the Abel map

$$(\gamma_1,\dots,\gamma_g)\to\phi_j=\sum_{i=1}^{g}\int_{P_1}^{\gamma_i}d\omega_j. \tag{3.26}$$

More generally, we can construct the bundles $\mathcal{N}_g^k(n,m)$ and $\mathcal{N}_g^k(n)$ with fiber $S^k(\Gamma)$ over the bases $\mathcal{M}_g(n,m)$ and $\mathcal{M}_g(n)$ respectively

$$\begin{array}{ccc} S^k(\Gamma) & \longrightarrow & \mathcal{N}_g^k(n,m)\\ & & \downarrow\\ & & \mathcal{M}_g(n,m)\end{array}\qquad \begin{array}{ccc} S^k(\Gamma) & \longrightarrow & \mathcal{N}_g^k(n)\\ & & \downarrow\\ & & \mathcal{M}_g(n)\end{array} \tag{3.27}$$

Thus the bundles $\mathcal{N}_g^{k=1}(n,m)\equiv\mathcal{N}_g(n,m)$, and $\mathcal{N}_g^{k=1}(n)\equiv\mathcal{N}_g(n)$ are the analogues in the our context of the universal curve. Returning to soliton equations, the geometric map $\mathcal{G}$ of (3.25) can now be succinctly described as a map from the fibrations $\mathcal{N}_g^g(n)$ and $\mathcal{N}_g^g(n,m)$ into the spaces respectively of operators L and pairs (L,A) of operators

$$\begin{aligned}\mathcal{G}&:\mathcal{N}_g^g(n)\to(L),\\ \mathcal{G}&:\mathcal{N}_g^g(n,m)\to(L,A).\end{aligned} \tag{3.28}$$

We emphasize that, although the operators in its image are not all periodic operators, the ones arising later upon restriction of $\mathcal{G}$ to suitable subvarieties of $\mathcal{N}_g^g(n,m)$ and $\mathcal{N}_g^g(n)$ with integral periods (c.f. Section III.C) will be.

We conclude this section by observing that, in the preceding construction, L and A depend only the singular part of Q, and hence are unaffected if dQ is shifted by a holomorphic differential. As we shall see below, the appropriate normalization in soliton theory is the *real normalization* by which we require that

$$\mathrm{Re}\oint_C dQ=0. \tag{3.29}$$

In the study of N=2 supersymmetric gauge theories, holomorphicity is a prime consideration, and we shall rather adopt in this context the *complex normalization*

$$\oint_{A_j}dQ=0,\ 1\le j\le g. \tag{3.30}$$

We note that each normalization provides an imbedding of $\mathcal{M}_g(n)$ into $\mathcal{M}_g(n,1)$, by making the choice $Q_+(K)=K$, with periods satisfying either (3.29) or (3.30), so that the operator A is just $A=\partial_x$ in either case. The image of $\mathcal{M}_g(n)$ in $\mathcal{M}_g(n,1)$ does depend on the normalization, however.

C. Geometric Symplectic Structures

We begin by discussing the basic local geometry of the moduli spaces $\mathcal{M}_g(n)$ and $\mathcal{M}_g(n,m)$. They are complex manifolds with only orbifold singularities, of dimensions

$$\begin{aligned} \dim \mathcal{M}_g(n) &= 4g - 3 + 2N + \sum_{\alpha=1}^{N} n_\alpha, \\ \dim \mathcal{M}_g(n,m) &= 5g - 3 + 3N + \sum_{\alpha=1}^{N} (n_\alpha + m_\alpha). \end{aligned} \tag{3.31}$$

Indeed, the number of degrees of freedom of an Abelian integral E with poles of order $n = (n_\alpha)$ is $1 + \sum_{\alpha=1}^{N}(n_\alpha + 1) - 1 + g = N + g + \sum_{\alpha=1}^{N} n_\alpha$, where the first 1 corresponds to the additive constant, and the remaining integer on the left is the dimension of meromorphic differentials with poles of order $\leq n_\alpha + 1$ at each P_α. For $g > 1$, the dimension of the moduli space of Riemann surfaces with N punctures is $3g-3+N$, which leads immediately to (3.31). For $g \leq 1$, it is easily verified that the same formula (3.31) holds, although the counting has to incorporate holomorphic vector fields and is slightly different in intermediate stages.

We can introduce explicit local coordinates on $\mathcal{M}_g(n,m)$. To obtain well-defined branches of Abelian integrals, we cut apart the Riemann surface Γ along a canonical homology basis A_i, B_j, $i,j = 1,\ldots,g$, and along cuts from P_1 to P_α for each $2 \leq \alpha \leq N$. Locally on $\mathcal{M}_g(n,m)$, this construction can be carried out continuously, with paths homotopic by deformations not crossing any of the poles. Denote the resulting surface by Γ_{cut}. On Γ_{cut}, the Abelian integrals E and Q become single-valued holomorphic functions, and we can introduce the one-form $d\lambda$ by

$$d\lambda = Q\, dE. \tag{3.32}$$

We observe that $d\lambda$ has a singularity of order $n_\alpha + m_\alpha + 1$ at each puncture P_α. Now the Abelian integral E defines a coordinate system z_α near each P_α by

$$E = z_\alpha^{-n_\alpha} + R_\alpha^E \log z_\alpha, \tag{3.33}$$

when n_α is strictly positive. When $n_\alpha = 0$, we write instead

$$E = R_\alpha^E \log z_\alpha. \tag{3.34}$$

The coordinate z_α can be used to fix the additive normalization of the Abelian integral λ, and to describe its Laurent expansion near each puncture. Thus we fix the additive constant in λ by demanding that its expansion in z_1 near P_1 have no constant term. The parameters $T_{\alpha,i}$, $1 \leq i \leq n_\alpha + m_\alpha$, R_α^λ, $2 \leq \alpha \leq N$, can then be defined by

$$\begin{aligned} T_{\alpha,i} &= -\frac{1}{i} \operatorname{Res}_{P_\alpha}(z_\alpha^i\, d\lambda),\ 1 \leq \alpha \leq N,\ 1 \leq i \leq n_\alpha + m_\alpha \\ R_\alpha^\lambda &= \operatorname{Res}_{P_\alpha}(d\lambda). \end{aligned} \tag{3.35}$$

The parameters $T_{\alpha,i}$ $(1 \leq \alpha \leq N)$ and R_α^λ $(2 \leq \alpha \leq N)$ account for $\sum_{\alpha=1}^N (n_\alpha + m_a) + N - 1$ parameters. The remaining parameters needed to parametrize $\mathcal{M}_g(n,m)$ consist of the $2N-2$ residues of dE and dQ

$$R_\alpha^E = \mathrm{Res}_{P_\alpha}\, dE, \; R_\alpha^Q = \mathrm{Res}_{P_\alpha}\, dQ, \; \alpha = 2, \ldots, N, \tag{3.36}$$

and the following $5g$ parameters which account for the presence of non-trivial topology

$$\tau_{A_i,E} = \oint_{A_i} dE, \; \tau_{B_i,E} = \oint_{B_i} dE, \tag{3.37}$$

$$\tau_{A_i,Q} = \oint_{A_i} dQ, \; \tau_{B_i,Q} = \oint_{B_i} dQ, \tag{3.38}$$

$$a_i = \oint_{A_i} Q\, dE, \; i = 1, \ldots, g. \tag{3.39}$$

Theorem 13. *Let $\mathcal{D}$ be the open set in $\mathcal{M}_g(m,n)$ where the zero divisors of dE and dQ, namely the sets $\{z; dE(z) = 0\}$ and $\{z; dQ(z) = 0\}$, do not intersect. Then*

(a) *Near each point in $\mathcal{D}$, the $5g - 3 + 3N + \sum_{\alpha=1}^N (n_\alpha + m_\alpha)$ parameters R_α^E, R_α^Q, R_α^λ, $T_{\alpha,k}$, $\tau_{A_i,E}$, $\tau_{B_i,E}$, $\tau_{A_i,Q}$, $\tau_{B_i,Q}$, a_i have linearly independent differentials, and thus define a local holomorphic coordinate system for $\mathcal{M}_g(n,m)$;*

(b) *The joint level sets of the set of all parameters except a_i define a smooth g-dimensional foliation of $\mathcal{D}$, independent of the choices we made to define the coordinates themselves.*

Theorem 10 is proved in [39]. Since $\mathcal{M}_g(n)$ can be imbedded in $\mathcal{M}_g(n,m)$ by choosing Q to have Laurent expansion $Q_+(K) = K$ and fixing its A_k periods, Theorem 1 also provides local coordinates for $\mathcal{M}_g(n)$. Specifically, the coordinates of $\mathcal{M}_g(n)$ which arise this way are

- $T_{\alpha,i}$ with $1 \leq i \leq n_\alpha$ for $\alpha \geq 2$, $T_{1,i}$ with $1 \leq i \leq n_1 - 1$ (since the normalization $Q = K^{-1} + O(K)$ fixes the coefficients of the two leading terms T_{1,n_1+1} and T_{1,n_1} in the singularity expansion of λ near P_1 to be $T_{1,n_1+1} = \frac{n_1}{n_1+1}$, $T_{1,n_1} = 0$);
- the $2N-2$ residues of dE and $d\lambda$ at P_α for $\alpha \geq 2$;
- the B_k periods of dQ, the A_k and B_k periods of dE, and the A_k periods of $d\lambda$,

for a total of $4g - 3 + 2N + \sum_{\alpha=1}^N n_\alpha$.

The intrinsic foliation obtained in Theorem 10 is central to our considerations, and we shall refer to it as the *canonical foliation*. Our goal is to construct now a symplectic form ω on the complex $2g$-dimensional space obtained by restricting the fibration $\mathcal{N}_g^g(n,m)$ to a g-dimensional leaf $\mathcal{M}$ of the canonical foliation of $\mathcal{M}_g(n,m)$. For this, we need to extend the differentials dE and dQ to one-forms on the whole fibration, and distinguish between the two ways this can be done.

One way is to consider the Abelian integrals E and Q as multi-valued functions on the fibration. Despite their multivaluedness, their differentials along any leaf of the canonical fibration are well-defined. In fact, E and Q are well-defined in a small neighborhood of the puncture P_1. The ambiguities in their values anywhere on each Riemann surface consist only of integer combinations of their residues or periods along closed cycles. Thus they are constant along any leaf of the canonical foliation, and disappear upon differentiation. The differentials along the fibrations obtained this way will be denoted by δE and δQ. Restricted to vectors tangent to the fiber, they reduce of course to the differentials dE and dQ.

The other way is to trivialize the fibration by using an Abelian integral, say E, as local coordinate in Γ (c.f. (3.21) and (3.22)). Equivalently, we note that at any point of $\mathcal{N}$, the varieties $E =$ *constant* are intrinsic and transversal to the fiber. Thus any one-form dQ on the fiber can be extended to a one-form on the total space of the fibration, by making it zero on vectors tangent to these varieties. We still denote this one-form by dQ. In the case of the Abelian integral E, we have $\delta E = dE$, but this is not true in general. To compare dQ with δQ, let $a_1, \dots, a_g$ be local coordinates for the leaf. Then $a_1, \dots, a_g, E$ are local coordinates for the whole fibration, $da_1, \dots, da_g$, dE are a basis of one-forms, and $dQ = \frac{dQ}{dE} dE$. On the other hand,

$$\delta Q = dQ + \sum_{i=1}^{g} \frac{\partial Q}{\partial a_i} da_i \equiv dQ + \delta^E Q. \tag{3.40}$$

Similarly, the full differential $\delta(Q\,dE)$ on the total space of the fibration is well-defined, despite the multivaluedness of Q. The partial derivatives $\partial_{a_i}(Q\,dE)$ along the base are all holomorphic, since both the singularities and the ambiguities in the differential are constant, and disappear upon differentiation. Recalling that $d\omega_j$ denotes the basis of holomorphic differentials dual to the homology basis A_k, B_k, we can then write

$$\frac{\partial}{\partial a_i}(Q\,dE) = d\omega_i. \tag{3.41}$$

By extending this construction to the fibration of Jacobians over a leaf $\mathcal{M}$ of the canonical foliation, we obtain the desired geometric symplectic form. Furthermore, this symplectic form coincides with the symplectic form constructed in Section II.E, upon imbedding the fibration of Jacobians in the space of soliton solutions of the equation $[\partial_y - L, \partial_t - A] = 0$ by the geometric map $\mathcal{G}$ of (3.28) [39]:

Theorem 14. *(a) The following two-form on the fibration* $\mathcal{N}^g(n,m)$ *restricted* *to a leaf $\mathcal{M}$ of the canonical foliation of $\mathcal{M}_g(n,m)$*

$$\omega_{\mathcal{M}} = \delta\left(\sum_{i=1}^{g} Q(\gamma_i)dE(\gamma_i)\right) = \sum_{i=1}^{g} \delta Q(\gamma_i)\wedge dE(\gamma_i) = \sum_{i=1}^{g} da_i \wedge d\omega_i \qquad (3.42)$$

defines a symplectic form; (b) Under the geometric correspondence $\mathcal{G}$, we have

$$\omega_{\mathcal{M}} = \mathcal{G}^*(\omega_m) \qquad (3.43)$$

where ω_m is the symplectic form constructed in Theorem 7.

It is instructive enough to present a proof of Theorem 11 for the basic symplectic structure defined by (2.62) on a space of linear operators L with scalar coefficients. In this case $Q = p$.

Let Ψ, Ψ^+ be the Baker-Akhiezer and the dual Baker-Akhiezer functions corresponding to the algebraic-geometric data (3.25) (c.f. Sections III.A and III.B). Then

$$\mathcal{G}^*(\omega) = \mathrm{Res}_P\langle\Psi^*\delta L\wedge\delta\Psi\rangle dp,\ \Psi^* = \Psi^+(\langle\Psi^+\Psi\rangle)^{-1}.$$

The differential

$$d\Omega^* = \langle\Psi^*\delta L\wedge\delta\Psi\rangle dp$$

is meromorphic. Thus its residue at P is the opposite of the sum of its other residues. The residues of $d\Omega^*$ at the poles γ_s of Ψ are given by

$$\mathrm{Res}_{\gamma_s} d\Omega^* = \langle\Psi^*\delta L\Psi\rangle\wedge\delta p(\gamma_s) = \delta E(\gamma_s)\wedge\delta p(\gamma_s).$$

In general, poles of $d\Omega^*$ may also arise from the zeroes q_i of dp, $dp(q_i) = 0$, which are singular points of the connection on $\mathcal{N}$ defined by dp. However,

$$\mathrm{Res}_{q_i} d\Omega^* = \mathrm{Res}_{q_i}\langle\Psi^*\delta L\, d\Psi\rangle\wedge\delta p(q_i) = 0,$$

in view of the relations $\Psi^*\delta L = \delta\Psi^*(\partial_y - L)$ and $(\partial_y - L)d\Psi = 0$. Theorem 11 is proved.

Remark. In general, when the Lax operator L explicitly depends on a spectral parameter (as in the example of the Calogero-Moser system below), a special gauge should be chosen in order for $d\Omega^*$ to be regular at the branching points of the connection.

D. Generalizations and Extensions

The above set-up can be easily adapted to a variety of important equations.

For the 2D Toda lattice, the differential operator ∂_x is replaced by a difference operator. Thus for the geometric data, we consider the moduli space of Riemann surfaces Γ with two punctures $P_\pm$, and Abelian differentials E, Q which are real-normalized, with E having a pole of order 1 at P_+, while dQ has simple poles at $P_\pm$. This is a leaf in the foliation of $\mathcal{M}_g(1,0;0,0)$. The discrete analogue of the Baker-Akhiezer function $\Psi(z;t)$ is now given by a sequence $\Psi_n(t_+,t_-;z)$ of Baker-Akhiezer functions characterized by

$$\Psi_n(t_+,t_-;z) = z_\pm^{\pm n}\left(\sum_{s=0}^{\infty}\xi_s(n,t)z_\pm^s\right)\exp(t_+z_+^{-1}+t_-z_-^{-1}) \tag{3.44}$$

As in Section II.I, we define the operators L and A as difference operators acting on sequences ψ_n which satisfy $(\partial_{t_+}-L)\Psi = (\partial_{t_-}-A)\Psi = 0$. Their coefficients c_n, v_n, ϕ_n are now functions of the geometric data. In analogy with (3.43), we have then

$$\begin{aligned}\omega_{\mathcal{M}} = \mathcal{G}^*(\omega_{\mathcal{M}}) &= -\sum_{\alpha=\pm}\mathrm{Res}_{P_\pm}\langle\Psi_n^*\delta L\wedge\delta\Psi_n\rangle dp = \sum_n\langle\delta\phi_n\wedge\delta v_n\rangle,\\ dp &= \mp\frac{dz}{z},\ z\to P_\pm.\end{aligned} \tag{3.45}$$

Note that the difference between this formula and the one in Section II.J is due to the choice t_+ instead of t_++t_- as the second variable.

In the case of $N\times N$ matrix equations $[\partial_y-L,\partial_t-A]=0$, we take the geometric data to consist of surfaces Γ with N punctures P_α, and E, Q to be real-normalized Abelian integrals with poles of order n and m at P_α. Let a_α be given coefficients, with $a_1=1$. The leaf $\mathcal{M}$ corresponding to the above specifications for E, Q, combines with the space of parameters a_α to a product space $\mathcal{M}\times\mathbf{C}^{N-1}$ of dimension $g+N-1$. On the fibration $\mathcal{N}^{g+N-1}$ above the product $\mathcal{M}\times\mathbf{C}^{N-1}$, we can construct the symplectic form

$$\omega_{\mathcal{M}} = \sum_{s=1}^{g+N-1}Q(\gamma_s)dE(\gamma_s). \tag{3.46}$$

Now corresponding to the preceding geometric data are local coordinates z_α near each puncture P_α given in analogy with (3.21) and (3.22) by $E=a_\alpha z_\alpha^{-n}$, polynomials $Q_{\alpha,+}(z_\alpha^{-1})$ which are the singular parts of Q near P_α, and thus a vector Baker-Akhiezer function $\Psi(z;x,y,t) = (\Psi_\alpha(z;x,y,t))_{\alpha=1}^N$ with the following essential singularity near P_β

$$\Psi_\alpha(z;x,y,t) = (\delta_{\alpha\beta}+\sum_{s=1}^{\infty}\xi_s^{\alpha\beta}(x,y,t)z_\beta^s)\exp(z_\beta^{-1}x+a_\beta z_\beta^{-n}y+Q_{\beta,+}(z_\beta^{-1})t). \tag{3.47}$$

As in Section II.I, there exist then unique matrix operators L,A so that $(\partial_y-L)\Psi=(\partial_t-A)\Psi=0$. They have expressions of the form in (2.93) and subsequent equations. We observe that the case where Q has only simple poles ($m_i=1$, $Q_{\beta,+}=z_\beta^{-1}$) is the matrix generalization of the scalar case considered earlier, where Q still has the interpretation of a quasi-momentum.

On the space of such operators, we had defined in Theorem 8 a symplectic form ω. Here again, the geometric and the formal symplectic forms (3.46) and

(2.96) correspond to one another under the basic geometric correspondence $\mathcal{G}$. More precisely,

$$\omega_{\mathcal{M}} = \sum_{\alpha=1}^{N} \mathrm{Res}_{P_\alpha} \langle \Psi^* (A_m^{(1)} \delta L - L^{(1)} \delta A_m) \wedge \delta \Psi \rangle dp = \mathcal{G}^*(\omega_m), \tag{3.48}$$

where

$$\omega_m = \mathrm{Res}_\infty \,\mathrm{Tr} \langle \Psi_0^* (A_m^{(1)} \delta L - L^{(1)} \delta A_m) \wedge \delta \Psi_0 \rangle dk. \tag{3.49}$$

In the case $m_i = 1$, we have $A = \partial_x$, and this relation reduces to

$$\omega_{\mathcal{M}} = - \sum_{\alpha=1}^{N} \mathrm{Res}_{P_\alpha} \langle \Psi^* \delta L \wedge \delta \Psi \rangle dp = \mathcal{G}^*(\omega).$$

with ω given by (2.96).

Similarly, our formalism can easily identify the action coordinates for the elliptic Calogero-Moser system, an issue which had been resolved only relatively recently [29]. We recall that the elliptic Calogero-Moser system is a system of N identical particles on a line, interacting with each other via the potential $V(x) = \wp(x)$

$$\ddot{x}_i = 4 \sum_{j \neq i} \wp'(x_i - x_j), \quad \wp(x) = \frac{d\wp(x)}{dx}. \tag{3.50}$$

Here $\wp(x) = \wp(x|\omega, \omega')$ is the Weierstrass elliptic function with periods 2ω, $2\omega'$, and ω, ω' are fixed parameters. The complete solution of the elliptic Calogero-Moser system was constructed by geometric methods in [35]. There an explicit Lax pair (L, M) was found, depending on a spectral parameter z varying on the torus $\mathbf{C}/(2\omega\mathbf{Z} + 2\omega'\mathbf{Z})$. Thus the dynamical system (3.50) is equivalent to the Lax equation $\dot{L} = [M, L]$, with L and M $N \times N$ matrices given by

$$\begin{aligned} L_{ij}(z) &= p_i \delta_{ij} + 2(1 - \delta_{ij}) \Phi(x_i - x_j, z), \; p_i = \dot{x}_i \\ M_{ij}(z) &= 2\delta_{ij} \sum_{k \neq i} \wp(x_i - x_j) + 2(1 - \delta_{ij}) \Phi'(x_i - x_j) \end{aligned} \tag{3.51}$$

and $\Phi(x, z)$ is the function

$$\Phi(x, z) = \frac{\sigma(z - x)}{\sigma(z)\sigma(x)} e^{\zeta(z)x} \tag{3.52}$$

with $\sigma(z)$, $\zeta(z)$ the usual Weierstrass elliptic functions. In view of the Lax equation, the characteristic polynomial $R(k, z) = \det(2k + L(z))$ is time-independent, and defines a time-independent spectral curve Γ

$$R(k, z) \equiv \sum_{i=0}^{N} r_i(z) k^i = 0, \tag{3.53}$$

where the $r_i(z)$ are elliptic functions of z. The Jacobians of the spectral curves Γ are levels of the involutive integrals of the system. In particular, we obtain *angle variables* φ_i by choosing 2π-periodic coordinates on them. However, as noted earlier, the identification of the canonically conjugate *action* variables is more difficult and has been carried out only recently [29]. In this work, it was shown that the Calogero-Moser sytem can be obtained through a Hamiltonian reduction of a Hitchin system, and as a result, the action-variables a_i are the periods of the differential $k\,dz$ along the A-cycles of the spectral curve Γ.

We can derive this result directly from our approach. From our viewpoint, the leaf $\mathcal{M}$ corresponding to the elliptic Calogero-Moser system is given by (Γ, k, z), where Γ is a Riemann surface of genus $g = N$, k is a function with simple poles at g points $P_1, \dots, P_g$, dz is a holomorphic Abelian differential whose periods as well as integrals $\int_{P_1}^{P_\alpha} dz$, $2 \leq \alpha \leq g$, are all on the lattice spanned by 2ω and $2\omega'$, and the residues of $k\,dz$ at P_α, $2 \leq \alpha \leq N$ are given by

$$\mathrm{Res}_{P_\alpha}(k\,dz) = 1, \; 2 \leq \alpha \leq N. \tag{3.54}$$

The analogues of the Baker-Akhiezer function and its dual are respectively the column vector $C(P) = (c_1, \dots, c_N)$ and the row vector $C^+(P) = (c_1^+, \dots, c_N^+)$, satisfying

$$\begin{aligned} &(2k + L(z))C = 0, \qquad C^+(2k + L(z)) = 0, \\ &\sum_i c_i \Phi(-x_i, z) = 1, \qquad \sum_{i=1} c_i^+ \Phi(x_i, z) = 1. \end{aligned} \tag{3.55}$$

Here $P = (k, z)$ is a point of the spectral curve Γ. The vectors $C(P)$ and $C^+(P)$ are meromorphic functions on Γ outside the points P_α on Γ corresponding to $z = 0$, and have each N poles. We denote these poles by $\gamma_1, \dots, \gamma_N$ and $\gamma_1^+, \dots, \gamma_N^+$, respectively. Near the points P_α, these vectors have the form

$$\begin{aligned} &c_i(z) = z(N^{-1} + O(z))e^{x_i z^{-1}}, \qquad c_i^+(z) = z(N^{-1} + O(z))e^{-x_i z^{-1}}, \qquad \text{for } \alpha = 1, \\ &c_i(z) = (c_i^\alpha + O(z))e^{x_i z^{-1}}, \qquad c_i^+(z) = (c_i^{\alpha,+} + O(z))e^{-x_i z^{-1}}, \qquad \text{for } \alpha > 1, \end{aligned} \tag{3.56}$$

where the coefficients c_i^α for $\alpha > 1$ satisfy

$$\sum_i c_i^\alpha = \sum_i c_i^{\alpha,+} = 0. \tag{3.57}$$

Let us make the gauge transformation

$$\tilde{L} = gLg^{-1}, \; \tilde{C} = gC, \; \tilde{C}^+ = C^+ g^{-1}, \; g_{ij} = e^{kx_i}\delta_{ij}.$$

The geometric symplectic form $\omega_{\mathcal{M}}$ constructed in Theorem 11 becomes in this case[1]

$$\omega_{\mathcal{M}} = \delta\left(\sum_{s=1}^N k(\gamma_s)dz\right) = \frac{1}{2}\sum_{\alpha=1}^N \mathrm{Res}_{P_\alpha} \frac{\langle \delta\tilde{C}^+ \wedge \delta\tilde{L}\tilde{C}\rangle}{\langle \tilde{C}^+\tilde{C}\rangle} dz, \tag{3.57}$$

[1]In [39] the operators L and C in equation (70) should be replaced by their gauge equivalent counterparts $\tilde{L}$ and $\tilde{C}$.

where $\langle f^+ g \rangle$ denotes the usual pairing between column vectors and row vectors. Substituting in the expansion (3.56), we obtain

$$\delta\left(\sum_{s=1}^{N} k(\gamma_s)dz\right) = \frac{1}{2}\sum_{i}^{N} \delta p_i \wedge \delta x_i. \tag{3.59}$$

This identifies our geometric symplectic form with the canonical symplectic form for the Calogero-Moser dynamical system. Since by construction (c.f.(3.42)), the geometric symplectic form admits the periods a_i of $k\,dz$ around A_i cycles as action variables dual to the angle variables on the Jacobian, our argument is complete.

IV. Whitham Equations

A. Non-linear WKB Methods in Soliton Theory

We have seen that soliton equations exhibit a unique wealth of exact solutions. Nevertheless, it is desirable to enlarge the class of solutions further, to encompass broader data than just rapidly decreasing or quasi-periodic functions. Typical situations arising in practice can involve Heaviside-like boundary conditions in the space variable x, or slowly modulated waves which are not exact solutions, but can appear as such over a small scale in both space and time.

The non-linear WKB method (or, as it is now also called, the Whitham method of averaging) is a generalization to the case of partial differential equations of the classical Bogolyubov-Krylov method of averaging. This method is applicable to nonlinear equations which have a moduli space of exact solutions of the form $u_0(Ux + Wt + Z|I)$. Here $u_0(z_1, \ldots, z_g|I)$ is a periodic function of the variables z_i; $U = (U_1, \ldots, U_g)$, $W = (W_1, \ldots, W_g)$ are vectors which like u itself, depend on the parameters $I = (I_1, \ldots, I_N)$, i.e. $U = U(I)$, $V = V(I)$. (A helpful example is provided by the solutions (3.10) of the KP equation, where I is the moduli of a Riemann surface, and U, V, W are the B_k-periods of its normalized differentials $d\Omega_1$, $d\Omega_2$, and $d\Omega_3$.) These exact solutions can be used as a leading term for the construction of asymptotic solutions

$$u(x,t) = u_0(\varepsilon^{-1}S(X,T) + Z(X,T)|I(X,T)) + \epsilon u_1(x,t) + \epsilon^2 u_2(x,t) + \ldots, \tag{4.1}$$

where I depend on the *slow* variables $X = \epsilon x, T = \epsilon t$ and and ϵ is a small parameter. If the vector-valued function $S(X,T)$ is defined by the equations

$$\partial_X S = U(I(X,T)) = U(X,T), \quad \partial_T S = W(I(X,T)) = W(X,T), \tag{4.2}$$

then the leading term of (4.1) satisfies the original equation up to first order one in ϵ. All the other terms of the asymptotic series (4.1) are obtained from the non-homogeneous linear equations with a homogeneous part which is just the linearization of the original non-linear equation on the background of the exact solution u_0. In general, the asymptotic series becomes unreliable on scales of the original variables x and t of order ϵ^{-1}. In order to have a reliable approximation, one needs to require a special dependence of the parameters $I(X,T)$. Geometrically, we note that $\epsilon^{-1}S(X,T)$ agrees to first order with $Ux+Vy+Wt$, and x,y,t are the fast variables. Thus $u(x,t)$ describes a motion which is to first order the original *fast* periodic motion on the Jacobian, combined with a slow drift on the moduli space of exact solutions. The equations which describe this drift are in general called *Whitham equations*, although there is no systematic scheme to obtain them.

One approach for obtaining these equations in the case when the original equation is Hamiltonian is to consider the Whitham equations as also Hamiltonian, with the Hamiltonian function being defined by the average of the original one. In the case when the phase dimension g is bigger than one, this approach does not provide a complete set of equations. If the original equation has a number of integrals one may try to get the complete set of equations by averaging all of them. This approach was used in [62] where Whitham equations were *postulated* for the finite-gap solutions of the KdV equation. The geometric meaning of these equations was clarified in [26]. The Hamiltonian approach for the Whitham equations of (1+1)-dimensional systems was developed in [23] where the corresponding bibliography can also be found.

In [36] a general approach for the construction of Whitham equations for finite-gap solutions of soliton equations was proposed. It is instructive enough to present it in case of the zero curvature equation (2.1) with scalar operators.

Recall from Sections III.A and III.B that the coefficients $u_i(x,y,t)$, $v_j(x,y,t)$ of the finite-gap operators L_0 and A_0 satisfying (2.1) are of the form (c.f. (3.10))

$$u_i = u_{i,0}(Ux+Vt+Wt+Z|I), \quad v_j = v_{j,0}(Ux+Vt+Wt+Z|I), \tag{4.3}$$

where $u_{i,0}$ and $v_{j,0}$ are differential polynomials in θ-functions and I is any coordinate system on the moduli space $\mathcal{M}_g(n,m)$.

We would like to construct operator solutions of (2.1) of the form

$$L = L_0 + \epsilon L_1 + \dots, \quad A = A_0 + \epsilon A_1 + \dots, \tag{4.4}$$

where the coefficients of the leading terms have the form

$$\begin{aligned} u_i &= u_{i,0}(\epsilon^{-1}S(X,Y,T) + Z(X,Y,T)|I(X,Y,T)), \\ v_j &= v_{j,0}(\epsilon^{-1}S(X,Y,T) + Z(X,Y,T)|I(X,Y,T)) \end{aligned} \tag{4.5}$$

From Section III.B, we also recall that $\mathcal{N}_g^1(n,m)$ is the bundle over $\mathcal{M}_g(n,m)$ with the corresponding curve Γ as fiber. If I is a system of coordinates on $\mathcal{M}_g(n,m)$, then we may introduce a system of coordinates $(z,\ I)$ on $\mathcal{N}_g^1(n,m)$ by choosing a coordinate along the fiber Γ. The Abelian integrals p,E,Q are multi-valued functions of $(\lambda,\ I)$, i.e. $p=p(\lambda,I)$, $E=E(\lambda,I)$, $Q=Q(\lambda,I)$. If we describe a drift on the moduli space of exact solutions by a map $(X,Y,T)\to I=I(X,Y,T)$, then the Abelian integrals p,E,Q become functions of z,X,Y,T, via

$$\begin{aligned} p(z,X,Y,T) &= p(z,I(X,Y,T)), \\ E(z,X,Y,T) &= E(z,I(X,Y,T)), \\ Q(z,X,Y,T) &= Q(z,I(X,Y,T)). \end{aligned}$$

The following was established in [36]:

Theorem 15. *A necessary condition for the existence of the asymptotic solution (4.4) with leading term (4.3) and bounded terms L_1 and A_1 is that the equation*

$$\frac{\partial p}{\partial z}\left(\frac{\partial E}{\partial T}-\frac{\partial Q}{\partial Y}\right)-\frac{\partial E}{\partial z}\left(\frac{\partial p}{\partial T}-\frac{\partial Q}{\partial X}\right)+\frac{\partial Q}{\partial z}\left(\frac{\partial p}{\partial Y}-\frac{\partial E}{\partial X}\right)=0 \qquad (4.6)$$

be satisfied.

The equation (4.6) is called the Whitham equation for (2.1). It can be viewed as a generalized dynamical system on $\mathcal{M}_g(n,m)$, i.e., a map $(X,Y,T)\to\mathcal{M}_g(n,m)$. Some of its important features are:

• Even though the original two-dimensional system may depend on y, Whitham solutions which are Y-independent are still useful. As we shall see later, this particular case has deep connections with topological field theories. If we choose the local coordinate z along the fiber as $z=E$, then the equation simplifies in this case to

$$\partial_T p = \partial_X Q. \qquad (4.7)$$

We note that it followed immediately from the consistency of (4.2) that we must have

$$\partial_T \oint_{B_k} dp = \partial_X \oint_{B_k} dQ.$$

Thus (4.7) is a strengthening of this requirement which encodes also the solvability term by term of the linearized equations for all the successive terms of the asymptotic series (4.3).

• Naively, the Whitham equation seems to impose an infinite set of conditions, since it is required to hold at every point of the fiber Γ. However, the functions involved are all Abelian integrals, and their equality over the whole of Γ can actually be reduced to a finite set of conditions. To illustrate this point, we consider the Y-independent Whitham equation on the moduli space of curves of the form

$$y^2=\prod_{i=1}^{3}(E-E_i)\equiv R(E).$$

Then the differentials dp and dQ are given by

$$dp = \frac{dE}{\sqrt{R(E)}} \left(E - \frac{\int_{E_2}^{E_3} \frac{E\,dE}{\sqrt{R}}}{\int_{E_2}^{E_3} \frac{dE}{\sqrt{R}}} \right),$$

$$dQ = \frac{dE}{\sqrt{R(E)}} \left(E^2 - \frac{1}{2}(E_1 + E_2 + E_3)E - \frac{\int_{E_2}^{E_3} \frac{(E - \frac{1}{2}\sum_{i=1}^{3} E_i)E\,dE}{\sqrt{R}}}{\int_{E_2}^{E_3} \frac{dE}{\sqrt{R}}} \right).$$

We view p and Q as functions of $(E; E_i)$, with E the coordinate in the fiber Γ, and E_i the coordinates on the moduli space of curves. Near each branch point E_i, $\sqrt{E - E_i}$ is a local coordinate and we may expand

$$\begin{aligned} p &= p(E_i) + \alpha\sqrt{E - E_i} + O(E - E_i), \\ Q &= Q(E_i) + \beta\sqrt{E - E_i} + O(E - E_i). \end{aligned} \tag{4.8}$$

Differentiating with respect to X and T, keeping E fixed, we find that the leading singularities of $\partial_T p$ and $\partial_X Q$ are respectively $-\frac{\alpha}{2\sqrt{E-E_i}}\partial_T E_i$ and $-\frac{\beta}{2\sqrt{E-E_i}}\partial_X E_i$. Since $\frac{\beta}{\alpha} = \frac{dQ}{dp}$, we see that the equation (4.7) implies

$$\partial_T E_i = \left(\frac{dQ}{dp} \right)\bigg|_{E_i} \partial_X E_i. \tag{4.9}$$

Conversely, if the equation (4.9) is satisfied, then $\partial_T p - \partial_X Q$ is regular and normalized, and hence must vanish. Thus the equation (4.7) is actually equivalent to the set of differential equations (4.9).

• The equation (4.7) can be represented in a manifestly invariant form, without explicit reference to any local coordinate system z. Given a map $(X, Y, T) \to \mathcal{M}_g(n, m)$, the pull back of the bundle $\mathcal{N}_g^1(n, m)$ defines a bundle over a space with coordinates X, Y, T. The total space $\mathcal{N}^4$ of the last bundle is 4-dimensional. Let us introduce on it the one-form

$$\alpha = p\,dX + E\,dY + Q\,dT, \tag{4.10}$$

Then (4.7) is equivalent to the condition that the wedge product of $d\alpha$ with itself be zero (as a 4-form on $\mathcal{M}^4$)

$$d\alpha \wedge d\alpha = 0. \tag{4.11}$$

• It is instructive to present the Whitham equation (4.7) in yet another form. Because (4.7) is invariant with respect to a change of local coordinate we may use $p = p(z, I)$ by itself as a local coordinate. With this choice we may view E and Q as functions of p, X, Y and T, i.e. $E = E(p, X, Y, T)$, $Q = Q(p, X, Y, T)$. With this choice of local coordinate (4.7) takes the form

$$\partial_T E - \partial_Y Q + \{E, Q\} = 0, \tag{4.12}$$

where $\{\cdot,\cdot\}$ stands for the usual Poisson bracket of two functions of the variables p and X, i.e.

$$\{f,g\} = f_p g_X - g_p f_X.$$

• In Theorem 12 we had focused on constructing an asymptotic solution for a single equation. This corresponds to a choice of A, and thus of an Abelian differential Q, and the Whitham equation is an equation for maps from (X,Y,T) to $\mathcal{M}_g(n,m)$. As in the case of the KP and other hierarchies, we can also consider a whole hierarchy of Whitham equations. This means that the Abelian integral Q is replaced by the really normalized Abelian integral Ω_i which has the following form

$$\Omega_i = K^i + O(K^{-1}), \quad K^n = E, \tag{4.13}$$

in a neighborhood of the puncture P (compare with (2.48)). The result is a hierarchy of equations on maps of the form

$$(X = T_1, Y = T_2, T = T_3, T_4, \ldots) \to \mathcal{M}_g(n). \tag{4.13}$$

The whole hierarchy may be written in the form (4.11) where we set now

$$\alpha = \sum_i \Omega_i dT_i. \tag{4.14}$$

B. Exact Solutions of Whitham Equations

In [38] a construction of exact solutions to the Whitham equations (4.7) was proposed. In the following section, we shall present the most important special case of this construction, which is also of interest to topological field theories and supersymmetric gauge theories. It should be emphasized that for these applications, the definition of the hierarchy should be slightly changed. Namely, the Whitham equations describing modulated waves in soliton theory are equations for Abelian differentials with a real normalization (3.29). In what follows we shall consider the same equations, but where the real-normalized differentials are replaced by differentials with the complex normalization (3.30). As discussed in Section III.A, the two types of normalization coincide on the subspace corresponding to M-curves, which is essentially the space where all solutions are regular and where the averaging procedure is easily implemented. Thus the two forms of the Whitham hierarchy can be considered as different extensions of the same hierarchy. The second one is an analytic theory, and we shall henceforth concentrate on it.

In this subsection and in the rest of the paper, we shall restrict ourselves to the hierarchy of "algebraic geometric solutions" of Whitham equations, that is, solutions of the following stronger version of the equations (4.11)

$$\frac{\partial}{\partial T_i} E = \{\Omega_i, E\}, \tag{4.15}$$

We note that the original Whitham equations can actually be interpreted as consistency conditions for the existence of an E satisfying (4.15). Furthermore, the solutions of (4.15) can be viewed in a sense as "Y-independent" solutions of Whitham equations, since the equation (4.12) reduces to $\partial_T E + \{E, Q\} = 0$ for Y-independent solutions. They play the same role as Lax equations in the theory of (2+1)-dimensional soliton equations. As stressed earlier, Y-independent solutions of the Whitham hierarchy can be considered even for two-dimensional systems where the y-dependence is non-trivial in general.

Our first step is to show that (4.15) defines a system of commuting flows on $\mathcal{M}_g(n)$. For the sake of simplicity, we assume that there is only one puncture. Let us start with a more detailed description of this space which is a complex manifold with only orbifold singularities. Its complex dimension is equal to (c.f. (3.31)) $\dim \mathcal{M}_g(n) = 4g + n - 1$, and we had constructed a set of local coordinates for it in Theorem 10 and subsequent discussion. Here we require the following slightly different set of coordinates (details can be found in [38]). The first $2g$ coordinates are still the periods of dE,

$$\tau_{A_i,E} = \oint_{A_i} dE, \ \tau_{B_i,E} = \oint_{B_i} dE. \tag{4.16}$$

The differential dE has $2g + n - 1$ zeros (counting multiplicities). When all zeroes are simple, we can supplement (4.16) by the $2g+n-1$ critical values E_s of the Abelian differential E, i.e.

$$E_s = E(q_s), \quad dE(q_s) = 0, \quad s = 1, \dots, 2g + n - 1, \tag{4.17}$$

In general, dE may have multiple zeroes, and we let $D = \sum \mu_s q_s$ be the zero divisor of dE. The degree of this divisor is equal to $\sum_s \mu_s = 2g-1+n$. Consider a small neighborhood of q_s, viewed as a point of the fibration $\mathcal{N}_g^1(n)$, above the original data point m_0 in the moduli space $\mathcal{M}_g(n)$. Viewed as a function on the fibration, E is a deformation of its value $E(z, m_0)$ above the original data point, with multiple critical points q_s. Therefore, on each of the corresponding curve, there exists a local coordinate w_s such that

$$E = w_s^{\mu_s+1}(z, m) + \sum_{i=0}^{\mu_s-1} E_{s,i}(m) w_s^i(z, m). \tag{4.18}$$

The coefficients $E_{s,i}(m)$ of the polynomial (4.18) are well-defined functions on $\mathcal{M}_g(n)$. Together with $\tau_{A_i,E}$, $\tau_{B_i,E}$, they define a system of local coordinates on $\mathcal{M}_g(n)$.If $\mu_s = 1$, then $E_{s,0}$ clearly coincides with the critical value $E(q_s)$ from (4.17).

Let $\mathcal{D}'$ be the open set in $\mathcal{M}_g(n)$ where the zero divisors of dE and dp, namely the sets $\{z; dE(z) = 0\}$ and $\{z; dp(z) = 0\}$, do not intersect and let $\mathcal{D}^0$ be the open set in $\mathcal{M}_g(n)$ where all zeros of dE are simple.

Theorem 16. *The Whitham equations (4.15) define a system of commuting meromorphic vector fields (flows) on $\mathcal{M}_g(n)$ which are holomorphic on*

$\mathcal{D}' \subset \mathcal{M}_g(n)$. On the open set $\mathcal{D}' \cap \mathcal{D}^0$, the equations (4.15) have the form

$$\frac{\partial}{\partial T_j}\tau_{A_i,E} = 0, \quad \frac{\partial}{\partial T_j}\tau_{B_i,E} = 0, \tag{4.18}$$

$$\partial_{T_j} E_s = \frac{d\Omega_j}{dp}(q_s)\partial_X E_s. \tag{4.19}$$

Indeed, the equations (4.15) are fulfilled at each point of Γ, and thus

$$\frac{\partial}{\partial T_i}\oint_C dE = \frac{d\Omega_i}{dp}(z)\partial_X\left(\oint_C dE\right) - \frac{dE}{dp}(z)\partial_X\left(\oint_C d\Omega_i\right) \tag{4.20}$$

The functions dE/dp and $d\Omega_i/dp$ are linearly independent. It follows that the periods of dE are constants. The equations (4.19) coincide with (4.15) at the point q_s (where dE equals zero).

In order to complete the proof of Theorem 13, it suffices to show that (4.18–4.19) imply (4.15). The equation (4.19) implies that the difference between the left and right hand sides of (4.15) is a meromorphic function $f(z)$ on Γ. This function is holomorphic outside the puncture P and the zeros of dp. At the puncture P, the function $f(z)$ has a pole of order less or equal to $(n-2)$. However, $f(z)$ equals zero at zeros of dE. Hence, the function $g(z) = f(z)\frac{dp}{dE}$ is holomorphic on Γ and equals zero at P. Therefore, $f(z) = 0$ identically. Theorem 13 is proved.

An important consequence of Theorem 13 is that the space $\mathcal{M}_g(n)$ admits a natural foliation, namely by the joint level sets of the functions $\tau_{A_i,E}$, $\tau_{B_i,E}$, which are smooth $(2g+n-1)$-dimensional submanifolds, and which are invariant under the flows of the Whitham hierarchy (4.15). We shall sometimes refer to the leaves $\hat{\mathcal{M}}$ of this foliation as *large leaves*, to stress their distinction from the g-dimensional leaves $\mathcal{M}$ of the canonical foliation of $\mathcal{M}_g(n,m)$.

The special case of the construction of exact solutions to (4.15) in [38] may now be described as follows: the moduli space $\mathcal{M}_g(n,m)$ provides the solutions of the first $n+m$-flows of (4.15) parametrized by $3g$ constants, which are the set of the last three coordinates (3.38–3.39) on $\mathcal{M}_g(n,m)$.

Theorem 17. *Let T_i, $i = 1,\dots,n+m$, $\tau_{A_i,E}$, $\tau_{B_i,E}$, $\tau_{A_i,Q}$, $\tau_{A_i,Q}$, a_i be the coordinates on $\mathcal{M}_g(n,m)$ defined in Theorem 10. Then the projection*

$$\begin{aligned}\mathcal{M}_g(n,m) &\to \mathcal{M}_g(n)\\ (\Gamma,E,Q) &\to (\Gamma,E)\end{aligned} \tag{4.21}$$

defines (Γ,E) as a function of the coordinates on $\mathcal{M}_g(n,m)$. For each fixed set of parameters $\tau_{A_i,E}$, $\tau_{B_i,E}$, $\tau_{A_i,Q}$, $\tau_{A_i,Q}$, a_i, the map $(T_i)_{i=1}^{i=n+m} \to \mathcal{M}_g(n)$ satisfies the Whitham equations (4.15).

Proof. Let us use $E(z)$ as a local coordinate on Γ. Then as we saw earlier, the equations (4.15) are equivalent to the equations $\partial_{T_i}p(E,T) = \partial_X\Omega_i(E,T)$.

These are the compatibility conditions for the existence of a generating function for all the Abelian differentials $d\Omega_i$. In fact, if we set

$$d\lambda = Q\,dE, \tag{4.22}$$

then it follows from the definition of the coordinates that

$$\partial_{T_i} d\lambda = d\Omega_i,\ \partial_x d\lambda = dp, \tag{4.23}$$

(For the proof of (4.23), it is enough to check that the right and the left hand sides of it have the same analytical properties.) □

Theorem 18. *We consider the same parametrization of $\mathcal{M}_g(n,m)$ as in Theorem 14. Then as a function of the parameters T_i, $1 \le i \le n+m$, the second Abelian integral $Q(p,T)$ satisfies the same equations as E, i.e.*

$$\partial_{T_i} Q = \{\Omega_i, Q\}. \tag{4.24}$$

Furthermore

$$\{E, Q\} = 1. \tag{4.25}$$

We note that (4.25) can be viewed as a Whitham version of the so-called string equation (or Virasoro constraints) in a non-perturbative theory of 2-d gravity [19][66].

C. The τ-Function of the Whitham Hierarchy

The solution of the Whitham hierarchy can be succinctly summarized in a single τ-function defined as follows. The key underlying idea is that suitable submanifolds of $\mathcal{M}_g(n,m)$ can be parametrized by $2g+N-1+\sum_{\alpha=1}^{N}(n_\alpha+m_\alpha)$ Whitham times T_A, to each of which is associated a "dual" time T_{DA}, and an Abelian differential $d\Omega_A$. We begin by discussing the simpler case of one puncture, $N=1$. Recall that the coefficients of the pole of $d\lambda$ has provided $n+m$ Whitham times $T_j = -\frac{1}{j}\,\mathrm{Res}(z^j d\lambda)$. Their "dual variables" are

$$T_{Dj} = \mathrm{Res}(z^{-j} d\lambda), \tag{4.26}$$

and the associated Abelian differential are the familiar $d\Omega_i$ of (3.4) (complex normalized). When $g>0$, the moduli space $\mathcal{M}_g(n,m)$ has in addition $5g$ more parameters. We consider only the foliations for which the following $3g$ parameters are fixed

$$\oint_{A_k} dE,\ \oint_{B_k} dE,\ \oint_{A_k} dQ. \tag{4.27}$$

Thus the case $g>0$ leads to two more sets of g Whitham times each

$$a_k = \oint_{A_k} d\lambda,\ T_k^E = \oint_{B_k} dQ. \tag{4.28}$$

Their dual variables are

$$a_{Dk} = -\frac{1}{2\pi i}\oint_{B_k} d\lambda,\ T^E_{Dk} = \frac{1}{2\pi i}\oint_{A_k^-} E\, d\lambda. \tag{4.29}$$

The corresponding Abelian differentials are respectively the holomorphic differentials $d\omega_k$ and the differentials $d\Omega^E_k$, defined to be holomorphic everywhere on Γ except along the A_j cycles, where they have discontinuities

$$d\Omega^{E+}_k - d\Omega^{E-}_k = \delta_{jk} dE. \tag{4.30}$$

We denote the collection of all $2g+n+m$ times by $T_A = (T_j, a_k, T^E_k)$. In the case of $N > 1$ punctures, we have $2g + \sum_\alpha (n_\alpha + m_\alpha)$ times $(T_{\alpha,j}, a_k, T^E_k)$ and $3N-3$ additional parameters for $\mathcal{M}_g(n,m)$, namely the residues of dQ, dE, and $d\lambda$ at the punctures P_α, $2 \le \alpha \le N$ (c.f. (3.35–3.36)). For simplicity, we only consider the leaves of $\mathcal{M}_g(n,m)$ where

$$\mathrm{Res}_{P_\alpha}(dQ) = 0,\ \mathrm{Res}_{P_\alpha}(dE) = \textit{fixed},\ \ 2 \le \alpha \le N, \tag{4.31}$$

and incorporate among the T_A the residues of $d\lambda$ at P_α, $2 \le \alpha \le N$, $R^\lambda_\alpha = \mathrm{Res}_{P_\alpha}(d\lambda)$ (c.f.(3.35)). The dual parameters to these $N-1$ additional Whitham times are then the regularized integrals

$$R^\lambda_{D\alpha} = -\int_{P_1}^{P_\alpha} d\lambda,\ 2 \le \alpha \le N. \tag{4.32}$$

More precisely, recall that the Abelian integral λ has been fixed by the condition that its expansion near P_1, in terms of the local coordinate z_α defined by E, have no constant term. Near each P_α, $2 \le \alpha \le N$, in view of (4.31), it admits an expansion of the form

$$\lambda(z_\alpha) = \sum_{j=1}^{n_\alpha+m_\alpha} \frac{T_{\alpha,j}}{z_\alpha^j} + \mathrm{Res}_{P_\alpha}(d\lambda)\log z_\alpha + \lambda_\alpha + O(z_\alpha). \tag{4.33}$$

For each α, $2 \le \alpha \le N$, we define the right hand side of (4.32) to be λ_α. The Abelian differential $d\Omega^{(3)}_\alpha$ associated with R^λ_α is the Abelian differential of the third kind, with simple poles at P_1 and P_α, and residue 1 at P_α. In summary, we have the following table

Times	**Dual Times**	**Differentials**
$-\frac{1}{j}\mathrm{Res}_{P_\alpha}(z^j d\lambda)$	$\mathrm{Res}_{P_\alpha}(z^{-j} d\lambda)$	$d\Omega^0_{\alpha,j}$
$\oint_{A_k} d\lambda$	$-\frac{1}{2\pi i}\oint_{B_k} d\lambda$	$d\omega_k$
$\mathrm{Res}_{P_\alpha} d\lambda$	$-\int_{P_1}^{P_\alpha} d\lambda$	$d\Omega^{(3)}_\alpha$
$\oint_{B_k} dQ$	$\frac{1}{2\pi i}\oint_{A_k} E d\lambda$	$d\Omega^E_k$

(4.34)

We can now define the τ-function of the Whitham hierarchy by

$$\tau(T) = e^{\mathcal{F}(T)},$$
$$\mathcal{F}(T) = \frac{1}{2}\sum_{A} T_A T_{DA} + \frac{1}{4\pi i}\sum_{k=1}^{g} a_k T_k^E E(A_k \cap B_k), \tag{4.35}$$

where $A_k \cap B_k$ is the point of intersection of the A_k and B_k cycles. In the case $\mathrm{Res}_{P_\alpha} dE = 0$ we have then (see [38])

Theorem 19. *The derivatives of $\mathcal{F}$ with respect to the $2g+\sum_\alpha(n_\alpha+m_\alpha)+N-1$ Whitham times T_A are given by*

$$\begin{aligned}
\partial_{T_A}\mathcal{F} &= T_{DA} + \frac{1}{2\pi i}\sum_{k=1}^{g} \delta_{a_k,A} T_k^E E(A_k \cap B_k),\\
\partial^2_{T_{\alpha,i},T_{\beta,j}}\mathcal{F} &= \mathrm{Res}_{P_\alpha}(z_\alpha^i d\Omega_{\beta,j}),\\
\partial^2_{a_j,A}\mathcal{F} &= \frac{1}{2\pi i}\left(E(A_k \cap B_k)\delta_{(E,k),A} - \oint_{B_k} d\Omega_A\right),\\
\partial^2_{(E,k)A}\mathcal{F} &= \frac{1}{2\pi i}\oint_{A_k} E\, d\Omega_A,\\
\partial^3_{ABC}\mathcal{F} &= \sum_{q_s}\mathrm{Res}_{q_s}\left(\frac{d\Omega_A d\Omega_B d\Omega_C}{dE\, dQ}\right).
\end{aligned} \tag{4.36}$$

These formulae require some modifications when $\mathrm{Res}_{P_\alpha} dE \neq 0$, which is a case of particular interest in supersymmetric QCD (see Section VI below). In particular, the first derivatives with respect to R_α^λ become [17]

$$\partial_{R_\alpha^\lambda}\mathcal{F} = T_{D\alpha}^\lambda + \frac{1}{2}\pi i \sum_{\beta} c_{\alpha,\beta} R_\beta^\lambda,$$

where $c_{\alpha,\beta}$ is an integer which is antisymmetric in α and β. It is then easy to see that the second derivatives are modified accordingly by constant shifts, while the third derivatives remain unchanged.

We would like to point out that the $2g+\sum_\alpha(n_\alpha+m_\alpha)+N-1$ submanifolds of $\mathcal{M}_g(n,m)$ defined by fixing (4.27) as well as the residues of dE and dQ are yet another version of the $2g+n-1$ large leaf $\hat{\mathcal{M}}$ of the foliation of $\mathcal{M}_g(n)$ encountered earlier in the case of one puncture. Indeed, imbedding $\mathcal{M}_g(n)$ in $\mathcal{M}_g(n,1)$ by choosing $Q_+ = k$ would fix two Whitham times T_n and T_{n+1}, as we saw after Theorem 10. This reduces the dimension $2g+n+1$ to the desired dimension $2g+n-1$.

We observe that the first derivatives of $\mathcal{F}$ give the coefficients of the Laurent expansions and the periods of the form λ. The second derivatives give the coefficients of the expansions and the periods of the differentials $d\Omega_A$. In that sense the function $\mathcal{F}$ encodes all the information on the Whitham hierarchy. The formulae for the first and the second derivatives with respect to the variables T_i are the analogues in the case of the averaged equations of the corresponding formulae for the τ-function of the KP hierarchy. The formulae for the third derivatives are specific to the Whitham theory. As we shall see later, they are reminiscent rather of marginal deformations of topological or conformal field theories and of special geometry.

Finally, it may be worth noting that the expression (4.35) for $\mathcal{F}$ can be elegantly summarized as

$$\mathcal{F} = \int_\Gamma \bar{d}\lambda \wedge d\lambda. \tag{4.37}$$

V. Topological Landau-Ginzburg Models on a Riamman Surface

In this section, we shall show that each $2g + n - 1$ leaf of the foliation of $\mathcal{M}_g(n)$ (or, equivalently, of the foliation of $\mathcal{M}_g(n, 1)$ upon imbedding $\mathcal{M}_g(n)$ into $\mathcal{M}_g(n, 1)$) actually parametrizes the marginal deformations of a topological field theory on a surface of genus g. Furthermore, the free energy of these theories coincides with the restriction to the leaf of the exponential of the τ-function for the Whitham hierarchy. We begin with a brief discussion of some key features of topological field theories [63–65][11–13].

A. Topological Field Theories

In general, a two-dimensional quantum field theory is specified by the correlation functions $\langle\phi(z_1)\ldots\phi(z_N)\rangle_g$ of its local physical observables $\phi_i(z)$ on any surface Γ of genus g. Here $\phi_i(z)$ are operator-valued tensors on Γ. The operators act on a Hilbert space of states with a designated vacuum state $|\Omega\rangle$. The correlators $\langle\phi(z_1)\ldots\phi(z_N)\rangle$ usually depend on the background metric on Γ and on the location of the insertion points z_i. In particular, they may develop singularities as z_i approaches z_j. Equivalently, the operator product $\phi_i(z_i)\phi_j(z_j)$ may develop singularities. For example, in a conformal field theory, $\phi_i(z_i)\phi_j(z_j)$ will have an operator product expansion of the form

$$\phi_i(z_i)\phi_j(z_j) = \sum_k c_{ij}^k (z_i - z_j)^{h_i + h_j - h_k}\phi_k(z_j) + \textit{descendants}, \tag{5.1}$$

where h_i is the conformal dimension of ϕ_i. If we let $\phi_0(z)$ be the field corresponding to $|\Omega\rangle$, under the usual *states* $\leftrightarrow$ *fields* correspondence, then we obtain a metric by setting

$$\eta_{ij} = c_{ij}^0. \tag{5.2}$$

Using η_{ij} to raise and lower indices, and noting that $\langle\phi_0(z)\rangle_0 = 1$, we can easily recognize $c_{ijk} = c_{ij}^k\eta_{kl}$ as the three-point function on the sphere

$$c_{ijk} = \langle\phi_i(z_i)\phi_j(z_j)\phi_k(z_k)\rangle_0, \tag{5.3}$$

which is actually independent of the insertion points z_i, z_j, z_k by SL(2,**C**) invariance.

Topological field theories are theories where the correlation functions are actually independent of the insertion points z_i. Thus they depend only on the labels of the fields ϕ_i and the genus g of Γ. This independence implies that the OPE of (5.1) contains no singularity, and that for all practical purposes, the operator product $\phi_i(z_i)\phi_j(z_j)$ can be replaced by the formal operator algebra

$$\phi_i\phi_j = \sum_k c_{ij}^k \phi_k. \tag{5.4}$$

The associativity of operator compositions translates into the associativity of the operator algebra (5.4). Furthermore, the operator algebra is commutative, since factorization of the 4-point function in the s and t channels must give the same answer. If we assume that in a topological field theory, physical correlation functions can be factored through only physical states (as is the case, if the Hilbert space of physical states arises as the cohomology of a nilpotent BRST operator Q acting on a larger Hilbert space containing spurious states), then the correlation functions of the theory on surfaces of all genera can all be expressed explicitly, through factorization, in terms of the structure constants c_{ijk}.

B. Deformations of a Topological Field Theory

We shall be particularly interested in the case where the topological field theory arises as a deformation by parameters t_i of a fixed theory. In a topological field theory, the physical fields ϕ_i have clearly scaling dimension 0. In presence of a BRST symmetry, they actually fit in a multiplet $\Phi_i = (\phi_i^{(0)} = \phi_i, \phi_i^{(1)}, \phi_i^{(2)})$, where the "descendants" $\phi_i^{(d)}$ are tensors of scaling dimension d which satisfy the descent equations $d\phi^{(d)} = \{Q, \phi^{(d+1)}\}$. In particular, $\phi_i^{(2)}$ can be integrated on the surface Γ. If we let $I(0)$ be the Lagrangian of the original theory, we can deform the theory to another theory with Lagrangian

$$I(t) = I(0) - \sum_i t_i \int_\Gamma \phi_i^{(2)}. \tag{5.5}$$

The structure constants $c_{ijk}(t)$ of the new theory are then given by

$$c_{ijk}(t) = \left\langle \phi_i\phi_j\phi_k \exp\left(\sum_i t_i \int_\Gamma \phi_i^{(2)}\right)\right\rangle_0 . \tag{5.6}$$

They define in turn a topological field theory. For deformations around a topological conformal field theory, i.e., a topological field theory whose stress tensor is already traceless before restricting to physical states, the structure constants $c_{ijk}(t)$ are known to satisfy the key compatibility condition $\partial_l c_{ijk} = \partial_k c_{ijl}$

[11–13]. This means that we can find a function $\mathcal{F}$, called the free energy and formally denoted by

$$\mathcal{F}(t) = \left\langle \exp\left(\sum_i t_i \int \Phi_i\right)\right\rangle_0 , \tag{5.7}$$

which satisfies the third derivative condition

$$\partial^3_{ijk}\mathcal{F}(t) = c_{ijk}. \tag{5.8}$$

In terms of $\mathcal{F}$, the commutativity and associativity conditions of the operator algebra with structure constants $c_{ijk}(t)$ become a system of differential equations of third order, called the Witten-Dijkgraaf-Verlinde-Verlinde (WDVV) equation.

C. The Framework of Solitons

In their original work [11–12], Dijkgraaf, Verlinde, and Verlinde derived an explicit expression for the free energy $\mathcal{F}(t)$ in the case of topological Landau-Ginzburg theories. These are the topological theories arising from twisting an N=2 superconformal field theory, which is itself obtained by following the renormalization group flow to the fixed point of a Landau-Ginzburg model. Although the renormalization group flow modifies the kinetic terms in the Landau-Ginzburg action, the superpotential $W(x)$ remains unchanged, and thus characterizes both the associated superconformal and the topological models [30][61][67]. Our goal in this section is to show how this theory can fit in the framework of solitons, and to exhibit the natural emergence of the differential $Q\,dE$ and Whitham times.

We consider first the case of genus 0, with $\Gamma = \{z \in \mathbf{C} \cup \infty\}$. The role of the superpotential $W(x)$ is played in our context by the Abelian integral E with a unique pole of order n at ∞. We consider then a leaf in the space $\mathcal{M}_0(n,1)$, characterized by the condition that $Q = z$, and E is of the form

$$E = z^n + \sum_{i=0}^{n-2} u_i z^i + O(z^{-1}). \tag{5.9}$$

This is of dimension $n-1$, and can be parametrized by the $n-1$ Whitham times T_A, $A = 1, \ldots, n-1$, with the other times fixed to $T_n = 0$, $T_{n+1} = \frac{n}{n+1}$. At each point E on the leaf, the primary fields ϕ_i can be identified with $d\Omega_i/dQ$. The structure constants are defined by

$$\langle \phi_i\phi_j\rangle = \mathrm{Res}_\infty\left(\frac{d\Omega_{(i)}d\Omega_{(j)}}{dE}\right), \quad \langle \phi_i\phi_j\phi_k\rangle = \mathrm{Res}_\infty\left(\frac{d\Omega_{(i)}d\Omega_{(j)}d\Omega_{(k)}}{dE\,dQ}\right). \tag{5.10}$$

We note that dQ corresponds to the field defined by the vacuum. The following can be derived from Theorem 16:

Theorem 20. *Let $\mathcal{F}(T)$ be the τ-function of the (complex normalized) Whitham hierarchy, restricted to the $(n-1)$ dimensional leaf described above. Then*

(i) *the fields ϕ_i anti-diagonalize the pairing η_{ij}, i.e., $\langle\phi_i\phi_j\rangle = \delta_{i+j,n}$;*

(ii) *$\mathcal{F}(T)$ is the free energy of the theory, i.e., $\partial^3_{T_iT_jT_k}\mathcal{F}(T) = c_{ijk}$. In particular, $\mathcal{F}(T)$ satisfies the system of WDVV equations.*

More generally, the case of Γ of genus g (with one puncture to simplify the notation) has been treated in [22][38]. In this case, the relevant leaf within $\mathcal{M}_g(n,1)$ is of dimension $n-1+2g$ and is given by the constraints

$$\begin{aligned} &T_n = 0,\ T_{n+1} = \frac{n}{n+1}, \\ &\oint_{A_k} dE = 0,\ \oint_{B_k} dE = fixed, \\ &\oint_{A_k} dQ = 0. \end{aligned} \tag{5.11}$$

Thus the leaf is parametrized by the $(n-1)$ Whitham times T_A, $A = 1, \ldots, n-1$, and by the periods a_j and T_j^E of (4.28). The fields ϕ_A of the theory need to be augmented accordingly. We take the $2g$ additional ones to be given by $d\omega_j/dQ$ and $d\Omega_j^E/dQ$, where the differentials $d\omega_j$ and $d\Omega_j^E$ are the ones associated to a_j and T_j^E, as described earlier in (4.30).

Theorem 21. *Let η_{AB} and c_{ABC} be defined as*

$$\eta_{A,B} = \sum_{q_s} \mathrm{Res}_{q_s} \frac{d\Omega_A d\Omega_B}{dE}, \tag{5.12}$$

$$c_{ABC} = \sum_{q_s} \mathrm{Res}_{q_s} \frac{d\Omega_A d\Omega_B d\Omega_C}{dEdQ}, \tag{5.13}$$

with q_s the zeroes of dE, and the indices A,B,C running this time through the augmented set of $n-1+2g$ indices given by $T_A = (T_i, a_j, T_{E,j})$. Then

(i) $\eta_{i,j} = \delta_{i+j,n}$, $\eta_{a_j,(E,k)} = \delta_{j,k}$. *All other pairings vanish;*

(ii) *Let $\mathcal{F}(T_i, a_j, T_{E,j})$ be the τ-function of the Whitham hierarchy restricted to the leaf (5.11). Then $\partial^3_{ABC}\mathcal{F}(T) = c_{ABC}$, where A runs through the augmented set of $n-1+2g$ indices.*

Note that in the genus 0 case when $Q = z$, the sum in (5.12–5.13) over the residues at the zeros of dE reduces to the residue at infinity.

Remarkably, the larger spaces $\mathcal{M}_g(n,m)$ can accommodate the gravitational descendants of the fields ϕ_A. More precisely, consider for $g = 0$ the leaf of the space $\mathcal{M}_0(n, mn+1)$ given by the following evident modification of the earlier normalization (5.11)

$$T_{in} = 0,\ i = 1, \ldots, m, \quad T_{nm+1} = \frac{nm}{nm+1}. \tag{5.14}$$

The space of Whitham times is automatically increased to the correct number by taking all the coefficients of $Q\,dE$. The additional $m(n-1)$ fields may be identified with the first m gravitational descendants of the primary fields. Namely, the p-th descendant $\sigma_p(\phi_i)$ of the primary field ϕ_i is just $\frac{d\Omega_{pn+i}}{dQ}$. This statement is a direct corollary of the following theorem.

Theorem 22. *The correlation functions given by* $\langle\phi_A\phi_B\phi_C\rangle = \partial^3_{ABC}\mathcal{F}$ *with* $\sigma_p(\phi_i) = d\Omega_{i+pn}/dQ$ *satisfy the factorization properties for descendant fields*

$$\langle\sigma_p(\phi_i)\phi_B\phi_C\rangle = \langle\sigma_{p-1}(\phi_i)\phi_j\rangle\eta^{jk}\langle\phi_k\phi_B\phi_C\rangle, \tag{5.15}$$

where $\phi_i, i = 1, \ldots, n-1$ *are primary fields, and* ϕ_A *are all fields (including descendants).*

Factorization properties for descendant fields were derived by Witten [63–66]. We note that the completeness of the operator algebra requires a larger set of fields that just g, which is the dimension of the small leaves of the canonical foliation of $\mathcal{M}_g(n,m)$, and which will be shown in the next section to be the dimension of the moduli space of vacua of certain supersymmetric gauge theories. This is one of the difficulties in establishing direct contact between topological field theories and supersymmetric gauge theories, although there has been progress in this direction [6][44][45].

VI. Seiberg-Witten Solutions of N=2 Supersymmetric Gauge Theories

Moduli spaces of geometric structures are appearing increasingly frequently as the key to the physics of certain supersymmetric gauge or string theories. One recurrent feature is a moduli space of degenerate vacua in the physical theory. The physics of the theory is then encoded in a Kähler geometry on the space of vacua, or, in presence of powerful constraints such as N=2 supersymmetry, in an even more restrictive *special geometry*, where the Kähler potential is dictated by single holomorphic function $\mathcal{F}$, called the prepotential. This was the case for Type IIA and Type IIB strings, where the vacua corresponding to compactifications on Calabi-Yau threefolds [31][68] produce effective N=2 four dimensional supergravity theories. The massless scalars of such theories (in this case, the moduli of the Calabi-Yau threefold) must parametrize a manifold equipped with special geometry [9][51][57]. More recently, a similar phenomenon has been brought to light by Seiberg-Witten [52][53] for N=2 supersymmetric gauge theories. Remarkably, the space of vacua of these theories, which is classically just a space of diagonalizable and traceless matrices, becomes upon quantization a moduli space of Riemann surfaces. The prepotential $\mathcal{F}$ for the quantum effective theory can then be derived from a meromorphic one-form $d\lambda$ on each surface. A particularly striking feature of these effective theories, noticed by many authors [24][28][42–43][49][56], is a strong but as yet ill-understood similarity with the

Whitham theory of solitons. Indeed, the quantum spaces of vacua for many N=2 SUSY theories actually coincide with certain leaves of the canonical foliation of $\mathcal{M}_g(n,m)$ [39], the Seiberg-Witten form $d\lambda$ with the one-form $Q\,dE$ central to Whitham theory, and the effective prepotential of the gauge theories with the exponential of the τ-function of the Whitham theory! The purpose of this section is to review some of these developments.

A. N=2 Supersymmetric Gauge Theories

We begin with a brief account of N=2 SUSY Yang-Mills theories in four dimensions with gauge group G [2]. The Yang-Mills gauge field $A = A_\mu dx^\mu$ is imbedded in an N=2 gauge multiplet consisting of A, left and right Weyl spinors λ_L and λ_R, and a complex scalar field ϕ, with all fields valued in the adjoint representation of G. The requirement of N=2 SUSY and renormalizability fixes uniquely the action

$$I = \int_{M^4} d^4x\, Tr\big[\frac{1}{4g^2}F\wedge F^* + \frac{\theta}{8\pi^2}F\wedge F + D\phi^\dagger\wedge * D\phi + [\phi,\phi^\dagger]^2\big] + fermions, \quad (6.1)$$

where g is the coupling constant, θ is the instanton angle, and we have written explicitly only the bosonic part of the action. The classical vacua are given by the critical points of the action. In this case, they work out to be $A=0$, ϕ is constant (up to a gauge transformation), and

$$[\phi,\phi^\dagger] = 0. \quad (6.2)$$

Thus ϕ must lie in the Cartan subalgebra. For G=SU(N_c) (N_c is commonly referred to as number of "colors"), we set

$$\phi = \begin{pmatrix} \bar a_1 & & & & \\ & \bar a_2 & & & \\ & & \ddots & & \\ & & & \ddots & \\ & & & & \bar a_{N_c} \end{pmatrix}, \quad \sum_{k=1}^{N_c} \bar a_k = 0. \quad (6.3)$$

Thus the classical space of vacua is parametrized by the $\bar a_k$, up to a Weyl permutation.

For a generic configuration $\bar a_k$, we have $\bar a_j \neq \bar a_k$ for any $j \neq k$, and the gauge group SU(N_c) is spontaneously broken down to $U(1)^{N_c-1}$. At the quantum level, we expect then the space of inequivalent vacua to be parametrized by $N_c - 1$ parameters a_k (thought of as renormalizations of the $\bar a_k$, $\sum_{k=1}^{N_c} a_k = 0$), with each vacuum corresponding to a theory of $N_c - 1$ interacting U(1) gauge fields A_j, i.e., $N_c - 1$ copies of electromagnetism. In the weak coupling regime, we expect singularities at $a_k = a_l$, where the gauge symmetry is suddenly enhanced. Since N=2 SUSY remains unbroken, each gauge field A_j is part of an N=2 SUSY U(1) gauge multiplet $(A_j, \lambda_{Lj}, \lambda_{Rj}, \phi_j)$, all in the adjoint representation of U(1). Again, the action for a theory with such a field content is fixed by N=2 supersymmetry. To leading order in the low momentum expansion for these fields, it must be of the form

$$I_{\mathit{eff}} = \frac{1}{8\pi}\int_{M^4} d^4x\big[(\mathrm{Im}\,\tau^{jk})F_j\wedge *F_k + (\mathrm{Re}\,\tau^{jk})F_j\wedge F_k + d\phi^{\dagger i}\wedge d\phi^i_D\big] + \mathit{fermions} \tag{6.4}$$

where

$$\tau^{jk} = \frac{\partial^2 \mathcal{F}}{\partial a_j \partial a_k},\ \phi^j_D = \frac{\partial \mathcal{F}}{\partial a_j}(\phi) \tag{6.5}$$

for a suitable complex and analytic function $\mathcal{F}(a,\Lambda)$, called the *prepotential.* We note that the prepotential $\mathcal{F}$ is a function not just of the vacua parameters a_k, but also of a scale Λ introduced by renormalization.

Thus the physics of the quantum theory is encoded in a single function $\mathcal{F}$. What is known about $\mathcal{F}$? To insure the positivity of the kinetic energy, we must have

$$\mathrm{Im}\frac{\partial^2\mathcal{F}}{\partial a_j\partial a_k} > 0. \tag{6.6}$$

Geometrically, $\mathcal{F}$ defines then a Kähler metric on the quantum moduli space by

$$ds^2 = \sum \mathrm{Im}\left(\frac{\partial^2\mathcal{F}}{\partial a_j\partial a_k}\right) da_j\overline{da_k}. \tag{6.7}$$

Furthermore, at weak-coupling $\Lambda \ll 1$, $\mathcal{F}$ can be evaluated in perturbation theory. For pure SU(N_c) Yang-Mills, one finds

$$\mathcal{F}(a,\Lambda) = \frac{2N_c}{2\pi i}\sum_{k=1}^{N_c} a_k^2 - \frac{1}{8\pi i}\sum_{j,k=1}^{N_c}(a_k-a_j)^2\log\frac{(a_k-a_j)^2}{\Lambda^2} + \sum_{d=1}^{\infty}\mathcal{F}_d\Lambda^{2N_c d}. \tag{6.8}$$

The first term on the right hand side is the classical prepotential. The second term is the perturbative one-loop quantum correction. In view of N=2 non-renormalization theorems, it is known that higher loops do not contribute. The third term is the instanton contribution, consisting of d-instanton processes for all d. We observe that the expansion (6.8) implies in particular that $\mathcal{F}$ has non-trivial monodromy around $a_j = a_k$ in the $\Lambda \ll 1$ regime. The exact solution of N=2 Yang-Mills theories is reduced in this way to finding a holomorphic $\mathcal{F}$ satisfying the constraints (6.6) and (6.8).

We have just described the main problem for N=2 SUSY pure SU(N) Yang-Mills theories. However, the same problem should be addressed for general N=2 SUSY gauge theories with gauge group G, with matter fields ("hyper-multiplets") in a representation R of G. As in the case of pure Yang-Mills, the Wilson effective Lagrangian of these theories is dictated by a prepotential $\mathcal{F}_{G,R}(a,\Lambda)$, and the problem is to determine $\mathcal{F}_{G,R}(a,\Lambda)$, subject to the constraints (6.6) and (6.8), where the right hand side of (6.8) has been modified to incorporate the contributions of the hypermultiplets. For example, in presence of N_f hypermultiplets in the fundamental representation of bare masses m_i, $1 \leq i \leq N_f$, the one-loop correction to the prepotential for the SU(N_c) theory contains the additional term

$$\sum_{k=1}^{N_c} \sum_{i=1}^{N_f} (a_k + m_j)^2 \log \frac{(a_k + m_j)^2}{\Lambda^2}.$$

B. The Seiberg-Witten Ansatz

The requirements that $\mathcal{F}$ have monodromy and a Hessian with positive definite imaginary part, suggest an underlying non-trivial geometry on the quantum space of vacua. In [52][53], Seiberg and Witten made the fundamental Ansatz that

- For each Λ, the quantum moduli space should parametrize a family of Riemann surfaces $\Gamma(a, \Lambda)$ of genus $g = N_c - 1$, now known as the spectral curves of the theory;
- on each $\Gamma(a, \Lambda)$, there is a meromorphic one-form $d\lambda$;
- $\mathcal{F}$ is determined by the periods of $d\lambda$ [2]

$$a_k = \frac{1}{2\pi i} \oint_{A_k} d\lambda, \qquad a_{D,k} = \frac{1}{2\pi i} \oint_{B_k} d\lambda, \qquad \frac{\partial \mathcal{F}}{\partial a_k} = a_{D,k}. \tag{6.9}$$

The gauge theories under consideration contain dyons, i.e., particles which carry both electric and magnetic charges. Let $(n, m) \in \mathbf{Z}^{N_c - 1} \times \mathbf{Z}^{N_c - 1}$ be their charges, with (n_i, m_i) the charge with respect to the i-th U(1) factor. The N=2 SUSY algebra implies the bound $M^2 \geq 2|an + a_D m|^2$ from below for their masses. Thus the states saturating this bound, known as Bogomolny-Prasad-Sommerfeld or BPS states, are described by the lattice spanned by the periods of $d\lambda$. The singular locus of the fibration $\Gamma(a, \Lambda)$, namely the points where the curve degenerates and a period a_j or $a_{D,j}$ vanishes, corresponds then to vacua where one or several dyons become massless.

For pure SU(2) Yang-Mills, the monodromy prescription at ∞ is restrictive enough to suggest the identification of the quantum moduli space of vacua with $H/\Gamma(2)$ (H denotes the upper half space, and $\Gamma(2)$ the subgroup of SL(2,$\mathbf{Z}$) matrices congruent to 1 mod 2), assuming the minimal number two of singularities in the interior of the quantum moduli space. Since then, spectral curves have been proposed for a variety of gauge theories with matter, based on physical considerations such as decoupling, or analogies with singularity theory or soliton theory (see e.g. [40] and references therein). However, at the present

[2]In this section, we adopt the normalization (6.9) for the periods a_i of $d\lambda$ rather than (3.39), in keeping with the literature on Seiberg-Witten theory. Similarly, the present $\mathcal{F}$ differs from the earlier τ-function $\mathcal{F}_{\text{Whitham}}$ of soliton theory (c.f. (4.35)) by $\mathcal{F} = -\frac{1}{2\pi i}\mathcal{F}_{\text{Whitham}}$.

time, we still do not have a complete correspondence between the group theoretic characterization of the gauge theory, consisting of the group G and the representation R for the hypermultiplets, and the fibration of spectral curves which characterizes its geometric and physical content.

C. The Framework of the Theory of Solitons

Nevertheless, an intriguing feature of most of the spectral curves for N=2 SUSY gauge theories known so far, is that they, together with the one-form $d\lambda$, fit exactly in the framework of the foliation on $\mathcal{M}_g(n,m)$ with $d\lambda = Q\,dE$. In particular,

• The spectral curves for SU(N_c) theories with $N_f < 2N_c$ hypermultiplets in the fundamental representation of bare masses m_i, $1 \leq i \leq N_f$, are given by the leaf (Γ, E, Q) with the following properties.

- dE has simple poles, at points P_+, P_-, P_i,with residues $-N_c$, $N_c - N_f$, and 1 ($1 \leq i \leq N_f$) respectively. Its periods around homology cycles are integer multiples of $2\pi i$;
- Q is a well-defined meromorphic *function*, with simple poles only at P_+ and P_-;
- The other parameters of the leaf are fixed by the following normalization of the one-form $d\lambda = Q\,dE$

$$\begin{aligned}\mathrm{Res}_{P_i}(d\lambda) &= -m_i,\\ \mathrm{Res}_{P_+}(z\,d\lambda) &= -N_c 2^{-1/N_c}, \mathrm{Res}_{P_-}(z\,d\lambda)\\ &= (N_c - N_f)\left(\frac{\Lambda^{2N_c-N_f}}{2}\right)^{1/(N_c-N_f)}\\ \mathrm{Res}_{P_+}(d\lambda) &= 0.\end{aligned} \tag{6.10}$$

  Here $z = E^{-1/N_c}$ or $z = E^{1/(N_c-N_f)}$ is as usual the holomorphic coordinate system near P_+ or near P_- adapted to the Abelian integral E.

These conditions imply that Γ is hyperelliptic, and admits an equation of the form (see [39])

$$y^2 = \prod_{k=1}^{N_c}(Q - \bar{a}_k)^2 - \Lambda^{2N_c-N_f}\prod_{j=1}^{N_f}(Q + m_j) \equiv A(Q)^2 - B(Q). \tag{6.11}$$

Strictly speaking, the parameters $\bar{a}_k$ of (6.11) agree with the classical vacua in (6.3) only when $N_c < N_f$. For $N_f \geq N_c$, there are $O(\Lambda)$ corrections, which can be absorbed in a reparametrization leaving the prepotential $\mathcal{F}$ invariant [15]. Thus we may view the $\bar{a}_k$ of (6.3) and (6.11) as identical. If we represent the Riemann surface (6.11) by a two-sheeted covering of the complex plane, then the meromorphic function Q on Γ in $d\lambda = Q\,dE$ is just the coordinate in each sheet, while the Abelian integral E is given by $E = \log(y + A(Q))$. The points $P_\pm$ correspond to the points at infinity, with the two possible choices of signs $\pm$ for $y = \pm\sqrt{A^2 - B}$. To choose a canonical homology basis, we let $x_k^\pm$, $1 \leq k \leq N_c$, be the branch points $A(x_k^\pm)^2 - B(x_k^\pm)$, A_k be a simple contour enclosing the slit from x_k^- to x_k^+ for $2 \leq k \leq N_c$, and B_k be the curve going from x_k^- to x_1^- on each sheet. We can now give a preliminary and easy check that the curve (6.11) is consistent with the expected behavior of the theory at weak-coupling. Consider for simplicity the case of pure Yang-Mills, $N_f = 0$. Then as $\Lambda \to 0$, the discriminant of the curve behaves as $\Lambda^{N_c} \prod_{j<k}(\bar{a}_j - \bar{a}_k)^2$, and the singularities are at the expected location. Furthermore, in this limit, $d\lambda \sim Q\frac{A'(Q)}{A(Q)}dQ + \ldots$, and the cycles A_k are just contours in the complex plane around a slit which shrinks to a single point $\bar{a}_k$. The residue formula gives at once

$$a_k = \frac{1}{2\pi i}\oint_{A_k} Q\frac{A'(Q)}{A(Q)}dQ + O(\Lambda^{N_c}) = \bar{a}_k + O(\Lambda^{N_c}), \tag{6.12}$$

identifying $\bar{a}_k$ as a classical order parameter.

• The spectral curves for the other classical gauge groups with matter in the fundamental representation are restrictions of the ones for SU(N_c) [4][16];

• The SU(N_c) theory with matter in the adjoint representation is of particular interest. For massless matter, the theory has actually an N=4 supersymmetry, and is conformally invariant. As the hypermultiplet acquires mass, the N=4 SUSY is broken down to an N=2 SUSY. In [18], Donagi and Witten argued that the spectral curves for the theory are then given by Hitchin systems. Expressed in terms of elliptic Calogero-Moser systems, the curves they proposed are given precisely by the leaf (Γ, k, z) in Section III.D. Here the hypermultiplet mass has been scaled to 1, and the moduli $\tau = \omega_2/\omega_1$ of the torus is the microscopic gauge coupling.

Although this suggests a deep relation between N=2 gauge theories and integrable models, such a relation is still not fully understood at the present time. Nevertheless, the parallelism between the two fields allows us to apply to the study of the prepotential $\mathcal{F}$ of gauge theories the methods developed in the theory of solitons. Thus Theorem 16 implies readily [17]

Theorem 23. *The prepotential $\mathcal{F}$ for* SU*(N_c) gauge theories with $N_f < 2N_c$ hypermultiplets of masses m_j in the fundamental representation, satisfies the following differential equation*

$$\sum_{j=1}^{N_c} a_j \frac{\partial \mathcal{F}}{\partial a_j} + \sum_{j=1}^{N_f} m_j \frac{\partial \mathcal{F}}{\partial m_j} - 2\mathcal{F} = \\ - \frac{1}{2\pi i}[\mathrm{Res}_{P_+}(z\,d\lambda)\,\mathrm{Res}_{P_+}(z^{-1}d\lambda) + \mathrm{Res}_{P_-}(z\,d\lambda)\,\mathrm{Res}_{P_-}(z^{-1}d\lambda)] \tag{6.13}$$

We observe that there is a slight abuse of language here, since in the case of the effective prepotential of gauge theories, $\mathcal{F}$ is only fixed up to a_k independent terms by (6.9). This is consistent with the fact that only derivatives of $\mathcal{F}$ with respect to a_k occur in the effective action. Thus the a_k independent terms on the right hand side of (6.13) can be ignored by adjusting $\mathcal{F}$. The prepotential (4.35) (restricted to the leaf corresponding to the SU(N_c) theory) is one choice of $\mathcal{F}$. Another choice suggested by dimensional analysis (c.f. (6.8)) is the prepotential $\mathcal{F}$ satisfying the homogeneity condition $(\frac{\partial}{\partial\Lambda}+\mathcal{D})\mathcal{F}=0$. In this case, we recognize (6.13) as a renormalization group equation, with the beta function given by the right hand side of (6.13). Earlier versions of (6.13) appear in [24][46][47][56].

To illustrate the power of Theorem 20, we shall show how it can generate explicit expressions for the contributions of instanton processes to any order. Thus we consider the regime where Λ is small and all the A_k-cycles degenerate simultaneously. A fundamental observation is that in this regime, the quantum order parameters a_k are perturbations of their classical counterparts $\bar{a}_k$ which can be determined explicitly to any order. In fact, as noted in the arguments leading to (6.12), the A_k cycles are simple contours shrinking to a point in one sheet of the Riemann surface and residue formulae apply. The approximation (6.12) can be improved to

$$\begin{aligned} a_k &= \bar{a}_k + \sum_{m=1}^{\infty} \frac{\Lambda^{(2N_c-N_f)m}}{2^{2m}(m!)^2} (\frac{\partial}{\partial \bar{a}_k})^{2m-1} \bar{S}_k(\bar{a}_k), \\ \bar{S}_k(x) &= \frac{\prod_{j=1}^{N_f}(x+m_j)}{\prod_{l\neq k}(x-\bar{a}_l)^2}. \end{aligned} \tag{6.14}$$

The evaluation of the dual periods a_{Dk} is of course more difficult. We need to show that the prepotential $\mathcal{F}$, as defined by the B_k-periods, reproduces the classical prepotential $\mathcal{F}^{(0)}$ in (6.8) (with hypermultiplets) and satisfies the non-renormalization theorem. This requires an analytic continuation in an auxiliary parameter ξ, as explained in [15]. However, once this is established, the difficult instanton contributions can be derived from the renormalization group equation. Setting $\mathcal{F}=\mathcal{F}^{(0)}+\mathcal{F}^{(1)}+\mathcal{F}^{(2)}+\ldots$, we have, say up to 2-instanton order and using Euler's homogeneity relation,

$$\sum_{j=1}^{N_c} a_j \frac{\partial \mathcal{F}}{\partial a_j} + \sum_{j=1}^{N_f} m_j \frac{\partial \mathcal{F}}{\partial m_j} = (N_f - 2N_c)\left(\frac{1}{4\pi i}\sum_{k=1}^{N_c} a_k^2 + \mathcal{F}^{(1)} + 2\mathcal{F}^{(2)}\right). \tag{6.15}$$

(We note the overall factor $N_f - 2N_c$, which confirms the known conformal invariance of the theory with $N_f = 2N_c$. For the spectral curves of this theory, we refer to [4]). On the other hand, dropping all a_k-independent terms, the right hand side of (6.13) is easily found

$$\sum_{j=1}^{N_c} a_j \frac{\partial \mathcal{F}}{\partial a_j} + \sum_{j=1}^{N_f} m_j \frac{\partial \mathcal{F}}{\partial m_j} - 2\mathcal{F} = \frac{1}{4\pi i}(N_f - 2N_c)\sum_{k=1}^{N_c} \bar{a}_k^2. \tag{6.16}$$

We can now use (6.14) to reexpress the right hand side of (6.16) in terms of the quantum order parameters a_k. The instanton contributions can then be read off after suitable rearrangements [15][17]

$$\begin{aligned}\mathcal{F}^{(1)} &= \frac{1}{8\pi i}\Lambda^{2N_c-N_f}\sum_{k=1}^{N_c} S_k(a_k),\\ \mathcal{F}^{(2)} &= \frac{1}{32\pi i}\Lambda^{2(2N_c-N_f)}\left(\sum_{k\neq l}\frac{S_k(a_k)S_l(a_l)}{(a_k-a_l)^2}+\frac{1}{4}\sum_{k=1}^{N_c} S_k(a_k)\partial_{a_k}^2 S_k(a_k)\right).\end{aligned} \tag{6.17}$$

Here the function $S_k(x)$ is defined in analogy with (6.14) by

$$S_k(x) = \frac{\prod_{j=1}^{N_f}(x+m_j)}{\prod_{j\neq k}^{N_c}(x-a_j)^2}.$$

We turn next to a determination of the effective prepotential at strong coupling. In general, when a single cycle A_k or B_k degenerates, we expect the effective prepotential to be expressible in terms of functions on the resulting surface of lower genus. A particularly interesting case is the behavior of $\mathcal{F}$ near a point on the quantum moduli space of maximum degeneracy, where all B_k cycles degenerate simultaneously, and the spectral curve degenerates to two spheres connected by thin tubes. Physically, this means that a maximum number of mutually local dyons become simultaneously massless. As shown in [20], the points of maximum degeneracy occur at the curves $y^2 = A_0(Q)^2 - 4\Lambda^{2N_c}$, where $A_0(Q)$ is given by the N-th Chebyshev polynomial

$$A_0 = 2\Lambda^{N_c} C_0\left(\frac{Q}{2\Lambda}\right), \quad C_0(z) = \cos(N\arccos(z)). \tag{6.18}$$

A neighborhood of the maximum degeneracy point on the quantum moduli space is parametrized by polynomials $P(Q)$ of degree $N-2$, with the spectral curve $y^2 = A(Q)^2 - 4\Lambda^{2N_c}$, $A(Q) = A_0(Q) + 2\Lambda^{N_c}P(\frac{Q}{2\Lambda})$. Since it is the B_k cycle which degenerates this time, it is more convenient to express the prepotential $\mathcal{F}(a_{Dk})$ and the beta function in terms of the dual variables a_{Dk}, which can then be evaluated to an arbitrary order of accuracy by residue methods. The renormalization group equation remains the same under interchange of the dual variables $a_k \leftrightarrow a_{Dk}$

$$\sum_{k=1}^{N_c} a_{Dk}\frac{\partial\mathcal{F}}{\partial a_{Dk}} - 2\mathcal{F} = \frac{2N_c}{2\pi i}u(a_D), \tag{6.19}$$

where u is the coefficient of Q^{N_c-2} in $A(Q)$. Residue calculations show next

a_{Dk} is of first order in P and that

$$a_{Dk} = i(-)^k \frac{s_k}{N_c} P(c_k) + \sum_{m=1}^{\infty} a_{Dk}^{(m)},$$
$$s_k = \sin\left(\frac{k\pi}{N_c}\right), \quad c_k = \cos\left(\frac{k\pi}{N_c}\right), \quad k = 1, \dots, N_c, \tag{6.20}$$

where the $a_{Dk}^{(m)}$ are of order $O^m(a_D)$ and can be evaluated explicitly. To third order in a_{Dk}, we find for u

$$u = 2i\Lambda \sum_{k=1}^{N_c-1} s_k a_{Dk} + \frac{1}{4N_c} \sum_{k=1}^{N_c-1} a_{Dk}^2 + \frac{i}{32N_c^2\Lambda}\left[\sum_{k=1}^{N_c-1} \frac{a_{Dk}^2}{s_k^3} - 4 \sum_{k\neq l}^{N_c-1} \frac{a_{Dk}^2 a_{Dl} s_l}{(c_k - c_l)^2}\right]. \tag{6.21}$$

Solving the renormalization group equation, we obtain [14]

Theorem 24. *Near the point of maximum degeneracy on the quantum moduli space given by (6.19), the prepotential $\mathcal{F}$ is given by the following expression*

$$\mathcal{F}(a_D) = -\frac{2N_c\Lambda}{\pi} \sum_{k=1}^{N_c-1} s_k a_{Dk} - \frac{i}{4\pi} \sum_{k=1}^{N_c-1} a_{Dk}^2 \log\frac{a_{Dk}}{\Lambda_k} + \frac{1}{32\pi N_c \Lambda}\left[\sum_{k=1}^{N_c-1} \frac{a_{Dk}^3}{s_k^3} - 4 \sum_{k\neq l}^{N_c-1} \frac{a_{Dk}^2 a_{Dl} s_l}{(c_k - c_l)^2}\right] \tag{6.22}$$

up to third order in the order parameters a_{Dk}. Here Λ_k is determined by $\log\frac{\Lambda_k}{\Lambda} = \frac{3}{2} + \log s_k$.

We should mention that there is by now an extensive literature on Seiberg-Witten theories, and we refer to [40] for a description of other recent advances and for a more complete list of references.

Bibliography

[1] Adler, M., *On a trace functional for formal pseudo-differential operators and the symplectic structure for Korteweg-de Vries type equations*, Invent. Math. **50** (1979) 219–248.

[2] Amati, D., Konishi, K., Y. Meurice, G.C. Rossi, and G. Veneziano, *Non-perturbative aspects of supersymmetric gauge theories*, Physics Reports **162** (1988) 169–248.

[3] Arbarello, E. and C. DeConcini, *On a set of equations characterizing Riemann matrices*, Ann. of Math. **120** (1984) 119–140.

[4] Argyres, P. and A.P. Shapere, *The vacuum structure of N=2 super-QCD with classical gauge groups*, Nucl. Phys. **B 461** (1996) 437–459, hep-th/9508175.

[5] Bogoyavlenski, O. and S.P. Novikov, *On the connection between Hamiltonian formalisms of stationary and non-stationary problems*, Funct. Anal. and Appl. **10** (1976) 9–13.

[6] G. Bonelli and M. Matone, *Non-perturbative relations in N=2 SUSY Yang-Mills and WDVV equations*, Phys.Rev.Lett. **77** (1996) 4712–4715, hep-th/9605090.

[7] Cherednik, I., *Differential equations for the Baker-Akhiezer functions of algebraic curves*, Funct. Anal. and Appl. **12** (1978) 45–52.

[8] Date, E., M. Jimbo, M. Kashiwara, and T. Miwa, *Transformation groups for soliton equations*, in M. Jimbo and T. Miwa (eds) *Non-linear integrable systems—classical and quantum theory*, Proc. RIMS Symposium, Singapore, 1983.

[9] de Wit, B., P.G. Lauwers, and A. van Proeyen, *Lagrangians of N=2 supergravity-matter systems*, Nucl. Phys. **B 255** (1985) 569–608.

[10] Dickey, L.A., *Soliton equations and Hamiltonian systems*, Advanced Series in Mathematical Physics, Vol. 12 (1991) World Scientific, Singapore.

[11] Dijkgraaf, R., E. Verlinde, and H. Verlinde, *Topological strings in $d < 1$*, Nucl. Phys. **B 352** (1991) 59–86.

[12] Dijkgraaf, R., E. Verlinde, and H. Verlinde, *Notes on topological string theory and 2D quantum gravity*, in *String theory and quantum gravity*, Proceedings of the Trieste Spring School 1990, M. Green (ed), World-Scientific, 1991.

[13] Dijkgraaf, R. and E. Witten, *Mean field theory, topological field theory, and multi-matrix models*, Nucl. Phys. **B 342** (1990) 486–522.

[14] D'Hoker, E. and D.H. Phong, *Strong coupling expansions in $SU(N_c)$ Seiberg-Witten theory*, Phys. Lett. **B 397** (1997) 94–103.

[15] D'Hoker, E., I.M. Krichever, and D.H. Phong, *The effective prepotential for N=2 supersymmetric SU(N) gauge theories*, Nucl. Phys. **B 489** (1997) 179–210.

[16] D'Hoker, E., I.M. Krichever, and D.H. Phong, *The effective prepotential for N=2 supersymmetric SO(N) and Sp(2N) gauge theories*, Nucl. Phys. **B 489** (1997) 211–222.

[17] D'Hoker, E., I.M. Krichever, and D.H. Phong, *The renormalization group equation for N=2 supersymmetric gauge theories*, Nucl. Phys. **B494** (1997) 89–104, hep-th/9610156.

[18] Donagi, R. and E. Witten, *Supersymmetric Yang-Mills theory and integrable systems*, Nucl. Phys. **B 460** (1996) 299–334, hep-th/9610101.

[19] Douglas, M. and S. Shenker, *Strings in less than one dimension*, Nucl. Phys. **B 335** (1990) 635–654.

[20] Douglas, M. and S. Shenker, *Dynamics of SU(3) gauge theories*, Nucl. Phys. **B 447** (1995) 271–296, hep-th/9503163.

[21] Dubrovin, B.A., *Integrable systems in topological field theory*, Nucl. Phys. **B 379** (1992) 627–689.

[22] Dubrovin, B.A., *Hamiltonian formalism of Whitham-type hierarchies and topological Landau-Ginzburg models*, Comm. Math. Phys. **145** (1992) 195–207.

[23] Dubrovin, B.A. and S.P. Novikov, *The Hamiltonian formalism of one-dimensional systems of the hydrodynamic type, and the Bogoliubov-Whitham averaging method*, Sov. Math. Doklady **27** (1983) 665–669.

[24] Eguchi, T. and S.K. Yang, *Prepotentials of N=2 supersymmetric gauge theories and soliton equations*, Mod.Phys.Lett. **A 11** (1996) 131–138, hep-th/9510183.

[25] Faddeev, L. and L. Takhtajan, *Hamiltonian methods in the theory of soliton*, Springer-Verlag, 1987.

[26] Flaschka, H., Forrest, M.G., and D. McLaughlin, *Multiphase averaging and the inverse spectral solution of the Korteweg-de Vries equation*, Comm. Pure Appl. Math. **33** (1980) 739–784.

[27] Gelfand, I.M. and L.A. Dickey, *Fractional powers of operators and Hamiltonian systems*, Funct. Anal. and Appl. **10** (1976) 13–29.

[28] Gorsky, A., I.M. Krichever, A. Marshakov, A. Mironov, and A. Morozov, Phys. Lett. **B 355** (1995) 466–474.

[29] Gorsky, A. and N. Nekrasov, *Elliptic Calogero-Moser system from two-dimensional current algebra*, preprint ITEP-NG/1-94, hep-th/9401021.

[30] Greene, B., C. Vafa, and N. Warner, *Calabi-Yau manifolds and renormalization group flows*, Nucl. Phys. **B 324** (1989) 371–390.

[31] Greene, B. and S.T. Yau, eds., *Mirror Symmetry II*, Studies in Advanced Mathematics, American Mathematical Society-International Press (1997), Rhode Island and Hong Kong.

[32] Klemm, A., W. Lerche, and S. Theisen, *Non-perturbative actions of N=2 supersymmetric gauge theories*, Int. J. Mod. Phys. **A 11** (1996) 1929–1974, hep-th/9505150.

[33] Krichever, I.M., *The algebraic-geometric construction of Zakharov-Shabat equations and their solutions*, Doklady Akad. Nauka USSR **227** (1976) 291–294.

[34] Krichever, I.M., *Methods of algebraic geometry in the theory of non-linear equations*, Russian Math Surveys **32** (1977) 185–213.

[35] Krichever, I.M., *Elliptic solutions of the Kadomtsev-Petviashvili equation and integrable systems of particles*, Funct. Anal. Appl. **14** (1980) 282–290.

[36] Krichever, I.M., *Averaging method for two-dimensional integrable equations*, Funct. Anal. Appl. **22** (1988) 37–52.

[37] Krichever, I.M., *Spectral theory of two-dimensional periodic operators and its applications*, Russian Math. Surveys (1989) **44** (1989) 145–225.

[38] Krichever, I.M., *The τ-function of the universal Whitham hierarchy, matrix models, and topological field theories*, Comm. Pure Appl. Math. **47** (1994) 437–475.

[39] Krichever, I.M. and D.H. Phong, *On the integrable geometry of soliton equations and N=2 supersymmetric gauge theories*, J. Differential Geometry **45** (1997) 349–389.

[40] Lerche, W., *Introduction to Seiberg-Witten theory and its stringy origins*, Proceedings of the *Spring School and Workshop in String Theory*, ICTP, Trieste (1996), hep-th/9611190.

[41] Magri, F., *A simple model of the integrable Hamiltonian equation*, J. Math. Phys. **19** (1978) 1156–1162.

[42] Martinec, E., *Integrable structures in supersymmetric gauge and string theory*, Phys. Lett. **B 367** (1996) 91–96, hep-th/9510204.

[43] Martinec, E. and N. Warner, *Integrable models and supersymmetric gauge theory*, Nucl. Phys. **B 459** (1996) 97–112, hep-th/9609161.

[44] Marshakov, A., *Non-perturbative quantum theories and integrable equations*, Int. J. Mod. Phys. **A 12** (1997) 1607–1650, hep-th/9610242.

[45] Marshakov, A., A. Mironov, and A. Morozov, *WDVV-like equations in N=2 SUSY Yang-Mills theory*, hep-th/9607109.

[46] Matone, M., *Instantons and recursion relations in N=2 Susy gauge theory*, Phys. Lett. **B 357** (1995) 342–348.

[47] Minahan, J. and D. Nemeschansky, *N=2 Super Yang-Mills and Subgroups of* $SL(2, Z)$, Nucl. Phys. B 468 (1996) 73–84, hep-th/9601059.

[48] Mumford, D., *Theta Functions I, II*, Birkhäuser (1980) Boston.

[49] Nakatsu, T. and K. Takasaki, *Whitham-Toda hierarchy and N=2 supersymmetric Yang-Mills theory*, Mod. Phys. Lett. **A 11** (1996) 157–168, hep-th/9509162.

[50] Novikov, S.P. and A. Veselov, *On Poisson brackets compatible with algebraic geometry and Korteweg-deVries dynamics on the space of finite-zone potentials*, Soviet Math. Doklady **26** (1982) 357–362.

[51] Periwal, V. and A. Strominger, *Kähler geometry of the space of N=2 supersonformal field theories*, Phys. Lett. **B 235** (1990) 261–267.

[52] Seiberg, N. and E. Witten, *Electric-magnetic duality, Monopole Condensation, and Confinement in N=2 Supersymmetric Yang-Mills Theory*, Nucl. Phys. **B 426** (1994) 1952, hep-th/9407087.

[53] Seiberg, N. and E. Witten, *Monopoles, Duality and Chiral Symmetry Breaking in N=2 Supersymmetric QCD*, Nucl. Phys. **B 431** (1994) 484–550, hep-th/9410167.

[54] Sato, M., *Soliton equations and the universal Grassmann manifold*, Math. Notes. Series 18 (1984) Sophia University, Tokyo.

[55] Shiota, T., *Characterization of Jacobian varieties in terms of soliton equations*, Invent. Math. **83** (1986) 333–382.

[56] Sonnenschein, J., S. Theisen, and S. Yankielowicz, *On the relations between the holomorphic prepotential and the quantum moduli in SUSY gauge theories*, Phys. Lett. **B 367** (1996) 145–150, hep-th/9510129.

[57] Strominger, A., *Special Geometry*, Comm. Math. Phys. **133** (1990) 163–180.

[58] Takebe, T., *Representation theoretic meaning for the initial-value problem for the Toda lattice hirearchy*, Lett. Math. Phys. **21** (1991), 77–84.

[59] Terng, C.L., *Soliton equations and differential geometry*, J. Differential Geometry **45** (1997) 407–445.

[60] Terng, C.L. and K. Uhlenbeck, to appear.

[61] Vafa, C., *Topological Landau-Ginzburg models*, Mod. Phys. Lett. **A 6** (1991) 337–346.

[62] Whitham, G., *Linear and non-linear waves*, Wiley (1974) New York.

[63] Witten, E., *Topological Quantum Field Theory*, Comm. Math. Phys. **117** (1988) 353–386.

[64] Witten, E., *Topological sigma models*, Comm. Math. Phys. **118** (1988) 411–449.

[65] Witten, E., *Mirror manifolds and topological field theory*, in *Essays on Mirror Manifolds*, ed. by S.T. Yau, International Press (1992), Hong-Kong, 120–159.

[66] Witten, E., *Two-dimensional gravity and intersection theory on moduli space*, Surveys in Differential Geometry **1** (1991) 281–332.

[67] Witten, E., *Phases of N=2 theories in two dimensions*, Nucl. Phys. **B 403** (1993) 159–222; also reprinted in [31].

[68] Yau, S.T., ed., *Essays on Mirror Manifolds*, International Press (1992) Hong-Kong.

Poisson Actions and Scattering Theory for Integrable Systems

Chuu-Lian Terng* and Karen Uhlenbeck†

Abstract

Conservation laws, heirarchies, scattering theory and Bäcklund transformations are known to be the building blocks of integrable partial differential equations. We identify these as facets of a theory of Poisson group actions, and apply the theory to the ZS-AKNS nxn heirarchy (which includes the non-linear Schrödinger equation, modified KdV, and the n-wave equation). We first find a simple model Poisson group action that contains flows for systems with a Lax pair whose terms all decay on R. Bäcklund transformations and flows arise from subgroups of this single Poisson group. For the ZS-AKNS nxn heirarchy defined by a constant $a \in u(n)$, the simple model is no longer correct. The a determines two separate Poisson structures. The flows come from the Poisson action of the centralizer H_a of a in the dual Poisson group (due to the behavior of $e^{a\lambda x}$ at infinity). When a has distinct eigenvalues, H_a is abelian and it acts symplectically. The phase space of these flows is the space S_a of left cosets of the centralizer of a in D_-, where D_- is a certain loop group. The group D_- contains both a Poisson subgroup corresponding to the continuous scattering data, and a rational loop group corresponding to the discrete scattering data. The H_a-action is the right dressing action on S_a. Bäcklund transformations arise from the action of the simple rational loops on S_a by right multiplication.

*Research supported in part by NSF Grant DMS 9626130

†Research supported in part by Sid Richardson Regents' Chair Funds, University of Texas system

Various geometric equations arise from appropriate choice of a and restrictions of the phase space and flows. In particuar, we discuss applications to the sine-Gordon equation, harmonic maps, Schrödinger flows on symmetric spaces, Darboux orthogonal coordinates, and isometric immersions of one space-form in another.

Table of Contents

1 Introducton

Soliton theory is an enticingly elegant part of modern mathematics. It has a multitude of interpretations in geometry, analysis and algebra. The main goal of this paper is to relate loop groups actions, scattering theory, and Bäcklund transformations within the same narrative, via Poisson actions. Our work is motivated by Beals and Coifman's rigorous and beautiful treatment of scattering and inverse scattering theory of the first order systems ([BC 1, 2, 3]). An expository version of our main result on scattering theory is contained in lecture notes by Richard Palais [Pa]. In retrospect,, we also find that many of our results in the $su(2)$ case are contained in the book by Faddeev and Takhtajian ([FT]). Throughout the paper, the matrix non-linear Schrödinger equation is used as a motivating example. In the final section we discuss a number of applications in geometry, including Darboux orthogonal coordinates and isometric immerisons of space forms in space forms. We begin the paper with a survey.

- **Finite dimensional mechanics**

To give some perspective, we start with a short review of finite dimensional Hamiltonian systems, complete integrability, symplectic actions, Poisson actions, and moment maps. A more detailed review of Poisson actions is given in section 2. A *symplectic structure* on a $2n$-dimensional manifold M is a closed, non-degenerate two form w on M. Since w is non-degenerate, it induces an isomorphism $J : T^*M \to TM$. A Hamiltonian on M is a smooth funtion $f : M \to R$. The Hamiltonian vector field X_f corresponding to f is the symplectic dual of df, i.e.,

$$X_f = J(df), \qquad \text{or} \qquad i_{X_f} w = df.$$

It follows from this definition that X_f is symplectic, i.e., $L_{X_f} w = 0$, or equivalently the one parameter subgroup generated by X_f preserves w.

A Poisson structure on M is a Lie bracket $\{\ ,\ \}$ on $C^\infty(M, R)$, which satisfies the Leibnitz rule

$$\{fg, h\} = f\{g, h\} + g\{f, h\}.$$

A symplectic form w induces a natural Poisson structure on M by

$$\{f, g\} = w(X_f, X_g) = df(X_g).$$

Then the map from $C^\infty(M, R)$ to the Lie algebra of vector fields on M defined by $f \mapsto X_f$ is a Lie algebra homomorphism, i.e.,

$$[X_f, X_g] = X_{\{f,g\}}.$$

Two Hamiltonians f, g are said to be *in involution* if $\{f, g\} = 0$. In this case the corresponding Hamiltonian flows commute, and g is a *conservation law* for the Hamiltonian system

$$\frac{dx}{dt} = X_f(x(t)), \tag{1.1}$$

i.e., g is constant on the integral curves of X_f.

The Hamiltonian system (1.1) on M^{2n} is called *completely integrable* if there exists n conservation laws $f_1 = f, f_2, \ldots, f_n$ that are in involution and $df_1, \ldots, df_n$ are linearly independent. For example, the Hamiltonian systems given by the Kowalevsky top, the Toda system and the geodesic flow on an ellipsoid are completely integrable.

Suppose f is completely integrable and $f = f_1, \ldots, f_n$ are in involution. Then $X_{f_1}, \ldots, X_{f_n}$ generate an action of R^n on M. If the map $\mu = (f_1, \ldots, f_n) : M \to R^n$ is proper, then $X_{f_1}, \ldots, X_{f_n}$ generate an action of the n-torus T^n on each R^n-orbit of M. Let $\theta_1, \ldots, \theta_n$ denote the angular coordinates on the torus orbit. Then $(f_1, \ldots, f_n, \theta_1, \ldots, \theta_n)$ is a coordinate system on M and the system

(1.1) is linearized in these coordinates. These are the *action-angle coordinates* in Liouville's Theorem (for detail see [AbM], [Ar]).

The notion of complete integrability can be extended to the notion of symplectic action of a Lie group. An action of G on (M, w) is *symplectic* if the action preserves w. A symplectic action of a Lie group G on M is *Hamiltonian* if there exists a map $\mu : M \to \mathcal{G}^*$ such that the infinitesimal vector field on M corresponding to $\xi \in \mathcal{G}$ is the Hamiltonian vector field of the function f_ξ defined by $f_\xi(x) = \mu(x)(\xi)$. Such μ is called a *moment map*. When G is abelian, the flows generated by the action commute. In particular, the study of completely integrable systems on M^{2n} is the same as the study of Hamiltonian actions of R^n on M^{2n}. When G is non-abelian, the flows generated by η in the centralizer of ξ $\mathcal{G}_\xi = \{\eta \in \mathcal{G} \mid [\xi, \eta] = 0\}$ commute with the flow generated by ξ. In other words, f_η is a conservation law for the flow generated by ξ. But the flows generated by $\eta_1, \eta_2 \in \mathcal{G}_\xi$ in general do not commute.

- **Poisson groups**

Given two Poisson manifolds $(M_1, \{\,,\,\}_1)$ and $(M_2, \{\,,\,\}_2)$, the *product Poisson structure* on $M_1 \times M_2$ is defined by

$$\{f, g\}(x, y) = \{f(\cdot, y), g(\cdot, y)\}_1(x) + \{f(x, \cdot), g(x, \cdot)\}(y).$$

A map $\phi : M_1 \to M_2$ is Poisson if ϕ preserves the Poisson structure, i.e.,

$$\{f, g\}_2 \circ \phi = \{f \circ \phi, g \circ \phi\}_1.$$

A Poisson group is a Poisson manifold $(G, \{\,,\,\})$ such that G is a Lie group and the multiplication map $m : G \times G \to G$ is a Poisson map, where $G \times G$ is equipped with the product Poisson structure. The modern study of Poisson groups was initiated by Drinfeld in [Dr] and there are several good articles by Lu and Weinstein [LW] and Semenov-Tian-Shanksy [Se1, 2]. Given a Poisson group G, there is a canonical construction of a dual Poisson group G^* (cf. [LW]). The simplest Poisson group is a Lie group G with the trivial Poisson structure, and its dual Poisson group is the dual $\mathcal{G}^*$ of Lie algebra $\mathcal{G}$ with the standard Lie Poisson structure and viewed as an abelian Lie group. In general, Poisson groups are best understood as part of a Manin triple. A Manin triple is a triple $(\mathcal{G}, \mathcal{G}_+, \mathcal{G}_-)$, where $\mathcal{G}$ is a Lie algebra with a non-degenerate bi-linear form $\langle\,,\rangle$, $\mathcal{G}_+, \mathcal{G}_-$ are Lie subalgebras of $\mathcal{G}$, $\mathcal{G} = \mathcal{G}_+ + \mathcal{G}_-$ as direct sum of vector spaces, and $\langle \mathcal{G}_+, \mathcal{G}_+\rangle = \langle \mathcal{G}_-, \mathcal{G}_-\rangle = 0$. Then the corresponding Lie group G_+ is Poisson and G_- is its Poisson dual. The triple (G, G_+, G_-) is called a *double group* in the literature. In this paper, we will call this triple a *Manin triple group* to avoid confusion with the completely different concept of a double loop group. For example, $(SL(n), SU(n), B_n)$ is a Manin triple group, where B_n is the subgroup of upper triangular matrices in $SL(n, C)$ with real diagonal entries and $\langle x, y\rangle = \mathrm{Im}(\mathrm{tr}(xy))$ is the non-degenerate bi-linear form. For the trivial Poisson structure on G, the Manin triple group is $(G \ltimes_{\mathrm{ad}} \mathcal{G}^*, G, \mathcal{G}^*)$, where $G \ltimes_{\mathrm{ad}} \mathcal{G}^*$ is the semi-direct product.

- **Adler-Kostant-Symes Theorem**

Let $\mathcal{G}$ be a Lie algebra equipped with an ad-invariant, non-degenerate bilinear form $\langle\ ,\rangle$. Suppose $\mathcal{K}$ and $\mathcal{N}$ are Lie subalgebras of $\mathcal{G}$ such that $\mathcal{G}$ is the direct sum of $\mathcal{K}$ and $\mathcal{N}$ as vector space. Then the space $\mathcal{K}^\perp$ perpendicular to $\mathcal{K}$ in $\mathcal{G}$ with respect to $\langle\ ,\rangle$ can be identified as the dual $\mathcal{N}^*$ of $\mathcal{N}$. Let $M \subset \mathcal{K}^\perp$ be a coadjoint N-orbit equipped with the standard co-adjoint orbit symplectic structure. The Adler-Kostant-Symes theorem ([AdM], [Kos]) states that if f and g are Ad-invariant function from $\mathcal{G}$ to R then $f\,|\,M$ and $g\,|\,M$ are commuting Hamiltonians. For example, let $\langle x, y\rangle = \mathrm{tr}(xy)$ and $sl(n,R) = \mathcal{K} + \mathcal{N}$, where $\mathcal{K} = so(n)$ and $\mathcal{N}$ is the subalgebra of real, trace zero, upper triangular matrices. Then $\mathcal{K}^\perp$ is the space of real, symmetric, trace zero matrices, and the coadjoint N-orbit M at $x_0 = \sum_{i=1}^{n-1}(e_{i,i+1} + e_{i+1,i})$ is the set of all tridiagonal matrices $z = \sum_{i=1}^{n} x_i e_{ii} + \sum_{i=1}^{n-1} y_i(e_{i,i+1} + e_{i+1,i})$ such that all $y_i > 0$ and $\sum_i x_i = 0$. Note that $f_k(x) = \mathrm{tr}(x^k)$ is Ad-invariant function on $sl(n,R)$. So the Hamiltonians $H_2 = f_2\,|\,M,\ \ldots,\ H_n = f_n\,|\,M$ are commuting, and the Hamiltonian system on M corresponding to H_2 is the Toda lattice.

Adler and van Moerbeke [AdM] have shown that many finite dimensional completely integrable systems can be obtained by applying the Adler-Kostant-Symes theorem to suitable Lie algebras. For more examples, see also the paper by Reyman [R].

- **Poisson actions and dressing actions**

An action of a Poisson group G on a Poisson manifold M is Poisson if the action $G \times M \to M$ is a Poisson map. When G is equipped with the trivial Poisson structure, a G-action on a symplectic manifold is Poisson if and only if it is symplectic. The coadjoint action of G on $\mathcal{G}^*$ is Poisson in this trivial structure. In general, if (G, G_+, G_-) is a Manin triple group such that the multiplication map from $G_- \times G_+$ to G is an isomorphism, then the action of G_+ on G_- defined by $g_+ * g_- = \tilde{g}_+$, where $\tilde{g}_+$ is obtained from the factorization

$$g_+ g_- = \tilde{g}_- \tilde{g}_+ \in G_- \times G_+,$$

is Poisson. This action of G_+ on G_- is called the *dressing action*. To construct a global dressing action, every element in G must be factored as a product $g_+ g_- \in G_+ \times G_-$. For example, in reference to the example in the previous paragraph, the factorization of $g \in GL(n)$ as $g_+ g_- \in U(n) \times B_n$ can be obtained by applying the Gram-Schmidt process to the columns of g. In general, this factorization cannot be carried out in the entire group G.

- **Birkhoff decompositions theorems**

We remark here that all of the definitions and results in symplectic and Poisson geometry mentioned above make sense in infinite dimensions.

Two typical examples of infinite dimensional Manin triple groups are given by:

(1) G = the loop group of smooth maps from S^1 to $GL(n, C)$, G_+ is the subgroup of $g \in G$ such that g is the boundary value of a holomorphic map in the disk $|\lambda| < 1$, and G_- is the subgroup of $g \in G$ such that g is the boundary value of a holomorphic maps in $|\lambda| > 1$ and $g(1) = I$.

(2) G and G_+ are the same as in example (1), and G_- is the subgroup of $g \in G$ such that $g(S^1) \subset U(n)$ and $g(1) = I$.

The two Birkhoff decomposition theorems, which are carefully explained by Pressley and Segal ([PrS]) state that the multiplication map from $G_+ \times G_-$ to G is injective onto an open dense subset of G in example (1) and is a diffeomorphism in example (2). We also need a third analytic theorem on how the decomposition depends on a parameter $x \in (x_0, \infty)$ (Theorem 7.14).

- **Soliton equations and inverse scattering**

Infinite dimensional completely integrable systems are defined in terms of the existence of action-angle coordinates. All interesting infinite dimensional completely integrable Hamiltonian systems seem to be generalizations of the "classical" soliton equations. To set the stage, we give a brief, biased history of some of the work of these equations that is directly related to our paper. It is impossible to give a full history here and we have omitted many major developments.

Solitons were first observed by J. Scott Russel in 1834 while riding on horseback following the bow-wave of a barge along a narrow canal. In 1895, Korteweg and de Vries [KdV] derived the equation

$$q_t = -q_{xxx} - 6qq_x \qquad (KdV)$$

to model the wave propagation in a shallow channel of water, and obtained solitary wave solutions, i.e., $q(x,t) = f(x-ct)$ and f decays at $\pm\infty$. The modern theory of soliton equations started with the famous numerical computation of the interaction of solitary waves of the KdV equation by Zabusky and Kruskal ([ZK]) in 1965. In 1967, Gardner, Green, Kruskal, and Miura [GGKM] used a method called inverse scattering of the one-dimensional linear Schrödinger operator to solve the Cauchy problem for rapidly decaying initial data for the KdV equation. In 1968, Lax ([La]) introduced the concept of Lax-pair for KdV and wrote KdV as the condition for a pair of commuting linear operators. Zakharov and Faddeev ([ZF] 1971) gave a Hamiltonian formulation of KdV, and proved that KdV is completely integrable by finding action-angle variables. Zakharov and Shabat ([ZS1] 1972) found a Lax pair of 2×2 first order differential operators for the non-linear Schrödinger equation:

$$q_t = \frac{i}{2}(q_{xx} + 2|q|^2 q) \qquad (NLS)$$

and solved the Cauchy problem via a similar inverse scattering. Wadati ([Wa] 1973) used the same inverse scattering transform for the Modified KdV equation

$$q_t = -(q_{xxx} + 6q^2 q_x). \qquad (mKdV)$$

Ablowitz, Kaup, Newell and Segar ([AKNS1] 1973) again used this same inverse scattering transform for the sine-Gordon equation

$$q_{xt} = \sin q, \qquad (SGE),$$

and also observed ([AKNS2]) that all these equations have a Lax pair of 2×2 linear operators. In 1973, Zakharov and Manakov ([ZMa1], [ZMa2]) "solved" the 3-wave equation

$$(u_{ij})_t = \frac{b_i - b_j}{a_i - a_j}(u_{ij})_x + \sum_{k \neq i,j} \left(\frac{b_k - b_j}{a_k - a_j} - \frac{b_i - b_k}{a_i - a_k} \right) u_{ik} u_{kj}, \qquad i \neq j.$$

using a Lax pair of 3×3 first order linear operators. In 1976, Gelfand and Dikii [GDi] found an evolution equation on the space of n-th order differential operators on the line with a Lax pair (this generalizes the KdV equation). In 1978, Zakharov and Mikhailov studied harmonic maps from $R^{1,1}$ to Lie group G using a Lax pair of $\mathcal{G}$-valued first order linear operators ([ZMi1], [ZMi2]).

The scattering theory of the $n \times n$ linear system was studied by Shabat [Sh], and Beals and Coifman [BC1], [BC2]. In a series of papers, Beals and Coifman studied the scattering and inverse scattering theory of the first order $n \times n$ linear operator:

$$\begin{cases} \frac{\partial \psi}{\partial x} = (a\lambda + u)\psi, \\ \lim_{x \to -\infty} e^{-a\lambda x} \psi(x, \lambda) = I, \\ e^{-a\lambda x} \psi(x, \lambda) \text{ is bounded in } x. \end{cases} \qquad (1.2)$$

Here $a = \operatorname{diag}(a_1, \ldots, a_n)$ is a constant diagonal matrix with distinct eigenvalues $a_1, \ldots, a_n$, and u lies in the space $\mathcal{S}(R, gl_*(n))$ of Schwartz maps from R to the space $gl_*(n)$ of all $y \in gl(n)$ with zero diagonal entries. The "scattering data", S, of u is defined in terms of the singularity of $e^{-a\lambda x}\psi(x, \lambda)$ in λ. Assume $b \in gl(n)$ is a diagonal matrix. Beals and Coifman defined an evolution equation, the j-th flow associated to a, b on $\mathcal{S}(R, gl_*(n))$, such that if $u(x,t)$ is a solution of this equation then the scattering data $S(\cdot, t)$ of $u(\cdot, t)$ is a solution of the following linear equation:

$$\frac{\partial S}{\partial t} = [S, \lambda^j b],$$

i.e., $S(\lambda, t) = e^{-b\lambda^j t} S(\lambda, 0) e^{b\lambda^j t}$. Then by the inverse scattering transform they solved the Cauchy problem for the j-th flow equation globally. When $n = 2$ and $u \in su(2)$, the second flow defined by $a = b = \operatorname{diag}(i, -i)$ is the non-linear Schrödinger equation, and when $n = 3$ and $u \in su(3)$, the first flow defined by a, b with $b \neq a$ is the 3-wave equation. They prove that

$$w = \operatorname{Re} \int_{-\infty}^{\infty} \operatorname{tr}(-\operatorname{ad}(a)^{-1}(v_1)v_2)\, dx \qquad (1.3)$$

is a symplectic structure on $\mathcal{S}(R, gl_*(n))$ and all the j-th flows are commuting Hamiltonian flows. In 1991, Beals and Sattinger ([BS]) proved that the j-th flow equation is completely integrable by finding action-angle variables. A good survey on recent results is contained in the article by Beals, Deift and Zhou [BDZ].

- **Soliton equations and geometry**

Even earlier than they appeared in applied problems, soliton equations occurred in classical differential geometry. It was known in the mid 19th century that solutions of the sine-Gordon equation correspond to surfaces in R^3 with constant Gaussian curvature -1. The Bäcklund transformations (cf. [Ba]) of surfaces in R^3 generate families of new surfaces with -1 curvature from a given one, and hence give a method of generating new solutions of the sine-Gordon equation from a given one by solving two compatible ordinary differential equations. Inspired by this classical result, Bäcklund transformations have been constructed for a large class of the equations already mentioned (cf. [Mi], [SZ1, 2], [GZ], [TU1]). Many more equations in differential geometry possesses Bäcklund transformations. For example, equations for submanifolds with constant curvature in Euclidean space ([TT], [Ten]), Darboux orthogonal coordinate systems ([Da2]), and harmonic maps from $R^{1,1}$ into a Lie group ([U1]).

Another interesting soliton equation made its appearance in differential geometry at the beginning of the twentieth century. Da Rios, a student of Levi-Civita, studied the free motion of a thin vortex tube in a liquid medium in his master degree thesis ([dR]). He modeled this motion using the evolution of curves in R^3 (the *vortex filament equation* or the *smoke ring equation*):

$$\gamma_t = \gamma_x \times \gamma_{xx}, \tag{1.4}$$

i.e., γ evolves along the direction of the binormal with curvature as speed. The corresponding evolution of the geometric quantity $q = k\exp(-i\int \tau\, dx)$ satisfies the non-linear Schrödinger equation, where $k(\cdot, t)$ and $\tau(\cdot, t)$ are the curvature and torsion of the curve $\gamma(\cdot, t)$. This is the Hasimoto transformation of the vortex filament equation to the non-linear Schrödinger equation. For an interesting historical account of the multiple rediscoveries of the non-linear Schrödinger equation for vortex tubes see an article by Ricca [Ri].

Recently, techniques developed in soliton theory have also been used successfully in several geometric problems whose differential equations are elliptic. For example, the studies of harmonic maps from S^2 to a compact Lie group by Uhlenbeck ([U1]), harmonic maps from a torus to S^3 by Hitchen ([Hi1]), into a symmetric space by Burstall, Ferus, Pedit and Pinkall ([BFPP]), constant mean curvature tori in S^3 by Pinkall and Sterling ([PiS]), constant mean curvature tori in 3-dimensional space forms by Bobenko ([Bo]), and minimal tori in S^4 by Ferus, Pedit, Pinkall, and Sterling ([FPPS]). For a detailed and beautiful account of these developments see the survey book by Guest [Gu].

- **Main goal of the paper**

The main point of our paper is to show how Bäcklund transformations, scattering theory, and the hierarchy of flows can be obtained in a uniform and natural way from "dressing actions" of suitable infinite dimensional Manin triple groups.

- **Decay case**

Most of the soliton equations considered in our paper are evolution equations on the space $\mathcal{S}_{1,a}$ of $gl(n)$-valued connections with one complex parameter:

$$\frac{d}{dx} + a\lambda + u.$$

Here $a \in u(n)$ is a fixed diagonal element and $u \in \mathcal{S}(R, \mathcal{U}_a^\perp)$, where $\mathcal{U}_a^\perp$ is the orthogonal complement of the centralizer $\mathcal{U}_a = \{y \in u(n) \mid [a, y] = 0\}$ of a.

To help explain the basic Poisson group action for soliton equations, we study a simpler case first: flows on the space $\mathcal{S}_+$ of $gl(n)$-valued connections of the form

$$\frac{d}{dx} + A(x, \lambda),$$

where $A(x, \lambda) = \sum_{j=0}^{k} \alpha_j(x)\lambda^j$ for some k and $\alpha_j \in \mathcal{S}(R, u(n))$ for all $0 \leq j \leq k$. Hence $A(x, \lambda)$ is rapidly decaying in x for each $\lambda \in C$. Consider now the infinite dimensional Manin triple group (G, G_+, G_-), where G is the group of holomorphic maps g from $\mathcal{O} \setminus \{\infty\}$ satisfying the condition $g(\bar{\lambda})^* g(\lambda) = I$, G_+ is the subgroup of $g \in G$ that are holomorphic in C and G_- is the subgroup of $g \in G$ that is holomorphic in $\mathcal{O}$, where $\mathcal{O}$ is a small neighborhood of ∞ in the Riemann sphere $C \cup \{\infty\}$. Since $\mathcal{S}_+$ can be identified as a subspace of the dual of the Lie algebra $C(R, \mathcal{G}_-)$, and is invariant under the coadjoint action, $\mathcal{S}_+$ is a Poisson manifold with the standard Lie-Poisson structure. Here we use λ to denote the loop variable for G and x for the variable $x \in R$. The trivialization F of $D = \frac{d}{dx} + A(x, \lambda)$ in $\mathcal{S}_+$ is the solution of

$$\begin{cases} F^{-1}F_x = A(x, \lambda), \\ \lim_{x \to -\infty} F(x, \lambda) = I, \end{cases} \tag{1.5}$$

and the *monodromy* of D is

$$F_\infty(\lambda) = \lim_{x \to \infty} F(x, \lambda).$$

Since $A(x, \lambda)$ is decaying in x, it follows that the linear system (1.5) has a unique global solution. This identifies $\mathcal{S}_+$ as a subset of $C(R, G_+)$. Moreover, the monodromy F_∞ exists and is an element in G_+. The group G_- acts on $C(R, G_+)$ by the pointwise dressing action of G_- on G_+, hence it induces an action $*$ of G_- on $\mathcal{S}_+$. In fact, given $g \in G_-$, factor

$$g(\cdot)F(x, \cdot) = \tilde{F}(x, \cdot)\tilde{g}(x, \cdot) \in G_+ \times G_-,$$

then $g * D = \frac{d}{dx} + \tilde{F}^{-1}\tilde{F}_x$. The fundamental theorem for the decay case appears in section 4. We show that the action of G_- on $\mathcal{S}_+$ is Poisson with the monodromy on the line as the moment map. We call the flows generated by the action of G_- on $\mathcal{S}_+$ the *negative flows*.

- **Rational loop group action**

For a fixed constant $a \in u(n)$, the phase space $\mathcal{S}_{1,a}$ is a coadjoint orbit of $C(R, G_-)$, and the w defined by formula (1.3), is the Kostant-Kirillov symplectic form. Since $a\lambda + u$ does not decay in x, the monodromy of the connection $\frac{d}{dx} + a\lambda + u(x) \in \mathcal{S}_{1,a}$ on the line is not defined. The "action" of G_- on $\mathcal{S}_{1,a}$ can still be defined formally by the dressing action. In fact, first we identify A with its trivialization $E \in C(R, G_+)$ of A normalized at $x = 0$, i.e., E is the solution of

$$\begin{cases} E^{-1}E_x = a\lambda + u, \\ E(0, \lambda) = I. \end{cases}$$

Let $\tilde{E}(x, \cdot)$ denote the dressing action of g on $E(x, \cdot)$ for each $x \in R$. Then

$$\tilde{E}^{-1}\tilde{E}_x = a\lambda + \tilde{u}$$

for some smooth $\tilde{u}$. In general, $\tilde{u}$ does not decay at $\pm\infty$ for $g \in G_-$. So the action of G_- on $\mathcal{S}_{1,a}$ does not exist. But $\tilde{u}$ does belong to the Schwartz class for $g \in G_-^m$, the subgroup of rational maps in G_-. Hence the subgroup G_-^m does act on $\mathcal{S}_{1,a}$. Moreover, if $g \in G_-^m$ is a linear fractional transformation then $\tilde{u}$ can be obtained by solving an ordinary differential equation or by an algebraic formula in terms of g and E. These results are proved in section 6.

- **Homogeneous structure of scattering data**

We are motivated by results from scattering theory to choose the group D_- of meromorphic maps f from $C \setminus R$ to $GL(n, C)$, which satisfy the following conditions:

(i) $f(\bar{\lambda})^* f(\lambda) = I$,

(ii) f has a smooth extension to the closure $\bar{C}_\pm$,

(iii) f has an asymptotic expansion at ∞,

(iv) $f_\pm(r) = \lim_{s \searrow} f(r \pm is)$ such that

$f_+ = h_+ v_+$ factors with v_+ unitary and h_+ upper triangular and $h_+ - I$ in the Schwartz class.

Note that G_-^m is a subgroup of D_-. But D_- is not a subgroup of G_- because we do not assume f is holomorphic at $\lambda = \infty$. Let D_-^c denote the subgroup of $f \in D_-$ such that f is holomorphic in $C \setminus R$. In section 7, we prove that D_- is diffeomorphic to $G_-^m \times D_-^c$ by translating the Birkhoff decomposition theorems for maps from the unit circle to maps from the real line using a linear fractional transformation. We identify the space $\mathcal{S}_{1,a}$ as the homogeneous space D_-/H_- of left cosets of H_- in D_- (scattering cosets), where H_- is the subgroup of $f \in D_-$ that commutes with a. In fact, given $f \in D_-$, we use the Birkhoff decomposition

$$f(\lambda)^{-1}e^{a\lambda x} = E(\lambda, x)M(\lambda, x)^{-1}, \tag{1.6}$$

with $E \in C(R, G_+)$ and $M \in C(R, D_-)$. Then $E^{-1}E_x$ is of the form $a\lambda + u$ for some decay map u, and the map sending the left coset H_-f to $\frac{d}{dx} + E^{-1}E_x$ gives the identification of D_-/H_- and $\mathcal{S}_{1,a}$. Moreover, the right action of D_- on D_-/H_- induces an action of D_- on $\mathcal{S}_{1,a}$, which extends the action $*$ of the G^m_- on $\mathcal{S}_{1,a}$ defined in section 6.

- **Poisson structure of positive flows**

Let H_+ denote the subgroup of G_+ generated by $\{e^p \,|\, p$ is a polynomial in λ which commutes with $a\}$. In section 8, we construct an action of H_+ on D_-/H_- by the "dressing action" of H_+ on D_-. Hence it induces an action of H_+ on $\mathcal{S}_{1,a}$. If a is regular (i.e., a has distinct eigenvalues) then H_+ is abelian, the action of H_+ on $\mathcal{S}_{1,a}$ is Hamiltonian, and the flows generated by H_+ are the commuting hierarchy of the j-th flows. If a is singular, then H_+ is a non-abelian Poisson group and H_+ contains a distinguished infinite dimensional abelian subgroup generated by polynomials in a. Although the action of H_+ is not symplectic, we prove the action of H_+ on $\mathcal{S}_{1,a}$ is Poisson by constructing a moment map. Here we need to prove the difficult result that $M_{\pm\infty} \in H_-$, where $M_{\pm\infty}(\lambda) = \lim_{x\to\pm\infty} M(x, \lambda)$. Then $M_{-\infty}^{-1}M_\infty$ is a moment map for the H_+-action. We also show that the pull back of the symplectic form w to the space of continuous scattering cosets $D^c_-/(H_- \cap D^c_-)$ is non-degenerate. We believe the restriction of w to each algebraic component of the space of discrete scattering cosets $G^m_-/(H_- \cap G^m_-)$ is also non-degenerate, and we prove this in one case.

- **Bäcklund transformations**

Since G^m_- acts on the phase space $\mathcal{S}_{1,a}$, it induces an action of G^m_- on the space of solutions of the j-th flow. In general, if G acts on M, the induced action of G on the space of solutions of a dynamical system is not easy to write down. In section 10, we prove that the induced action of G^m_- on the space of solutions of the j-th flow on $\mathcal{S}_{1,a}$ can be constructed again by dressing action as done in section 6. In fact, if $a\lambda + u$ is a solution of the j-th flow and $g \in G^m_-$ is a linear fractional transformation, then $g \sharp (a\lambda + u)$ can be obtained by solving two compatible ordinary differential equations. The action of such g gives the classical Bäcklund transformation for the sine-Gordon equation. The orbit of the rational negative loop group G^m_- through the vacuum (trivial) solution can be computed explicitly, and is the space of pure solitons. Using the action of G^m_-, we are also able to construct periodic (breather) solutions for the harmonic map equation and the j-th flow equation with $j \geq 2$.

- **Geometric Non-linear Schrödinger equation**

In section 11, we apply soliton theory to the Schrödinger flow on $Gr(k, C^n)$. Suppose (M, g, J) is a complex Hermitian manifold. The geometric non-linear Schrödinger equation (GNLS) is the following evolution equation of curves on M:

$$J\phi_t = \Delta\phi = \nabla_{\phi_x}\phi_x, \qquad (GNLS)$$

where ∇ is the Levi-Civita connection of the metric g. When $M = S^2$, this equation is equivalent via the Hasimoto transformation to the non-linear Schrödinger equation. In fact, if γ evolves according to the vortex filament equation (1.4) and x is the arc length parameter, then γ_x satisfies the geometric non-linear Schrödinger equation on S^2. When M is the complex Grassmannian manifold $Gr(k, C^n)$, the GNLS gives the matrix non-linear Schrödinger equation (MNLS) studied by Fordy and Kulish [FK] for maps q from R^2 to the space $\mathcal{M}_{k\times(n-k)}$ of $k \times (n-k)$ complex matrices:

$$q_t = \frac{i}{2}(q_{xx} + 2qq^*q), \qquad (MNLS)$$

where $q^* = \bar{q}^t$. The MNLS is the second flow on $\mathcal{S}_{1,a}$ defined by

$$a = \begin{pmatrix} iI_k & 0 \\ 0 & -iI_{n-k} \end{pmatrix}.$$

This flow has a Lax pair:

$$\left[\frac{\partial}{\partial x} + \begin{pmatrix} i & 0 \\ 0 & -i \end{pmatrix}\lambda + \begin{pmatrix} 0 & q \\ -q^* & 0 \end{pmatrix},\right.$$
$$\left.\frac{\partial}{\partial t} + \begin{pmatrix} i & 0 \\ 0 & -i \end{pmatrix}\lambda^2 + \begin{pmatrix} 0 & q \\ -q^* & 0 \end{pmatrix}\lambda + \frac{1}{2i}\begin{pmatrix} qq^* & -q_x \\ -q_x & -q^*q \end{pmatrix}\right] = 0.$$

By applying soliton theory to the MNLS, we can solve the Cauchy problem globally with decay initial data, and obtain a Poisson action of H_+ on $\mathcal{S}_{1,a}$ such that the flow generated by $a\lambda^2$ is the MNLS. The flow generated by $b\lambda^j$ with $b \in \mathcal{U}_a$ commutes with MNLS. If $n = 2$ then H_+ is abelian and the action is symplectic. If $n > 2$ then H_+ is non-abelian and the flows generated by $b\lambda^k$ with $b \in \mathcal{U}_a$ commute with the MNLS. But the flows generated by $b_1\lambda^j$ and $b_2\lambda^s$ with $b_1, b_2 \in \mathcal{U}_a$ do not commute if $[b_1, b_2] \neq 0$. Not all these flows are described by differential equations. The flow generated by $b\lambda^k$, $b \neq a$ and $k \neq 1$, are mixed integral-differential flows.

- **Restriction of the phase space by an automorphism**

The phase space of the modified KdV (mKdV) equation is the following subspace of $\mathcal{S}_{1,a}$:

$$\mathcal{S}'_{1,a} = \left\{\frac{d}{dx} + \begin{pmatrix} i & 0 \\ 0 & -i \end{pmatrix}\lambda + \begin{pmatrix} 0 & q \\ -q & 0 \end{pmatrix} \,\middle|\, q \in \mathcal{S}(R, R)\right\}.$$

The third flow defined by $b = a = \text{diag}(i, -i)$ leaves $\mathcal{S}'_{1,a}$ invariant and is the modified KdV flow. While all the even flows vanish on $\mathcal{S}'_{1,a}$, all odd flows leave $\mathcal{S}'_{1,a}$ invariant. This is a special case of restrictions given by finite order automorphisms. To explain this in a more general context, we let $\mathcal{U}$ be a semi-simple Lie algebra (not necessary a subalgebra of $su(n)$), and let $\langle\ ,\rangle$ denote the Killing form. Given $a \in \mathcal{U}$, let $\mathcal{S}_{1,a}$ denote the space of all connections of the form $\frac{d}{dx} + a\lambda + u$, where u is a Schwartz class map from R to the orthogonal complement $\mathcal{U}_a^\perp$ of the centralizer $\mathcal{U}_a$ of a in $\mathcal{U}$. Then $\text{ad}(a)$ maps $\mathcal{U}_a^\perp$ isomorphically onto $\mathcal{U}_a^\perp$. Hence

$$w(v_1, v_2) = \text{Re} \int_{-\infty}^{\infty} \langle -\,\text{ad}(a)^{-1}(v_1), v_2\rangle dx$$

still defines a symplectic structure on $\mathcal{S}_{1,a}$.

Suppose σ is an order k Lie algebra automorphism of $\mathcal{U}$ such that there is an eigendecomposition of σ

$$\mathcal{U} = \mathcal{U}_0 + \ldots + \mathcal{U}_{k-1},$$

where $\mathcal{U}_j$ is the eigenspace with eigenvalues $e^{2(j-1)\pi i/k}$ with $1 \le j \le k$. Assume $a \in \mathcal{U}_1$, and consider the following subspace of $\mathcal{S}_{1,a}$:

$$\mathcal{S}_{1,a}^\sigma(\mathcal{U}) = \left\{ \frac{d}{dx} + a\lambda + u \;\middle|\; u \in \mathcal{U}_0 \cap \mathcal{U}_a^\perp . \right\}$$

Note that when $\mathcal{U} = su(2)$, $\sigma(x) = \bar{x}$, and $a = \text{diag}(i, -i)$, we have $\mathcal{S}_{1,a}^\sigma = \mathcal{S}'_{1,a}$.

It was shown by the first author [Te2] that there exist a sequence of symplectic structures w_r such that $w_{-1} = w$ and all positive flows are Hamiltonian with respect to w_r. In section 9, we study the restriction of the sequence w_r of symplectic forms and the hierarchy of flows to the subspace $\mathcal{S}_{1,a}^\sigma$. We generalize results proved in [Te2] when σ is of order 2 and a result for the generalized modified KdV equation proved by Kupershmidt and Wilson [KW] when $\mathcal{U} = gl(n, C)$, σ is the order n automorphism defined by the conjugation of the operator $c \in GL(n)$ that permutes the standard basis of C^n cyclically, and $a = \text{diag}(1, \alpha, \cdot, \alpha^{n-1})$ with $\alpha = \exp(2\pi i/n)$. In fact, we prove:

(i) If $j \not\equiv 1 \pmod{k}$, then the j-th flow vanishes on $\mathcal{S}_{1,a}^\sigma$, and if $j \equiv 1$ (mod k) then the j-th flow leaves $\mathcal{S}_{1,a}^\sigma$ invariant.

(ii) The restriction of w_r on $\mathcal{S}_{1,a}^\sigma$ is zero if $r \not\equiv 0 \pmod{k}$, and is non-degenerate if $r \equiv 0 \pmod{k}$.

(iii) Let J_{rk} denote the Poisson structure corresponding to w_{rk}, and F_{jk+1} the Hamiltonian for the $(jk+1)$-th flow with respect to J_0. Then the $(k+1)$-th flow satisfies the Lenard relation

$$u_t = J_0(\nabla F_{k+1}) = J_k(\nabla F_1).$$

We should point out that when $\mathcal{U} \subset su(n)$, σ must have order 2. So the order k automorphisms occur in a more general context, in situations for which the scattering theory is considerably more difficult than the case we have discussed. This leads us to the question of other algebraic situations.

- **Other semi-simple Lie algebras**

In this paper we have proved that all rational factorizations can be carried out, and all the formal scattering coset data yield actual geometric flows when $\mathcal{U} = su(n)$. It follows that any problem for a Lie algebra $\mathcal{U} \subset su(n)$ becomes purely an algebraic subproblem. However, many interesting equations in differential geometry arise as flows on a twisted space $\mathcal{S}^{\sigma}_{1,a}(\mathcal{U})$, where $\mathcal{U} \not\subset su(n)$. We believe that some form of the discrete factorization theory and construction of scattering coset can be carried out for many real semi-simple Lie algebras. However, one normally expects a certain number of the factorization theorems to fail off a "big cell". Even more complications arise in trying to handle systems which lie properly in the full complex group. For example, the Gelfand-Dikii hierarchy for a k-th order differential operators is linked to a restriction by an order k automorphism (k-twist) in the full $gl(k, C)$, and the scattering theory is along rays in the directions of k-th roots of unity. Our formal observations about twists apply, and can help understand pure soliton solutions, but do not address the scattering theory difficulties.

- **First flows and flat metrics**

A symmetric space U/K is formed by a splitting of the Lie algebra $\mathcal{U} = \mathcal{K} + \mathcal{P}$, where

$$[\mathcal{K}, \mathcal{K}] \subset \mathcal{K}, \quad [\mathcal{K}, \mathcal{F}] \subset \mathcal{P}, \quad [\mathcal{P}, \mathcal{P}] \subset \mathcal{K}.$$

The rank of a symmetric space is the maximal number of linearly independent commuting elements in $\mathcal{P}$, i.e., the dimension of a maximal abelian subalgebra $\mathcal{T}$ in $\mathcal{P}$. Choose a basis $b_1, \ldots, b_k$ of $\mathcal{T}$. Then for each element $[f]$ of the scattering coset, from our point of view (at least formally, rigorously if $\mathcal{U} \subset su(n)$), there are k commuting first flows in variables we call $x_1, \ldots, x_k$. This yields a flat connection

$$\frac{\partial}{\partial x_i} + b_i \lambda + u_i$$

of k variables for each scattering coset $[f]$. For example, Darboux orthogonal coordinates in R^n ([Da2]), isometric immersions of R^n into R^{2n} with flat normal bundle and maximal rank ([Te2]), equations of hydrodynamic types ([DN1], [DN2], [Dub1], [Ts]) and Frobenius manifolds ([Dub2], [Hi2]) are of this type. In the appendix, we apply some of the soliton theory to these examples.

The authors would like to thank Mark Adler, Percy Deift, Gang Tian and Pierre Van Moerbeke for many helpful discussions. We are grateful to Dick Palais and Gudlaugur Thorbergsson for reading a draft of this paper.

2 Review of Poisson Actions

In this section, we review basic definitions and theorems on Poisson Lie groups and Poisson actions. Two good introductions for this material are articles by Lu and Weinstein [LW] and Semenov-Tian-Shansky [Se1].

A *Poisson structure* on a smooth manifold M is a smooth section π of $L(T^*M, TM)$ such that the bilinear map

$$\{\,,\,\} : C^\infty(M,R) \times C^\infty(M,R) \to C^\infty(M,R)$$

defined by $\{f,g\} = dg(\pi(df))$ is a Lie bracket and satisfies the condition

$$\{fg,h\} = f\{g,h\} + g\{f,h\}, \quad \text{for all}\, f,g,h \in C^\infty(M,R).$$

We will refer to either $\{\,,\,\}$ or π as the Poisson structure on M. The section π can also be viewed as a section of $(T^*M \otimes T^*M)^*$ or a section of $TM \otimes TM$, which will still be denoted by π. Symplectic manifolds are well-known examples of Poisson manifolds.

Let $(M, \{\,,\,\}_M)$ and $(N, \{\,,\,\}_N)$ be two Poisson manifolds. A smooth map $\phi : M \to N$ is called a *Poisson map* if $\{f_1 \circ \phi, f_2 \circ \phi\}_M = \{f_1, f_2\}_N \circ \phi$. *The product Poisson structure* on $M \times N$ is defined by

$$\{f,g\}(x,y) = \{f(*,y), g(*,y)\}_M(x) + \{f(x,*), g(x,*)\}_N(y).$$

A submanifold N of M is a *Poisson submanifold* if there exists a Poisson structure on N such that the inclusion map $i : (N, \{\,,\,\}_N) \to (M, \{\,,\,\}_M)$ is Poisson.

The dual $\mathcal{G}^*$ of a Lie algebra has a natural Lie-Poisson structure by

$$\pi_\ell(x,y) = \ell([x,y]), \qquad \ell \in \mathcal{G}^*, x, y \in \mathcal{G} = (\mathcal{G}^*)^*,$$

with coadjoint orbits as its symplectic leaves. If $\mathcal{G}$ has a non-degenerate ad-invariant form $(\,,\,)$, then by identifying $\mathcal{G}^*$ with $\mathcal{G}$ via $(\,,\,)$, the Lie-Poisson structure on $\mathcal{G}$ is $\pi_x(y,z) = (x, [y,z])$ for all $x, y, z \in \mathcal{G}$.

2.1 Definition. A *Poisson group* is a Lie group G together with a Poisson structure π such that the multiplication map $m : G \times G \to G$ is a Poisson map, where $G \times G$ is equipped with the product Poisson structure.

Note that $\pi(e) = 0$ when π is viewed as a map from $G \to TG \times TG$. Moreover, the dual of $d\pi_e$ is a map from $\mathcal{G}^* \times \mathcal{G}^* \to \mathcal{G}^*$, which defines a Lie bracket on $\mathcal{G}^*$. The corresponding simply connected Lie group G^* has a natural Poisson structure π^* such that the dual of $d(\pi^*)_e$ is the Lie bracket on $\mathcal{G}$. We will call (G^*, π^*) the *dual Poisson group* of (G, π). This pair often fits into a larger group and we call the collection of three groups a Manin triple group. We first explain the Manin triple at the level of Lie algebras.

2.2 Definition. A *Manin triple* is a collection of three Lie algebras $(\mathcal{G}, \mathcal{G}_+, \mathcal{G}_-)$ and an ad-invariant non-degenerate bilinear form $\langle\,,\,\rangle$ on $\mathcal{G}$ with the properties:

(1) $\mathcal{G}_+, \mathcal{G}_-$ are subalgebras of $\mathcal{G}$ and $\mathcal{G} = \mathcal{G}_+ + \mathcal{G}_-$ as direct sum of vector spaces,

(2) $\mathcal{G}_+, \mathcal{G}_-$ are isotropic, i.e., $\langle \mathcal{G}_+, \mathcal{G}_+ \rangle = \langle \mathcal{G}_-, \mathcal{G}_- \rangle = 0$.

Let $(\mathcal{G}, \mathcal{G}_+, \mathcal{G}_-)$ be a Manin triple with respect to $\langle\ ,\rangle$. Then $\mathcal{G}_+ \simeq \mathcal{G}_-^*$ and the infinitesimal vector field corresponding to $x_- \in \mathcal{G}_-$ for the coadjoint action of G_- on $\mathcal{G}_+$ is

$$v_{x_-}(y_+) = [x_-, y_+]_+.$$

The Lie Poisson structure on $\mathcal{G}_+$ is

$$(\pi_+)_{x_+}(y_-) = [x_+, y_-]_+.$$

If there are corresponding Lie groups (G, G_+, G_-) we call this a *Manin triple group*. If (G, G_+, G_-) is a Manin triple group, then G_+ and G_- have natural Poisson group structures. To describe the Poisson structures on G_+ and G_-, we first set up some notation: Given $x_\pm \in \mathcal{G}_\pm$, let $\ell_{x_\pm}, \tau_{x_\pm}$ denote the 1-forms on $G_\mp$ defined by

$$\ell_{x_-}(y_+g_+) = \langle x_-, y_+ \rangle, \qquad \tau_{x_-}(g_+y_+) = \langle x_-, y_+ \rangle,$$
$$\ell_{x_+}(y_-g_-) = \langle x_+, y_- \rangle, \qquad \tau_{x_+}(g_-y_-) = \langle x_+, y_- \rangle.$$

Then the Poisson structures on $G_\pm$ are given explicitly:

$$(\pi_+)_{g_+}(\ell_{x_-}, \ell_{y_-}) = \langle (g_+^{-1}x_-g_+)_+, g_+^{-1}y_-g_+ \rangle$$
$$(\pi_-)_{g_-}(\tau_{x_+}, \tau_{y_+}) = \langle (g_-x_+g_-^{-1})_-, g_-y_+g_-^{-1} \rangle.$$

This is equivalent to

$$(\pi_+)_{g_+}(\ell_{x_-}) = g_+(g_+^{-1}x_-g_+)_+, \qquad (\pi_-)_{g_-}(\tau_{x_+}) = (g_-x_+g_-^{-1})_-g_-,$$

where $g_\pm \in G_\pm$, $x_\pm \in \mathcal{G}_\pm$ and $y_\pm$ denotes the projection of $y \in \mathcal{G}$ onto $\mathcal{G}_\pm$ with respect to the decomposition $\mathcal{G} = \mathcal{G}_+ + \mathcal{G}_-$. Here we identify $\mathcal{G}_-$ as $\mathcal{G}_+^*$, $\mathcal{G}_+$ as $\mathcal{G}_-^*$ via $\langle\ ,\rangle$, and use the matrix convention $gx = (\ell_g)_*(x)$, $gxg^{-1} = \mathrm{Ad}(g)(x)$, and so forth. Since

$$\ell_{x_+}(g_-) = \tau_{(g_-^{-1}x_+g_-)_+}(g_-),$$

we have

$$\begin{aligned}
&(\pi_-)_{g_-}(\ell_{x_+}, \ell_{y_+}) \\
&\qquad = \langle (g_-(g_-^{-1}x_+g_-)_+g_-^{-1})_-,\ g_-(g_-^{-1}y_+g_-)_+g_-^{-1} \rangle \\
&\qquad = \langle (g_-(g_-^{-1}x_+g_- - (g_-^{-1}x_+g_-)_-)g_-^{-1})_-,\ g_-(g_-^{-1}y_+g_-)_+g_-^{-1} \rangle \\
&\qquad = -\langle g_-(g_-^{-1}x_+g_-)_-g_-^{-1},\ g_-(g_-^{-1}y_+g_-)_+g_-^{-1} \rangle \\
&\qquad = -\langle (g_-^{-1}x_+g_-)_-,\ (g_-^{-1}y_+g_-)_+ \rangle \\
&\qquad = -\langle (g_-^{-1}x_+g_-)_-,\ g_-^{-1}y_+g_- \rangle.
\end{aligned}$$

Hence (G_+, π_+) is the dual Poisson group of (G_-, π_-). Conversely, if K is a Poisson group and K^* is its dual Poisson group, then there exist an Ad-invariant form $\langle\ ,\rangle$ and a Lie bracket on $\mathcal{G} = \mathcal{K} + \mathcal{K}^*$ such that $(\mathcal{G}, \mathcal{K}, \mathcal{K}^*)$ is a Manin triple. Hence there is a bijective correspondence between the Manin triples and simply connected Poisson groups. The Manin triple group (G, G_+, G_-) is called a *double group* in the literature. In some cases, multiplication in G can not be globally defined. In this case, we call (G, G_+, G_-) a *local Manin triple group*.

2.3 Example. Let $G = SL(n, C)$, $G_+ = SU(n)$, G_- the subgroup of upper triangular matrices with real diagonal entries, and $\langle x, y\rangle = \mathrm{Im}(\mathrm{tr}(xy))$ the non-degenerate bi-invariant form on $\mathcal{G}$. Then (G, G_+, G_-) is a Manin triple group, and the multiplication map $G_+ \times G_- \to G$ and $G_- \times G_+ \to G$ are isomorphisms. The decomposition of $g \in SL(n, C)$ as $g = g_+ g_- \in G_+ \times G_-$ and $g = h_- h_+ \in G_- \times G_+$ are obtained by applying the Gram-Schmidt process to the columns and rows of g respectively.

2.4 Examples. The type of Poisson groups we need in this paper are generally credited to Cherednik ([Ch]). Let Ω_+ and Ω_- be two domains of $S^2 = C \cup \{\infty\}$ such that $S^2 = \Omega_+ \cup \Omega_-$ and both Ω_+ and Ω_- are invariant under complex conjugation. Let $\mathcal{O} = \Omega_+ \cap \Omega_-$. A map $g : \mathcal{O} \to SL(n, C)$ is called $su(n)$*-holomorphic* if g is holomorphic and satisfies the reality condition $g(\bar{\lambda})^* g(\lambda) = I$ for all $\lambda \in \mathcal{O}$. Let

$$G = \{g : \mathcal{O} \to GL(n, C) \mid g \text{ is } su(n)\text{-holomorphic}\}.$$

Now we fix a normalization point $\lambda_o \in C \cup \{\infty\}$. If $\lambda_o \in \Omega_+$, define

$$\begin{aligned} G_+ &= \{g \in G \mid g \text{ extends } su(n)\text{-holomorphically to } \Omega_+\ g(\lambda_o) = I\}, \\ G_- &= \{g \in G \mid g \text{ extends } su(n)\text{-holomorphically to } \Omega_-\}. \end{aligned}$$

Similarly, if $\lambda_o \in \Omega_-$, we define

$$\begin{aligned} G_+ &= \{g \in G \mid g \text{ extends } su(n)\text{-holomorphically to } \Omega_+\}, \\ G_- &= \{g \in G \mid g \text{ extends } su(n)\text{-holomorphically to } \Omega_-\ g(\lambda_o) = I\}. \end{aligned}$$

The normalization point λ_o determines an Ad-invariant bilinear form $\langle\ ,\rangle$ on $\mathcal{G} = \mathcal{G}_+ + \mathcal{G}_-$ such that $(\mathcal{G}, \mathcal{G}_+, \mathcal{G}_-)$ is a Manin triple. In fact,

$$\langle u, v\rangle = \begin{cases} \dfrac{1}{2\pi i} \displaystyle\oint_\gamma \frac{\mathrm{tr}(u(\lambda)v(\lambda))}{(\lambda - \lambda_o)^2} d\lambda, & \text{if } \lambda_o \in C, \\ \dfrac{1}{2\pi i} \displaystyle\oint_\gamma \mathrm{tr}(u(\lambda)v(\lambda)) d\lambda, & \text{if } \lambda_o = \infty, \end{cases}$$

where $\gamma = \partial\mathcal{O}$. Note that if $u(\lambda) = \sum_k u_k (\lambda - \lambda_o)^k$ and $v = \sum_k v_k (\lambda - \lambda_o)^k$, then

$$\langle u, v\rangle_{\lambda_o} = \begin{cases} \sum_k \operatorname{tr}(u_k v_{-k+1}), & \text{if } \lambda_o \in C, \\ \sum_k \operatorname{tr}(u_k v_{-k-1}), & \text{if } \lambda_o = \infty. \end{cases}$$

The main examples we use in this paper are:

(i) $\Omega_+ = C$, $\Omega_- = \mathcal{O}_\infty$ a neighborhood of ∞,

(ii) $\Omega_+ = C \setminus \{0\}$, $\Omega_- = \mathcal{O}_0 \cup \mathcal{O}_\infty$.

It follows from the Birkhoff Decomposition Theorem (cf. p. 120 Theorem 8.1.2 in the book by Pressley and Segal ([PrS])) that the multiplication map $G_+ \times G_- \to G$ for example (i) is injective and maps onto an open dense subset of G. McIntosh shows that the multiplication map for example (ii) is a diffeomorphism [Mc].

Now suppose (G, G_+, G_-) is a Manin triple group, and the multiplication map $G_+ \times G_- \to G$ is a diffeomorphism. Then given $g_\pm \in G_\pm$, we decompose

$$g_+ g_- = f_- f_+ \in G_- G_+, \qquad g_- g_+ = h_+ h_- \in G_+ G_-.$$

Define

$$g_+ \# g_- = f_-, \qquad g_- \# g_+ = h_+.$$

Then $\#$ defines the dressing action of G_+ on G_- on the left, and the dressing action of G_- on G_+ on the left respectively. Let $x_- \in \mathcal{G}_-$, and $\tilde{x}_-$ denote the infinitesimal vector field of the action of G_- on G_+. Then

$$\tilde{x}_-(g_+) = g_+(g_+^{-1} x_- g_+)_+, \qquad \tilde{x}_+(g_-) = g_-(g_-^{-1} x_+ g_-)_-.$$

There are clearly also corresponding dressing action of G_- on G_+ and G_+ on G_- on the right.

Since the image of the multiplication map is an open dense subset for Example 2.4 (i) and the whole group G for Example 2.4 (ii), the dressing actions for the corresponding Manin triple groups are local and global respectively. However, the Lie algebra actions are defined for all elements in both cases.

2.5 Definition. An action of a Poisson group G on a Poisson manifold P is *Poisson* if the action $G \times P \to P$ is a Poisson map.

It is clear that if the G-action on P is Poisson, M is a Poisson submanifold of P, and M is invariant under G, then the G-action on M is also Poisson. Here one must be careful as the requirement that $M \subset P$ is Poisson is quite restrictive.

A *symplectic structure* on P is a Poisson structure π such that $\pi_x : TP_x^* \to TP_x$ is injective for all $x \in P$. This definition agrees with the standard one when P is finite dimensional, and is the definition of a weak symplectic structure defined in the lecture notes of Chernoff and Marsden [CM] when P is of infinite dimension. For simplicity of notation, we still call such structure a symplectic structure. A G-action on P is called symplectic if $g_*(\pi) = \pi$ for all $g \in G$. If G is equipped with the trivial Poisson structure ($\pi_G = 0$), then an action of G on a symplectic manifold P is Poisson if and only if it is symplectic. However, in general these two notions of actions are different on symplectic manifolds.

A *moment map* of a symplectic action of G on a symplectic manifold P is a G-equivariant map $\mu : P \to \mathcal{G}^*$ such that $\pi_P(df_\xi)$ is the infinitesimal vector field $\tilde{\xi}$ associated to ξ, where f_ξ is the function on P defined by $f_\xi(x) = \mu(x)(\xi)$. When the action is Poisson, we can not expect to define a Poisson map $\mu : P \to \mathcal{G}^*$. The following theorem gives a natural generalization of moment map for Poisson actions.

2.6 Theorem ([Lu]). *Suppose the Poisson group (G, π) acts on the Poisson manifold (P, π_P), and there exists a G-equivariant Poisson map*

$$m : (P, \pi_P) \to (G^*, \pi^*)$$

such that

$$\pi_P(((dm)m^{-1})(\xi)) = \tilde{\xi}, \quad \forall\, \xi \in \mathcal{G},$$

where $\tilde{\xi}$ is the infinitesimal vector field on P associated to ξ and (G^*, π^*) is the dual Poisson group of (G, π). Then the action of (G, π) on (P, π_P) is Poisson.

2.7 Definition. A *moment map* for a Poisson action of a Poisson group G on a Poisson manifold P is a map $m : P \to G^*$ which satisfies the assumptions in the above theorem.

2.8 Example. Suppose (G, G_+, G_-) is a Manin triple group, and the multiplication maps $G_+ \times G_- \to G$ and $G_- \times G_+ \to G$ are diffeomorphisms. Then the dressing action of (G_-, π_-) on (G_+, π_+) is Poisson and the identity map $\mathrm{id} : G_+ \to G_-^* = G_+$ is a moment map. To see this, note first that the identity map is Poisson and equivariant. So by Theorem 2.6 it suffices to check

$$(\pi_+)_{g_+}(dg_+g_+^{-1}(x_-)) = (\pi_+)_{g_+}(\ell_{x_-}) = g_+(g_+^{-1}x_-g_+)_+ = \tilde{x}_-(g_+).$$

Similarly, the dressing action of (G_+, π_+) on (G_-, π_-) is Poisson.

3 Negative flows in the decay case

Our starting point is the Manin triple $(\mathcal{G}, \mathcal{G}_+, \mathcal{G}_-, \langle\ ,\ \rangle)$ of Cherednik type (Example 2.4 (i)) with $\Omega_+ = C$, $\Omega_- = \mathcal{O}_\infty$ and

$$\langle u, v\rangle = \frac{1}{2\pi i}\oint_{\gamma_\infty} \mathrm{tr}(u(\lambda), v(\lambda))d\lambda,$$

where $\gamma_\infty = \partial\mathcal{O}_\infty$ is a contour around ∞. The basic geometric object is a $\mathcal{G}_+$-valued connection on the real line R of the form

$$D = \frac{d}{dx} + A(x,\lambda) = \frac{d}{dx} + \alpha_k(x)\lambda^k + \alpha_{k-1}(x)\lambda^{k-1} + \ldots + \alpha_0(x).$$

From the analytic point of view there are three distinct theories which have very different algebraic structures:

(1) Asymptotically constant cases—the leading term α_k is a constant $a \in gl(n,C)$ and $\alpha_j(x)$ decays in x for $0 \le j < k$.

(2) Decay case—$\alpha_j(x)$ decays in x for all $0 \le j \le k$.

(3) Periodic case—$\alpha_j(x)$ is periodic in x for all j.

Most of the classical scattering theory deals with the asymptotically constant case, which is the case we discuss in most of the paper. For the periodic case we refer the readers to papers by Krichever [Kr1], [Kr2]. We start with the decay case, as a warm-up for the asymptotically constant case.

Fix an element $\alpha_k \in L^1(R)$. An important example would be $\alpha_k(x) = \rho(x)a$ for $\rho \in L^1(R)$ and $a \in gl(n,C)$. If $\rho = dy/dx$, then we can rewrite the connection in y as

$$\frac{dy}{dx}\left(\frac{d}{dy} + a\lambda^k\right) = \frac{d}{dx} + \rho(x)a\lambda^k.$$

Hence the decay case is in reality the case of a "finite interval". However, we use the parametrization of the infinite interval to demonstrate structural relationships with the asymptotically constant case.

Let $C(R,G_\pm)$ be a linear subspace of maps from R to $G_\pm$, that has a formal Lie group structure with Lie algebra $C(R,\mathcal{G}_\pm)$, where $C(R)$ consists of functions which decay at least as fast as those in $L^1(R)$. Identify a map $A \in C(R,\mathcal{G}_+)$ with an element in $C(R,\mathcal{G}_-)^*$ via the pairing

$$\langle\langle A,T\rangle\rangle = \int_{-\infty}^{\infty} \langle A,T\rangle\, dx.$$

(Note that if $A\,dx$ is thought as a one form, then the above formulation is coordinate invariant). Let $\mathcal{S}$ be a subset of $C(R,\mathcal{G}_+)$ that is invariant under the coadjoint action of $C(R,G_-)$. The infinitesimal vector fields for the coadjoint action of $C(R,G_-)$ on $\mathcal{S}$ are

$$v_T(A)(x) = [A(x),T_-(x)]_+,$$

where $+$ indicates the orthogonal projection from $\mathcal{G}$ onto $\mathcal{G}_+$. The Poisson structure on $\mathcal{S}$ is given by

$$\pi_A(T_-) = -[A,T_-]_+,$$

and gives rise to a symplectic structure on the coadjoint orbits of $C(R, G_-)$ on $\mathcal{S}$.

The coadjoint orbit of $\alpha(x)\lambda^k$ under $C(R, G_-)$ is clearly contained in the set of polynomials of degree k of the form

$$A = \alpha(x)\lambda^k + \alpha_{k-1}(x)\lambda^{k-1} + \ldots + \alpha_1(x)\lambda + \alpha_0(x)$$

with the condition that $\alpha_{k-1}(x)$ is of the form $[\alpha(x), v(x)]$ for some v. For many choice of α, this will be the only constraint. The vector field v_T for $T = \sum_{k=1}^{\infty} T_j(x)\lambda^{-j}$ is

$$v_T(A) = [A, T]_+ = \sum_{j=0}^{k-1} \left(\sum_{i=j+1}^{k} [\alpha_i(x), T_{i-j}(x)] \right) \lambda^j,$$

where $\alpha_k = \alpha$.

The negative flows in the decay case can be easily described. Let $\mathcal{P}(R, \mathcal{G}_+)$ denote the Lie algebra of maps $A : R \to \mathcal{G}_+$ such that $A(x)(\lambda)$ is a polynomial in λ and decay in x. Let $\mathcal{P}_k$ denote the set of all $A \in \mathcal{P}(R, \mathcal{G}_-)$ of degree k, and $\mathcal{P}_{k,\alpha}$ the set of all $A \in \mathcal{P}(R, \mathcal{G}_-)$ whose leading term is $\alpha\lambda^k$. Then $\mathcal{P}(R, \mathcal{G}_+)$, $\mathcal{P}_k$ and $\mathcal{P}_{k,\alpha}$ are invariant under the coadjoint action of $C(R, G_-)$, and

$$\pi_A(T_-) = -[A, T_-]_+$$

gives the Poisson structure.

3.1 Definition. The *trivialization of* $A = \sum_{j=0}^{k} \alpha_j(x)\lambda^j$ *normalized at* $x = -\infty$ is the solution $F(A) \in C(R, G_+)$ of

$$F^{-1}F_x = A, \qquad \lim_{x \to -\infty} F(x, \lambda) = I.$$

Given $b \in su(n)$ and $A = \sum_{j=0}^{k} \alpha_j(x)\lambda^j$, then $F(A)^{-1}(x)bF(A)(x) \in \mathcal{G}_+$. Write the expansion of $F(A)^{-1}bF(A)$ at $\lambda = 0$ to get

$$F(A)^{-1}bF(A) = \beta_0 + \beta_1\lambda + \beta_2\lambda^2 + \ldots. \tag{3.1}$$

The β_j's can be computed explicitly from A. Since

$$(F^{-1}bF)_x + [A, F^{-1}bF] = 0, \tag{3.2}$$

we can compare coefficients of λ^j in equation (3.2) to get

$$\begin{cases} (\beta_0)_x + [\alpha_0, \beta_0] = 0, \\ (\beta_j)_x + [\alpha_0, \beta_j] + \sum_{i=1}^{\min\{j,k\}} [\alpha_i, \beta_{j-i}] = 0. \end{cases}$$

The β_j's can be solved explicitly from $\alpha_0, \ldots, \alpha_k$ as follows: Let $g : R \to GL(n, C)$ be the solution to

$$\begin{cases} g^{-1}g_x = \alpha_0 \\ \lim_{x \to -\infty} g(x) = I. \end{cases}$$

Then

$$\begin{cases} \beta_0 = g^{-1}bg, \\ \beta_j(x) = -g^{-1}(x)\sum_{i=1}^{\min\{j,k\}}\left(\int_{-\infty}^{x} g(y)[\alpha_i(y), \beta_{j-i}(y)]g^{-1}(y)dy\right) g(x). \end{cases} \tag{3.3}$$

Hence we have obtained a family of integral equations to describe the β_j's.

The Lax pair for this system is written

$$\left[\frac{\partial}{\partial x} + A, \ \frac{\partial}{\partial t} + (F^{-1}b\lambda^{-m}F)_-\right] = 0. \tag{3.4}$$

It follows from the definition of β's that the coefficient of λ^j with $j < 0$ in the left hand side of equation (3.4) is automatically zero. Setting the coefficients of λ^j ($j \geq 0$) in equation (3.4) to zero gives a system of equations describing a flow on $\mathcal{P}_k$:

$$\begin{cases} \dfrac{d\alpha_k}{dt} = 0, \\ \dfrac{d\alpha_j}{dt} = \sum_{i=j+1}^{\min\{k,m+j\}} [\alpha_i, \beta_{m+j-i}]. \end{cases} \tag{3.5}$$

We call this flow the *$-m$-flow on $\mathcal{P}_{k,\alpha}$ defined by b.* Equation (3.4) also gives

$$\begin{aligned} A_t &= ((\lambda^{-m}F^{-1}bF)_x)_- + [A, (\lambda^{-m}F^{-1}bF)_-] \\ &= [(\lambda^{-m}F^{-1}bF), A]_- + [A, (\lambda^{-m}F^{-1}bF)_-] \\ &= [A, (F^{-1}b\lambda^{-m}F)_-]_+. \end{aligned}$$

So the $-m$-th flow can also be written as

$$A_t = [A, (F^{-1}b\lambda^{-m}F)_-]_+. \tag{3.6}$$

Since the vector field

$$\xi_{b,j}(A) = [A, (F^{-1}b\lambda^{-m}F)_-]_+$$

is bounded in L^1, it is not difficult to see that the $-m$-flow is global. We will prove these flows generate a natural Poisson group action on $\mathcal{P}_{k,\alpha}$ in the next section.

For our basic model, $k = 1$, we have

$$\begin{aligned} &A = \alpha(x)\lambda + u(x), \\ &v_T(u) = [\alpha(x), T_1(x)], \\ &\{T, V\}_A = \int_{-\infty}^{\infty} \operatorname{tr}(\alpha(x)[T_1(x), V_1(x)])\, dx, \end{aligned}$$

where $T = \sum_{j=1}^{\infty} T_j\lambda^{-j}$ and $V = \sum_{j=1}^{\infty} V_j\lambda^{-j}$. This gives our next proposition.

3.2 Proposition. *The $-m$-th flow on $\mathcal{P}_{1,\alpha}$ defined by b is*

$$u_t = [\alpha, \beta_{m-1}],$$

where β_j is defined inductively by

$$\begin{cases} \beta_0 = g^{-1} b g, \\ \beta_j(x) = -g^{-1}(x) \left(\int_{-\infty}^{x} g[\alpha, \beta_{j-1}] g^{-1} dy \right) g(x), \end{cases}$$

and g is the solution to $g^{-1} g_x = u$ and $\lim_{x \to -\infty} g(x) = I$.

A simple change of gauge (cf. [Te2]) implies that the -1-flow describes the geometric equation for harmonic maps from $R^{1,1}$ into $U(n)$ in characteristic coordinates:

3.3 Proposition. *Fix a smooth L^1-map $\alpha : R \to u(n)$ and $b \in u(n)$. Suppose $u(x,t)$ is a solution of the -1-flow equation on $\mathcal{P}_{1,\alpha}$ defined by b:*

$$u_t = [\alpha, g^{-1} b g], \qquad \text{where } g^{-1} g_x = u, \quad \lim_{x \to -\infty} g(x) = I. \tag{3.7}$$

Then there exists a unique solution $E(x,t,\lambda)$ for

$$\begin{cases} E^{-1} E_x = \alpha \lambda + u, \\ E^{-1} E_t = \lambda^{-1} g^{-1} b g, \\ E_\lambda(0,0) = I. \end{cases}$$

Set $s(x,t) = E(x,t,-1) E(x,t,1)^{-1}$. Then $s : R^{1,1} \to U(n)$ is harmonic, $(s^{-1} s_x)(x,t)$ is conjugate to $-2\alpha(x)$, and $(s^{-1} s_t)(x,t)$ is conjugate to $-2b$ for all $t \in R$.

Harmonic maps into a symmetric space are obtained by restriction ([Te2]). This is discussed in section 9. Also, a more elaborate choice of Cherednik splittings allows more complicated examples like the harmonic map equation in space-time (laboratory) coordinates.

4 Poisson structure for negative flows (decay case)

The dressing action defines a local action of G_- on $\mathcal{P}(R, \mathcal{G}_+)$ which is Poisson and generates the negative flows. The notation is the same as in section 3.

4.1 Theorem. *For $A \in \mathcal{P}(R, \mathcal{G}_+)$, let $F(A) : R \to G_+$ denote the trivialization of A normalized at $x = -\infty$. Given $g_- \in G_-$, let $\tilde{F}(x) = g_- \sharp (F(A)(x))$, where $\sharp$ denotes the dressing action of G_- at G_+ for each $x \in R$. Define*

$$g_- * A = \tilde{F}^{-1} \tilde{F}_x.$$

Then $g_- * A$ defines a local action of G_- on $\mathcal{P}(R, \mathcal{G}_+)$. Moreover, the infinitesimal vector field $\tilde{\xi}_-$ associated to $\xi_- \in \mathcal{G}_-$ for this action is

$$\tilde{\xi}_-(A) = -[A, (F^{-1}(A)\xi_- F(A))_-]_+. \tag{4.1}$$

Proof. It is clear that $(g_- * A)$ defines a local action of G_- on $\mathcal{C}(R, \mathcal{G}_+)$. Now we compute the infinitesimal vector field $\tilde{\xi}_-$ on $\mathcal{C}(R, \mathcal{G}_+)$. Write $g_- F = \tilde{F} f_-$, and let δ denote the tangent variation. Then $(\delta g_-)F = \delta\tilde{F} + F\delta f_-$, which implies that

$$F^{-1}(\delta g_-)F = F^{-1}\delta\tilde{F} + \delta f_-.$$

If $\xi_- = \delta g_-$, then we have

$$\delta f_- = (F^{-1}\xi_- F)_-, \qquad F^{-1}\delta\tilde{F} = (F^{-1}\xi_- F)_+. \tag{4.2}$$

Since $g_- * A = \tilde{F}^{-1}\tilde{F}_x$, we obtain

$$\begin{aligned}
\tilde{\xi}_-(A) &= -F^{-1}(\delta\tilde{F})F^{-1}F_x + F^{-1}(\delta\tilde{F})_x \\
&= -(F^{-1}\xi_- F)_+ A + F^{-1}(F(F^{-1}\xi_- F)_+)_x \\
&= -(F^{-1}\xi_- F)_+ A + A(F^{-1}\xi_- F)_+ + ((F^{-1}\xi_- F)_x)_+ \\
&= [A, (F^{-1}\xi_- F)_+] + [F^{-1}\xi_- F, A]_+ \\
&= -[A, (F^{-1}\xi_- F)_-]_+.
\end{aligned}$$

Since $x \mapsto A(x)(\lambda)$ is in $L^1(R) \cap C^\infty(R)$ and $x \mapsto F(x, \lambda)$ is bounded for all λ, we have $\tilde{\xi}_-$ is tangent to $\mathcal{P}(R, \mathcal{G}_+)$. □

4.2 Corollary. *The local action of G_- on $\mathcal{P}(R, \mathcal{G}_+)$ leaves $\mathcal{P}_{k,\alpha}$ invariant, and the flow generated by $\xi_- = -b\lambda^{-m}$ is the $-m$-flow on $\mathcal{P}_{k,\alpha}$ defined by b.*

4.3 Theorem. *The local action of G_- on $\mathcal{P}(R, \mathcal{G}_+)$ is Poisson. The infinitesimal vector field corresponding to ξ_- is $\tilde{\xi}_-(A) = -[A, (F^{-1}\xi_- F)_-]_+$, where F is the trivialization of A normalized at $x = -\infty$. In fact, the map $\phi : \mathcal{P}(R, \mathcal{G}_+) \to G_+ = G_-^*$ defined by $\phi(A) = \lim_{x\to\infty} F(A)(x)$ is a moment map for this action.*

To prove the theorem, we first need a lemma:

4.4 Lemma. $d\phi_A(B) = \left(\int_{-\infty}^{\infty} F(A)BF(A)^{-1}dx\right)\phi(A)$.

Proof. Let F denote $F(A)$, and $\delta F = dF_A(B)$. Taking the variation of the equation $F^{-1}F_x = A$, we get $(F^{-1}\delta F)_x + [A, F^{-1}\delta F] = B$. This implies that

$$F^{-1}\delta F = F(A)^{-1}\left(\int_{-\infty}^{x} F(A)(y,\lambda)B(y,\lambda)F^{-1}(A)(y,\lambda)dy\right)F(A).$$

Then the lemma follows from taking the limit as $x \to \infty$. □

4.5 Proof of Theorem 4.3. It suffices to prove that ϕ satisfies the assumption in Theorem 2.6. First we prove that ϕ is G_--equivariant. Taking the limit of $g_-F = \tilde{F}f_-$ as $x \to -\infty$, we get

$$\lim_{x\to-\infty} \tilde{F}(\lambda, x) = I, \qquad \lim_{x\to-\infty} f_-(\lambda, x) = g_-(\lambda).$$

So $F(g_- * A) = \tilde{F}$ and

$$\phi(g_- * A) = \lim_{x\to\infty} \tilde{F} = g_-\phi(A)(\lim_{x\to\infty} f_-)^{-1} = g_-\#\phi(A).$$

This proves that ϕ is G_--equivariant.

Given $\xi_- \in \mathcal{G}_-$ and $B \in \mathcal{P}(R, \mathcal{G}_+)$, using Lemma 4.4 we get

$$\begin{aligned}\langle\langle d\phi_A(B)(\phi(A))^{-1}, \xi_-\rangle\rangle &= \langle\langle F(A)BF(A)^{-1}, \xi_-\rangle\rangle \\ &= \langle\langle B, F(A)^{-1}\xi_-F(A)\rangle\rangle \\ &= \langle\langle B, (F^{-1}\xi_-F)_-\rangle\rangle.\end{aligned}$$

So $(\Pi_+)_A(d\phi_A\phi(A)^{-1}, \xi_-) = \tilde{\xi}_-(A)$.

It remains to prove that ϕ is a Poisson map. Given $\xi_-, \eta_- \in \mathcal{G}_-$, let $g_+ = \phi(A)$, and ℓ_i the linear functional on $T(G_+)_{g_+}$ defined by

$$\ell_1(z_+g_+) = \langle \xi_-, z_+\rangle, \qquad \ell_2(z_+g_+) = \langle \eta_-, z_+\rangle.$$

It follows form Lemma 4.4 that

$$\ell_1 \circ d\phi_A(B) = \langle B, (F^{-1}\xi_-F)_-\rangle, \qquad \ell_2 \circ d\phi_A(B) = \langle B, (F^{-1}\eta_-F)_-\rangle.$$

But $(F^{-1}\xi_-F)_x + [A, F^{-1}\xi_-F] = 0$. So we get

$$\begin{aligned}&\Pi_+(\ell_1 \circ d\phi_A, \ell_2 \circ d\phi_A) \\ &\quad = -\langle\langle [A, (F^{-1}\xi_-F)_-], (F^{-1}\eta_-F)_-\rangle\rangle \\ &\quad = -\langle\langle [A, (F^{-1}\xi_-F) - (F^{-1}\xi_-F)_+], (F^{-1}\eta_-F)_-\rangle\rangle \\ &\quad = \langle\langle (F^{-1}\xi_-F)_x + [A, (F^{-1}\xi_-F)_+], (F^{-1}\eta_-F)_-\rangle\rangle \\ &\quad = \langle\langle (F^{-1}\xi_-F)_x, (F^{-1}\eta_-F)_-\rangle\rangle + \langle\langle [A, (F^{-1}\xi_-F)_+], (F^{-1}\eta_-F)_-\rangle\rangle \\ &\quad = \langle F^{-1}\xi_-F, (F^{-1}\eta_-F)_-\rangle \mid_{x=-\infty}^{x=\infty} - \langle\langle (F^{-1}\xi_-F), ((F^{-1}\eta_-F)_-)_x\rangle\rangle \\ &\qquad + \langle\langle [A, (F^{-1}\xi_-F)_+], (F^{-1}\eta_-F)_-\rangle\rangle.\end{aligned}$$

The first term is equal to

$$\begin{aligned}\langle g_+^{-1}\xi_-g_+, (g_+^{-1}\eta_-g_+)_-\rangle - \langle \xi_-, \eta_-\rangle &= \langle g_+^{-1}\xi_-g_+, (g_+^{-1}\eta_-g_+)_-\rangle \\ &= (\pi_+)_{g_+}(\xi_-g_+, \eta_-g_+) = (\pi_+)_{g_+}(\ell_1, \ell_2),\end{aligned}$$

where $\langle \xi_-, \eta_- \rangle = 0$ because $\mathcal{G}_-$ is isotropic with respect to (,). The second term is

$$\begin{aligned}\langle\langle (F^{-1}\xi_- F), ((F^{-1}\eta_- F)_-)_x \rangle\rangle &= \langle\langle (F^{-1}\xi_- F)_+, (F^{-1}\eta_- F)_x \rangle\rangle \\ &= \langle\langle (F^{-1}\xi_- F)_+, -[A, F^{-1}\eta_- F] \rangle\rangle \\ &= \langle\langle [A, (F^{-1}\xi_- F)_+], F^{-1}\eta_- F \rangle\rangle \\ &= \langle\langle [A, (F^{-1}\xi_- F)_+], (F^{-1}\eta_- F)_- \rangle\rangle,\end{aligned}$$

which cancels the third term. This proves that ϕ is Poisson. Since ϕ satisfies all assumptions of Theorem 4.3, the action of G_- on $\mathcal{L}_+$ is Poisson and ϕ is a moment map. □

5 Positive flows in the asymptotically constant case

In this section, we will use the same Manin triple as in section 3, and describe flows in the asymptotically constant case. We restrict our discussion to the simplest cases.

Fix $a \in su(n)$, and set

$$\begin{aligned}U_a &= \{g \in SU(n) \mid ga = ag\}, \\ \mathcal{U}_a &= \{y \in su(n) \mid [a, y] = 0\}, \\ \mathcal{U}_a^{\perp} &= \{z \in su(n) \mid \langle z, \mathcal{U}_a \rangle = 0\}.\end{aligned}$$

Given a vector space V, we let $\mathcal{S}(R, V)$ denote the space of all maps from R to V that are in the Schwartz class. Let $\mathcal{S}_{1,a}$ denote the space of all maps $A : R \to \mathcal{G}_+$ such that $A(x)(\lambda) = a\lambda + u(x)$ with $u \in \mathcal{S}(R, \mathcal{U}_a^{\perp})$. The basic symplectic structure on $\mathcal{S}_{1,a}$ is similar to what we have described already for the decay case. However, the structure of the natural flows is different because we may not normalize at $x = -\infty$. Integration as described in the negative flows will tend to destroy the decay condition. The -1-flow does in some sense exist: $\mathcal{S}_{1,a}$

$$\begin{cases} u_t = [a, g^{-1}bg], \\ g_x = gu, \\ \lim_{x \to -\infty} g = I. \end{cases} \tag{5.1}$$

However, the right-hand boundary at ∞ will not be under control and the symplectic structure does not make coherent sense.

Rather than identify A with the trivialization F normalized at $x = -\infty$, we use two different trivializations. For the purposes of constructing Bäcklund transformations, we identify A with the trivialization E normalized at $x = 0$, i.e.,

$$E^{-1}E_x = a\lambda + u, \qquad E(0,\lambda) = I.$$

When we describe the Poisson structure of the positive flows we use $M(x,\lambda)$, where

$$(e^{a\lambda x}M)^{-1}(e^{a\lambda x}M)_x = A, \qquad \lim_{x\to-\infty} M(x,\lambda) = I.$$

Since both E and $e^{a\lambda x}M$ solve the same linear equation, there exists $f(\lambda)$ such that

$$f(\lambda)E(x,\lambda) = e^{a\lambda x}M(x,\lambda).$$

Note that $f(\lambda) = M(0,\lambda)$ contains all the spectral information. The general condition is that f is not holomorphic at $\lambda = \infty$, but that both $f(\lambda)$ and $M(x,\lambda)$ have asymptotic expansion at $\lambda = \infty$. This is known to be the case in scattering theory, and we need our theory to mesh with this analysis.

The positive flows for the asymptotically constant case are defined in a similar fashion as the negative flows for the decay case with the restriction that the generators commute with a. The hierarchy of flows is now mixed ordinary differential and integral equations. Let $A = a\lambda + u$ with $u \in \mathcal{S}(R, \mathcal{U}_a^\perp)$, and M as above. Fix $b \in u(n)$ such that $[a,b] = 0$. Then $M^{-1}bM$ has an asymptotic expansion at $\lambda = \infty$ (cf. [BC1,2]):

$$M^{-1}bM \sim Q_{b,0} + Q_{b,1}\lambda^{-1} + Q_{b,2}\lambda^{-2} + \dots, \qquad Q_{b,0} = b.$$

Since

$$M^{-1}bM = E^{-1}f^{-1}e^{a\lambda x}be^{-a\lambda x}fE = E^{-1}f^{-1}bfE,$$

we get $(M^{-1}bM)_x + [a\lambda + u, M^{-1}bM] = 0$. So we have

$$(Q_{b,i})_x + [u, Q_{b,i}] + [a, Q_{b,i+1}] = 0. \tag{5.2}$$

This defines $Q_{b,i}$'s recursively.

An element $a \in u(n)$ is *regular* if a has distinct eigenvalues. Otherwise, a is *singular*. If a is regular, then it is known that $Q_{b,i}$'s are polynomial differential operators in u (cf [Sa]). But when a is singular, the $Q_{b,i}$'s are integral-differential operators in u. To be more precise, we decompose

$$Q_{b,j} = P_{b,j} + T_{b,j} \in \mathcal{U}_a + \mathcal{U}_a^\perp.$$

Using equation (5.2), P's and T's can be solved recursively. In fact,

$$\begin{cases} P_{b,0} = 0, \\ T_{b,0} = b, \\ P_{b,j+1} = -\operatorname{ad}(a)^{-1}((P_{b,j})_x + [u, Q_{b,j}]^\perp), \\ T_{b,j+1} = -\int_{-\infty}^{x}[u, P_{b,j+1}]^{\mathcal{T}}dy, \end{cases} \tag{5.3}$$

where $v^{\perp}$ and $v^{\mathcal{T}}$ denote the projection onto $\mathcal{U}_a^{\perp}$ and $\mathcal{U}_a$ respectively, and $-\operatorname{ad}(a)$ maps $\mathcal{U}_a^{\perp}$ isomorphically to $\mathcal{U}_a^{\perp}$. It follows from induction and formula (5.3) that the $T_{b,j}$ are bounded and the $P_{b,j}$ are in the Schwartz class.

Consider the Lax pair

$$\left[\frac{\partial}{\partial x}+A,\ \frac{\partial}{\partial t}+(M^{-1}b\lambda^j M)_+\right]=0. \tag{5.4}$$

Set the coefficient of λ^{j-k}, $0\le k<j$, in equation (5.4) equal to zero to get

$$\begin{cases}[a,Q_{b,0}]=0,\\ (Q_{b,k})_x+[u,Q_{b,k}]+[a,Q_{b,k+1}]=0, \quad 1\le k<j.\end{cases} \tag{5.5}$$

This defines the $Q_{b,j}$'s. The constant term gives

$$u_t=(Q_{b,j})_x+[u,Q_{b,j}]=[Q_{b,j+1},a] \tag{5.6}$$

which is called the *j-th flow equation on $\mathcal{S}_{1,a}$ defined by b*. Equation (5.4) can also be written as

$$\begin{aligned}A_t&=((M^{-1}b\lambda^jM)_+)_x+[A,(M^{-1}b\lambda^jM)_+]\\ &=[M^{-1}b\lambda^jM,M^{-1}M_x]_++[A,(M^{-1}b\lambda^jM)_+]\\ &=[M^{-1}b\lambda^jM,A-M^{-1}a\lambda M]_++[A,(M^{-1}b\lambda^jM)_+]\\ &=[M^{-1}b\lambda^jM,A]_++[A,(M^{-1}b\lambda^jM)_+]\\ &=[(M^{-1}b\lambda^jM)_-,A]_+=[Q_{b,j+1},a].\end{aligned}$$

It is clear that the following three statements are equivalent:

(i) $\left[\frac{\partial}{\partial x}+A,\frac{\partial}{\partial t}+B\right]=0$,

(ii) the connection 1-form $\theta=A\,dx+B\,dt$ is flat for all λ, i.e., $d\theta=-\theta\wedge\theta$,

(iii) $\begin{cases}E^{-1}E_x=A,\\ E^{-1}E_t=B,\end{cases}$ is solvable.

So we have

5.1 Proposition. *$A=a\lambda+u$ is a solution of the j-th flow (5.6) on $\mathcal{S}_{1,a}$ defined by b if and only if*

$$\theta(x,t,\lambda)=(a\lambda+u)dx+(b\lambda^j+Q_{b,1}\lambda^{j-1}+\ldots+Q_{b,j})dt$$

is flat on the (x,t)-plane for each λ.

5.2 Definition. The one parameter family of connection 1-form θ defined in Proposition 5.1 is called the *flat connection associated to the solution* A of the j-th flow. The unique solution $E : R^2 \times C \to GL(n, C)$ of

$$\begin{cases} E^{-1}E_x = a\lambda + u, \\ E^{-1}E_t = b\lambda^j + Q_{b,1}\lambda^{j-1} + \ldots + Q_{b,j}, \\ E(0,0,\lambda) = I \end{cases}$$

is called *the trivialization of the flat connection* θ *normalized at the origin* or *the trivialization of the solution* A *at* $(x,t) = (0,0)$.

When a is regular, positive flows are the familiar hierarchy of commuting Hamiltonian flows described by differential equations. When a is singular, positive flows generate a non-abelian Poisson group action. This will be described in section 8.

5.3 Example. For $su(2)$ with $a = \mathrm{diag}(i, -i)$, $\mathcal{S}_{1,a}$ is the set of A of the form $a\lambda + u$, where

$$u = \begin{pmatrix} 0 & f \\ -\bar{f} & 0 \end{pmatrix}$$

and $f : R \to C$ is in the Schwartz class. The first flow is the translation $u_t = u_x$, the second flow defined by a is the non-linear Schrödinger equation (NLS)

$$q_t = \frac{i}{2}(q_{xx} + 2|q|^2 q), \tag{5.7}$$

and the positive flows are the hierarchy of commuting flows associated to the non-linear Schrödinger equation.

5.4 Example. For $a = \mathrm{diag}(a_1, \ldots, a_n) \in su(n)$ with $a_1 < \ldots < a_n$, $\mathcal{S}_{1,a}$ is the set of all $A = a\lambda + u$, where $u = (u_{ij}) \in su(n)$ and $u_{ii} = 0$ for all $1 \le i \le n$. The first flow on $\mathcal{S}_{1,a}$ defined by a is the translation

$$u_t = u_x.$$

The first flow on $\mathcal{S}_{1,a}$ defined by $b = \mathrm{diag}(b_1, \ldots, b_n)$ $(b \ne a)$ is the n-wave equation ([ZMa1, 2]) for u:

$$(u_{ij})_t = \frac{b_i - b_j}{a_i - a_j}(u_{ij})_x + \sum_{k \ne i,j} \left(\frac{b_k - b_j}{a_k - a_j} - \frac{b_i - b_k}{a_i - a_k} \right) u_{ik}u_{kj}, \qquad i \ne j.$$

If a is singular and $[b, a] = 0$, then the j-th flow on $\mathcal{S}_{1,a}$ defined by b is in general an integro-differential equation. But the j-th flow on $\mathcal{S}_{1,a}$ defined by a is again a differential operator:

5.5 Proposition. *$Q_{a,j}(u)$ is always a polynomial differential operator in u.*

Proof. It is easy to see that $Q_{a,1} = u$. We will prove this Proposition by induction. Suppose $Q_{a,i}$ is a polynomial differential operator in u for $i \le j$. Write

$$Q_{a,i} = P_{a,i} + T_{a,i} \in \mathcal{U}_a^{\perp} + \mathcal{U}_a$$

as before. Using formula (5.3), we see that $P_{a,j+1}$ is a polynomial differential operator in u. But we can not conclude from formula (5.3) that $T_{a,j+1}$ is a polynomial differential operator in u. Suppose a has k distinct eigenvalues $c_1, \ldots, c_k$. Then

$$f(t) = (t - c_1)(t - c_2) \ldots (t - c_k)$$

is the minimal polynomial of a. So $f(M^{-1}aM) = 0$, which implies that the formal power series

$$f(a + Q_{a,1}\lambda^{-1} + Q_{a,2}\lambda^{-2} + \ldots) = 0. \tag{5.8}$$

Notice that $f'(a)$ is invertible and $T_{a,j+1}$ commutes with a. Now compare coefficient of $\lambda^{-(j+1)}$ in equation (5.8) implies that $T_{a,j+1}$ can written in terms of $a, Q_{a,1}, \ldots, Q_{a,j}$. This proves that $Q_{a,j+1}$ is a polynomial differential operator in u. □

5.6 Example. For $u(n)$ with

$$a = \begin{pmatrix} iI_k & 0 \\ 0 & -iI_{n-k} \end{pmatrix},$$

$$\mathcal{S}_{1,a} = \left\{ a\lambda + u \;\middle|\; u = \begin{pmatrix} 0 & X \\ -X^* & 0 \end{pmatrix}, X \in \mathcal{M}_{k\times(n-k)} \right\},$$

where $\mathcal{M}_{k\times(n-k)}$ is the space of $k \times (n-k)$ complex matrices. Identifying $\mathcal{S}_{1,a}$ as $\mathcal{S}(R, \mathcal{M}_{k\times(n-k)})$, then the bi-linear form

$$\langle u, v\rangle = \int_{-\infty}^{\infty} \operatorname{tr}(uv)\, dx$$

on $\mathcal{S}(R, \mathcal{U}_a^{\perp})$ induces the following bi-linear form on $\mathcal{S}(R, \mathcal{M}_{k\times(n-k)})$:

$$\langle X, Y\rangle = -\int_{-\infty}^{\infty} \operatorname{tr}(XY^* + X^*Y)\, dx.$$

The orbit symplectic structure on $\mathcal{S}_{1,a}$ induces the following symplectic structure on $\mathcal{S}(R, \mathcal{M}_{k\times(n-k)})$:

$$w(X, Y) = \left\langle \frac{i}{2}X, Y \right\rangle.$$

According to Propositions 5.5, the j-th flow defined by a can be written down explicitly. For

$$u = \begin{pmatrix} 0 & B^* \\ -B^* & 0 \end{pmatrix},$$

we have $Q_{a,0} = a$, $Q_{a,1} = u$,

$$Q_{a,2} = \begin{pmatrix} \frac{1}{2i}BB^* & \frac{i}{2}B_x \\ \frac{i}{2}B_x^* & -\frac{1}{2i}B^*B \end{pmatrix}.$$

The first three flows on $\mathcal{S}(R, \mathcal{M}_{k\times(n-k)})$ are

$$\begin{aligned} B_t &= B_x \\ B_t &= \frac{i}{2}(B_{xx} + 2BB^*B) \\ B_t &= -\frac{1}{4}B_{xxx} - \frac{3}{4}(B_xB^*B + BB^*B_x). \end{aligned}$$

Notice that the second flow is the matrix non-linear Schrödinger equation associated to $Gr(k, C^n)$ by Fordy and Kulish [FK]. By Proposition 5.1, B is a solution of the second flow if and only if

$$(a\lambda + u)\, dx + \left(a\lambda^2 + u\lambda + Q_{a,2}\right) dt$$

is flat for all λ.

6 Action of the rational loop group

The rational loop group is used to construct the soliton data for the positive flows discussed in section 5. We first define a local action $\sharp$ of G_- on $C(R, \mathcal{G}_+)$ via the dressing action. In general the G_--action does not preserve the space $\mathcal{S}_{1,a}$ (because the Schwartz condition on u for $A = a\lambda + u$ is not preserved even locally). However, we prove that the action $\sharp$ of the subgroup G_-^m of rational maps in G_- leaves $\mathcal{S}_{1,a}$ invariant. We also show that the factorization can be done explicitly. In particular, the action $g_- \sharp A$ can be computed explicitly in terms of the trivialization $E(A)$ of A normalized at $x = 0$. In fact, $g_- \sharp A$ is given by an algebraic formula in terms of $E(A)$ and g.

Let $A \in C(R, \mathcal{G}_+)$, and $E(x, \lambda)$ denote the trivialization of A normalized at $x = 0$. Then the map $A \mapsto E$ identifies $C(R, \mathcal{G}_+)$ with a subset of $C(R, G_+)$. (We write $E(x)(\lambda) = E(x, \lambda)$).

Given $f_- \in G_-$ and $A \in C(R, \mathcal{G}_+)$, define

$$f \sharp A = \tilde{E}^{-1}(\tilde{E})_x,$$

where $\tilde{E}(x) = f_- \sharp E(x)$ is the dressing action of G_- on G_+ for each $x \in R$. In other words, we factor

$$f_-E(x) = \tilde{E}(x)\tilde{f}_-(x) \in G_+ \times G_-.$$

Clearly, this defines a local action of G_- on $C(R, \mathcal{G}_+)$. For $\xi_- \in \mathcal{G}_-$, the corresponding infinitesimal vector field on $C(R, \mathcal{G}_+)$ is

$$\tilde{\xi}_-(A) = -[A, (E(A)^{-1}\xi_- E(A))_-]_+.$$

6.1 Proposition. *Let a be a fixed diagonal element in $u(n)$, and $C_{1,a}$ the space of all $A \in C(R, \mathcal{G}_+)$ such that $A(x)(\lambda) = a\lambda + u(x)$ and $u : R \to \mathcal{U}_a^\perp$ is a smooth map. Then $\xi_- \mapsto \tilde{\xi}_-$ defines an action of $\mathcal{G}_-$ on $C_{1,a}$.*

In general, the action of G_- does not preserve the Schwartz condition for $\mathcal{S}_{1,a}$. So it does not define an action on $\mathcal{S}_{1,a}$. But the subgroup G_-^m of rational maps does preserve the decay condition.

6.2 Theorem. *Let G_-^m be the subgroup of rational maps $g \in G_-$. Then the $\sharp$ action of G_-^m on $C(R, \mathcal{G}_+)$ leaves the space $\mathcal{S}_{1,a}$ invariant. Moreover, let $g \in G_-^m$, $A \in \mathcal{S}_{1,a}$, and E the trivialization of A normalized at $x = 0$, then*

(i) we can factor $gE(x) = \tilde{E}(x)\tilde{g}(x) \in G_+ \times G_-^m$ and $g \sharp A = \tilde{E}^{-1}(\tilde{E})_x$,

(ii) $g \sharp A$ can be constructed algebraically from E and g.

To prove this theorem, we first recall the following result of the second author [U1]:

6.3 Proposition ([U1]). *Let $z \in C \setminus R$, V a complex linear subspace of C^n, π the projection of C^n onto V, and $\pi^\perp = I - \pi$. Set*

$$g_{z,\pi}(\lambda) = \pi + \frac{\lambda - z}{\lambda - \bar{z}}\pi^\perp. \tag{6.1}$$

Then

(i) $g_{z,\pi} \in G_-^m$,

(ii) G_-^m is generated by $\{g_{z,\pi} \mid z \in C \setminus R, \pi \text{ is a projection}\}$. ($g_{z,\pi}$ will be called a simple element).

6.4 Proposition.

(i) Let $g(\lambda) = \Pi_{j=1}^r \frac{\lambda - z_j}{\lambda - \bar{z}_j}$, and $A = a\lambda + u$. Then $g \in G_-^m$ and $g \sharp A = A$.

(ii) Let $v_1, \ldots, v_k$ be a unitary basis of the linear subspace V, π_j the projection of C^n onto Cv_j, and π the projection onto V. Then

$$\prod_{j=1}^k g_{z,\pi_j} = \left(\frac{\lambda - z}{\lambda - \bar{z}}\right)^{k-1} g_{z,\pi}.$$

Proof. Statement (i) follows from the fact that g commutes with G_+ and G_-. Statement (ii) follows from a direct computation. □

The above two Propositions imply that to prove Theorem 6.2 it suffices to prove $g_{z,\pi} \sharp A \in \mathcal{S}_{1,a}$, where π is the projections onto a one dimensional subspace. First, we give an explicit construction of $g_{z,\pi} \sharp A$.

6.5 Theorem. *Let $A = a\lambda + u \in \mathcal{S}_{1,a}$, and E the trivialization of A normalized at $x = 0$. Let $z \in C\backslash R$, V a complex linear subspace of C^n, and π the projection onto V. Set*

$$\begin{aligned}\tilde{V}(x) &= E(x,z)^*(V),\\ \tilde{\pi}(x) &= \text{the projection of } C^n \text{ onto } \tilde{V}(x),\\ \tilde{E}(x,\lambda) &= g_{z,\pi}(\lambda)E(x,\lambda)g_{z,\tilde{\pi}(x)}{}^{-1}\\ &= \left(\pi + \frac{\lambda - z}{\lambda - \bar{z}}\pi^{\perp}\right) E(x,\lambda)\left(\tilde{\pi}(x) + \frac{\lambda - \bar{z}}{\lambda - z}\tilde{\pi}(x)^{\perp}\right).\end{aligned}$$

Then:

(i) $g_{z,\pi} \sharp E = \tilde{E}$.

(ii) $\tilde{\pi}^{\perp}(\tilde{\pi}_x + (a\bar{z} + u)\tilde{\pi}) = 0$.

(iii) If $v : R \to C^n$ is a smooth map such that $v(x) \in \tilde{V}(x)$ for all $x \in R$, then $v_x(x) + (a\bar{z} + u)v(x) \in \tilde{V}(x)$ for all x.

(iv) $g_{z,\pi} \sharp A = A + (z - \bar{z})[\tilde{\pi}, a]$.

Proof. First we claim that $\tilde{E}(x,\lambda)$ is holomorphic for $\lambda \in C$. By definition, $\tilde{E}$ is holomorphic in $\lambda \in C\backslash\{z,\bar{z}\}$ and has possible poles at $z, \bar{z}$ with order one. The residues of $\tilde{E}$ at these two points can be computed easily:

$$\begin{aligned}\text{Res}(\tilde{E}, z) &= (z - \bar{z})\pi E(x,z)\tilde{\pi}^{\perp}(x),\\ \text{Res}(\tilde{E}, \bar{z}) &= (\bar{z} - z)\pi^{\perp}E(x,\bar{z})\tilde{\pi}(x).\end{aligned}$$

Since $A(x,\bar{z})^* + A(x,z) = 0$ and $E(0,\lambda) = I$, $E(x,\bar{z})^*E(x,z) = I$. This implies that

$$\tilde{V}(x) = E(x,z)^*(V) = E(x,\bar{z})^{-1}(V).$$

So both residues are zero, and the claim is proved. In particular, we have $g_{z,\pi}E(x) = \tilde{E}(x)g_{z,\tilde{\pi}(x)} \in G_+ \times G_-$. This implies (i).

By Proposition 6.1, $\tilde{E}^{-1}(\tilde{E})_x = a\lambda + \tilde{u}(x)$ for some smooth $\tilde{u} : R \to \mathcal{U}_a^{\perp}$. We get from the formula for $\tilde{E}$ that

$$\begin{aligned}a\lambda + \tilde{u} &= g_{z,\tilde{\pi}}(a\lambda + u)g_{z,\tilde{\pi}}^{-1} - (g_{z,\tilde{\pi}})_x g_{z,\tilde{\pi}}^{-1}\\ &= \left(\tilde{\pi} + \frac{\lambda - z}{\lambda - \bar{z}}\tilde{\pi}^{\perp}\right)(a\lambda + u)\left(\tilde{\pi} + \frac{\lambda - \bar{z}}{\lambda - z}\tilde{\pi}^{\perp}\right)\\ &\qquad - \left(\tilde{\pi}_x + \frac{\lambda - z}{\lambda - \bar{z}}\tilde{\pi}_x^{\perp}\right)\left(\tilde{\pi} + \frac{\lambda - \bar{z}}{\lambda - z}\tilde{\pi}^{\perp}\right).\end{aligned} \tag{6.2}$$

Since the left hand side is holomorphic at $\lambda = z$, the residue of the right hand side at $\lambda = z$ is zero. This gives $(\tilde{\pi}(az+u) - \tilde{\pi}_x)\tilde{\pi}^\perp = 0$, which is equivalent to (ii).

Statement (iii) follows from (ii) since

$$\begin{aligned} v_x + (a\bar{z}+u)v &= (\tilde{\pi}(v))_x + (a\bar{z}+u)v \\ &= \tilde{\pi}_x v + \tilde{\pi} v_x + (a\bar{z}+u)v \\ &= (\tilde{\pi}_x + (a\bar{z}+u)\tilde{\pi})(v) + \tilde{\pi}(v_x) \in \tilde{V}(x). \end{aligned}$$

To prove (iv), we multiply $g_{z,\tilde{\pi}}$ to both sides of equation (6.2) and get

$$\begin{aligned} ((\lambda - \bar{z})\tilde{\pi} + (\lambda - z)\tilde{\pi}^\perp)(a\lambda + u) - ((\lambda-\bar{z})\tilde{\pi}_x + (\lambda - z)\tilde{\pi}_x^\perp) \\ = (a\lambda + \tilde{u})((\lambda-\bar{z})\tilde{\pi} + (\lambda - z)\tilde{\pi}^\perp). \end{aligned}$$

Set $\lambda = z$ and $\lambda = \bar{z}$ in the above equation, we get

$$\begin{cases} \tilde{\pi}(az+u) - \tilde{\pi}_x = (az+\tilde{u})\tilde{\pi}, \\ \tilde{\pi}^\perp(a\bar{z}+u) - \tilde{\pi}_{\bar{x}}^- = (a\bar{z}+\tilde{u})\tilde{\pi}^\perp. \end{cases} \tag{6.3}$$

Add the two equations in (6.3) to get

$$\tilde{u} = u + (z - \bar{z})[\tilde{\pi}, a]. \tag{6.4}$$

□

6.6 Theorem. *The map $\tilde{\pi}$ in Theorem 6.5 is the solution of the following ordinary differential equation:*

$$\begin{cases} (\tilde{\pi})_x + [az+u, \tilde{\pi}] = (\bar{z} - z)[\tilde{\pi}, a]\tilde{\pi}, \\ \tilde{\pi}^* = \tilde{\pi}, \quad \tilde{\pi}^2 = \tilde{\pi}, \quad \tilde{\pi}(0) = \pi. \end{cases} \tag{6.5}$$

Moreover, if $\tilde{\pi}$ is a solution of this equation then $[\tilde{\pi}, a]$ is in the Schwartz class.

Proof. Substitute equation (6.4) into the first equation of (6.3) to get the equation (6.5).

By Proposition 6.4, to prove $[\tilde{\pi}, a]$ is in the Schwartz class it suffices to prove it for the case when V is of one dimensional. By Theorem 6.5 (iii) there exist smooth maps $v : R \to C^n$ and $\phi : R \to C$, such that $v(x)$ spans the linear subspace $\tilde{V}(x)$ and

$$v_x + (a\bar{z}+u)v = \phi v.$$

Set $w = \exp\left(-\int_{-\infty}^x \phi\right) v$. Then $w(x)$ generates $\tilde{V}(x)$ and

$$w_x + (a\bar{z}+u)w = 0. \tag{6.6}$$

We may assume that

$$a = \mathrm{diag}(ic_1, \dots, ic_n), \qquad c_1 \le \dots \le c_n.$$

Let $\psi_j : R \to C^n$ denote the solution of

$$\begin{cases} (\psi_j)_x + (a\bar{z} + u)\psi_j = 0, \\ \lim_{x\to-\infty} e^{-ic_j\bar{z}x}\psi_j(x) = e_j, \end{cases}$$

where $\{e_1, \dots, e_n\}$ is the standard basis of R^n. The construction of the ψ_j is a standard textbook part of the scattering theory. Then $\psi_1, \dots, \psi_n$ form a basis of the solution for equation (6.6). So there exist constants $b_1, \dots, b_n$ such that $w = \sum_{j=1}^n b_j\psi_j$. Let $\bar{z} = r + is$ with $s > 0$ and choose j to be the smallest integer such that $b_j \neq 0$. Then

$$e^{-ic_j\bar{z}x}w = \sum_{k\le j}^{n} e^{-ic_j\bar{z}x}b_k\psi_k = \sum_{k\le j}^{n} e^{i(-c_j+c_k)\bar{z}x}b_k(e^{-ic_k\bar{z}x}\psi_k)$$

Since $\lim_{x\to-\infty} e^{i(-c_j+c_k)\bar{z}x} = 0$ if $c_k < c_j$, we get

$$\lim_{x\to-\infty} e^{-ic_j\bar{z}x}w(x) = \sum_{c_k=c_j} b_k e_k,$$

which is an eigenvector for a. So $\lim_{x\to-\infty}[\tilde{\pi}(x), a] = 0$. Moreover,

$$e^{-ic_j\bar{z}x}w(x) = \sum_{c_k=c_j} b_k e_k + \sum_{c_j<c_k} e^{i(-c_j+c_k)\bar{z}x}b_k e^{-ic_k\bar{z}x}\psi_k.$$

Since $\lim_{x\to-\infty} e^{-ic_k\bar{z}x}\psi_k(x) = e_k$,

$$\left| e^{-ic_j\bar{z}x}w(x) - \sum_{c_k=c_j} b_k e_k \right| = e^{(-c_j+c_m)x\,\mathrm{Im}(z)}\mathcal{O}(1),$$

where c_m is the next non-zero term. Hence $\tilde{\pi}(x) - \lim_{x\to-\infty}\tilde{\pi}(x)$ decays exponentially, so $[\tilde{\pi}, a]$ also decays exponentially as $x \to -\infty$. Similarly, we can prove that $[\tilde{\pi}(x), a] = 0$ decays exponentially when $x \to \infty$.

From equation (6.5)

$$|\tilde{\pi}_x| \le 2|z|\,|[a, \tilde{\pi}]| + 4|u|.$$

So $\tilde{\pi}_x$ decays like u. Repeated differentiation of equation (6.5) gives the desired result. In fact, $\tilde{\pi}_x \in \mathcal{S}(R)$ as well. □

If U is a matrix whose columns form a basis of V, then the projection of C^n onto V is $\pi = U(U^*U)^{-1}U^*$. This follows from elementary linear algebra. So we have:

6.7 Corollary. *Let V be a k-dimensional linear subspace of C^n, and U a matrix whose columns form a basis of V. Then $g_{z,\pi} \sharp A = A + (\bar{z} - z)[\tilde{\pi}, a]$, where*

$$\tilde{\pi}(x) = E^*(x,z)U(U^*E(x,z)E^*(x,z)U)^{-1}U^*E(x,z). \qquad (6.7)$$

6.8 Proof of Theorem 6.2.

Given $g \in G_-^m$, write g as product of simple elements $\prod_{j=1}^k g_{z_j,\pi_j}$ (note that the factorization of g into simple elements is not unique, for example see Proposition 6.4 (ii)). Use Theorem 6.5 to see that $g \sharp A \in \mathcal{S}_{1,a}$, and $g \sharp E$ and $g \sharp A$ are obtained by algebraic formulae from E and g. □

7 Scattering data and Birkhoff decomposition

The asymptotically constant case is the case standardly treated in the soliton literature. We obtain many hints of how to describe the theory, since most of what we need is already contained in scattering theory literature. The main purpose of this section is to give a homogeneous structure for the space of scattering data, to obtain the Inverse Scattering Transform using the standard Birkhoff decompositions, and to relate the action of the rational loop group described in section 6 to the scattering data in a natural and simple way.

We first review results of Beals-Coifman ([BC1, 2]) and Zhou ([Zh1, 2]) on scattering theory for $n \times n$ first order linear system. Let

$$a = \text{diag}(ia_1, \dots, ia_n) \in u(n), \qquad a_1 \leq a_2 \leq \dots \leq a_n,$$

and $A = a\lambda + u \in \mathcal{S}_{1,a}$. Consider the linear system

$$\begin{cases} \psi_x = \psi(a\lambda + u), \\ \lim_{x \to -\infty} e^{-a\lambda x}\psi(x,\lambda) = I, \\ m(x,\lambda) = e^{-a\lambda x}\psi(x,\lambda) \text{ is bounded in } x. \end{cases} \qquad (7.1)$$

(m will be called the *normalized (matrix) eigenfunction* of A).

7.1 Theorem ([BC1, 2],[Zh2]). *Given $A = a\lambda + u \in \mathcal{S}_{1,a}$ there exists a bounded discrete subset D of $C \setminus R$ such that the normalized eigenfunction $m(x,\lambda) = e^{-a\lambda x}\psi(x,\lambda)$ is holomorphic in $\lambda \in C \setminus (R \cup D)$ and has poles at $z \in D$. Moreover, there exists a dense open subset $\mathcal{S}'_{1,a}$ of $\mathcal{S}_{1,a}$ such that for $A = a\lambda + u \in \mathcal{S}'_{1,a}$, the normalized eigenfunction $m(x,\lambda)$ satisfies the following conditions:*

(i) The subset D is finite, and m has only simple poles at $z \in D$,

(ii) The matrix function m can be extended smoothly to the real axis from the upper and lower half λ-plane,

(iii) As a function of λ, m has an asymptotic expansion at $\lambda = \infty$.

The open dense subset $\mathcal{S}'_{1,a}$ contains all $\{u \in \mathcal{S}_{1,a}$ such that the L^1-norm of u is less than 1 and all u with compact support.

7.2 Theorem ([BC1,2]). *Let m be the normalized eigenfunction of $A = a\lambda + u \in \mathcal{S}_{1,a}$, and $b \in u(n)$ such that $[a, b] = 0$. Set $Q_b = m^{-1}bm$. Then Q_b has an asymptotic expansion at $\lambda = \infty$:*

$$Q_{b,\lambda} \sim b + Q_{b,1}\lambda^{-1} + Q_{b,2}\lambda^{-2} + \dots .$$

Moreover,

(i) $(Q_{b,j})_x + [u, Q_{b,j}] = [Q_{b,j+1}, a]$.

(ii) The j-th flow $u_t = [Q_{b,j+1}, a]$ is symplectic with respect to the symplectic structure $w(v_1, v_2) = \langle -\operatorname{ad}(a)^{-1}(v_1), v_2\rangle$.

Recall

$$G_+ = \{g : C \to GL(n, C) \mid g \text{ is holomorphic, } g(\bar{\lambda})^* g(\lambda) = I\},$$
$$G_- = \{g : \mathcal{O}_\infty \to GL(n, C) \mid g \text{ is holomorphic, } g(\bar{\lambda})^* g(\lambda) = I, g(\infty) = I\}.$$

Since $m(x, \lambda)$ is not holomorphic at $\lambda = \infty$, we must change G_-, and restrict G_+ to have a singularity at $\lambda = \infty$ of the type exp(polynomial).

We are motivated by Theorem 7.1 to choose a different negative group D_-:

7.3 Definition. Let D_- denote the group of meromorphic maps f from $C \setminus R$ to $GL(n, C)$ satisfying the following conditions:

(i) $f(\bar{\lambda})^* f(\lambda) = I$.

(ii) f has a smooth extension to the closure $\bar{C}_\pm$, i.e., $f_\pm(r) = \lim_{s\searrow 0} f(r \pm is)$ exists and is smooth for $r \in R$, (since $f(\bar{\lambda})^* f(\lambda) = I$, we have $f_-(r) = (f_+(r)^*)^{-1}$).

(iii) f has an asymptotic expansion at ∞.

(iv) $f_+ - I$ lies in the Schwartz class modulo unitary maps. In other words, if we factor $f_+ = h_+ v_+$ with v_+ unitary and h_+ upper triangular then $h_+ - I$ is in the Schwartz class.

Let $m(x, \lambda)$ be the normalized eigenfunction for $A \in \mathcal{S}'_{1,a}$, and E the trivialization of A normalized at $x = 0$. Since both $e^{a\lambda x} m(x, \lambda)$ and $E(x, \lambda)$ satisfy the ordinary differential equation in x:

$$E^{-1}E_x = (e^{a\lambda x} m)^{-1} (e^{a\lambda x} m)_x = A,$$

there exists f such that

$$e^{a\lambda x} m(x,\lambda) = f(\lambda) E(x,\lambda).$$

In fact, $f(\lambda) = m(0,\lambda)$. By Theorem 7.1, $f \in D_-$.

Beals and Coifman [BC1,2] defined the scattering data of $A = a\lambda + u \in \mathcal{S}'_{1,a}$ to be the map $S : R \cup D \to GL(n)$: for $z \in D$, $S(z)$ is the element in $GL(n, C)$ such that

$$(I - (\lambda - z)^{-1} e^{azx} S(z) e^{-azx}) m(x,\lambda)$$

has a removable singularity at $\lambda = z$, and for $r \in R$, $S(r) = v_-^{-1}(r) v_+(r)$, where

$$\lim_{s \searrow 0} m(x, r + is) = e^{arx} v_\pm(r) e^{-arx} m(x,r).$$

They prove:

(i) The map sending A to S is injective.

(ii) If $u(x,t)$ is a solution of the j-th flow on $\mathcal{S}_{1,a}$ defined by b, $u_t = [Q_{b,j+1}, a]$, and $S(\lambda, t)$ is the corresponding scattering data, then

$$S_t(\lambda, t) = [S(\lambda, t), \lambda^j b].$$

In particular, $S(\lambda, t) = e^{-b\lambda^j t} S(\lambda, 0) e^{b\lambda^j t}$.

We note that scattering data S for A is determined by $f(\lambda) = m(0,\lambda)$. In fact, $S(r) = f_-(r)^{-1} f_+(r)$ for $r \in R$ and $S(z)$ can be obtained from the residue of $f(\lambda)$ at $z \in D$.

7.4 Remark. The rational group G_-^m defined in section 6 is a subgroup of D_-.

Instead of using S as the scattering data, we use the left coset $H_- f$ in D_-/H_- as the scattering data of A, where H_- is the subgroup of $h \in D_-$ that commutes with a. We will call $[f] = H_- f$ the *scattering coset of* A. One advantage of using the scattering cosets is that the inverse scattering transform can be obtained from the standard Birkhoff Decomposition Theorems. Another advantage is that the natural action of the subgroup G_-^m of rational maps on D_-/H_- on the right by multiplication induces the action of G_-^m on $\mathcal{S}_{1,a}$ defined in section 6. To explain this, we first prove a decomposition theorem.

7.5 Theorem. *Let D_-^c denote the subgroup of $v \in D_-$ such that v is holomorphic in $C \setminus R$. Then any $f \in D_-$ can be uniquely factored into*

$$f = gh = \tilde{h}\tilde{g},$$

where $g, \tilde{g} \in G_-^m$ and $h, \tilde{h} \in D_-^c$. Moreover, the multiplication map

$$G_-^m \times D_-^c \to D_-$$

is a diffeomorphism.

This theorem is the real line version of the Birkhoff decomposition theorem, which can be seen by transforming the domain C_+ to the unit disk and real axis to the unit circle S^1 by a linear fractional transformation. To be more precise, let $LGL(n, C)$ denote the loop group of smooth maps from S^1 to $GL(n, C)$, and $L^+GL(n, C)$ the group of maps $g \in LGL(n, C)$ such that g is the boundary value of a holomorphic map

$$g : \{z \,|\, |z| < 1\} \to GL(n, C)\}.$$

Let $\Omega U(n)$ denote the based loop group of maps $g : S^1 \to U(n)$ such that $g(-1) = I$, Recall that the standard Birkhoff Decomposition Theorem (cf. [PrS] p. 120, Theorem 8.1.1) is:

7.6 Birkhoff Decomposition Theorem. *Any $g \in LGL(n, C)$ can be factored uniquely as*

$$g = g_+ g_- = h_- h_+,$$

where $g_+, h_+ \in L^+GL(n, C)$ and $g_-, h_- \in \Omega U(n)$. In other words, the multiplication map

$$L^+GL(n, C) \times \Omega U(n) \to LGL(n, C)$$

is a diffeomorphism.

A direct computation shows:

7.7 Proposition. *Given $g : S^1 \to GL(n, C)$, define $\Phi(g) : R \to GL(n, C)$ by $\Phi(g)(r) = g\left(\frac{1+ir}{1-ir}\right)$. Then*

(i) g is smooth if and only if $\Phi(g)$ is smooth and has the same asymptotic expansions at $-\infty$ and ∞,

(ii) $g - I$ is infinitely flat at $z = -1$ if and only if $\Phi(g) - I$ is in the Schwartz class,

(iii) $g : C \to GL(n, C)$ satisfies the reality condition $g(1/\bar{z})^* g(z) = I$ if and only if $\tilde{g}(\lambda) = g(\frac{1+i\lambda}{1-i\lambda})$ satisfies the reality condition $\tilde{g}(\bar{\lambda})^* \tilde{g}(\lambda) = I$.

7.8 Corollary. *The group D_- is isomorphic to the group of smooth loops $g : S^1 \to GL(n, C)$ that are boundary values of meromorphic maps with finitely many poles in $|\, z \,| < 1$ and $g^* g - I$ is infinitely flat at $z = -1$.*

As a consequence of Theorem 7.6 and Proposition 7.7, we have

7.9 Corollary. *If $f : R \to GL(n, C)$ is smooth and has an asymptotic expansion at $\lambda = \infty$, then f can be factored*

$$f = vg,$$

where g is unitary and v is the boundary value of a holomorphic map on C_+.

7.10 Proof of Theorem 7.5.
It follows from Corollary 7.9 that given $f \in D_-$, we can factor $f_\pm$

$$f_\pm(r) = h_\pm(r) g_\pm(r), \qquad r \in R$$

where $h_\pm$ is the boundary value of a holomorphic map h on $C_\pm$ and $g_\pm$ is a smooth map from R to $U(n)$. It follows from $f_- = (f_+^*)^{-1}$ that we have $g_+ = g_-$ and $h(\bar{\lambda})^* h(\lambda) = I$. Write $f = hg$. Since f is meromorphic and h is holomorphic in C_+, g is meromorphic in C_+. However, $g(r)^* g(r) = I$ for $r \in R$ implies that g extends holomorphically across the real axis. So g is meromorphic in C and bounded near infinity. This implies that g is rational, i.e., $g \in G_-^m$. □

Recall that $a = \mathrm{diag}(ia_1, \ldots, ia_n) \in u(n)$ is a fixed diagonal matrix, and G_+ is the group of holomorphic maps $g : C \to GL(n, C)$.

7.11 Theorem. *Let $f \in D_-$, k a positive integer, and $b \in u(n)$ such that $[a, b] = 0$. Let $e_{b,k}(x)(\lambda) = e^{b\lambda^k x}$. Then there exists a unique $E(x, \lambda)$ and $M(x, \lambda)$ such that*

$$f^{-1} e_{b,k}(x) = E(x, \cdot) M^{-1}(x, \cdot) \in G_+ \times D_-.$$

Proof. Write $f = hg$ as in Theorem 7.5 with $h \in D_-^c$ and $g \in G_-^m$. Write $h = pv$, where p is upper triangular and v is unitary. By definition of D_-, $p - I$ is in the Schwartz class when restricted to the real axis in the λ-plane. So $e_{b,k}^{-1}(x) p^{-1} e_{b,k}(x)$ has an asymptotic expansion at $r = \pm\infty$ for each x. Write $e_{b,k}^{-1}(x) p^{-1} e_{b,k}(x) = \tilde{v}(x) \tilde{h}(x)$, where $\tilde{v}$ is unitary and $\tilde{h}$ is the boundary value of a holomorphic map on C_+. Notice f, p, v and h do not depend on x, whereas the rest of the matrix functions do depend on x. So

$$h^{-1} e_{b,k}(x) = v^{-1} e_{b,k}(x) \tilde{v}(x) \tilde{h}(x) = B(x) \tilde{h}(x),$$

where $B(x) = v^{-1} e_{b,k}(x) \tilde{v}(x)$ is unitary. Both $\tilde{h}(x)$ and h are holomorphic in $\lambda \in C_+$, and $e_{b,k}(x)$ is holomorphic in C_+. Hence $B(x)$ is holomorphic in $\lambda \in C_+$. However, $B(x)$ is unitary hence it is holomorphic in $\lambda \in C$.

Next we claim that we can factor $g^{-1} B(x) = E(x) g_1^{-1}(x)$ with E holomorphic in C and $g_1 \in G_-^m$. This can be proved exactly the same way as Theorem 6.2. Then

$$\begin{aligned} f^{-1} e_{b,k}(x) &= (hg)^{-1} e_{b,k}(x) = g^{-1} h^{-1} e_{b,k}(x) \\ &= g^{-1} B(x) \tilde{h}(x) = E(x) g_1^{-1}(x) \tilde{h}(x) = E(x) M^{-1}(x), \end{aligned}$$

which finishes the proof. □

7.12 Definition. A matrix q is called a-diagonal if $q_{jk} = 0$ whenever $a_j \neq a_k$, q is (strictly) upper a-triangular if $q_{jk} = 0$ whenever $a_j > a_k$ (and $q_{jk} = 0$ or I if $a_j = a_k$), and q is (strictly) lower a-triangular if $q_{jk} = 0$ whenever $a_j < a_k$ (and $q_{jk} = 0$ or I if $a_j = a_k$). Let q_d denote the *a-diagonal projection of* q, i.e.,

$$(q_d)_{ij} = \begin{cases} q_{ij}, & \text{if } a_i = a_j, \\ 0, & \text{if } a_j \neq a_k. \end{cases}$$

7.13 Proposition. *Any $f \in D_-$ can be factored uniquely as*

$$f = pv = q\tilde{v},$$

where p is upper a-triangular, q is lower a-triangular, v and $\tilde{v}$ are unitary, and the a-diagonal projections p_d, q_d are holomorphic in $C_\pm$.

Proof. Write $g = p_0 v_0$, where p_0 is upper a-triangular and v_0 is unitary. Such p_0, v_0 are not unique because an element in $U_a = \{y \in U(n) \mid ay = ya\}$ is both a-triangular and unitary. Write $p_0 = p_1 p_2$, where p_1 is strictly upper a-triangular and p_2 is in a-diagonal. Factor $p_2 = p_3 h$, where p_3 is holomorphic in C_+ and h is unitary. Then $g = p_1 p_3 h v_0 = pv$, where $p = p_1 p_3$ is upper a-triangular and $v = hv_0$ is unitary. Since p_1 is strictly upper a-triangular, $p_d = p_3$ is holomorphic. □

To study how the Birkhoff factorization of Theorem 7.11 depends on parameter x, we introduce the class of Schwartz maps from $[r_0, \infty)$ to a Hilbert space. Let H be a Hilbert space, a map $\phi : [r_0, \infty) \to H$ is in $\mathcal{S}([r_0, \infty), H)$ if for each pair of integers (m, s) there exists a constant $c_{m,s}$ such that

$$\left\| \left(\frac{d}{dx}\right)^m \phi(x) \right\| \leq \frac{c_{m,s}}{(1 + |\, x\,|)^s}.$$

Let H_1 denote the Sobolev space for maps from $R^+ = [0, \infty)$ to $\mathcal{U}_a^\perp$. In other words, $u \in H_1$ if

$$\begin{aligned} ||u||_1^2 &= \int_0^\infty \left(\left\| \frac{du}{dr} \right\|^2 + ||u||^2 \right) dr \\ &= \int_0^\infty (y^2 + 1) ||\hat{u}||^2 dy \quad < \infty, \end{aligned}$$

where $\hat{u}$ is the Fourier transform of u.

The following is a functional analytic extension of Birkhoff decomposition.

7.14 Theorem. *Let $I + D(x, \cdot) = (I + h(x, \cdot))V(x, \cdot)$ be the Birkhoff decomposition, where $h(x)(r)$ is the boundary value of a holomorphic map in the upper half plane and $V(x)(r)$ is unitary. If $D \in \mathcal{S}(R^+, H_1)$, then h and $V - I$ are in $\mathcal{S}(R^+, H_1)$.*

Proof. This should be regarded as an implicit function theorem. It is based on the two facts about the Sobolev space H_1. The first is that H_1 is an algebra under multiplication and $\exp : H_1 \to H_1$ is smooth. The second is that the linear Birkhoff decomposition can be defined using the Fourier transform $\mathcal{F}$. Let $\Psi : L^2(R) \to L^2(R)$ denote the linear operator defined by

$$\Psi(f)(y) = \begin{cases} f(y) + f(-y)^*, & \text{if } y \geq 0, \\ 0, & \text{otherwise,} \end{cases}$$

and let $\pi_+ : H_1 \to H_1$ be the bounded linear map

$$\pi_+ = \mathcal{F}^{-1}\Psi\mathcal{F}, \qquad \text{i.e.,}$$
$$\pi_+(f)(r) = \int_0^\infty (\hat{f}(y) + \hat{f}(-y)^*)e^{iry}\,dy.$$

We claim that $f = \pi_+(f) + (I - \pi_+)(f)$ is the linear Birkhoff Decomposition, or equivalently, $\pi_+(f)$ is the boundary value of a holomorphic map on C_+ and $(I - \pi_+)(f)$ is is in $u(n)$. To see this, we note that $\pi_+(f)$ is the boundary value of the holomorphic map

$$\lambda \in C_+ \mapsto \int_0^\infty (\hat{f}(y) + \hat{f}(-y)^*)e^{i\lambda y}\,dy.$$

Then

$$\begin{aligned}(I - \pi_+)(f) &= f(r) - \int_0^\infty (\hat{f}(y) + \hat{f}(-y)^*)e^{iry}\,dy \\ &= \int_{-\infty}^\infty \hat{f}(y)e^{iry}\,dy - \int_0^\infty (\hat{f}(y) + \hat{f}(-y)^*)e^{iry}\,dy \\ &= \int_{-\infty}^0 \hat{f}(y)e^{iry}\,dy - \int_0^\infty \hat{f}(-y)e^{iry}\,dy.\end{aligned}$$

It follows that $(I - \pi_+)(f)^* = -(I - \pi_+)(f)$.

Due to the linearity of π_+, it is easy to see that this extends to the parameter version in x. We write this as

$$\pi_+ : \mathcal{S}(R^+, H_1) \to \mathcal{S}(R^+, H_1).$$

Now the Birkhoff decomposition is a non-linear operator. However we are near the identity, so it can be regarded as a perturbation of the linear operation because the exponential map is smooth on $H_1(R)$.

Let $Y : \mathcal{S}(R^+, H_1) \to \mathcal{S}(R^+, H_1)$ be the map defined by

$$Y(f) = e^{\pi_+(f)}e^{(I-\pi_+)(f)}.$$

Given D, we wish to find $\tilde{D}$ such that

$$(I + D) = \exp(\pi_+(\tilde{D}))\exp((I - \pi_+)\tilde{D}) = Y(\tilde{D}).$$

Since $dY_0 = I$, for x sufficiently large

$$||\pi_+(\tilde{D})||_1 \leq ||\tilde{D}||_1 \leq O_s,$$

where $O_s(x) = c||D||_1 \leq cc_s/(1 + |x|)^s$. The estimate on derivatives in x is more difficult. Let $I + h = \exp(\pi_+(\tilde{D}))$. Then

$$(I + h)^{-1} D_x V^{-1} = (I + h)^{-1} h_x + V_x V^{-1}.$$

On the right, the first term is holomorphic in the upper half plane, the second term is unitary. Hence

$$(I + h)^{-1} h_x = \pi_+((I + h)^{-1} D_x V^{-1}) = \pi_+((I + h)^{-1} D_x (I + D)^{-1}(I + h)).$$

Or

$$h_x = (I + h)\pi_+((I + h)^{-1} D_x (I + D)^{-1}(I + h)).$$

Certainly, $D_x(I + D)^{-1} \in \mathcal{S}(R^+, H_1)$, π_+ is linear, and H_1 is an algebra. Using the Leibnitz rule repeatedly, we can obtain

$$\left\| \left(\frac{\partial}{\partial x}\right)^m h(x) \right\|_1 \leq C_m(h) \max_{j \leq i} \left\| \left(\frac{\partial^m D}{\partial x^m}(x)\right) \right\|_1,$$

where $C_m(h) = C(||h||, \ldots, ||(\partial/\partial x)^{m-1} h||)$. Estimates in the Schwartz topology follows by induction on m. □

7.15 Remark. The awkwardness of this proof reminds one that the classical use of the Schwartz space is probably not as natural for the analysis as various choices of Hilbert or Banach spaces in x would be. The above proof would then be a straight forward use of the usual implicit function theorem (rather than a reproof). Notice that in fact $H_1(R)$ could be replaced by any $H_k(R)$, $k > \frac{1}{2}$.

Recall that $e_{a,1}(x)(\lambda) = e^{a\lambda x}$.

7.16 Theorem. *In the Birkhoff factorization of Theorem 7.11*

$$f^{-1} e_{a,1}(x) = E(x) M^{-1}(x) \in G_+ \times D_-.$$

We have in addition the following properties

(i) $E^{-1} E_x = A$, *where* $A(x, \lambda) = a\lambda + u(x)$ *for some* $u \in \mathcal{S}(R, \mathcal{U}_a^\perp)$,

(ii) $M_{\pm\infty} \in H_-$, *where* $M_{\pm\infty}(\lambda) = \lim_{x \to \pm\infty} M(x, \lambda)$,

(iii) if λ *is not a pole of* f *then* $\tilde{M}_\pm(\cdot, \lambda) = M(\cdot, \lambda) - M_{\pm\infty}(\lambda)$ *is in the Schwartz class.*

To prove this theorem, we need the following Lemma:

7.17 Lemma. *Given $q \in \mathcal{S}(R)$ and $\beta < 0$ a constant, and set*

$$\Delta(x)(r) = \int_{-\infty}^{0} q(y+\beta x)e^{iry}\,dy,$$
$$\xi(x)(r) = \int_{0}^{\infty} q(y+\beta x)e^{iry}\,dy.$$

Then

(i) $\Delta \in \mathcal{S}(R^+, H_1)$,

(ii) *for any integer $m \geq 0$ there exists a constant c_m such that*

$$\left|\frac{\partial^m \xi}{\partial x^m}(x,r)\right| \leq c_m.$$

Proof. Write $\Delta(x,r) = \Delta(x)(r)$ and $\xi(x,r) = \xi(x)(r)$. Then

$$\left(\frac{\partial}{\partial x}\right)^m \Delta(x,r) = \beta^m \int_{-\infty}^{0} \frac{\partial^m q}{\partial y^m}(y+\beta x)e^{iry}\,dy.$$

Since $q \in \mathcal{S}(R^-)$,

$$\left|\left(\frac{\partial}{\partial y}\right)^m q(y)\right| \leq \frac{c_{m,s}}{(1+|y|)^s}.$$

But by the Plancherel Theorem

$$\begin{aligned}\left\|\left(\frac{\partial}{\partial x}\right)^m \Delta(x,\cdot)\right\|_1^2 &= \beta^{2m}\int_{-\infty}^{0}(1+y^2)\left(\frac{\partial^m q}{\partial y^m}(y+\beta x)\right)^2 dy\\ &\leq \beta^{2m}\int_{-\infty}^{0}\frac{(1+|y|^2)c_{m,s}^2}{(1+|y|+|\beta x|)^{2s}}dy\\ &\leq \frac{\beta^{2m}c_{m,s}^2}{(1+|\beta x|)^{2s-3}}\end{aligned}$$

.

An adjustment of the constants completes the proof of (i). A straight forward computation gives (ii). □

7.18 Proof of Theorem 7.16.

Take the variation with respect to $f \in D_-$ in the formulas

$$\begin{aligned} f^{-1}(\lambda)e^{a\lambda x} &= E(x,\lambda)M^{-1}(x,\lambda),\\ A(x,\lambda) &= E^{-1}(x,\lambda)E_x(x,\lambda).\end{aligned}$$

We get

$$-E^{-1}f^{-1}\delta f E = E^{-1}\delta E - M^{-1}\delta M,$$
$$\delta A = [A, (E^{-1}f^{-1}\delta f E)_-]_+.$$

For $f = I$, we have $A = a\lambda$. So $\delta A(x)$ is independent of λ and lies in $\mathcal{U}_a^\perp$ for all x. This implies that $E^{-1}E_x = a\lambda + u$ for some $u : R \to \mathcal{U}_a^\perp$. The fact that $u \in \mathcal{S}(R, \mathcal{U}_a^\perp)$ follows directly from (ii) and (iii). Write

$$u = M^{-1}M_x + M^{-1}[a\lambda, M].$$

Now $M(\cdot, \lambda) - M_\infty \in \mathcal{S}(R^+)$ and $[M_\infty, a] = 0$ imply that $u \,|\, R^+ \in \mathcal{S}(R^+)$. The corresponding argument gives $u \,|\, R^- \in \mathcal{S}(R^-)$.

We first prove the theorem for $f \in D_-^c$. Use Proposition 7.13 to write

$$f = p_d(I + p)v,$$

where p_d is a-diagonal and holomorphic in C_+, p is strictly upper a-triangular, and v is unitary. We will be looking at $x \to \infty$. Examine the formula for $\lambda \in C_+$:

$$(e^{-a\lambda x}p(\lambda)e^{a\lambda x})_{jk} = \begin{cases} 0, & \text{if } a_j \geq a_k, \\ p_{jk}(\lambda)e^{-i(a_j - a_k)\lambda x}, & \text{if } a_j < a_k. \end{cases} \tag{7.2}$$

Here $p_{jk} \,|\, R$ lies in the Schwartz space if $a_j < a_k$.

Use inverse Fourier transform to write

$$p_{jk}(r) = \int_{-\infty}^{\infty} \hat{p}_{jk}(y)e^{iry}\,dy.$$

So

$$p_{jk}(r)e^{-i(a_j - a_k)rx} = \int_{-\infty}^{\infty} \hat{p}_{jk}(y + (a_j - a_k)x)e^{iry}\,dy.$$

The piece $\int_0^\infty \hat{p}_{jk}(y + (a_j - a_k)x)e^{iry}\,dy$ is the boundary value of a holomorphic map in C_+, which can be written

$$\xi_{jk}(x, \lambda) = \int_0^\infty \hat{p}_{jk}(y + (a_j - a_k)x)e^{i\lambda y}\,dy.$$

So

$$p_{jk}(r)e^{-i(a_j - a_k)rx} = \xi_{jk}(r, x) + \triangle_{jk}(x, r), \qquad \text{where}$$
$$\triangle_{jk}(x, r) = \int_{-\infty}^{0} \hat{p}_{jk}(y + (a_j - a_k)x)e^{iry}\,dy.$$

It follows from Lemma 7.17 that $\triangle \in \mathcal{S}(R^+, H_1)$ and $\left\| \frac{\partial^m \xi}{\partial x^m(x,r)} \right\| \leq c_m$.

Now write

$$\begin{aligned} e_a^{-1} p_d (I+p) e_a &= p_d e_a^{-1}(I+p) e_a = p_d(I + \xi + \triangle) \\ &= p_d((I+\xi)(I+\triangle) - \xi\triangle) \\ &= p_d(I+\xi)(I + \triangle - (I+\xi)^{-1}\xi\triangle). \end{aligned}$$

We claim that $D = \triangle - (I+\xi)^{-1}\xi\triangle \in \mathcal{S}(R^+, H_1)$. Note that

$$(I+\xi)^{-1} = I - \xi + \xi^2 - \xi^3 + \ldots + \xi^n$$

is a finite series since ξ is strictly upper a-triangular. The rules of multiplication of $\mathcal{S}(R^+, H_1)$ by a smooth bounded function give the result that $D \in \mathcal{S}(R^+, H_1)$.

Let $(I + D) = (I + h)V$ be the Birkhoff decomposition. By Theorem 7.14, h and $V - I$ are in $\mathcal{S}(R^+, H_1)$. So

$$\begin{aligned} e_a^{-1} f = e_a^{-1} p_d (I+p) v &= e_a^{-1} p_d (I+p) e_a e_a^{-1} v \\ &= p_d(I+\xi)(I+D) e_a^{-1} v \\ &= p_d(I+\xi)(I+h) V e_a^{-1} v. \end{aligned}$$

By definition $M = p_d(I+\xi)(I+h)$, and

$$M - p_d(I+\xi) = p_d(I+\xi)h.$$

Since $h \in \mathcal{S}(R^+, H_1)$, and we have uniform estimates on all derivatives of $p_d(I + p)$, $M - p_d(I+\xi) \in \mathcal{S}(R^+, H_1)$. The same argument, in which a factorization $f = q\tilde{v}$ for q lower a-triangular and $\tilde{v}$ unitary, proves Schwartz space decay as $x \to -\infty$.

To complete the proof, given $f \in D_-$, write $f = hg \in D_-^c \times G_-^m$. Write

$$h^{-1} e_{a,1}(x) = E_0(x) M_0^{-1}(x) \in G_+ \times D_-.$$

By Theorem 6.5, we factor $g^{-1} E_0(x) = \tilde{E(x)} g(x) \in G_+ \times G_-^m$. Then

$$\begin{aligned} f^{-1} e_{a,1}(x) = g^{-1} h^{-1} e_{a,1}(x) &= g^{-1} E_0(x) M_0^{-1}(x) = E(x) \tilde{g}(x) M_0^{-1}(x) \\ &= E(x) M(x). \end{aligned}$$

By Theorem 6.6 $\tilde{g}$ satisfies condition (ii) and (iii). But we just proved that M_0 satisfies (ii) and (iii), so is $M = \tilde{g} M_0^{-1}$. □

Note the convergence is actually uniform in the argument in Theorem 7.16. So we have

7.19 Theorem. *As in Theorem 7.16 let $f \in D_-$, $f^{-1}e_{a,1}(x) = E(x)M(x)^{-1} \in G_+ \times D_-$, and $f_+ = \lim_{s\searrow 0} f(r+is)$. Factor $f_+ = Pv = Q\tilde{v}$, where $v, \tilde{v}$ are unitary, P is upper a-triangular, Q is lower a-triangular, and P_d, Q_d is holomorphic in C_+. Then*

$$\lim_{x\to\infty} e^{arx}M_+(x,r)e^{-arx} = P(r),$$
$$\lim_{x\to-\infty} e^{arx}M_+(x,r)e^{-arx} = Q(r).$$

7.20 Theorem. *Let $\Psi : D_- \to \mathcal{S}_{1,a}$ be the map defined by $\Psi(f) = E^{-1}E_x$, where E is obtained from f as in Theorem 7.16. Let H_- denote the subgroup of $f \in D_-$ such that $fa = af$. Then*

(i) $\mathcal{S}^0_{1,a} = \Psi(D_-)$ *is an open and dense subset of* $\mathcal{S}_{1,a}$,

(ii) $\Psi(f) = \Psi(g)$ *if and only if there exist $h \in H_-$ such that $g = hf$,*

(iii) $\mathcal{S}^0_{1,a}$ *is isomorphic to the homogeneous space D_-/H_- of left cosets of H_- in G_-,*

(iv) if $A = \Psi(f)$ and M is as in Theorem 7.16, then the normalized eigenfunction m in Theorem 7.1 of A is $M_{-\infty}^{-1}M$.

Proof. The first part (i) is a consequence of Theorem 7.1. Both (iii) and (iv) follow from (ii). To prove (ii), recall if

$$f^{-1}e_{a,1}(x) = E(x)M^{-1}(x) \in G_+ \times D_-,$$
$$g^{-1}e_{a,1}(x) = E(x)N^{-1}(x) \in G_+ \times D_-.$$

Then

$$M(x)N^{-1}(x) = e_{a,1}(x)^{-1}fg^{-1}e_{a,1}(x).$$

Suppose $\operatorname{Im}\lambda > 0$. Then the limit of the right hand side is upper a-triangular when $x \to \infty$, and the limit is lower a-triangular when $x \to -\infty$. So MN^{-1} is both upper and lower a-triangular. Hence it is a-diagonal, i.e., $MN^{-1} \in H_-$. So $fg^{-1} \in H_-$.

Conversely, if $g = hf$ for some $h \in H_-$ and $f^{-1}e_{a,1}(x) = E(x)M(x)^{-1} \in G_+ \times D_-$, then

$$\begin{aligned} g^{-1}e_{a,1}(x) &= f^{-1}h^{-1}e_{a,1}(x) = f^{-1}e_{a,1}(x)h^{-1} \\ &= E(x)(M(x)^{-1}h^{-1}) \in G_+ \times D_-. \end{aligned}$$

So $\Psi(f) = \Psi(g)$. □

In summary, we have shown that given $f \in D_-$, we can construct an $A \in \mathcal{S}_{1,a}$ such that $A = \Psi(f)$ by using various Birkhoff decomposition theorems repeatedly.

7.21 Theorem. *The natural right action of D_- on the space D_-/H_- of left cosets induces a natural action $*$ of D_- on $\mathcal{S}^0_{1,a}$ via the isomorphism $\bar{\Psi}$ from D_-/H_- to $\mathcal{S}^0_{1,a}$. Equivalently, if $A = \Psi(f)$ and $g \in D_-$ then $g * A = \Psi(fg^{-1})$. Moreover:*

(i) Let $g \in D_-$, $A \in \mathcal{S}^0_{1,a}$, and E the trivialization of A normalized at $x = 0$, then we can factor

$$gE(x) = \tilde{E}(x)\tilde{g}(x) \in G_+ \times D_-,$$

and $g * A = \tilde{E}^{-1}\tilde{E}_x$.

(ii) If $g \in G^m_-$, then $g * A = g \sharp A$, where $\sharp$ is the action of G^m_- on $\mathcal{S}_{1,a}$ defined in Theorem 6.2. In other words, if $A = \Psi(f)$ and $g \in G^m_-$, then $g \sharp A = \Psi(fg)$. Or equivalently, if H_-f is the scattering coset of A then H_-fg is the scattering coset of $g \sharp f$.

Proof. Given $f, g \in D_-$, we factor

$$f^{-1}e_{a,1}(x) = E(x)M^{-1}(x), \qquad (fg)^{-1}e_{a,1}(x) = \tilde{E}(x)\tilde{M}^{-1}(x).$$

Then $g^{-1}E(x) = \tilde{E}(x)(\tilde{M}^{-1}(x)M(x)) \in G_+ \times D_-$. This defines the action of D_- on $\mathcal{S}^0_{1,a}$, and it extends the action of G^m_- on $\mathcal{S}_{1,a}$ defined in Theorem 6.2. □

7.22 Remark. If the scattering data of A has k poles counted with multiplicity, then $g_{z,\pi} \sharp A$ typically has $k+1$ poles, but it may have k or $k-1$ poles for special choices of z and π. To see this, let $z \in C \setminus R$, and π a projection such that $\pi a \neq a\pi$. If $\bar{z}$ is not a pole of the scattering data of A then $g_{z,\pi} \sharp A$ add one pole $\bar{z}$ to the scattering data. Let $A = g_{\bar{z},\pi} \sharp A_0$, where A_0 is the vacuum solution. Then the scattering data of $g_{z,\pi_1} \sharp A$

(i) has no poles if $\pi_1 = \pi$,

(ii) has only one pole z if π_1 and π commute and $\pi + \pi_1 \neq I$.

8 Poisson structure for the positive flows

Let H_+ denote the subgroup of G_+ generated by

$$\{e^{p(\lambda)} \mid p(\lambda) \text{ is a polynomial } p(\lambda)a = ap(\lambda)\}.$$

In this section, we prove that the right dressing action of H_+ on D_- induces a Poisson group action of H_+ on $\mathcal{S}^0_{1,a}$ and show that it generates the positive flows defined in section 5. We also study the induced symplectic structure on the space of discrete scattering data $G^m_-/(G^m_- \cap H_-)$, and the space of the continuous scattering data $D^c_-/(D^c_- \cap H_-)$.

Set $e_{a,j,b}(x,t)(\lambda) = e^{a\lambda x + b\lambda^j t}$ and recall $e_{b,j}(x)(\lambda) = e^{b\lambda^j x}$.

8.1 Theorem. *Let $a, b \in u(n)$ such that $[a, b] = 0$. Then we can factor*

$$f^{-1}e_{a,j,b}(x,t) = E(x,t)M^{-1}(x,t) \in G_+ \times D_-.$$

Moreover, E and M satisfy the following conditions:

(i) $E^{-1}E_x = A$ is a solution of the j-th flow defined by b, where $A(x,\lambda) = a\lambda + u(x)$.

(ii) $E^{-1}E_t = B$, where

$$B(\cdot,\lambda) = b\lambda^j + Q_{b,1}(u)\lambda^{j-1} + \ldots + Q_{b,j}(u) = (M^{-1}b\lambda^j M)_+.$$

Proof. Since $[a, b] = 0$, $\exp(a\lambda x + b\lambda^j t) = e^{a\lambda x}e^{b\lambda^j t}$. Use Theorem 7.11 to factor

$$f^{-1}e_{a,j,b}(x,t) = f^{-1}e_{a,1}(x)e_{b,j}(t) = E_0(x)M_0^{-1}(x)e_{b,j}(t).$$

Use Theorem 7.11 again to factor

$$M_0^{-1}(x)e_{b,j}(t) = E_1(x,t)M(x,t) \in G_+ \times D_-.$$

The variational form of $f^{-1}e_{a,j,b} = EM^{-1}$ implies

$$\begin{cases} E^{-1}E_x = a\lambda + u, \\ E^{-1}E_t = b\lambda^j + q_1\lambda^{j-1} + \ldots + q_j. \end{cases}$$

So

$$\left[\frac{\partial}{\partial x} + a\lambda + u,\ \frac{\partial}{\partial t} + b\lambda^j + q_1\lambda^{j-1} + \ldots + q_j\right] = 0. \tag{8.1}$$

Compare coefficient of λ^i in equation (8.1) to get

$$\begin{cases} (q_i)_x + [u, q_i] = [q_{i+1}, a], & \text{if } 0 \le i < j, \\ u_t = (q_j)_x + [u, q_{j+1}]. \end{cases}$$

This is the same system as (5.2) defining the $Q'_{b,i}s$. Hence $q_i = Q_{b,i}$. □

8.2 Corollary. *The dressing action $\natural$ of H_+ on D_- on the right is well-defined and H_- is fixed under this action. Hence an action $\natural$ of H_+ on D_-/H_- is defined, which leads to an action on $\mathcal{S}^0_{1,a}$. In fact, this action is defined as follows: Write $A = \Psi(f)$, $f^{-1}e_{a,1}(x) = E(x)M(x)^{-1}$. For $h \in H_+$, we factor*

$$hM(x) = \tilde{M}(x)\tilde{h}(x) \in D_- \times G_+$$

to get

$$h \natural A = (e^{a\lambda x}\tilde{M})^{-1}(e^{a\lambda x}\tilde{M})_x.$$

8.3 Corollary. *If* $A_0 = \Psi(f_0)$*, then* $A(t) = \Psi(e_{b,j}(t) \natural f_0)$ *is the solution of the* j*-th flow on* $\mathcal{S}_{1,a}$ *defined by* b *with* $A(0) = A_0$*.*

8.4 Corollary. *Let* $a_1, \ldots, a_n$ *be a basis of the space* $\mathcal{T}$ *of diagonal matrices in* $u(n)$*,* $f \in D_-$*, and* $e_{a_1,\ldots,a_n}(x_1, \ldots, x_n)(\lambda) = \exp(\sum_{j=1}^n a_j x_j \lambda)$*. Factor*

$$f^{-1} e_{a_1,\ldots,a_n}(x) = E(x)M(x) \in G_+ \times D_-.$$

Then there exists $v : R^n \to \mathcal{T}^\perp$ *such that*

(i) $E^{-1}E_{x_j} = a_j\lambda + [a_j, v]$ *for all* $1 \le j \le n$*,*

(ii) v *is a solution of equation*

$$\left[a_i, \frac{\partial v}{\partial x_i}\right] - \left[a_j, \frac{\partial v}{\partial x_j}\right] = [[a_i, v], [a_j, v]]. \tag{8.2}$$

8.5 Remark. Equation (8.2) is the n-dimensional system associated to $U(n)$ constructed in the paper of the first author [Te2].

8.6 Theorem. *The action* $\natural$ *of* H_+ *on* $\mathcal{S}^0_{1,a}$ *is Poisson. Moreover, the map* $\mu : \mathcal{S}^0_{1,a} \to H_- = H_+^*$ *defined by* $\mu(A)(\lambda) = M_{-\infty}^{-1} M_\infty$ *is a moment map, where* $A = \Psi(f)$*,* $f^{-1}e_{a,1}(x) = E(x)M(x)^{-1} \in G_+ \times D_-$ *and* $M_{\pm\infty}(\lambda) = \lim_{x\to\pm\infty} M(x, \lambda)$*.*

Proof. Suppose $A = \Psi(f)$, i.e.,

$$\begin{cases} f^{-1}e_{a,1}(x) = E(x)M^{-1}(x) \in G_+ \times D_-, \\ A = E^{-1}E_x = (e^{a\lambda x}M)^{-1}(e^{a\lambda x}M)_x. \end{cases}$$

The second equation implies

$$M^{-1}M_x + \lambda M^{-1}aM = A. \tag{8.3}$$

Set $\eta = M^{-1}\delta M$, $B = \delta A$ and $\psi = e^{a\lambda x}M$. Compute the variation directly from equation (8.3) to derive

$$\eta_x + [A, \eta] = B, \qquad \lim_{x\to-\infty} \eta = 0,$$
$$\eta(x) = \psi(x)^{-1} \int_{-\infty}^x (\psi B \psi^{-1}) dy\, \psi(x).$$

For $\xi_+ \in \mathcal{H}_+$, since $[\xi_+, a] = 0$, we have $M^{-1}\xi_+ M = \psi^{-1}\xi_+\psi$.

$$
\begin{aligned}
&\langle d\mu_A(B)\mu(A)^{-1}\,,\,\xi_+\rangle \\
&\quad = \lim_{x\to\infty}\left\langle M(x)\psi^{-1}(x)\int_{-\infty}^{x}(\psi B\psi^{-1})dy\psi(x)M^{-1}(x)\,,\,\xi_+\right\rangle, \\
&\quad = \lim_{x\to\infty}\left\langle e^{-a\lambda x}\int_{-\infty}^{x}(\psi B\psi^{-1})dye^{a\lambda x}\,,\,\xi_+\right\rangle \\
&\quad = \lim_{x\to\infty}\left\langle \int_{-\infty}^{x}(\psi B\psi^{-1})dy\,,\,e^{a\lambda x}\xi_+e^{-a\lambda x}\right\rangle \\
&\quad = \lim_{x\to\infty}\left\langle \int_{-\infty}^{x}(\psi B\psi^{-1})dy\,,\,\xi_+\right\rangle = \lim_{x\to\infty}\int_{-\infty}^{x}\langle\psi B\psi^{-1}\,,\,\xi_+\rangle dy \\
&\quad = \lim_{x\to\infty}\int_{-\infty}^{x}\langle B\,,\,\psi^{-1}e^{-a\lambda y}\xi_+e^{a\lambda y}\psi\rangle dy \\
&\quad = \langle\langle B\,,\,(M^{-1}\xi_+M)_-\rangle\rangle.
\end{aligned}
$$

The rest of the proof goes exactly the same as for Theorem 4.3. □

8.7 Remark. Let $a = \mathrm{diag}(ia_1, \dots, ia_n)$, and $a_1 < \dots < a_n$. Then $\mathcal{U}_a$ is the set of all diagonal matrices in $u(n)$, $\mathcal{U}_a^\perp$ is the set of all matrices $u \in u(n)$ such that $u_{ii} = 0$ for all $1 \le i \le n$. So H_+ is abelian and the action of H_+ on $\mathcal{S}_{1,a}$ is in fact symplectic.

The following theorem was proved by Flaschke, Newell and Ratiu [FNR1, 2] for $n = 2$ and by one of us [Te2] for general n:

8.8 Theorem ([Te2]). *The Hamiltonian function on $\mathcal{S}_{1,a}$ corresponding to the j-th flow defined by a is:*

$$F_{a,j}(u) = -\frac{1}{j+1}\int_{-\infty}^{\infty}(Q_{a,j+2}, a)\,dx, \tag{8.4}$$

i.e., $\nabla F_{b,j}(u) = Q^\perp_{b,j+1}(u)$.

8.9 Remark. Let $b, c \in \mathcal{U}_a$, and $\xi_{b,j}$ and $\xi_{c,k}$ denote infinitesimal vector field for the H_+-action on $\mathcal{S}^0_{1,a}$ corresponding to $b\lambda^j$ and $c\lambda^k$ respectively. Then the bracket $[\xi_{b,j}, \xi_{c,k}]$ is equal to the infinitesimal vector field corresponding to $[b,c]\lambda^{k+j}$. Unless $[b,c] = 0$, these two flows do not commute.

8.10 Remark. If we replace the group $SU(n)$ by a simple compact Lie group, then what we have discussed still holds if appropriate algebraic conditions are prescribed.

In the end of this section, we will study the pull back of the symplectic structure w on $\mathcal{S}_{1,a}$ to D_-/H_- via the isomorphism Ψ. Note that $\Psi(G^m_-)$ ($\Psi(D^c_-)$ resp.) is the space of A's with only discrete (continuous resp.) scattering data.

We have been using the base point $x = 0$, i.e., $f(\lambda) = M(0, \lambda)$. But there is nothing special about $x = 0$. In the following, we choose a base point y and let $y \to -\infty$. The expression with base point 0, and with y, differ by a term which cancels out when we evaluate integrals at the end points.

We only deal with the symplectic structure on $\Psi(D_-^c)$. However, this set is pretty large. For example, Beals and Coifman and later Zhou show the following:

8.11 Theorem ([BDZ]). *Let B_1 denote the unit ball in $\mathcal{S}_{1,a}$ with respect to the L^1-norm, i.e., B_1 is set of all $A = a\lambda + u \in \mathcal{S}_{1,a}$ such that $\int_{-\infty}^{\infty} \|u\| dr < 1$. Then $B_1 \subset \Psi(D_-^c)$.*

Let S denote the scattering transform that maps $A \in \mathcal{S}'_{1,a}$ to its scattering data S (defined in section 7). The restriction of the symplectic form w on $\mathcal{S}_{1,a}$ to $S(B_1)$ was computed by Beals and Sattinger [BS]. We will compute the restriction of w to $\Psi(D_-^c)$ in terms of variations in D_-^c below. Let

$$\langle f(r), g(r) \rangle = \int_{-\infty}^{\infty} \mathrm{Im}(\mathrm{tr}(f(r)g(r))dr.$$

By the same computation as in Theorem 4.3, the Poisson bracket on $B_1 \subset \mathcal{S}_{1,a}$ is

$$\begin{aligned} \{\delta_1 A, \delta_2 A\} &= \lim_{x\to\infty} \langle (E^{-1}(x) f^{-1} \delta_1 f E(x))_-, E(x)^{-1} f^{-1} \delta_2 f E(x) \rangle \\ &\quad - \lim_{y\to-\infty} \langle (E^{-1}(y) f^{-1} \delta_1 f E(y))_-, E(y)^{-1} f^{-1} \delta_2 f E(y) \rangle \\ &= \lim_{x\to\infty} \langle E^{-1}(x) \delta_1 E(x), M^{-1} \delta_2 M(x) \rangle \\ &\quad - \lim_{y\to-\infty} \langle E^{-1}(y) \delta_1 E(y), M^{-1} \delta_2 M(y) \rangle. \end{aligned}$$

Now, let the vacuum be based at y, i.e., factor

$$f^{-1}(\lambda) e^{a\lambda(x-y)} = E(x, y, \lambda) M^{-1}(x, y, \lambda).$$

Hence $f(\lambda) = M(y, y, \lambda)$ and $\delta E(y, y, \lambda) = 0$. The y-term in the above description is now zero, and we have

$$\begin{aligned} &= \lim_{x\to\infty, y\to-\infty} -\langle M(x)^{-1} e^{-a\lambda(x-y)} \delta_1 f f^{-1} e^{a\lambda(x-y)} M(x),\ M^{-1}(x) \delta M(x) \rangle \\ &= \lim_{x\to\infty, y\to-\infty} -\langle e^{-a\lambda(x-y)} \delta_1 f f^{-1} e^{a\lambda(x-y)},\ \delta_2 M(x) M^{-1}(x) \rangle \\ &= \lim_{x\to\infty, y\to-\infty} -\langle e^{a\lambda y} \delta_1 M(y) M^{-1}(y) e^{-a\lambda y},\ e^{a\lambda x} \delta_2 M(x) M^{-1}(x) e^{-a\lambda x} \rangle. \end{aligned}$$

Now by Theorem 7.19, we get

8.12 Theorem. *The Poisson structure on the unit ball B_1 in $\mathcal{S}_{1,a}$ with respect to the L^1-norm, written in terms of variations in D_-^c, is*

$$\{\delta_1 A, \delta_2 A\} = \int_{-\infty}^{\infty} \langle \delta_1 P P^{-1}, \delta_2 Q Q^{-1} \rangle,$$

where $A = \Psi(f)$, $f_+ = Pv = Q\tilde{v}$ is the factorization of f_+ into upper a-triangular and lower a-triangular times unitary and P_d, Q_d are holomorphic as given in Proposition 7.13.

Next we study the pull back the symplectic form

$$w(q_1, q_2) = \int_{-\infty}^{\infty} \operatorname{tr}(\operatorname{ad}(a)^{-1}(q_1)(q_2))\, dx$$

on $\mathcal{S}_{1,a}$ to the space $\Psi(G_-^m)$. This space has many complicated algebraic components. For example the space of all $A \in \mathcal{S}_{1,a}$ whose scattering data have only one pole (or equivalently, $A = \Psi(g)$, where g is a simple element) can be parametrized by

$$\bigcup_{k=1}^{n} C_+ \times \{V \in Gr(k, C) \mid a(V) \not\subset V\}.$$

However, the space of A whose scattering data has only two poles immediately becomes complicated as the factorization of $g \in G_-^m$ as product of simple elements is not unique. The following Proposition gives the restriction of w to the simplest component of $\Psi(G_-^m)$. We believe that the restriction of w to each algebraic component should be symplectic, but we have not yet found an efficient way to compute the general case.

8.13 Proposition. *Let $a = \operatorname{diag}(-i, i, \dots, i)$. Then:*

(i) *The space of all $A = \Psi(g_{z,\pi})$, where π is the projection onto a one dimensional subspace Cv, is isomorphic to*

$$N = C_+ \times (C^{n-1} \setminus 0) = \{(z, v) \mid z \in C \setminus R, v = (v_2, \dots, v_n) \neq 0\}.$$

(ii) *The pull back of the symplectic form w to N is*

$$2\operatorname{Re}\left(d\bar{z} \wedge \partial \log(|v|^2) + (z - \bar{z})\partial\bar{\partial}\log(|v|^2)\right),$$

where $|v| = \sum_{j=2}^{n} |v_j|^2$.

Proof. Let π denote the projection of C^n onto the one dimensional subspace spanned by $(1, v)$, where $v = (v_2, \dots, v_n) \in C^{n-1}$. By Theorem 6.5 (vi) and formula (6.7), $g_{z,\pi} \sharp 0 = (u_{ij})$, where $(u_{ij}) \in u(n)$, $u_{ij} = 0$ if $2 \leq i, j \leq n$ and

$$u_{1j}(x) = \frac{2i(z - \bar{z})\bar{v}_j e^{i(z+\bar{z})x}}{e^{-i(z-\bar{z})x} + e^{i(z-\bar{z})}(\mid v_2 \mid^2 + \ldots + \mid v_n \mid^2)}.$$

Then the proposition follows from at least two separate computations, neither of which is very illuminating. We hope to provide the more general results in a future paper. □

8.14 Remark. Fix $z \in C_+$. Then the restriction of the symplectic form to the subset $\{(z,v) \mid v \in C^{n-1} \setminus \{0\}, \|v\| = 1\}$ of M in the theorem above gives the standard symplectic structure of CP^{n-2}.

9 Symplectic structures for the restricted case

Most of the interesting applications in geometry come from restrictions of the full flow equation to a smaller phase space satisfying additional algebraic conditions. This leads to a serious problem, not with the flows and the scattering cosets, but with the symplectic structure. Generally the original symplectic structure we have used to this point vanishes on the restricted submanifolds. In this section, we describe a typical restrictions and the construction of the hierarchy of symplectic structures. We give an outline of the theory, and explain how it can be applied. The details of this construction for involutions appear in [Te2].

Let U be a simple Lie group, $\langle\ ,\rangle$ a non-degenerate, ad-invariant bilinear form on the Lie algebra $\mathcal{U}$, and σ an order m automorphism of U. For simplicity, we denote the Lie algebra automorphism $d\sigma_e$ on $\mathcal{U}$ again by σ. Fix a primitive m-th root of unity α. Suppose σ has an eigendecomposition on $\mathcal{U}$:

$$\mathcal{U} = \sum_{j=0}^{m} \mathcal{U}_j,$$

where $\mathcal{U}_j$ is the eigenspace of σ on $\mathcal{U}$ with eigenvalue α^j. Then

$$\begin{cases} [\mathcal{U}_j, \mathcal{U}_k] \subset \mathcal{U}_{j+k}, & \text{for all } j,k, \\ \langle \mathcal{U}_j, \mathcal{U}_k \rangle = 0, & \text{if } j \neq k. \end{cases}$$

Here we use the convention $\mathcal{U}_j = \mathcal{U}_k$ if $j \equiv k \pmod m$.

Fix an element $a \in \mathcal{U}_1$. Let $\mathcal{U}_a$ denote the centralizer of a, and $\mathcal{U}_a^{\perp}$ the orthogonal complement of $\mathcal{U}_a$ in $\mathcal{U}$. Consider a subspace of $\mathcal{S}_{1,a}$:

$$\mathcal{S}_{1,a}^{\sigma} = \{A = a\lambda + u \in \mathcal{S}_{1,a} \mid u \in \mathcal{U}_0 \cap \mathcal{U}_a^{\perp}\} \quad \subset \mathcal{S}_{1,a}.$$

Then $A \in \mathcal{S}_{1,a}^{\sigma}$ satisfies the reality condition

$$\sigma(A(\alpha^{-1}\lambda)) = A(\lambda). \tag{9.1}$$

Hence the trivialization E of $A \in \mathcal{S}_{1,a}$ normalized at the origin satisfies the condition $\sigma(E(\alpha^{-1}\lambda)) = E(\lambda)$. Then

9.1 Proposition. *$\mathcal{S}_{1,a}^{\sigma}$ is invariant under the action $\sharp$ of $G_-^{m,\sigma}$ on $\mathcal{S}_{1,a}$ defined in section 6, where $G_-^{m,\sigma}$ is the subgroup of $g \in G_-^m$ such that $\sigma(g(\alpha^{-1}\lambda)) = g(\lambda)$.*

The trivialization M of $A \in \mathcal{S}^{\sigma}_{1,a}$ normalized at infinity also satisfies the same reality condition as E. So for $b \in \mathcal{U}_1 \cap \mathcal{U}_a$, we have

$$\sigma(M^{-1}(\alpha^{-1}\lambda)bM(\alpha^{-1}\lambda)) = \alpha M^{-1}(\lambda)bM(\lambda).$$

Since

$$M^{-1}bM \sim \sum_{j=0}^{\infty} Q_{b,j}\lambda^{-j},$$

we get

$$Q_{b,j} \in \mathcal{U}_{-j+1}, \qquad \text{for all } j \geq 0.$$

In particular, $[Q_{j+1}, a] \in \mathcal{U}_{-j+1}$. So $[Q_{b,j+1}(u), a]$ is normal to $\mathcal{S}^{\sigma}_{1,a}$ if $j \not\equiv 1$ (mod m), and is tangent to $\mathcal{S}^{\sigma}_{1,a}$ if $j \equiv 1$ (mod m). And we have

9.2 Proposition. *The j-th flow preserves $\mathcal{S}^{\sigma}_{1,a}$ for all j and any order m automorphism σ. If $j \not\equiv 1$ (mod m), then the flows are identically constant.*

However, the symplectic form

$$w(\delta_1 u, \delta_2 u) = \int_{-\infty}^{\infty} \operatorname{tr}(-\operatorname{ad}(a)^{-1}(\delta_1 u), \delta_2 u)\, dx$$

vanishes on $\mathcal{S}^{\sigma}_{1,a}$.

The sequence of symplectic structures constructed by Terng can be described using a sequence of coadjoint orbits, which arise using a shift in the bi-linear form $\langle\ ,\ \rangle$ on the loop algebra $\mathcal{G}$.

For $k \leq -1$, let M_k denote the coadjoint $C(R, G_-)$-orbit at $\left(\frac{d}{dx} + a\lambda\right)\nu_k^{-1}$ in $C(R, \mathcal{G}_+)$, where $\nu_k(\lambda) = \lambda^{k+1}$. Set

$$\mathcal{S}_{1,a,k} = (M_k \nu_k) \cap \mathcal{S}_{1,a}.$$

Then δu lies in the tangent space of $\mathcal{S}_{1,a,k}$ at $\frac{d}{dx} + a\lambda + u$ if and only if

$$\delta u(x) = \left(\left[\xi_-(x), \frac{d}{dx} + A(x)\right]\nu_k^{-1}\right)_+ \nu_k, \tag{9.2}$$

where formally $\xi_-(x) \in \mathcal{G}_-$. Here $(\cdot)_+$ is the projection into $\mathcal{G}_+$, and the construction is entirely algebraic. For $A = a\lambda + u$, write

$$\xi_-(\delta u) = \xi_{-1}(\delta u)\lambda^{-1} + \xi_{-2}(\delta u)\lambda^{-2} + \dots .$$

Then equation (9.2) gives

$$[\xi_{-1}, a] = \delta u,$$
$$[\xi_j, a] = \left[\frac{d}{dx} + u, \xi_{j+1}\right], \qquad k \leq j \leq -1.$$

This gives a recipe to compute $\xi_-(\delta u)$ explicitly via a mixed integro-differential operation:

$$\begin{aligned}\xi_{-1}^{\perp}(\delta u) &= J_a^{-1}(\delta u),\\ \xi_j^{\perp}(\delta u) &= (J_a^{-1}P_u)^{-j-1}J_a^{-1}(\delta u),\end{aligned}$$

where

$$\begin{aligned}J_a(v) &= [v,a],\\ P_u(v) &= v_x + [u,v]^{\perp} - [u,\eta_u(v)],\\ \eta_u(v) &= \int_{-\infty}^{x}[u(y),v(y)]^d dy.\end{aligned}$$

Here $v^{\perp}$ and v^d denote the projection onto a-off diagonal ($\mathcal{U}_a^{\perp}$) and a-diagonal ($\mathcal{U}_a$) respectively. Set

$$J_k = J_a(J_a^{-1}P_u)^{k+1}.$$

The natural shifted symplectic structure is given by

$$\begin{aligned}w_k(\delta_1 u,\delta_2 u) &= \int_{-\infty}^{\infty}\left\langle \frac{d}{dx}+A, \nu_k^{-1}[\xi_-(\delta_1 u),\xi_-(\delta_2 u)]\right\rangle dx\\ &= \int_{-\infty}^{\infty}\operatorname{tr}\left(\left(\frac{d}{dx}+a\lambda+u\right)([\xi_-(\delta_1 u),\xi_-(\delta_2 u)])\right)_k dx,\\ &= \int_{-\infty}^{\infty}\operatorname{tr}\left((\delta_1 u)\,\xi_k^{\perp}\,(\delta_2 u)\right)\,dx,\\ &= \int_{-\infty}^{\infty}\operatorname{tr}\left((\delta_1 u)\,J_k^{-1}\,(\delta_2 u)\right)\,dx,\end{aligned}$$

where $(\cdot)_k$ denote the coefficient of λ^k in $(\cdot)$. In particular, the first two in the series are:

$$\begin{aligned}w_{-1}(\delta_1 u,\delta_2 u) &= w(\delta_1 u,\delta_2 u) = \int_{-\infty}^{\infty}\operatorname{tr}((-\operatorname{ad}(a)^{-1}(\delta_1 u))\delta_2 u)\,dx,\\ w_{-2}(\delta_1 u,\delta_2 u) &= \int_{-\infty}^{\infty}\operatorname{tr}((\delta_1 u)(J_{-2})_u^{-1}(\delta_2 u))\,dx\\ &= \int_{-\infty}^{\infty}\operatorname{tr}((\delta_1 u)J_a^{-1}P_uJ_a^{-1}(\delta_2 u))\,dx.\end{aligned}$$

The natural coadjoint orbits require the relevant terms of ξ_- to lie in the Schwartz class. So the tangent space of the smaller submanifold $\mathcal{S}_{1,a,k}$ =

$(M_k\nu_k) \cap \mathcal{S}_{1,a}$ at u is

$$\{\delta u \mid \xi_{-j}(\delta u)(\infty) = 0, 1 \le j \le -k\}.$$

Hence $\mathcal{S}_{1,a,k}$ is a finite codimension submanifold of $\mathcal{S}_{1,a}$ and the formulas we write down for w_k are skew symmetric.

For $k \ge 0$, let M_k denote the coadjoint $C(R, G_+)$-orbit at $\left(\frac{d}{dx} + a\lambda\right)\nu_k^{-1}$ in $C(R, \mathcal{G}_-)$, and $\mathcal{S}_{1,a,k} = (M_k\nu_k) \cap \mathcal{S}_{1,a}$, where $\nu_k(\lambda) = \lambda^{k+1}$. Then δu lies in the tangent space of $\mathcal{S}_{1,a,k}$ at $\frac{d}{dx} + A$ if and only if

$$\delta u(x) = \left(\left[\xi_+(x), \frac{d}{dx} + A(x)\right]\nu_k^{-1}\right)_- \nu_k, \tag{9.3}$$

where formally $\xi_+(x) \in \mathcal{G}_+$. Here $()_-$ is the projection into $\mathcal{G}_-$. For $A = a\lambda + u$, write

$$\xi_+(\delta u) = \xi_0(\delta u) + \xi_1(\delta u)\lambda + \dots.$$

Then equation (9.3) gives

$$\begin{aligned} \left[\frac{d}{dx} + u, \xi_0\right] &= \delta u, \\ \left[\frac{d}{dx} + u, \xi_j\right] &= [\xi_{j-1}, a], \qquad 1 \le j \le k. \end{aligned}$$

Hence

$$\xi_j^\perp(\delta u) = (P_u^{-1}J_a)^j P_u^{-1}(\delta u) = J_j^{-1}(\delta u).$$

The natural shifted symplectic structure is given by

$$w_k(\delta_1 u, \delta_2 u) = \int_{-\infty}^{\infty} \operatorname{tr}((\delta_1 u) J_k^{-1}(\delta_2 u))\, dx.$$

In particular,

$$w_0(\delta_1 u, \delta_2 u) = \int_{-\infty}^{\infty} \operatorname{tr}((\delta_1 u) P_u^{-1}(\delta_2 u))\, dx.$$

If $a \in \mathcal{U}_1$, then $J_a = -\operatorname{ad}(a)$ maps $\mathcal{U}_j$ to $\mathcal{U}_{j+1}$. This implies that J_k maps $\mathcal{U}_j$ to $\mathcal{U}_{j-k}$. Thus we obtain:

9.3 Proposition ([Te2]). *w_k is a symplectic structure on $\mathcal{S}_{1,a,k}$. Moreover,*

(i) $w_k = 0$ on $\mathcal{S}_{1,a,k} \cap \mathcal{S}_{1,a}^\sigma$ if $k \not\equiv 0 \pmod m$,

(ii) w_k is non-degenerate on $\mathcal{S}_{1,a,k} \cap \mathcal{S}_{1,a}^\sigma$ if $k \equiv 0 \pmod m$.

Recall that $F_{b,j}$ defined by formula (8.4) is the Hamiltonian for the j-th flow on $\mathcal{S}_{1,a}$ defined by b with respect to the symplectic form w_{-1}, and $\nabla F_{b,j} = Q_{b,j+1}^\perp$. Since $P_u(Q_{b,j}^\perp) = [Q_{b,j+1}, a]$, we get

9.4 Theorem ([Te2]). *If a is regular, then*

(i) $J_r(\nabla F_{b,j}) = [Q_{b,j+r+2}, a]$,

(ii) the Hamiltonian flow corresponding to $F_{b,j}$ *on* $(\mathcal{S}_{1,a}, w_r)$ *is the* $(j+r+1)$*-th flow defined by* b.

9.5 Examples.

Example 1. Let σ denote the involution $\sigma(y) = -y^t$ of $SU(n)$, and $a = \text{diag}(i, -i, \dots, -i)$. Then $\mathcal{S}^{\sigma}_{1,a,0}$ is the set of all $A = a\lambda + u$ with

$$u = \begin{pmatrix} 0 & v \\ -v^t & 0 \end{pmatrix},$$

where $v : R \to \mathcal{M}_{1\times(n-1)}$ is a decay map from R to the space $\mathcal{M}_{1\times(n-1)}$ of real $1 \times (n-1)$ matrices. The even flows vanishes on $\mathcal{S}^{\sigma}_{1,a,0}$, and the odd flows are extensions of the usual hierarchy of flows for the modified KdV. The third flow written in terms of $v : R \to \mathcal{M}_{1\times(n-1)}$ is the matrix modified KdV equation:

$$v_t = -\frac{1}{4}(v_{xxx} + 3(v_x v^t v + v v^t v_x).$$

(When $n = 2$, v is a scalar function and the above equation is the classic modified KdV equation.) The 2-form w_0 gives the appropriate non-degenerate symplectic structure for the matrix modified KdV equation and the hierarchy of odd flows.

Example 2. It seems appropriate to mention the relation of the restriction to the sine-Gordon equation. The sine-Gordon equation is written in space time coordinates (τ, y) as

$$\frac{\partial^2 q}{\partial \tau^2} - \frac{\partial^2 q}{\partial y^2} + \sin q = 0,$$

or

$$q_{xt} = \sin q$$

in characteristic coordinates. This is the -1-flow on $\mathcal{S}^{\sigma}_{1,a}$ defined by b, where $\sigma(y) = -y^t$ is the involution on $su(2)$, $a = \text{diag}(i, -i)$ and $b = -a/4$. The Lax pair is best written in characteristic coordinates:

$$\left[\frac{\partial}{\partial x} + a\lambda + u, \frac{\partial}{\partial t} + \lambda^{-1}B\right] = 0,$$

where

$$a = \begin{pmatrix} i & 0 \\ 0 & -i \end{pmatrix}, \quad u = \begin{pmatrix} 0 & \frac{q_x}{2} \\ -\frac{q_x}{2} & 0 \end{pmatrix}, \quad B = i \begin{pmatrix} \cos q & \sin q \\ \sin q & -\cos q \end{pmatrix}.$$

The restriction is the same as for the modified KdV. The natural Cauchy problem is in space time coordinates (τ, y), but the scattering theory has been developed for characteristic coordinates. However, the classical Bäcklund transformations work well with either choice of coordinates , and preserve whatever decay conditions have been described in either coordinate systems.

Example 3. We obtain the Kupershmidt and Wilson equation ([KW]) in terms of a restriction by an order n automorphism of $sl(n)$. Let $\alpha = e^{2\pi i/m}$, and $p \in SL(n)$ the matrix representing the cyclic permutation $(12\dots n)$, i.e., $p(e_i) = e_{i+1}$ (here we use the convention that $e_i = e_j$ if $i \equiv j \pmod n$). Let $\sigma : sl(n) \to sl(n)$ be the order n automorphism defined by $\sigma(y) = p^{-1}yp$. Then $X \in \mathcal{U}_j$ if and only if $\sigma(X) = \alpha^j X$. Let

$$a = \mathrm{diag}(1, \alpha, \dots, \alpha^{n-1}) \in \mathcal{U}_1.$$

Note that $\mathcal{U}_a^\perp$ is the space of all matrices $X \in sl(n)$ such that $X_{ii} = 0$ for all $i = 1, \dots, n$. So

$$\mathcal{S}_{1,a}^{\sigma} = \begin{cases} \left\{ \begin{pmatrix} 1 & 0 \\ 0 & -1 \end{pmatrix}\lambda + \begin{pmatrix} 0 & v \\ v & 0 \end{pmatrix} \,\middle|\, v \in \mathcal{S}(R, C) \right\}, & \text{if } n = 2, \\ \left\{ \begin{pmatrix} 1 & 0 & 0 \\ 0 & \alpha & 0 \\ 0 & 0 & \alpha^2 \end{pmatrix}\lambda + \begin{pmatrix} 0 & v_1 & v_2 \\ v_2 & 0 & v_1 \\ v_1 & v_2 & 0 \end{pmatrix} \,\middle|\, v_1, v_2 \in \mathcal{S}(R, C) \right\}, & \text{if } n = 3. \end{cases}$$

In general, $A = a\lambda + u \in \mathcal{S}_{1,a}^{\sigma}$ is determined by $(n-1)$ functions (the first row of u). By Propositions 9.3, $\{w_{rn} \mid r \in Z\}$ is a sequence of symplectic forms on $\mathcal{S}_{1,a}^{\sigma}$. The $(n+1)$-th flow is the Kupershmidt-Wilson equation. By Theorem 9.4 it satisfies the Lenard relation:

$$u_t = [Q_{a,n+2}, a] = J_0(\nabla F_{b,n}) = J_n(\nabla F_{b,0}).$$

When $n = 2$, the third flow on $\mathcal{S}_{1,a}^{\sigma}$ defined by a gives the modified KdV equation:

$$v_t = \frac{1}{4}(v_{xxx} - 6v^2 v_x), \tag{9.4}$$

and all the odd flows are the hierarchy of commuting flows of the modified KdV equation. For $n > 2$, this gives another generalization of modified KdV equation.

10 Bäcklund transformations for j-th flows

This section contains a brief outline of ideas and results in [TU1]. The classical Bäcklund transformations are originally geometric constructions by which a two parameters family of constant Gaussian curvature -1 surfaces is obtained from a single surface of Gaussian curvature -1. This is accomplished by solving two ordinary differential equations with a parameter s. The second parameter is the initial data. Since surfaces of Gaussian curvature -1 are classically known to be equivalent to local solutions of the sine-Gordon equation ([Da1], [Ei])

$$q_{xt} = \sin q$$

this provides a method of deriving new solutions of a partial differential equation from a given solution via the solution of ordinary differential equations. Most of the known "integrable systems" possess transformations of this type, which are sometimes called *Darboux transformations*. Ribaucour and Lie transformations are other classical transformations that generate new solutions from a given one.

The action of the rational loop group we constructed in section 6 can be extended to an action which transforms solutions of the j-th flow equation. In this section we describe very briefly the results in [TU1], which will construct an action of the semi-direct product of $R^* \ltimes G_-^m$ on the solution space of the j-th flow. The construction of this loop group action is motivated by the construction given by the second author in [U1] for harmonic maps. We will see

(1) the action of a simple element $g_{z,\pi}$ corresponds to a Bäcklund transformation,

(2) the action of R^* corresponds to the Lie transformations,

(3) the Bianchi permutability formula arises from the various ways of factoring quadratic elements in the rational loop group into simple elements,

(4) the Bäcklund transformations arise from ordinary differential equations if one solution is known,

(5) once given the trivialization of the Lax pair corresponding to a given solution, the Bäcklund transformations become algebraic.

Since the sine-Gordon equation arises as part of the algebraic structure (the -1-flow for $su(2)$ with an involution constraint), we can check that we are generalizing the classical theory. The choice of group structure depends on the choice of the base point (just as the scattering theory depends on the choice of a vacuum, or the choice of $0 \in R$). Hence the group structure was not apparent to the classical geometers.

One of the most interesting observations is that appropriate choices of poles for the rational loop yield time periodic solutions. This yields an interesting insight into the construction of time-periodic solutions (or the classical breathers) to the sine-Gordon equation as explained in Darboux ([Da1]). For recent developments concerning breathers of the sine-Gordon equation see [BMW], [De], [SS]. There are no simple factors in the rational loop group corresponding to the placement of poles for time periodic solutions. However, there are quadratic elements, whose simple factors do not satisfy the algebraic constraints to preserve sine-Gordon, but which nevertheless generate the well-known breathers (one way to think of them is as the product of two complex conjugate Bäcklund transformations). The product of these quadratic factors generate arbitrarily complicated time periodic solutions.

The classical theory of Bäcklund transformations is based on ordinary differential equations.

10.1 Theorem ([Ei]). *Suppose q is a solution of the sine-Gordon equation, and $s \neq 0$ is a real number. Then the following first order system is solvable for q^*:*

$$\begin{cases} (q^* - q)_x = 4s \sin\left(\dfrac{q^* + q}{2}\right) \\ (q^* + q)_t = \dfrac{1}{s} \sin\left(\dfrac{q^* - q}{2}\right). \end{cases} \tag{10.1}$$

Moreover, q^* is again a solution of the sine-Gordon equation.

10.2 Definition. If q is a solution of the sine-Gordon equation, then given any $c_o \in R$ there is a unique solution q^* for equation (10.1) such that $q^*(0,0) = c_o$. Then $B_{s,c_o}(q) = q^*$ is a transformation on the space of solutions of the sine-Gordon equation, which will be called a *Bäcklund transformation* for the sine-Gordon equation.

10.3 Proposition ([Ei]). *Define $L_s(q)(x,t) = q(sx, s^{-1}t)$. Then q is a solution of the sine-Gordon equation if and only if $L_s(q)$ is a solution of the sine-Gordon equation. (L_s is called a Lie transformation).*

10.4 Proposition ([Ei]). *Bäcklund transformations and Lie transformations of the sine-Gordon equation are related by the following formula:*

$$B_{s,c_o} = L_s^{-1} B_{1,c_o} L_s.$$

There is also a Bianchi permutability theorem for surfaces with Gaussian curvature -1 in R^3, which gives the following analytical formula for the sine-Gordon equation:

10.5 Theorem ([Ei]). *Suppose q_0 is a solution of the sine-Gordon equation, $s_1^2 \neq s_2^2$, and $s_1 s_2 \neq 0$. Let $q_i = B_{s_i,c_i}(q_0)$ for $i = 1,2$. Then there exist $d_1, d_2 \in R$, which can be constructed algebraically, such that*

(1) $B_{s_1,d_1} B_{s_2,c_2} = B_{s_2,d_2} B_{s_1,c_1}$,

(2) let $q_3 = B_{s_1,d_1} B_{s_2,c_2}(q_0)$, *then*

$$\tan \frac{q_3 - q_0}{4} = \frac{s_1 + s_2}{s_1 - s_2} \tan \frac{q_1 - q_2}{4}. \tag{10.2}$$

This is called the Bianchi permutability formula for the sine-Gordon equation.

Next we describe the action of G_-^m on the spaces of solutions of the j-th flow ($j \geq -1$). This construction is again using dressing action as in section 6 for the action of G_-^m on $\mathcal{S}_{1,a}$. First we make some definitions:

10.6 Definition. Let $\mathcal{M}(j, a, b)$ denote the space of all solutions of the j-th flow (equation (5.6)) on $\mathcal{S}_{1,a}$ defined by b with $[a, b] = 0$ for $j = -1$ and $j \geq 1$ respectively.

Assume $j \geq 1$. Let $A = a\lambda + u \in \mathcal{M}(j, a, b)$, and $E(x, t, \lambda)$ the trivialization of A normalized at $(x, t) = 0$, i.e.,

$$\begin{cases} E^{-1}E_x = a\lambda + u, \\ E^{-1}E_t = b\lambda^j + Q_{b,1}\lambda^{j-1} + \ldots + Q_{b,j}, \\ E(0, 0, \lambda) = I. \end{cases}$$

Given $g \in G_-^m$, by exactly the same method as in section 6, we can factor

$$g(\lambda)E(x, t, \lambda) = \tilde{E}(x, t, \lambda)\tilde{g}(x, t, \lambda),$$

such that $E(x, t, \cdot) \in G_+$ and $\tilde{g}(x, t, \cdot) \in G_-^m$. Define

$$\begin{cases} g \bullet E = \tilde{E}, \\ g \bullet A = \tilde{E}^{-1}\tilde{E}_x \\ . \end{cases}$$

Then $g \bullet A \in \mathcal{M}(j, a, b)$, and $\bullet$ defines an action of G_-^m on $\mathcal{M}(j, a, b)$.

Recall that the $g \in G_-^m$ can be generated by simple elements $g_{z,\pi} \in G_-^m$, which are rational functions of degree 1. Choose a pole $z \in C \setminus R$, and a subspace $V \subset C^n$, which is identified with the Hermitian projection

$$\pi : C^n \to V.$$

Write

$$g_{z,\pi}(\lambda) = \pi + \frac{\lambda - z}{\lambda - \bar{z}}\pi^{\perp}$$

as in Proposition 6.3.

10.7 Definition. $A \mapsto g_{z,\pi} \bullet A$ is a Bäcklund transformation for the j-th flow.

Compute the action of $g_{z,\pi}$ explicitly as in section 6 to get:

10.8 Theorem. *Let $g_{z,\pi}$ be a generator in G_-^m, where π is the projection of C^n onto a k-dimensional complex linear subspace V. Let $A = a\lambda + u \in \mathcal{M}(j, a, b)$, and $E(x, t, \lambda)$ the trivialization of A normalized at $(x, t) = 0$. Set $\tilde{V}(x, t) = E(x, t, z)^*(V)$, and let $\tilde{\pi}(x, t)$ denote the projection of C^n onto $\tilde{V}(x, t)$. Set*

(10.3)

$$\tilde{\pi}(x, t) = E^*(x, t, z)U(U^*E(x, t, z)E^*(x, t, z)U)^{-1}U^*E(x, t, z),$$

$$g_{z,\tilde{\pi}(x,t)(\lambda)} = \tilde{\pi}(x, t) + \frac{\lambda - z}{\lambda - \bar{z}}\tilde{\pi}(x, t)^{\perp}$$

where U is a $n \times k$ matrix whose columns form a basis for V. Then

(i) $g_{z,\pi} \bullet E = g_{z,\pi} E g_{z,\tilde{\pi}}{}^{-1}$,

(ii) $g_{z,\pi} \bullet A = A + (z - \bar{z})[\tilde{\pi}, a]$.

10.9 Theorem. *The $\tilde{\pi}$ constructed in Theorem 10.8 is the solution of the following compatible first order system:*

$$\begin{cases} (\tilde{\pi})_x + [az + u, \tilde{\pi}] = (\bar{z} - z)[\tilde{\pi}, a]\tilde{\pi}, \\ (\tilde{\pi})_t = \sum_{k=0}^{j} [\tilde{\pi}, Q_{b,j-k}(u)](z + (\bar{z} - z)\tilde{\pi})^k, \\ \tilde{\pi}^* = \tilde{\pi}, \quad \tilde{\pi}^2 = \tilde{\pi}, \quad \tilde{\pi}(0,0) = \pi. \end{cases} \tag{10.4}$$

Moreover,

(i) *equation (10.4) is solvable for $\tilde{\pi}$ if and only if $A = a\lambda + u$ is a solution of the j-th flow on $\mathcal{S}_{1,a}$ defined by b,*

(ii) *if $A = a\lambda + u$ is a solution of the j-th flow and $\tilde{\pi}$ is a solution of equation (10.4), then $\tilde{A} = A + (z - \bar{z})[\tilde{\pi}, a]$ is again a solution of the j-th flow.*

10.10 Definition. Let $R^* = \{r \in R \mid r \neq 0\}$ denote the multiplicative group, and $R^* \ltimes G_-^m$ the semi-direct product of R^* and G_-^m defined by the homomorphism

$$\rho : R^* \to \mathrm{Aut}(G_-^m), \qquad \rho(r)(g)(\lambda) = g(r\lambda),$$

i.e., the multiplication in $R^* \ltimes \Omega(G)$ is defined by

$$(r_1, h_1) \cdot (r_2, h_2) = (r_1 r_2, h_1(\rho(r_1)(h_2))).$$

10.11 Theorem. *The action $\bullet$ of G_-^m extends to an action of $R^* \ltimes G_-^*$ on the space $\mathcal{M}(j, a, b)$ of solutions of the j-th flow on $\mathcal{S}_{1,a}$ defined by b. In fact, if $A = a\lambda + u \in \mathcal{M}(j, a, b)$ and E is the trivialization of A normalized at $(x,t) = 0$, then*

$$\begin{cases} (r \bullet E)(x, t, \lambda) = E(r^{-1}x, r^{-j}t, r\lambda), \\ (r \bullet A)(x, t, \lambda) = a\lambda + r^{-1}u(r^{-1}x, r^{-j}t). \end{cases}$$

Since $(r^{-1}, 1)(1, g_{e^{i\alpha},\pi})(r, 1) = (1, g_{re^{i\alpha},\pi})$, we have

10.12 Corollary. *If $A \in \mathcal{M}(j, a, b)$, then*

$$r^{-1} \bullet (g_{e^{i\alpha},\pi} \bullet (r \bullet A)) = g_{re^{i\alpha},\pi} \bullet A.$$

Next we state an analogue of the Bianchi Permutability Theorem for the positive flows:

10.13 Theorem. *Let $z_1 = r_1 + is_1, z_2 = r_2 + is_2 \in C \setminus R$ such that $r_1 \neq r_2$ or $s_1^2 \neq s_2^2$, and π_1, π_2 projections of C^n. Let $A_0 \in \mathcal{M}(j, a, b)$, and $A_i = g_{z_i,\pi_i} \bullet A_0$ for $i = 1, 2$. Set*

$$\xi_i = \big(-(z_1 - z_2)I + 2i(s_1\pi_1 - s_2\pi_2)\big)\pi_i\big((-(z_1 - z_2)I + 2i(s_1\pi_1 - s_2\pi_2)\big)^{-1},$$
$$\tilde{\xi}_i = (-(z_1 - z_2)I + 2i(s_1\tilde{\pi}_1 - s_2\tilde{\pi}_2))\tilde{\pi}_i(-(z_1 - z_2)I + 2i(s_1\tilde{\pi}_1 - s_2\tilde{\pi}_2))^{-1}, \tag{10.5}$$

for $i = 1, 2$, where $\tilde{\pi}_i$ is as in Theorem 10.8 and $A_i = A_0 + 2is[\tilde{\pi}_i, a]$. Then

(i) $g_{z_2,\xi_2} g_{z_1,\pi_1} = g_{z_1,\xi_1} g_{z_2,\pi_2}$,

(ii)

$$\begin{aligned} A_3 &= (g_{z_2,\xi_2} g_{z_1,\pi_1}) \bullet A_0 = A_0 + 2i[s_1\tilde{\pi}_1 + s_2\tilde{\xi}_2, a] \\ &= (g_{z_1,\xi_1} g_{z_2,\pi_2}) \bullet A_0 = A_0 + 2i[s_1\tilde{\xi}_1 + s_2\tilde{\pi}_2, a]. \end{aligned} \tag{10.6}$$

Note that if A lies in the orbit of $R^* \ltimes G_-^m$ through the vacuum solution $A_0 = a\lambda$ of the j-th flow then A has no continuous scattering data.

10.14 Definition. An element A in the orbit of $R^* \ltimes G_-^m$ through the vacuum solution $A_0 = a\lambda$ of the j-th flow is called a *k-soliton* if the normalized eigenfunction has k poles counted with multiplicities.

10.15 Remark. The trivialization of the vacuum solution $A_0 = a\lambda$ of the j-th flow on $\mathcal{S}_{1,a}$ defined by b is $E = \exp(a\lambda x + b\lambda^j t)$. So 1-soliton is

$$g_{z,\pi} \bullet A_0 = a\lambda + (z - \bar{z}) \left[e^{-a\bar{z}x - b\bar{z}^j t} U (U^* e^{a(z-\bar{z})x + b(z^j - \bar{z}^j)t} U)^{-1} U^* e^{azx + bz^j t},\ a\right],$$

where U is a matrix whose columns form a basis of $\mathrm{Im}(\pi)$. If $g = \prod_{i=1}^k g_{z_i,\pi_i}$, then $g \sharp A_0$ can be written in terms of 1-solitons $g_{z_1,\pi_1} \sharp A, \ldots, g_{z_k,\pi_k} \sharp A$ algebraically by applying the permutability formula (10.6) repeatedly.

10.16 Remark. If A is a soliton solution of the j-th flow on $\mathcal{S}_{1,a}$ defined by b, then by Corollary 8.3 there exists $g \in G_-^m$ such that

$$A(t) = \Psi(e_{b,j}(t) \natural g).$$

As noted in Remark 7.22, given $g_{z,\pi} \notin H_-$, $g_{z,\pi} \bullet A$ can be a $m+1, m$ or $m-1$-soliton if A is a m-soliton solution.

10.17 Proposition. *The set of all soliton solutions of the j-th flow on $\mathcal{S}_{1,a}$ defined by b is isomorphic to the space G_-^m / H_-^m of left cosets, where $H_-^m = G_-^m \cap H_-$.*

10.18 Remark. Although the space G^m_-/H^m_- of all multi-solitons does not have a manifold structure, the set $\mathcal{B}_m$ all m-solitons is the union of complex algebraic varieties. For example,

$$\mathcal{B}_1 = \bigcup_{k=1}^{n-1} (C_+ \times X_k), \text{ where}$$
$$X_k = Gr(k, C^n) \setminus \{V \in Gr(k, C^n) \mid a(V) = V\}.$$

But $\mathcal{B}_m$ with $m \geq 2$ is much more complicated because of the non-uniqueness of the factorization and the fact that generators of G^m_- have complicated relations such as the permutability formula given in Theorem 10.13 (i).

Next we apply the action of G^m_- on the first flows to get actions of G^m_- on the space $\mathcal{M}$ of solutions of the n-dimension system (8.2) associated to $U(n)$. Given $v \in \mathcal{M}$, the trivialization E of v normalized at the origin is the solution

$$\begin{cases} E^{-1}E_{x_j} = a_j\lambda + [a_j, v], & 1 \leq j \leq n \\ E(0, \lambda) = I. \end{cases}$$

10.19 Theorem. *The group $R^* \times G^m_-$ acts on $\mathcal{M}$, and the action $\bullet$ is constructed in the same manner as on the spaces of solutions of the first flow. In fact, given $g_{z,\pi} \in G^m_-$, the following initial value problem is solvable for $\tilde{\pi}$ and has a unique solution:*

$$\begin{cases} (\tilde{\pi})_{x_j} + [a_j z + [a_j, v], \tilde{\pi}] = (\bar{z} - z)[\tilde{\pi}, a_j]\tilde{\pi}, \\ \tilde{\pi}^* = \tilde{\pi}, \quad \tilde{\pi}^2 = \tilde{\pi}, \quad \tilde{\pi}(0) = \pi. \end{cases}$$

Moreover,

(i) $g_{z,\pi} \bullet v = v - ((z - \bar{z})\tilde{\pi})^\perp$, where $y^\perp$ denote the projection of $y \in \mathcal{U}$ onto $\mathcal{T}^\perp$,

(ii) the trivialization of $g_{z,\pi} \bullet v$ is $g_{z,\pi} E g^{-1}_{z,\tilde{\pi}}$,

(iii) $\tilde{\pi}$ is the projection onto the linear subspace $E^*_z(V)$, where V is the image of the projection π,

(iv) $(r \bullet v)(x) = r^{-1}v(r^{-1}x)$ for $r \in R^*$.

10.20 Remark. The permutability Theorem 10.13 holds for system (8.2) with the same formula.

There are also analogous results for the -1-flow:

10.21 Theorem. *The group $R^* \ltimes G^m_-$ acts on the space $\mathcal{M}(-1, a, b)$ of solutions of the -1-flow on $\mathcal{S}_{1,a}$ defined by b:*

$$\begin{cases} u_t = [a, g^{-1}bg], \\ g_t = gu, \qquad \lim_{x \to -\infty} g(x) = I. \end{cases} \tag{10.7}$$

Moreover, let $A \in \mathcal{M}(-1, a, b)$, E the trivialization of A normalized at $(x, t) = 0$, and $g_{z,\pi}$ a simple element of G_-^m, then

(i) Theorems 10.8, 10.13 and Corollary 10.12 hold with the same formulas,

(ii) $\tilde{\pi}$ is the solution to

$$\begin{cases} (\tilde{\pi})_x + [az + u, \tilde{\pi}] = (\bar{z} - z)[\tilde{\pi}, a]\tilde{\pi}, \\ (\tilde{\pi})_t = \dfrac{1}{|z|^2}\left((z - \bar{z})\tilde{\pi}g^{-1}bg\tilde{\pi} - zg^{-1}bg\tilde{\pi} + \bar{z}\tilde{\pi}g^{-1}bg\right), \\ \tilde{\pi}^* = \tilde{\pi}, \quad \tilde{\pi}^2 = \tilde{\pi}, \quad \tilde{\pi}(0,0) = \pi, \end{cases} \tag{10.8}$$

(iii) for $r \in R^*$ we have

$$\begin{cases} (r \bullet E)(x, t, \lambda) = E(r^{-1}x, rt, r\lambda), \\ (r \bullet A)(x, t, \lambda) = a\lambda + r^{-1}u(r^{-1}x, rt). \end{cases}$$

10.22 Remark. Let $A \in \mathcal{M}(-1, a, b)$, and E its trivialization normalized at the origin. Then $s(x, t) = E(x, t, -1)E(x, t, 1)^{-1}$ is a harmonic map from $R^{1,1}$ with the metric $2dx\,dt$ to $U(n)$ and $s^{-1}s_x$ is conjugate to a and $s^{-1}s_t$ is conjugate to b. In particular, this says that $\mathcal{M}(-1, a, b)$ is a subset of the space $\mathcal{F}$ of harmonic maps from $R^{1,1}$, to $U(n)$. The action of G_-^m on $\mathcal{F}$ constructed in [U1] leaves $\mathcal{M}(-1, a, b)$ invariant, and agrees with the action $\bullet$ we constructed here.

Recall that given an involution σ of $su(n)$, $G_-^{m,\sigma}$ is the subgroup of $g \in G_-^m$ such that $\sigma(g(-\lambda)) = g(\lambda)$ and $\mathcal{S}_{1,a}^\sigma = \{A \in \mathcal{S}_{1,a} \,|\, \sigma(A(-\lambda)) = A(\lambda)\}$. Since the action of $G_-^{m,\sigma}$ leaves $\mathcal{S}_{1,a}^\sigma$ invariant, we have

10.23 Theorem. *The space $\mathcal{M}^\sigma(2j-1, a, b)$ of solutions of the $(2j-1)$-th flow associated to U/K defined by b is a subset of $\mathcal{M}(2j-1, a, b)$ for $j \geq 0$. Moreover, the action $\bullet$ of $R^* \ltimes G_-^{m,\sigma}$ leaves $\mathcal{M}^\sigma(2j-1, a, b)$ invariant, and Theorems 10.8, 10.9, 10.11 and Corollary 10.12 hold for $\mathcal{M}^\sigma(2j-1, a, b)$ and $R^* \ltimes G_-^{m,\sigma}$.*

10.24 Theorem. *The space $\mathcal{M}^\sigma(-1, a, b)$ of solutions of the -1-th flow equation on $\mathcal{S}_{1,a}^\sigma$ defined by b is a subset of $\mathcal{M}(-1, a, b)$. Moreover, the action of $R^* \ltimes G_-^{m,\sigma}$ leaves $\mathcal{M}^\sigma(-1, a, b)$ invariant and Theorem 10.21 holds for $\mathcal{M}_{1,a}^\sigma$ and $R^* \ltimes G_-^{m,\sigma}$.*

10.25 Proposition. *Let σ denote the involution on $SU(n)$ defined by $\sigma(y) = \bar{y}$. Then*

(i) $g_{z,\pi} \in G_-^{m,\sigma}$ if and only if $z = -\bar{z}$ and $\bar{\pi} = \pi$,

(ii) if $\bar{\pi} = \pi$, then $g_{z,\pi}g_{-\bar{z},\pi} \in G_-^{m,\sigma}$.

Next we explain the relation between the classical Bäcklund transformations and the action of $G_-^{m,\sigma}$ on the space of solutions of the sine-Gordon equation (or the space $\mathcal{M}^\sigma(-1,a,b)$ with σ, a,b defined as in Example 9.5). If $s \in R$, $\tilde{\pi}^* = \tilde{\pi} = (\bar{\pi})^t$ and $(\bar{\pi})^* = \tilde{\pi}$, then by Proposition 10.25, $g_{is,\tilde{\pi}} \in G_-^{m,\sigma}$. Hence

$$\tilde{\pi} = \begin{pmatrix} \cos^2 \frac{f}{2} & \sin \frac{f}{2} \cos \frac{f}{2} \\ \sin \frac{f}{2} \cos \frac{f}{2} & \sin^2 \frac{f}{2} \end{pmatrix}$$

for some function f, i.e., $\tilde{\pi}$ is the projection onto $\begin{pmatrix} \cos \frac{f}{2} \\ \sin \frac{f}{2} \end{pmatrix}$. So the first order system (10.8) for $\tilde{\pi}$ becomes

$$\begin{cases} f_x = \frac{q_x}{2} + 2s \sin f, \\ f_t = \frac{1}{2s} \sin(f - q). \end{cases} \tag{10.9}$$

Write

$$\tilde{u} = g_{is,\beta} \bullet u = \begin{pmatrix} 0 & \tilde{q}_x/2 \\ -\tilde{q}_x/2 & 0 \end{pmatrix}.$$

But $\tilde{u} = u + 2is[\tilde{\pi}, a]$, hence we have $\tilde{q} = 2f - q$. Writing equation (10.9) in terms of $\tilde{q}$, we get

$$\begin{cases} (q^* - q)_x = 4s \sin(\frac{q^*+q}{2}) \\ (q^* + q)_t = \frac{1}{s} \sin(\frac{q^*-q}{2}), \end{cases}$$

which is the classical Bäcklund transformation for the sine-Gordon equation. So we have:

10.26 Proposition. *Let q be a solution of the sine-Gordon equation, and $0 < c_0 < \pi$. Set*

$$\begin{cases} A = a\lambda + \begin{pmatrix} 0 & \frac{q_x}{2} \\ -\frac{q_x}{2} & 0 \end{pmatrix}, \\ f_o = 1/2(q(0,0) + c_o) \\ \pi = \begin{pmatrix} \cos^2 \frac{f_0}{2} & \sin \frac{f_0}{2} \cos \frac{f_0}{2} \\ \sin \frac{f_0}{2} \cos \frac{f_0}{2} & \sin^2 \frac{f_0}{2} \end{pmatrix}. \end{cases}$$

Then

$$B_{s,c_o}(q) = g_{is,\pi} \bullet A.$$

(We will still use $\bullet$ to denote the induced action of $G_-^{m,\sigma}$ on the set of solutions of the sine-Gordon equation).

10.27 Proposition. *Let q is a solution of the sine-Gordon equation. Then:*

(i) $s \bullet q$ is the Lie transformation $L_s(q)$.

(ii) *Proposition 10.4 is a consequence of the following equality in the group* $R^* \ltimes G_-^m$:

$$(s^{-1}, I)(1, g_{e^{i\alpha},\beta})(s, I) = g_{se^{i\alpha},\beta}.$$

(iii) *The Permutability Formula (10.2) in Theorem 10.5 is the same formula (10.6) given in Theorem 10.13, which follows from the following relation of the generators of* G_-^m:

$$g_{z_1,\xi_1} g_{z_2,\pi_2} = g_{z_2,\xi_2} g_{z_1,\pi_1},$$

where π_i *and* ξ_i *are related by formula (10.5).*

10.28 Remark. It is noted by Xi Du [Du] that a classical Ribaucour transformations as defined in Darboux ([Da1]) for surfaces of $K = -1$ in R^3 correspond to the action of an element $g \in G_-^{m,\sigma}$, which is the product of two simple elements as in Proposition 10.25 (ii).

Using the action of G_-^m, we obtain many solutions of the j-th flow that are periodic in time. This is an algebraic calculation, which shows that when the poles are properly placed, the solutions are periodic. Multi-solitons will be time periodic if the periods of the component solitons are rationally related.

10.29 Theorem. *Let* $j > 1$ *be an integer,* $a = \text{diag}(ia_1, \ldots, ia_n)$, *and* $b = \text{diag}(ib_1, \ldots, ib_n)$. *If* $b_1, \ldots, b_n$ *are rational numbers. Then the* j*-th flow equation defined by* a, b:

$$u_t = [Q_{b,j+1}(u), a]$$

has infinitely many m*-soliton solutions that are periodic in* t.

The trivialization of the vacuum solution for the -1-flow defined by $a = \text{diag}(i, \ldots, i, -i, \ldots, -i)$ is $E(\lambda, x, t) = \exp(a(\lambda x + \lambda^{-1} t))$. By Theorem 10.21, the 1-soliton $g_{e^{i\theta},\pi} \bullet 0$ is a function of

$$\exp(i(\cos\theta(x+t) - i\sin(x-t))) = \exp(i\tau\cos\theta + y\sin\theta),$$

where $y = x - t$ and $\tau = x + t$ are the space-time coordinates. This gives

10.30 Theorem. *If* $z = e^{i\theta}$ *and* $a = \text{diag}(i, \ldots, i, -i, \ldots, -i)$, *then the* 1*-soliton* $g_{z,\pi} \bullet 0$ *for the* -1*-flow (harmonic maps from* $R^{1,1}$ *to* $SU(n)$*)) is periodic in time with period* $\frac{2\pi}{\cos\theta}$. *A multiple soliton generated by a rational loop with poles at* $z_1 = e^{i\theta_1}, \ldots, z_r = e^{i\theta_r}$ *will be periodic with period* τ *if there exists integers* $k_1, \ldots, k_r$ *such that*

$$\tau = \frac{2\pi k_j}{\cos\theta_j} \qquad \forall\, 1 \le j \le r.$$

The multi-solitons above satisfy the sine-Gordon equation if the rational loop satisfies $\overline{f(-\bar{\lambda})} = f(\lambda)$. This means the poles occur in pairs $(e^{i\theta_j}, -e^{-i\theta_j})$ and the projection matrices π_j must be real.

10.31 Corollary. *Multiple-breather solutions exists for the sine-Gordon equation.*

10.32 Example. If π is real, then

$$(g_{e^{i\theta},\pi} g_{-e^{-i\theta},\pi}) \bullet 0 = 4\tan^{-1}\left(\frac{\sin\theta\sin((x+t)\cos\theta)}{\cos\theta\cosh((x-t)\sin\theta)}\right)$$

is the classical breather solution for the sine-Gordon equation. Theorems 10.24 and 10.30 give m-breather solutions explicitly, although the computations are quite long.

10.33 Corollary. *There are infinitely many harmonic maps from $R^{1,1}$ to a symmetric space that are periodic in time.*

11 Geometric non-linear Schrödinger equation

Consider the evolution of curves in R^3

$$\gamma_t = \gamma_x \times \gamma_{xx}, \tag{11.1}$$

where $\times$ denote the cross-product in R^3. This equation is known as the *vortex filament equation* , and has a long and interesting history (cf. [Ri]). It is easy to see that $||\gamma_x||^2$ is preserved under the evolution. It follows that if $\gamma(\cdot,0)$ is parametrized by its arc length, then so are all $\gamma(\cdot,t)$ for all t. So equation (11.1) can also be viewed as the evolution of a curve that moves along the direction of binormal with the curvature as its speed. Let $k(\cdot,t)$ and $\tau(\cdot,t)$ be the curvature and torsion of the curve $\gamma(\cdot,t)$. Then there exists a unique $\theta(x,t)$ such that $\theta_x = \tau$ and

$$q(x,t) = k(x,t)e^{-i\int^x \tau(s,t)ds}$$

is a solution of the non-linear Schrödinger equation:

$$q_t = \frac{i}{2}\left(q_{xx} + 2|q|^2 q\right).$$

There is another interesting evolution of curves in S^2 that is also associated to the non-linear Schrödinger equation:

11.1 Proposition. *$\gamma(x,t)$ is a solution of equation (11.1) with x as the arc length parameter if and only if $\phi(x,t) = \gamma_x(x,t) : R^2 \to S^2$ satisfies the equation*

$$J(\phi_t) = \nabla_{\phi_x}\phi_x, \tag{11.2}$$

where ∇ is the Levi-Civita connection and J is the complex structure of the standard two sphere S^2.

Equation (11.2) is the geometric non-linear Schrödinger equation (GNLS) on S^2. Such equation can be defined on any complex Hermitian manifold (M, g, J). Consider the Schrödinger flow on the space $\mathcal{S}(R, M)$ of Schwartz maps, i.e., the equation for maps $\phi : R \times R \to M$:

$$J\left(\frac{\partial \phi}{\partial t}\right) = \triangle \phi,$$

where $\triangle \phi = \nabla_{\phi_x} \phi_x$ is the gradient of the energy functional on $\mathcal{S}(R, M)$, or the accelleration.

In this section, we give a brief outline of ideas and results in a forthcoming paper [TU3]. There is a Hasimoto type transformation that transforms the GNLS equation associated to $Gr(k, C^n)$ to the the second flow on $\mathcal{S}_{1,a}$ defined by

$$a = \begin{pmatrix} iI_k & 0 \\ 0 & -iI_{n-k} \end{pmatrix} \in u(k) \times u(n-k). \tag{11.3}$$

We have seen in Example 5.6 that identifying $\mathcal{S}_{1,a}$ as the space $\mathcal{M}_{k\times(n-k)}$ of $k \times (n-k)$ matrices, the second flow defined by a is the matrix non-linear Schrödinger equation:

$$B_t = \frac{i}{2}(B_{xx} + 2BB^*B). \tag{11.4}$$

Applying our theory to equation (11.4), we obtain many beautiful properties for the GNLS associated to $Gr(k, C^n)$. For example, we have

(i) a Hamiltonian formulation,

(ii) long time existence for the Cauchy problem,

(iii) a sequence of commuting Hamiltonian flows,

(iv) explicit soliton solutions,

(v) a non-abelian Poisson group action on the space of solutions of the GNLS,

(vi) a sequence of compatible symplectic structures on the space $\mathcal{S}(R, Gr(k, C^n))$ in which the GNLS is Hamiltonian and has a Lenard relation.

Let $U(n)$ be equipped with a bi-invariant metric. It is well-known that $Gr(k, C^n)$ can be naturally embedded as a totally geodesic submanifold M of $U(n)$. In fact, M is the set of all $X \in U(n)$ such that X is conjugate to a as described by formula (11.3). The invariant complex structure on M is given by

$$J_h(v) = [v, h].$$

Consider the following equation for maps $\phi : R^2 \to M$:

$$J_\phi(\phi_t) = \nabla_{\phi_x}(\phi_x), \qquad \phi : R^2 \to M \tag{11.5}$$

where ∇ is the Levi-Civita connection of the standard Kahler metric on M. A direct computation gives

$$\nabla_{\phi_x}(\phi_x) = \phi(\phi^{-1}\phi_x)_x.$$

So equation (11.5) becomes

$$\phi_t = -\frac{1}{2}(\phi^{-1}\phi_x)_x. \tag{GNLS}$$

Next we want to associate to each solution of equation (11.4) a solution of the GNLS. This is a generalization of the Hasimoto transformation of the vortex filament equation to non-linear Schrödinger equation. As noted in Example 5.6, $A = a\lambda + u \in$ is a solution of (11.4) if and only if

$$\theta_\lambda = (a\lambda + u)dx + (a\lambda^2 + u\lambda + Q_{a,2}(u))dt$$

is flat for all λ, where

$$u = \begin{pmatrix} 0 & B \\ -B^* & 0 \end{pmatrix}, \qquad Q_{a,2} = \begin{pmatrix} \frac{1}{2i}BB^* & \frac{i}{2}B_x \\ \frac{i}{2}B_x^* & -\frac{1}{2i}B^*B \end{pmatrix}.$$

In particular, $\theta_0 = u\,dx + Q_{a,2}(u)dt$ is flat. Let g be the trivialization of θ_0, i.e.,

$$\begin{cases} g^{-1}g_x = u, \\ g^{-1}g_t = Q_{a,2}(u). \end{cases}$$

Set $\phi = gag^{-1}$. Changing the gauge of θ_λ by g gives

$$\begin{aligned} \tau_\lambda &= g\theta_\lambda g^{-1} - dgg^{-1} = (gag^{-1}\lambda)dx + (gag^{-1}\lambda^2 + g_x g^{-1}\lambda)dt \\ &= \phi\lambda\, dx + (\phi\lambda^2 + gug^{-1}\lambda)dt. \end{aligned}$$

Since τ_λ is flat for all λ, we get

$$\begin{cases} \phi_t = (gug^{-1})_x, \\ \phi_x = -[\phi, gug^{-1}]. \end{cases} \tag{11.6}$$

But for $u \in \mathcal{U}_a^\perp$, we have $a^{-1}ua = -u$. Hence

$$\begin{aligned} \phi^{-1}\phi_x &= ga^{-1}g^{-1}(g_x ag^{-1} - gag^{-1}g_x g^{-1}) = ga^{-1}uag^{-1} - g_x g^{-1} \\ &= -gug^{-1} - g_x g^{-1} = -2g_x g^{-1} = -2gug^{-1}. \end{aligned} \tag{11.7}$$

So the first equation of (11.6) implies that ϕ is a solution of the GNLS.

Conversely, suppose $\phi : R^2 \to M$ is a solution of the GNLS. Then there exists $g : R^2 \to U(n)$ such that $\phi = gag^{-1}$ and $g^{-1}g_x(x,t) \in \mathcal{U}_a^\perp$ for all (x,t). Set

$$u = g^{-1}g_x, \qquad f = -\frac{1}{2}(\phi^{-1}\phi_x).$$

Then the equation (11.7) implies that $\phi^{-1}\phi_x = g_x g^{-1} = f$. Differentiate $\phi = gag^{-1}$ with respect to x to get

$$\phi_x = [g_x g^{-1}, \phi] = [f, \phi].$$

But the GNLS gives $\phi_t = f_x$. So

$$(\phi\lambda)dx + (\phi\lambda^2 + f\lambda)dt$$

is flat for all λ. Changing the gauge by g^{-1}, we find that

$$\beta_\lambda = (a\lambda + u)dx + (a\lambda^2 + u\lambda + h)dt$$

is flat, where $h = g^{-1}g_t$. Flatness of β_λ on the (x,t)-plane for all λ implies that $h = Q_{a,2}(u)$. So this proves that u is a solution of the second flow equation (11.4). To summarize,

11.2 Proposition. *If $B : R^2 \to \mathcal{M}(k \times (n-k))$ is a solution of the matrix NLS (11.4), then there is $g : R^2 \to U(n)$ such that*

$$g^{-1}g_x = \begin{pmatrix} 0 & B \\ -B^* & 0 \end{pmatrix}, \qquad g^{-1}g_t = \begin{pmatrix} \frac{1}{2i}BB^* & \frac{i}{2}B_x \\ \frac{i}{2}B_x^* & -\frac{1}{2i}B^*B \end{pmatrix}$$

and $\phi = gag^{-1}$ is a solution of the GNLS. Conversely, if ϕ is a solution of the GNLS, then there is $g : R^2 \to U(n)$ such that

$$\phi = g\begin{pmatrix} iI_k & 0 \\ 0 & -I_{n-k} \end{pmatrix} g^{-1}, \qquad g^{-1}g_x = \begin{pmatrix} 0 & B \\ -B^* & 0 \end{pmatrix},$$

and B is a solution of the matrix NLS equation (11.4).

To end this section, we will translate properties for the second flow (11.4) to properties of the GNLS equation.

When a is singular, formula (8.4) implies that the corresponding Hamiltonians for the first three flows on $\mathcal{S}_{1,a}$ defined by a are

$$\begin{aligned}
F_1(B) &= \frac{1}{4}\int_{-\infty}^{\infty} \operatorname{tr}(iB_xB^* - iBB_x)dx = \frac{1}{4}\langle iB_x, B\rangle, \\
F_2(B) &= \frac{1}{4}\int_{-\infty}^{\infty} \operatorname{tr}(-B_xB_x^* + B^*BB^*B)\,dx \\
&= \frac{1}{8}(-\langle B_x, B_x\rangle + \langle B^*B, B^*B\rangle), \\
F_3(B) &= \frac{i}{16}\int_{-\infty}^{\infty} \operatorname{tr}(-BB_{xxx}^* + B_{xxx}B^*) + 3\operatorname{tr}(-BB^*BB_x^* + B_xB^*BB^*)dt \\
&= -\frac{1}{16}(\langle iB_{xxx}, B\rangle + 3\langle iB_x, BB^*B\rangle).
\end{aligned}$$

11.3 Remark. If $b \in \mathcal{U}_a$, i.e., $[a,b] = 0$, and $b \neq a$, then

(i) $Q_{b,j}(u)$ in general is not a local operator in u,

(ii) the flow generated by $b\lambda^n \in \mathcal{H}_+$ commutes with the flow (11.4),

(iii) given $b_1, b_2 \in \mathcal{U}_a$, the flow generated by $b_1\lambda^k$ and $b_2\lambda^j$ need not commute and

$$[\xi_{b_1,k}, \xi_{b_2,j}] = \xi_{[b_1,b_2],k+j},$$

where $\xi_{b,m}$ denotes the infinitesimal vector fields corresponding to $b\lambda^m$,

(iv) the action of $H_+ = \{g \in G_+ \mid ga = ag\}$ on $\mathcal{S}_{1,a}$ is Poisson.

It follows from the discussion in section 7 that the Cauchy problem for equation (11.4) with initial condition $A_0 = a\lambda + u_0 \in \mathcal{S}'_{1,a}$ can be solved by using factorizations. First we use the direct scattering of $A_0 = a\lambda + u_0$ on the line, i.e., solve equation (7.1). Set $f(\lambda) = \psi(0, \lambda)$. Then $f \in D_-$. Decompose

$$f(\lambda)e^{a\lambda x + b\lambda^j t} = E(x,t,\lambda)M(x,t,\lambda)^{-1}$$

as in section 7 by applying Birkhoff decompositions repeatedly. Then $A = E^{-1}E_x$ is the solution of the Cauchy problem.

The rational group G_-^m acts on the space of solutions of the GNLS, and soliton solutions can be calculated explicitly using the formulas in Theorem 10.8.

11.4 Proposition. *Let $a = \mathrm{diag}(-i, i, \dots, i)$, and choose a pole $z = r + is \in C \setminus R$. Let π be the projection on the subspace spanned by $(1, v)^t = (1, v_2, \dots, v_n)^t$. Then the one-solitons for the j-th flow defined by a generated by Bäcklund transformations from the vacuum solution $A_0(x,t,\lambda) = a\lambda$ are of the form*

$$A(x,t,\lambda) = a\lambda + u(x,t),$$
$$u(x,t) = \begin{pmatrix} 0 & B(x,t) \\ -B^*(x,t) & 0 \end{pmatrix},$$

where

$$B(x,t) = \frac{4se^{-2i(rx + \mathrm{Re}(z^j)t)}}{e^{-2(sx + \mathrm{Im}(z^j)t)} + e^{2(sx + \mathrm{Im}(z^j)t)}\|v\|^2}\,\bar{v}.$$

Proof. We use Theorem 10.8 to make our computations. We start with $A_0 = a\lambda$. According to Theorem 10.8,

$$A_0 \mapsto a\lambda + (z - \bar{z})[\tilde{\pi}, a],$$

where $a = \mathrm{diag}(-i, i, \dots, i)$. Here $\tilde{\pi}$ is the projection on $(1, \tilde{v})^t$, where

$$\hat{v} = (1, \tilde{v}) = (1, e^{2i(\bar{z}x + \bar{z}^j t)}v).$$

Let $z = r + is$. Then

$$\tilde{\pi}(x,t) = \frac{1}{e^{-2(sx+\mathrm{Im}(z^j)t)} + e^{2(sx+\mathrm{Im}(z^j)t)}\|v\|^2}\,\hat{v}^*\hat{v}.$$

The formula for B follows. □

12 First flows and flat metrics

The integrable equations of evolution we have been describing up to this point have at most two independent variables. The flow of the first variable, regarded as a spacial variable, is used to construct the initial Cauchy data from the scattering coset (hence the "first flow" terminology). The second variable is considered to be the time variable, and the flow in this variable is the evolution. Many authors consider a commuting heirarchy of flows to generate functions of an infinite sequence of time variables. However, the physical and geometric applications do not require this consideration.

We turn our attention to a family of geometric problems in n spacial variables, which we shall call *n-dimensional systems* or *n-dimensional flows*. In the applications, the n variables are on an equal footing, and the flows in each variable is a first flow. The flows commute, and hence the resulting geometric object is always a flat connection on a region of R^n with special properties. From our viewpoint, the natural parameter (moduli) space of solutions is a coset space of the sort we have just described. In many cases, we have obtained global results on connections in R^n via the decay theorems in section 7.

The n commuting first flows associated to a rank n symmetric space have been discussed in a paper by the first author ([Te2]). We outline the general theory and give some of the basic examples. The results on coset spaces and Bäcklund transformations apply naturally to these systems.

12.1 Definition ([Te2]). Let U be a rank n Lie group, $\mathcal{T}$ a maximal abelian subalgebra of the Lie algebra $\mathcal{U}$, and $a_1, \ldots, a_n$ a basis of $\mathcal{T}$. *The n-dimensional system associated to U* is the following first order system:

$$[a_i, v_{x_j}] - [a_j, v_{x_i}] = [[a_i, v], [a_j, v]], \qquad v : R^n \to \mathcal{T}^\perp. \tag{12.1}$$

12.2 Definition ([Te2]). Let U/K be a rank n symmetric space, $\sigma : \mathcal{U} \to \mathcal{U}$ the corresponding involution, $\mathcal{U} = \mathcal{K} + \mathcal{P}$ the Cartan decomposition, $\mathcal{A}$ a maximal abelian subalgebra in $\mathcal{P}$, and $a_1, \ldots, a_n$ a basis of $\mathcal{A}$. *The n-dimensional system associated to U/K* is the first order system:

$$[a_i, v_{x_j}] - [a_j, v_{x_i}] = [[a_i, v], [a_j, v]], \qquad v : R^n \to \mathcal{P} \cap \mathcal{A}^\perp. \tag{12.2}$$

12.3 Theorem ([Te2]). *The following conditions are equivalent:*

(i) v is a solution of equation (12.1) (or (12.2))

(ii) *the connection* 1*-form* $\theta = \sum_{j=1}^{n}(a_j\lambda + [a_j, v])\, dx_j$ *is left flat, i.e.,* $d\theta = -\theta \wedge \theta$,

(iii) $\left[\frac{\partial}{\partial x_i} + (a_i\lambda + [a_i, v]), \frac{\partial}{\partial x_j} + (a_j\lambda + [a_j, v])\right] = 0$ *for all* $i \neq j$.

12.4 Theorem ([Te2]). *Let* $\mathcal{T}$ *be a maximal abelian subalgebra of* $\mathcal{U}$, $\mathcal{T}^{\perp}$ *the orthogonal complement of* $\mathcal{T}$ *in* $\mathcal{U}$, *and let* $\mathcal{S}(R^n, \mathcal{T}^{\perp})$ *denote the space of Schwartz maps from* R^n *to* $\mathcal{T}^{\perp}$. *Let* $a_1, \dots, a_n$ *be regular elements and form a basis of* $\mathcal{T}$. *Then there exists a dense open subset* $\mathcal{S}_0$ *of* $\mathcal{S}(R^n, \mathcal{T}^{\perp})$ *such that given* $v_0 \in \mathcal{S}_0$, *the following Cauchy problem for equation (12.1) has a unique solution:*

$$\begin{cases} [a_i, v_{x_j}] - [a_j, v_{x_i}] = [[a_i, v], [a_j, v]], & \text{if } i \neq j, \\ v(t, 0, \dots, 0) = v_0(t). \end{cases}$$

At this point, it is important to give some explaination and application. Because these flows all commute, a change of basis in the abelian subalgebra $\mathcal{T}$ or $\mathcal{A}$ can be represented by composition with an element of $GL(n, R)$. So we might as well assume that $a_1, \dots, a_n$ are generic or regular (have distinct eigenvalues). Starting at $0 \in R^n$, given an element in the coset space D_-/H_-, we can solve for $E(x_1, 0, \dots, 0, \lambda)$ and find $\frac{d}{dx} + \lambda a_1 + u(x_1, 0, \dots, 0)$ as if we were solving for initial Cauchy data. Instead of going to one of the heirarchy of flows, we solve for the entire family of first flows in variables $(x_1, \dots, x_n)$. This gives us a map

$$D_-/H_- \mapsto \{\text{flat connections on a region of } R^n\}.$$

In the case that our flows can be embedded in the unitary flows,

$$D_-/H_- \mapsto \{\text{flat connections on } R^n \text{ decaying at } \infty\}$$

(by Theorem 7.16). This gives a proof of Theorem 12.4.

12.5 Remark. Note that we are constructing a more rigid structure then a flat connection. We are actually constructing special connection one-forms, and we do not allow arbitray coordinate change or gauge changes in the theory.

By expressing the parameter space in this form, we have made a beginning towards thinking about the natural symplectic structure on this solution space. In the case of one-dimensional Cauchy data, the symplectic structures were averaged out over the one-parameter flow. Here we need to average them out over an n-dimensional flow to obtain a natural structure.

The canonical examples of these flat connections are quite easy to describe.

12.6 Examples.

Example (i) Let $U = GL(n, R)$, $\mathcal{T}$ the maximal abelian subalgebra of diagonal matrices, and $\{e_{11}, \dots, e_{nn}\}$ a basis of $\mathcal{T}$, where e_{ij} denote the matrix in $gl(n)$ all whose entries are zero except that the ij-th entry is equal to 1. Then the n-dimensional system associated to $GL(n,R)$ is the system for

$$f = (f_{ij}) : R^n \to gl(n,R), \quad f_{ii} = 0,\ 1 \le i \le n$$

$$\begin{cases} (f_{ij})_{x_i} + (f_{ij})_{x_j} + \sum_k f_{ik} f_{kj} = 0, & \text{if } i \ne j, \\ (f_{ij})_{x_k} = f_{ik} f_{kj}, & \text{if } i,\ j,\ k \text{ are distinct.} \end{cases} \tag{12.3}$$

Example (ii) Let $U/K = GL(n,R)/SO(n)$, and $\mathcal{U} = \mathcal{K} + \mathcal{P}$ the corresponding Cartan decomposition. Then $\mathcal{P}$ is the set of all real symmetric $n \times n$ matrices, the space $\mathcal{A}$ of all diagonal matrices is a maximal abelian subalgebra in $\mathcal{P}$, $e_{11}, \dots, e_{nn}$ form a basis of $\mathcal{A}$, and $\mathcal{P} \cap \mathcal{A}^\perp$ is the space $gl_s(n)$ of all symmetric $n \times n$ matrices whose diagonal entries are zero. The n dimensional system (12.2) associated to $GL(n,R)/SO(n)$ is the system for

$$F = (f_{ij}) : R^n \to gl(n,R), \quad f_{ij} = f_{ji}, \quad f_{ii} = 0 \text{ if } 1 \le i \le n$$

$$\begin{cases} (f_{ij})_{x_i} + (f_{ij})_{x_j} + \sum_k f_{ik} f_{kj} = 0, & \text{if } i \ne j, \\ (f_{ij})_{x_k} = f_{ik} f_{kj}, & \text{if } i,\ j,\ k \text{ are distinct.} \end{cases} \tag{12.4}$$

Note that system (12.4) is the system (12.3) restricted to maps $f = (f_{ij})$ that are symmetric.

Example (iii) Let $U/K = U(n)/SO(n)$, and $u(n) = so(n) + \mathcal{P}$ a Cartan decomposition. Then $\mathcal{P}$ is the set of all symmetric pure imaginary $n \times n$ matrices and the space $\mathcal{A}$ of all diagonal matrices in $\mathcal{P}$ is a maximal abelian algebra. Let $ia_1, \dots, ia_n$ be a basis of $\mathcal{A}$. Write $v : R^n \to \mathcal{P} \cap \mathcal{A}^\perp$ as $v = -iF$, where F is a real $n \times n$ symmetric matrix. Then equation (12.2) for v is the equation (12.4) for F. This is a special case of the general fact that the n-dimensional system associated to a compact symmetric space is the same as that associated to its non-compact dual.

Example (iv) Let $U/K = SO(2n)/S(O(n) \times O(n))$, and $\mathcal{U} = \mathcal{K} + \mathcal{P}$ the corresponding Cartan decomposition. Then

$$\mathcal{K} = so(n) \times so(n) = \left\{ \begin{pmatrix} B & 0 \\ 0 & C \end{pmatrix} \,\middle|\, B, C \in so(n) \right\},$$

$$\mathcal{P} = \left\{ \begin{pmatrix} 0 & F \\ -F^t & 0 \end{pmatrix} \,\middle|\, F \in gl(n) \right\},$$

and

$$\mathcal{A} = \left\{ \begin{pmatrix} 0 & -D \\ D & 0 \end{pmatrix} \,\middle|\, D \text{ is diagonal} \right\}$$

is a maximal abelian subalgebra of $\mathcal{P}$. Let $a_i = \begin{pmatrix} 0 & -e_{ii} \\ e_{ii} & 0 \end{pmatrix}$. Then $a_1, \dots, a_n$ form a basis of $\mathcal{A}$, and $\mathcal{P} \cap \mathcal{A}^\perp$ is the set of matrices of the form $\begin{pmatrix} 0 & X \\ -X^t & 0 \end{pmatrix}$

such that $X = (x_{ij})$ is $n \times n$ matrix with $x_{ii} = 0$ for all i. Then equation (12.2) for $v = \begin{pmatrix} 0 & F \\ -F^t & 0 \end{pmatrix}$, with $F = (f_{ij}) : R^n \to gl(n, R)$ and $f_{ii} = 0$ for all $1 \leq i \leq n$, is

$$\begin{cases} (f_{ij})_{x_i} + (f_{ji})_{x_j} + \sum_k f_{ki} f_{kj} = 0, & \text{if } i \neq j \\ (f_{ij})_{x_j} + (f_{ji})_{x_i} + \sum_k f_{ik} f_{jk} = 0, & \text{if } i \neq j \\ (f_{ij})_{x_k} = f_{ik} f_{kj}, & \text{if } i,\ j,\ k \text{ are distinct.} \end{cases} \tag{12.5}$$

Up to now, the flat connections were not constructed to relate to Riemannian geometry. To explain the relation of these flat connections to geometry, we need to set up some notations. A *diagonal metric* is a metric of the form

$$ds^2 = \sum_j b_j(x)^2 dx_j^2.$$

If this diagonal metric is flat, then $(x_1, \ldots, x_n)$ is an orthogonal coordinate system on R^n in the sense of Darboux ([Da2]). These examples arise in the study of isometric immersions of constant sectional curvature n-manifolds into Euclidean space.

On the other hand, to study Lagrangian flat submanifolds in C^n or Frobenius manifolds (used in quantum cohomology), we consider *Egoroff metrics*. These are metrics of the form

$$ds^2 = \sum_j \phi_{x_j} dx_j^2$$

for some function ϕ.

The Levi-Civita connection 1-form for the diagonal metric $\sum_{i=1}^n b_i(x)^2 dx_i^2$ is

$$w = (w_{ij}) = (-f_{ij} dx_i + f_{ji} dx_j), \qquad f_{ij} = \frac{(b_i)_{x_j}}{b_j}.$$

Or equivalently,

$$w = -\delta F + F^t \delta, \qquad \text{where } \delta = \operatorname{diag}(dx_1, \ldots, dx_n).$$

Hence we are looking for flat connections o this special form.

The Levi-Civita connection of a Egoroff metric is $w = [F, \delta]$ with $F = F^t$. It is easy to see that a diagonal metric is *Egoroff* if and only if $f_{ij} = f_{ji}$.

12.7 Definition. A *Darboux connection* is a connection of the form $-\delta F + F^t \delta$, and an *Egoroff connection* is a connection of the form $[F, \delta]$ with F symmetric, where $\delta = \operatorname{diag}(dx_1, \ldots, dx_n)$.

By definition of flatness, we get

12.8 Proposition. *A Darboux connection $-\delta F + F^t\delta$ is flat if and only if $F = (f_{ij})$ satisfies*

$$\begin{cases} (f_{ij})_{x_j} + (f_{ji})_{x_i} + \sum_k f_{ik}f_{jk} = 0, & \text{if } i \neq j \\ (f_{ij})_{x_k} = f_{ik}f_{kj}, & \text{if } i,\ j,\ k \text{ are distinct.} \end{cases} \tag{12.6}$$

An Egoroff connection $[F, \delta]$ (with $F = F^t$) is flat if and only if $F = (f_{ij})$ is a solution of equation (12.4), the n-dimensional system associated to the symmetric space $GL(n, R)/SO(n)$.

Let $w = -\delta F + F^t\delta$ be a flat Darboux connection. Then a metric $ds^2 = \sum_j b_j^2(x)dx_j^2$ has w as its Levi-Civita connection if and only if $(b_1, \dots, b_n)$ is a solution of

$$\frac{(b_i)_{x_j}}{b_j} = f_{ij}, \quad i \neq j. \tag{12.7}$$

In general, given a solution $F = (f_{ij})$ of equation (12.6), equation (12.7) has infinitely many local solutions parametrized by n functions b_i defined on the line $x_j = 0$ for $j \neq i$. These are used as the initial conditions for the ordinary differential equations (12.7).

Next we will explain the relation between the space of solutions of equation (12.6) and the set of flat n-submanifolds in R^{2n} with flat normal bundle and maximal rank. First we need the following definition:

12.9 Definition. The rank of a submanifold M^n of R^m at $x \in M$ is the dimension of the space of shape operators at $x \in M$. M is said to have constant rank k if the rank of M at x is equal to k for all $x \in M$. In general, the rank k of M at x is less than or equal to the codimension of M in R^m.

Using the local theory of submanifolds, it is easy to see that (cf. [Te2]) if M^n is a flat submanifold of R^{2n} with flat normal bundle and constant rank n, then locally there exist a coordinate system $x : \mathcal{O} \to M \subset R^{2n}$, $A = (a_{ij}) : \mathcal{O} \to O(n)$, $b_i : \mathcal{O} \to R$ and parallel normal frame $\{e_{n+1}, \dots, e_{2n}\}$ such that the two fundamental forms are:

$$\begin{cases} I = \sum_i b_i(x)^2 dx_i^2, \\ II = \sum_{i,j=1}^n a_{ji} b_i dx_i^2 e_{n+j}. \end{cases}$$

The coordinate system x is unique up to permutation and changing x_i to $-x_i$ (i.e., the action of the Weyl group B_n). Such coordinates will be called *principal curvature coordinates*, and (b, A) will be called the *fundamental data* of M.

12.10 Theorem ([Te2]). *Suppose M^n is flat submanifold of R^{2n} with flat normal bundle and constant rank n, x is a principal curvature coordinate system, and (b, A) is the fundamental data of M. Set*

$$f_{ij} = \begin{cases} (b_i)_{x_j}/b_j, & \text{if } i \neq j, \\ 0, & \text{if } i = j. \end{cases}$$

Then $F = (f_{ij})$ is a solution of equation (12.5), the system associated to the rank n symmetric space $SO(2n)/S(O(n) \times O(n))$. Conversely, if $F = (f_{ij})$ is a solution of equation (12.5), then there exist an open subset $\mathcal{O}$ of R^n, $b : \mathcal{O} \to R^n$, $A : \mathcal{O} \to O(n)$ and an immersion $X : \mathcal{O} \to R^{2n}$ such that

$$\begin{cases} dA = A(-F\delta + \delta F^t), & \text{where } \delta = \operatorname{diag}(dx_1, \dots, dx_n) \\ (b_i)_{x_j} = f_{ij} b_j, & \text{if } i \neq j, \end{cases} \tag{12.8}$$

and

(i) *the immersion X is flat, has flat normal bundle and constant rank n,*

(ii) *x is a principal curvature coordinate system for $X(\mathcal{O})$ and (b, A) is its fundamental data,*

(iii) *given any constants $c_1, \dots, c_n$ and set $b_i = \sum_j c_j a_{ji}$ for $1 \leq i \leq n$. Then $(b_1, \dots, b_n)$ is a solution of the second equation of system (12.8),*

(iv) *let $b = (a_{11}, \dots, a_{1n})$, then $X(\mathcal{O}) \subset S^{2n-1}$.*

12.11 Remark. If $F = F^t$ is a solution of equation (12.4), then F is a solution of equation (12.5). Let A be as in Theorem 12.10 and $b = (a_{11}, \dots, a_{1n})$, and $X : \mathcal{O} \to R^{2n}$ the corresponding immersion. It was observed by Dajczer and Tojeiro [DaR2] that the condition $F = F^t$ is equivalent to the condition that $X(\mathcal{O})$ is a Lagrangian flat submanifold of $R^{2n} = C^n$.

12.12 Remark. Let $N^n(c)$ denote the n-dimensional space form of sectional curvature c. It was proved by Cartan ([Ca]) that if $c < c'$ then $N^n(c)$ can not be locally isometrically embedded in $N^m(c')$ when $m < 2n - 1$, but can if $m \geq 2n - 1$. An analogue of Bäcklund's theorem for immersions of $N^n(c)$ into $N^{2n-1}(c')$ was constructed by Tenenblat and the first author [TT] for $c = -1$ and by Tenenblat [Ten] for $c = 0$. The corresponding Gauss-Codazzi equations for these immersions are called the *generalized sine-Gordon equation (GSGE)* and *generalized wave equation GWE* respectively. GSGE and GWE arise as the n-dimensional system associated to the symmetric spaces $SO(2n,1)/SO(n) \times SO(n,1)$ and $SO(2n)/S(O(n) \times O(n))$ respectively (cf. [Te2]). Du ([Du]) noted that the equation for isometric immersions of $N^k(c)$ in $N^m(c')$ is the k-dimensional system associated to a suitable rank k symmetric space. For example, the equation for immersions of R^k into S^n, $n \geq 2k - 1$, is the k-dimensional system associated to $Gr(k, R^{n+1})$. Du also proved that the Bäcklund transformations constructed in [TT], [Ten], and Ribaucour transformations constructed in [DaR1] are given by actions of certain order two elements in G_-^m.

Darboux' orthogonal coordinate systems arises naturally in the work of Dubrovin and Novikov ([DN1, 2]) and Tsarev ([Ts]) on Hamiltonian system of hydrodynamic type. A brief review of their results follows. Given a smooth section P of $L(TR^n, TR^n)$, the following first order quasi-linear system for $\gamma : R^2 \to R^n$

$$\frac{\partial \gamma}{\partial t} = P(\gamma)\left(\frac{\partial \gamma}{\partial x}\right) \tag{12.9}$$

is called a *hydrodynamic system*. If $(u_1, \ldots, u_n)$ is a local coordinate system on R^n, then $P(u)\left(\frac{\partial}{\partial u_i}\right) = \sum_{j=1}^n v_{ij}(u)\frac{\partial}{\partial u_j}$ for some smooth map $v = (v_{ij}) : R^n \to gl(n)$. System (12.9) is said to be *diagonalizable* if given any point $x \in R^n$ there is a local coordinate u around x such that the corresponding matrix map v for the smooth section P is diagonal. Let ds_0^2 denote the standard flat metric on R^n, and ∇ its Levi-Civita connection. Given two functionals F and G on $\mathcal{S}(R, R^n)$,

$$\{F, G\}(\gamma) = \int_{-\infty}^{\infty} (\delta F(\gamma), \nabla_{\gamma_x}(\delta G(\gamma))\, dx$$

defines a Poisson structure on $\mathcal{S}(R, R^n)$. Dubrovin and Novikov ([DN1], [DN2]) proved that this is the only Poisson structure on $\mathcal{S}(R, R^n)$ that is given by a first order differential operator. Given a zero order Lagrangian F with density $f : R^n \to R$, i.e.,

$$F(\gamma) = \int_{-\infty}^{\infty} f(\gamma(x))\, dx,$$

the Hamiltonian equation with respect to the Poisson structure defined above is

$$\frac{\partial \gamma}{\partial t} = \nabla_{\gamma_x}(\nabla f(\gamma)). \tag{12.10}$$

Such system is called *Hamiltonian system of hydrodynamic type*. Novikov conjectured that if system (12.10) is diagonalizable then it is completely integrable. This conjecture is proved by Tsarev in [Ts]. In these results, the boundary conditions for the Poisson bracket are ignored and equation (12.10) is defined on an open subset of R^n. In other words, this is the local theory of Hamiltonian hydrodynamic systems. Below we state some of Tsarev's results:

12.13 Theorem ([Ts]). *Suppose the Hamiltonian system for*

$$F(\gamma) = \int_{-\infty}^{\infty} f(\gamma(x))\, dx$$

is diagonalizable with respect to a local coordinate system $(u_1, \ldots, u_n)$. *Then* $\nabla^2 f = \sum_i v_i(u) du_i \otimes \frac{\partial}{\partial u_i}$, *and the Hamiltonian system (12.10) is*

$$\frac{\partial u_i}{\partial t} = v_i(u)\frac{\partial u_i}{\partial x}. \tag{12.11}$$

Moreover:

(i) $(u_1, \dots, u_n)$ *is a local orthogonal coordinate system on* R^n*, i.e., the standard metric* ds_0^2 *on* R^n *is* $\sum_{i=1}^n b_i(u)^2 du_i^2$ *for some smooth functions* $b_1, \dots, b_n$*. Moreover,*

$$\frac{1}{v_i - v_j}\frac{\partial v_i}{\partial u_j} = -\frac{1}{b_j}\frac{\partial b_i}{\partial u_j}. \tag{12.12}$$

(ii) If $(\tilde{v}_1, \dots, \tilde{v}_n)$ *is a solution of system (12.12), then* $\frac{\partial u_i}{\partial t} = \tilde{v}_i(u)\frac{\partial u_i}{\partial x}$ *is also a Hamiltonian system of hydrodynamic type.*

(iii) Suppose $\sum_{i=1}^n \tilde{b}_i^2(u)du_i^2$ *and* $\sum_{i=1}^n b_i(u)^2 du_i^2$ *have the same connection 1-form, i.e.,* $(b_i)_{u_j}/b_j = (\tilde{b}_i)_{u_j}/\tilde{b}_j$ *for all* $i \neq j$*. Set* $h_i = \tilde{b}_i/b_i$*. Then*

$$\frac{\partial u_i}{\partial t} = h_i(u)\frac{\partial u_i}{\partial x} \tag{12.13}$$

is a Hamiltonian system of hydrodynamic type and commutes with system (12.11).

(iv) If $v_i, \dots, v_n$ *are distinct, then system (12.11) is completely integrable.*

To end this section, we review some of the elementary relations between Dubrovin's Frobenius manifolds ([Dub2]) and the n-dimensional system (12.4) associated to the symmetric space $GL(n)/SO(n)$. For more deep and detailed results of Frobenius manifolds, we refer the reader to Dubrovin's article [Dub2].

12.14 Definition ([Dub2], [Hi2]). A *Frobenius manifold* of degree m (not necessarily an integer) is a quintuple (R^n, x, g, θ, ξ), where x is a coordinate system on R^n, $g = \sum_j \phi_{x_j} dx_j^2$ a flat Egoroff metric, $\theta = \sum_j \phi_{x_j} dx_j$ and $\xi = \sum_j \phi_{x_j} dx_j^3$ for some function ϕ such that ϕ, g, θ, ξ satisfy the following conditions:

(i) $\nabla\theta = 0$, where ∇ is the Levi-Civita connection of g,

(ii) $\nabla\xi$ is a symmetric 4 tensor,

(iii) ϕ_{x_j} is homogeneous of degree m for all j, i.e., $\phi_{x_j}(rx) = r^m \phi_{x_j}(x)$ for all $r \in R^*$ and $x \in R^n$.

The coordinate system x is called a *canonical coordinate system*.

Each tangent plane of the Frobenius manifold has a natural multiplication defined as follows: Set $v_i = \partial/\partial x_i$. Then

$$v_i v_j = v_j v_i = \delta_{ij} v_i, \quad \forall\ 1 \leq i, j \leq n$$

defines a multiplication on the tangent plane of R^n. Moreover, $T(R^n)_x$ is a commutative algebra and

$$\theta(uv) = g(u, v), \qquad \xi(u, v, w) = g(uvw). \tag{12.14}$$

The dual of the 1-form θ is $e = \sum_j \partial/\partial x_j$, which is the identity, i.e., $ve = ev = v$ for all $v \in T(R^n)_x$.

The following Proposition gives the relation between Frobenius manifolds and solutions of the n-dimensional system associated to $GL(n)/SU(n)$.

12.15 Proposition ([Dub2], [Hi2]). *Let (R^n, x, g, θ, ξ) be a Frobenius manifold of degree m, $g = \sum_j b_j^2 dx_j^2$, and $w_{ij} = f_{ij}(-dx_i + dx_j)$ the Levi-Civita connection of g, i.e., $b_i^2 = \phi_{x_i}$ and $f_{ij} = (b_i)_{x_j}/b_j$. Set*

$$F = (f_{ij}), \qquad S(x) = (S_{ij}(x)) = (f_{ij}(x)(x_i - x_j)).$$

Then

(i) $F = (f_{ij})$ is a solution of equation (12.4), the n-dimensional system associated to $GL(n)/SO(n)$,

(ii) F is invariant under the action of R^* defined in section 9, i.e.,

$$r \cdot F(x) = r^{-1} F(r^{-1}x), \text{for all } r \neq 0.$$

(iii) $\frac{\partial S}{\partial x_i} = [[F, e_{ii}], S]$, where e_{ii} is the diagonal matrix with all entries zero except the ii-th entry is 1,

(iv) $(b_1, \ldots, b_n)^t$ is an eigenvector of the matrix (S_{ij}) with eigenvalue $m/2$.

Since Frobenius manifolds are flat, there are also coordinate systems such that all the coordinate vector fields are covariant constant. A coordinate system $(t_1, \ldots, t_n)$ on a Frobenius manifold (R^n, x, g, θ, ξ) is called a *flat coordinate system* if g has constant coefficients with respect to the t-coordinates. Since $\nabla\theta = 0$ and e is the dual of θ, we have $\nabla e = 0$. So there exists a flat coordinates $(t_1, \ldots, t_n)$ such that $e = \partial/\partial t_1$. It follows from the condition that $\nabla\xi$ is symmetric, there exists a function $h(t)$ such that

$$\begin{cases} g = \sum_{jk} h_{t_1 t_j t_k} dt_j dt_k, \\ \theta = dt_1, \\ \xi = \sum_{ijk} h_{t_i t_j t_k} dt_i dt_j dt_k. \end{cases}$$

Using condition (12.14) and the fact that $\partial/\partial t_1$ is the identity, the multiplication can be written down explicitly in terms of h. In order for this multiplication to be associative, h has to satisfy a complicated non-linear equation, which is the WDVV (Witten-Dijkgraaf-Verlinde-Verlinde) equation. The WDVV equation arises in the study of Gromov-Witten invariants, and we refer the readers to work of Dubrovin ([Dub2]) and Ruan and Tian ([RT]).

Bibliography

[AC] Ablowitz, M.J.,Clarkson, P.A., *Solitons, non-linear evolution equations and inverse scattering.* Cambridge Univ. Press, 1991.

[AKNS1] Ablowitz, M.J., Kaup, D.J., Newell, A.C. and Segur, H., Method for solving the sine-Gordon equation, *Phy. Rev. Lett.* **30** (1973), 1262-1264.

[AKNS2] Ablowitz, M.J., Kaup, D.J., Newell, A.C. and Segur, H., The inverse scattering transform - Fourier analysis for nonlinear problems, *Stud. Appl. Math.* **53** (1974), 249-315.

[AbM] Abraham, R., Marsden, J.E., *Foundations of mechanics.* Benjamin/Cummings, 1978.

[Ad] Adler, M., On a trace functional for formal pseudo-differential operators and the symplectic structure of the Korteweg-de Vries Type Equations, *Invent. Math.* **50** (1979), 219-248.

[AdM] Adler, M., van Moerbeke, P., Completely integrable systems, Euclidean Lie algebras and curves, *Adv. Math.* **38** (1980), 267-317.

[Ar] Arnold, V.I., *Mathematical methods of classical mechanics.* Springer-Verlag, 1978.

[Ba] Bäcklund, A.V., *Concerning surfaces with constant negative curvature.* New Era Printing Co., Lancaster, PA, . original 1883, translation 1905

[BC1] Beals, R., Coifman, R.R., Scattering and inverse scattering for first order systems, *Commun. Pure Appl. Math.* **37** (1984), 39-90.

[BC2] Beals, R., Coifman, R.R., Inverse scattering and evolution equations, *Commun. Pure Appl. Math.* **38** (1985), 29-42.

[BC3] Beals, R., Coifman, R.R., Linear spectral problems, non-linear equations and the $\bar{\partial}$-method, *Inverse Problems* **5** (1989), 87-130.

[BDZ] Beals, R., Deift, P., Zhou, X., The inverse scattering transform on the line, *Important development in soliton theory* (1993), 7-32. Springer Ser. Nonlinear Dynam.

[BS] Beals, R., Sattinger. D.H., On the complete integrability of complete integrable systems, *Commun. Math. Phys.* **138** (1991), 409-436.

[BMW] Birnir, B., McKean, H., Weinstein, A., The rigidity of sine-Gordon breathers, *Comm. Pure Appl. Math.* **47** (1994), 1043-1051.

[Bo] Bobenko, A.I., All constant mean curvature tori in R^3, S^3, H^3 in terms of theta functions, *Math. Ann.* **290** (1991), 209-245.

[Bu] Budagov, Completely integrable model of classical field theory with nontrivial particle interaction in two-dimensional space-time, *Dokl. Akad. Nauk SSSR* **235** (1977), 24-56. (in Russian)

[BuC] Bullough, R.K., Caudrey, P.J., *Solitons.* Topics in Current Physics, vol. 117, Springer-Verlag, 1980.

[BFPP] Burstall, F.E., Ferus, D., Pedit, F., Pinkall, U., Harmonic tori in symmetric spaces and commuting Hamiltonian systems on loop algebras, *Annals of Math.* **138** (1993), 173-212.

[BG] Burstall, F.E., Guest, M.A., Harmonic two-spheres in compact symmetric spaces, *Math. Ann.* **309** (1997), 541-572.

[Ca] Cartan, E., Sur les variétés de courbure constante d'un espace euclidien ou non-euclidien, *Bull. Soc. Math. France* **47** (1919), 132-208.

[Ch] Cherednik, I.V., Definition of τ-functions for generalized affine Lie algebras, *Funct. Anal. Appl.* **17** (1983), 243-244.

[CM] Chernoff, P., Marsden, J., *Properties of infinite dimensional Hamiltonian system.* Lecture Notes in Math., vol. 425, Springer-Verlag, Berlin and New York, 1974.

[dR] Da Rios, , *Rend. Circ. Mat. Palermo* **22** (1906), 117-135.

[DaR1] Dajczer,M., Tojeiro, R., An extension of the classical Ribaucour transformation, preprint

[DaR2] Dajczer,M., Tojeiro, R., The Ribaucour transformation for flat totally real submanifolds of complex space forms, preprint

[Da1] G. Darboux, *Leçons sur la théorie générale des surfaces.* Chelsea, 1972. 3rd edition

[Da2] G. Darboux, *Lecon sur les Systèmes Orthogonaux.* Gauthier-Villars, 1910. 2nd edition

[De] Denzler, J., Nonpersistence of breather families for the perturbed sine-Gordon equation, *Commun. Math. Phys.* **158** (1993), 397-430.

[DPW] Dorfmeister, J., Pedit, F., Wu, H., Weierstrass type representation of harmonic maps into symmetric spaces, preprint

[Dr] Drinfel'd, V.G., Hamiltonian structures on Lie groups, Lie bialgebras and the geometrical meaning of Yang-Baxter equations, *Sov. Math. Doklady* **27** (1983), 68-71.

[DS] Drinfel'd, V.G., and Sokolov, V.V., Equations of Korteweg-de Vries type and simple Lie algebras, *Dokl. Akad. Nauk SSSR* **258** (1981), 11-16. (Trans. as Soviet Math. Dokl. 23, 457-462

[Du] Du, X., Isometric immersions of space forms in space forms, thesis, Northeastern University, 1997

[Dub1] Dubrovin, B.A., Differential geometry of strongly integrable systems of hydrodynamic type, *Funct. Anal. Appl.* **24** (1990), 280-285.

[Dub2] Dubrovin, B.A., Geometry of 2D topological field theories, *Lecture Notes in Mathematics, vol. 1620, Springer-Verlag* (1996), .

[DN1] Dubrovin, B.A., Novikov, S.P., Hamiltonian formalism of one-dimensional systems of hydrodynamic type, and the Bogolyubov-Whitham averaging method, *Soviet Math. Dokl.* **27** (1983), 781-785.

[DN2] Dubrovin, B.A., Novikov, S.P., On Poisson brackets of hydrodynamic type, *Soviet Math. Dokl.* **30** (1984), 294-297.

[Ei] Eisenhart, L.P., *A treatise on the differential geometry of curves and surfaces.* Ginn, 1909.

[FT] Faddeev, L.D., Takhtajan, L.A., *Hamiltonian Methods in the theory of Solitons.* Springer-Verlag, 1987.

[FPPS] Ferus, D., Pedit, F., Pinkall, U., Sterling, I., Minimal tori in S^4, *J. reine angew. Math.* **429** (1992), 1-47.

[FNR1] Flaschka, H., Newell, A.C., Ratiu, T., Kac-Moody Lie algebras and soliton equations, II. Lax equations associated with $A_1^{(1)}$, *Physica* **9D** (1983), 303-323.

[FNR2] Flaschka, H., Newell, A.C., Ratiu, T., Kac-Moody Lie algebras and soliton equations, IV. Lax equations associated with $A_1^{(1)}$, *Physica* **9D** (1983), 333-345.

[FK] Fordy, A.P., Kulish, P.P., Nonlinear Schrödinger equations and simple Lie algebra, *Commun. Math. Phys.* **89** (1983), 427-443.

[FW] Fordy, A.P., Wood, J.C., *Harmonic maps and integrable systems.* Vieweg, 1994. editors

[FRS] Frenkel, I..E., Reiman, A.G., Semenov-Tian-Shansky, M.A., Graded Lie algebras and completely integrable dynamical systems, *Soviet Math. Dokl.* **20** (1979), 811-814.

[GGKM] Gardner, C.S., Greene, J.M., Kruskal, M.D., Miura, R.M., Method for solving the Korteweg-de Vries equation, *Physics Rev. Lett.* **19** (1967), 1095-1097.

[GDi] Gel'fand, I.M., Dikii, L.A., Fractional Powers of Operators and Hamiltonian Systems, *Funkcional'nyi Analiz i ego Prilozenjia* **10** (1976), 13-29.

[GDo] Gel'fand, I.M., Dorfman, I. Ya, Hamiltonian operators and algebraic structures related to them, *Functional Anal. Appl.* **13** (1979), 248-261.

[GZ] Gu, C.H., Zhou, Z.X., On Darboux transformations for soliton equations in high dimensional spacetime, *Letters in Math. Phys.* **32** (1994), 1-10.

[Gu] Guest, M., *Integrable systems and harmonic maps.* Cambridge University Press, 1997.

[Ha] Hasimoto, H., Motion of a vortex filament and its relation to elastic, *J. Phys. Soc. Jap.* **31** (1971), 293-.

[Hi1] Hitchen, N.J., Harmonic maps from a 2-torus to the 3-sphere, *J. Differential Geom.* **31** (1990), 627-710.

[Hi2] Hitchen, N.J., Frobenius manifolds, *Nato Adv. Sci. Inst. Ser. C. Math. Phys. Sci. (Gauge theory and symplectic geometry)* **488** (1997), 61-112.

[Hi3] Hitchen, N.J., Integrable systems in Riemannian geometry, preprint

[Ka] Kac, V.G., *Infinite Dimensional Lie Algebras.* Cambridge University Press, 1985.

[Kon] Konopelchenko, B.G., *Nonlinear Integrable Equations.* Lecture Notes in Physics, vol. 270, 1987.

[Kos] Kostant, B., The solution to a generalized Toda lattice and representation theory, *Adv. Math.* **34** (1979), 195-338.

[KdV] Koteweg, D.J., de Vries, G., On the change of form of long waves advancing in a rectangular canal, and on a new type of long stationary waves, *Philos. Mag. Ser. 5* **39** (1895), 422-443.

[Kr1] Krichever, I.M., Integration of non-linear equations by methods of algebraic geometry, *Funct. Anal. Appl.* **11** (1977), 12-26.

[Kr2] Krichever, I.M., Methods of algebraic geometry in the theory of non-linear equations, *Russian Math. Surveys* **32** (1977), 185-213.

[KW] Kuperschmidt, B.A., Wilson, G., Modifying Lax equations and the second Hamiltonian structure, *Invent. Math.* **62** (1981), 403-436.

[LP] Langer, J., Perline, R., Local geometric invariants of integrable evolution equations, *J. Math. Phys.* **35** (1994), 1732-1737.

[La] Lax, P.D., Integrals of nonlinear equations of evolution adn solitary waves, *Comm. Pure. Appl. Math.* **31** (1968), 467-490.

[Lu] Lu, J.-H., Momentum mappings and reduction of Poisson actions, *Symplectic geometry, groupoid, and integrable systems, Math. Sci. Res. Inst. Publ.* **20** (1991), 209-226.

[LW] Lu, J.-H., Weinstein, A., Poisson Lie groups, dressing actions, and Bruhat decompositions, *J. Differential Geom.* **31** (1990), 501-526.

[M] Magri, F., A simple model of the integrable Hamiltonian equation, *J. Math. Phys.* **19** (1978), 1156-1162.

[Ma] Manakov, S.V., An example of a completely integrable nonlinear wave field with nontrivial dynamics (Lee model), *Teor. Mat. Phys.* **28** (1976), 172-179.

[Mc] McIntosh, I., Infinite dimensional Lie groups and the two dimensional Toda lattice, *Harmonic maps and integrable systems* (1994), 205-220. Fordy, A.P., Wood, J.C. editors, Vieweg

[Mi] Miura, R.M., *Bäcklund transformations.* Lecture Notes in Mathematics, vol. 515, Springer-Verlag, 1976. editor

[N] Newell, A.C., *Solitons in Mathematics and Physics.* SIAM, CBMS-NSF vol. 48, 1985.

[NMPZ] Novikov, S., Manakov, S., Pitaevskii, L.B. Zakharov, V.E., *Theory of Solitons.* Plenum. New York, 1984.

[Pa] Palais, R.S., The symmetries of solitons, *Bulletin AMS* **34** (1997), 339-403.

[PT] Palais R.S., and Terng C.L., *Critical Point Theory and Submanifold Geometry.* Lecture Notes in Math., vol. 1353, Springer-Verlag, Berlin and New York, 1988.

[PiS] Pinkall, U., Sterling, I., On the classification of constant mean curvature tori, *Ann. of Math.* **130** (1989), 407-451.

[PrS] Pressley, A. and Segal, G. B., *Loop Groups.* Oxford Science Publ., Clarendon Press, Oxford, 1986.

[R] Reyman, A.G., Integrable Hamiltonian systems connected with graded Lie algebras, *J. Sov. Math.* **19** (1982), 1507-1545.

[RS] Reyman, A.G., Semenov-Tian-Shansky, Current algebras and non-linear partial differential equations, *Sov. Math., Dokl.* **21** (1980), 630-634.

[Ri] Ricca, R.L., Rediscovery of Da Rios equations, *Nature* **352** (1991), 561-562.

[RT] Ruan, Y., G. Tian, A mathematical theory of quantum cohomology, *J. Differential Geom.* **42** (1995), 259-367.

[Sa] Sattinger, D.H., Hamiltonian hierarchies on semi-simple Lie algebras, *Stud. Appl. Math.* **72** (1984), 65-86.

[SZ1] Sattinger, D.H.,Zurkowski, V.D., Gauge theory of Bäcklund transformations.I, *Dynamics of infinite dimensional systems, Nato Sci. Inst. Ser. F. Comput. Systems Sci.* **37** (1987), 273-300. Springer-Verlag

[SZ2] Sattinger, D.H.,Zurkowski, V.D., Gauge theory of Bäcklund transformations.II, *Physica* **26D** (1987), 225-250.

[SW] Segal, G., Wilson, G., Loop groups and equations of KdV type, *Publ. Math. IHES* **61** (1985), 5-65.

[Se1] Semenov-Tian-Shansky, M.A., Dressing transformations and Poisson group actions, *Publ. RIMS Kyoto Univ.* **21** (1985), 1237-1260.

[Se2] Semenov-Tian-Shansky, M.A., Classical r-matrices, Lax equations, Poisson Lie groups, and dressing transformations, *Lecture Notes in Physics, Springer-Verlag* **280** (1986), 174-214.

[Sh] Shabat, A.B., An inverse scattering problem, *Diff. Equ.* **15** (1980), 1299-1307.

[SS] Shatah, J., Strauss, W., Breathers as homoclinic geometric wave maps, *Physics D.* **99** (1996), 113-133.

[Sy] Symes, W.W., Systems of Toda type, Inverse spectral problems, and representation theory, *Inventiones Math.* **59** (1980), 13-51.

[Ten] Tenenblat, K., Bäcklund's theorem for submanifolds of space forms and a generalized wave equation, *Boll. Soc. Brasil. Mat.* **16** (1985), 67-92.

[TT] Tenenblat, K., Terng, C.L., Bäcklund's theorem for n-dimensional submanifolds of R^{2n-1}, *Ann. Math.* **111** (1980), 477-490.

[Te1] Terng, C.L., A higher dimensional generalization of the sine-Gordon equation and its soliton theory, *Ann. Math.* **111** (1980), 491-510.

[Te2] Terng, C.L., Soliton equations and differential geometry, *J. Differential Geom* **45** (1997), 407-445.

[TU1] Terng, C.L., Uhlenbeck, K., Bäcklund transformations and loop group actions, preprint

[TU2] Terng, C.L., Uhlenbeck, K., Homoclinic wave maps into compact symmetric spaces, in preparation

[TU3] Terng, C.L., Uhlenbeck, K., Hamiltonian theory of the geometric non-linear Schrödinger equation on a Hermitian symmetric space, in preparation

[Ts] Tsarëv, S.P., The geometry of Hamiltonian systems of hydrodynamic type. The generalized Hodograph method, *Math. USSR Izvestiya* **37** (1991), 397-419.

[U1] Uhlenbeck, K., Harmonic maps into Lie group (classical solutions of the Chiral model), *J. Differential Geometry* **30** (1989), 1-50.

[U2] Uhlenbeck, K., On the connection between harmonic maps and the self-dual Yang-Mills and the sine-Gordon equations, *J. Geom. Phys.* **8** (1992), 283-316.

[Wa] Wadati, M., The modified Korteweg-de Vries equation, *J. Phys. Soc. Japan* **34** (1973), 380-384.

[Wi] Wilson, G., The modified Lax equations and two dimensional Toda lattice equations associated with simple Lie algebras, *Ergodic Theory and Dynamical Systems I* **30** (1981), 361-380.

[ZF] Zakharov, V.E., Faddeev, L.D., Korteweg-de Vries equation, a completely integrable Hamiltonian system, *Func. Anal. Appl.* **5** (1971), 280-287.

[ZMa1] Zakharov, V.E., Manakov, S.V., On resonant interaction of wave packets in non-linear media, *JETP Letters* **18** (1973), 243-247.

[ZMa2] Zakharov, V.E., Manakov, S.V., The theory of resonant interaction of wave packets in non-linear media, *Sov. Phys. JETP* **42** (1974), 842-850.

[ZMi1] Zakharov, V.E., Mikhailov, A.V., Example of nontrivial interaction of solitons in two-dimensional classical field theory, *JETP Letters* **27** (1978), 42-46.

[ZMi2] Zakharov, V.E., Mikhailov, A.V., Relativistically invariant 2-dimensional models of field theory which are integrable by means of the inverse scattering problem method, *Soviet Physics JETP* **47** (1978), 1017-1027.

[ZS1] Zakharov, V.E., Shabat, A.B., Exact theory of two-dimensional self-focusing and one-dimensional of waves in nonlinear media, *Sov. Phys. JETP* **34** (1972), 62-69.

[ZS2] Zakharov, V.E., Shabat, A.B., Integration of non-linear equations of mathematical physics by the inverse scattering method, II, *Funct. Anal. Appl.* **13** (1979), 166-174.

[ZK] Zabusky, N.J., Kruskal, M.D., Interaction of solitons in a collisionless plasma and the recurrence of initial states, *Physics Rev. Lett.* **15** (1965), 240-243.

[Zh1] Zhou, X., Riemann Hilbert problem and inverse scattering, *SIAM J. Math. Anal.* **20** (1989), 966-986.

[Zh2] Zhou, X., Direct and inverse scattering transforms with arbitrary spectral singularities, *Comm. Pure. Appl. Math.* **42** (1989), 895-938.

Chuu-lian Terng
Department of Mathematics
Northeastern University
Boston, MA 02115
email: terng@neu.edu

Karen Uhlenbeck
Department of Mathematics
The University of Texas at Austin
RLM8.100 Austin, Texas 78712
email:uhlen@math.utexas.edu

Loop Groups and Equations of KdV Type

Graeme Segal and George Wilson
Mathematical Institute,
University of Oxford,
24–29 St. Giles,
Oxford OX1 3LB, UK

The purpose of this paper is to work out some of the implications of recent ideas of M. and Y. Sato about the Korteweg-de Vries (KdV) equation and related non-linear partial differential equations. We learned of these ideas from the papers [5] of Date, Jimbo, Kashiwara and Miwa (the original work of M. and Y. Sato appears to be available only in Japanese). We shall describe a construction which assigns a solution of the KdV equation to each point of a certain infinite dimensional Grassmannian. The class of solutions obtained in this way, which is misleadingly referred to as "the general solution" in [5], includes the explicit algebro-geometric solutions of Krichever [10, 11]; among these are the well known "n-soliton" and rational solutions.

Our main aims are to determine what class of solutions is obtained by the method, to illustrate in detail how the geometry of the Grassmannian is reflected in properties of the solutions, and to show how the algebro-geometric solutions fit into the picture. We have also tried to explain the geometric meaning of the "τ-function", which plays a fundamental role in the papers [5]. But above all we have endeavoured to present a clear and self-contained account of the theory, and hope to have elucidated a number of points left obscure in the literature.

1 Introduction

The KdV equation

$$\frac{\partial u}{\partial t} = \frac{\partial^3 u}{\partial x^3} + 6u\frac{\partial u}{\partial x}$$

This article is reprinted by permission from Communications on Pure and Applied Mathematics 37 (1984), pp. 39 - 90.

describes the time-evolution of a function u of the variable x: we think of the equation geometrically as defining a flow on a suitable space of functions u. It is well known that the theory of the equation is closely connected with that of the linear differential operator $L_u = D^2 + u$, where $D = \partial/\partial x$, which is to be regarded as an operator on functions of x which varies with time. In fact the KdV equation can be written in the "Lax form"

$$\frac{\partial L_u}{\partial t} = 4[P_u, L_u],$$

where P_u is the operator $D^3 + \frac{3}{4}(uD + Du)$.

The operator P_u is almost characterized by the fact that—for any function u—the commutator $[P_u, L_u]$ is a multiplication operator. More precisely, for given L_u there is a canonical sequence of operators

$$P_u^{(k)} = D^k + p_{k2}D^{k-2} + \ldots + p_{kk}$$

such that each $[P_u^{(k)}, L_u]$ is a multiplication operator, and any operator P with the same property is a constant linear combination of the $P_u^{(k)}$. It turns out that the coefficients of such as operator P must be differential polynomials in u, i.e. polynomials in u and its x-derivatives $\partial^j u/\partial x^j$. For each k the equation

$$\frac{\partial L_u}{\partial t} = [P_u^{(k)}, L_u] \tag{1.1}$$

defines a flow on the space of functions of x. These flows are called the "KdV hierarchy". The case $k = 3$ is the original KdV equation (apart from the factor 4). When $k = 1$ we have $P_u^{(1)} = D$, and the corresponding flow is just uniform translation of u. When k is even we have $P_u^{(k)} = (L_u)^{k/2}$, so that the corresponding flow is stationary. It is a fundamental theorem of the subject that the flows given by (1.1) for various k *commute* among themselves.

In this paper we shall describe the KdV flows on a certain class $\mathcal{C}^{(2)}$ of functions u. Our approach is in terms of the geometry of an infinite dimensional manifold which is of considerable interest in its own right. It has two alternative descriptions. The first is as the space ΩU_2 of loops in the unitary group U_2. The second, more immediately relevant, description is as the Grassmannian $\mathrm{Gr}^{(2)}$ of all closed subspaces W of the Hilbert space $H = L^2(S^1)$ of square-summable complex-valued functions on the circle $S^1 = \{z \in \mathbf{C} : |z| = 1\}$ which satisfy the two conditions

(i) $z^2 W \subset W$, and

(ii) W is comparable with H_+.

Here z denotes the operator $H \to H$ given by multiplication by the function z on S^1, and H_+ is the closed subspace of H spanned by $\{z^k\}$ for $k \geq 0$, i.e. the boundary values of holomorphic functions in $|z| < 1$. The meaning of "comparable" is explained in §2.

Our basic construction associates to each point W in a connected component of $\mathrm{Gr}^{(2)}$ a meromorphic function u_W on the line, belonging to the class $\mathcal{C}^{(2)}$. The group Γ_+ of holomorphic maps $D_0 \to \mathbb{C}^\times$, where D_0 is the disc $\{z \in \mathbb{C} : z \leq 1\}$, acts by multiplication operators on H, and hence acts on $\mathrm{Gr}^{(2)}$. The action of Γ_+ induces the KdV flows on $\mathcal{C}^{(2)}$ in the following sense: if

$$g = \exp \sum t_k z^k \in \Gamma_+,$$

where $(t_1, t_2, \dots)$ are real numbers almost all zero, then u_{gW} is the function obtained from u_W by letting it flow for time t_k along the k-th KdV flow, for each k. (This makes sense precisely because the KdV flows commute.)

The meromorphic function u_W is obtained from the so-called "τ-function" τ_W of W by the formula

$$u_W(x) = 2\left(\frac{d}{dx}\right)^2 \log \tau_W(x);$$

$\tau_W(x)$ is the determinant of the orthogonal projection $e^{-xz}W \to H_+$. Of course the determinant needs to be suitably interpreted. To define it one must choose bases in W and H_+, and accordingly $\tau_W(x)$ is defined only up to a multiplier independent of x. The determinant $\tau_W(x)$ vanishes, and hence $u_W(x)$ has a pole, precisely when $e^{-xz}W$ intersects $H_+^\perp$.

For certain particular subspaces W belonging to the Grassmannian it turns out that the τ-function is a Schur function. This was discovered by Sato, and it was, we have been told, the observation that led him to develop his theory. In general a point of the Grassmannian can be described by its Plücker coordinates, and (as we shall prove in §8) the corresponding τ-function is an infinite linear combination of Schur functions with the Plücker coordinates as coefficients.

It is not practical, however, in developing the theory, to pass directly from the τ-function to the function u_W. Instead, one introduces an intermediate object, the *Baker function* ψ_W. This is an eigenfunction of the operator $D^2 + u_W$:

$$(D^2 + u_W)\psi_W(x, z) = z^2 \psi_W(x, z);$$

on the other hand for each fixed x it is the unique element of W which is of the form

$$e^{xz}(1 + a_1(x)z^{-1} + a_2(x)z^{-2} + \dots). \tag{1.2}$$

Finding the formula ((5.14) below) for the Baker function in terms of the τ-function was one of the most important contributions of the Japanese school. The formula is a precise analogue of a formula known earlier, in the case of solutions arising from an algebraic curve, expressing the Baker function in terms of θ-functions.

At this point we should say something about the class $\mathcal{C}^{(2)}$ of solutions u_W which we obtain. Suppose to begin with that u is a $\mathbb{C}^\infty$ function defined in an interval of $\mathbb{R}$. Then the eigenvalue problem $L_u\psi = z^2\psi$ has a *formal* solution of the form (1.2). The coefficients a_i in the formal series are $\mathbb{C}^\infty$ functions determined recursively by

$$-2a_i' = L_u a_{i-1},$$

with $a_0 = 1$. Each successive a_i involves a new constant of integration: this means that ψ is determined up to multiplication by an arbitrary power series in z^{-1} with constant coefficients. The series (1.2) will usually not converge for any values of z. The class of functions $\mathcal{C}^{(2)}$ is, roughly speaking, those such that it can be chosen convergent in a neighbourhood of $z = \infty$. To see how restrictive this is, consider the case of functions u which are rapidly decreasing as $x \to \pm\infty$. Then there are unique genuine solutions $\psi_+(x,z)$ and $\psi_-(x,z)$ of $L_u\psi = z^2\psi$, defined and holomorphic in z for $\mathrm{Re}(z) > 0$ and $\mathrm{Re}(z) < 0$ respectively, characterized by the properties

$$\psi_+(x,z) \sim e^{xz} \text{ as } x \to -\infty,$$
$$\psi_-(x,z) \sim e^{xz} \text{ as } x \to +\infty,$$

These solutions both extend to the axis $\mathrm{Re}(z) = 0$, but unless u belongs to the exceptional class of so-called "reflectionless potentials" or "multisolitons" they will be linearly independent functions of x, and then no genuine solution of the form (1.2) can exist. The situation is similar if we consider the case where u is a real $\mathbb{C}^\infty$ periodic function: of these, our class $\mathcal{C}^{(2)}$ contains only the "finite gap" potentials u. The periodic KdV flows have been described by McKean and Trubowitz [25] in terms of Riemann surfaces of (in general) infinite genus: the finite gap potentials are precisely those for which the Riemann surface involved is of finite genus. The corresponding solutions to the KdV hierarchy are then included in the class obtained by Krichever's method.

We next explain how Krichever's construction is included in ours. Krichever associates a function of x, say $u_{X,\mathcal{L}}$, to an algebraic curve X with a distinguished point x_∞ and a line bundle $\mathcal{L}$ (and also some additional data which we shall overlook in this introduction). A solutions of the KdV equation is obtained by letting $\mathcal{L}$ move along a straight line in the Jacobian of X. We shall see that a space $W \in \mathrm{Gr}^{(2)}$ is naturally associated to Krichever's data. Think of the circle S^1 as a small circle around the point x_∞ of X; then W consists of those functions on S^1 which are boundary values of holomorphic sections of $\mathcal{L}$ outside S^1—we suppose that $\mathcal{L}$ has been trivialized near x_∞. Krichever's solution $u_{X,\mathcal{L}}$ is simply u_W. When the curve X is non-singular, we shall show in §9 that the τ-function τ_W is essentially the θ-function of X.

The algebro-geometric solutions u are precisely those such that the operator L_u commutes with an operator of odd order. There is a very elegant theory, due essentially to Burchnall and Chaundy [4], relating commutative rings of differential operators to algebraic curves. A modern treatment of the subject has been given by Mumford [16]; but as it fits very naturally into our framework we have included a short self-contained account in §6.

We shall describe in particular detail the KdV flows on the two dense subspaces $\mathrm{Gr}_0^{(2)}$, and $\mathrm{Gr}_1^{(2)}$ of the Grassmannian corresponding respectively to polynomial and rational loops in U_2. The first spaces corresponds exactly to the rational solutions of the KdV equations which are zero at ∞. It is a beautiful fact that the orbits of the group Γ_+ of KdV flows on $\mathrm{Gr}_0^{(2)}$ form a cell decomposition of $\mathrm{Gr}_0^{(2)}$, with one cell of each complex dimension. (The n-th cell is the orbit of the function $-n(n+1)/x^2$.)

The points of $\mathrm{Gr}_1^{(2)}$ are those that arise by Krichever's construction from rational curves with singularities. For any $W \in \mathrm{Gr}_1^{(2)}$ the orbit of W under Γ_+ can be identified with the Jacobian of the corresponding curve.

The KdV hierarchy has fairly obvious generalizations in which the operator D^2+u is replaced by an operator of order n: these hierarchies are related to the loop space of U_n in the same way that the KdV equations are related to ΩU_2. For simplicity of explanation we have restricted ourselves in the introduction to the case $n = 2$, but in the body of the paper we shall always treat the general case, which presents no additional difficulty. In fact we shall treat a more general hierarchy still, that of the "Kadomtsev-Petviashvili" (KP) equations; the hierarchies already mentioned are all specializations of this. Less obvious are the generalizations of the KdV hierarchy due to Drinfel'd and Sokolov [6], in which, roughly speaking, U_n is replaced by an arbitrary compact Lie group; more precisely, Drinfel'd and Sokolov associate several "KdV" hierarchies to each affine Kac-Moody algebra. Some of these hierarchies are discussed in [5], though no general theory is developed there. The key step in [6] which is missing from [5] is to view the KdV flows as quotients of certain simpler ones, the "modified KdV" flows [12, 20]: the generalization of the latter is fairly evident. We refer to [35] for a brief account of how the present theory generalizes to the equations of Drinfel'd and Sokolov: here we just mention that from this point of view our main construction appears as a special case of a well known procedure ("dressing") of Zakharov and Shabat [23].

We end with a technical remark. In this introduction we have been considering u_W and τ_W as functions of the single variable x. In the body of the paper, however, it will be more convenient to think of them as functions of an infinite sequence of variables $(x, t_2, t_3, \dots)$, or alternatively as functions of the element

$$g = \exp(xz + t_2 z^2 + t_3 z^3 + \dots)$$

of the group Γ_+. To do this we define

$$\begin{aligned} u_W(x, t_2, t_3, \dots) &= u_{g^{-1}W}(0), \\ \tau_W(x, t_2, t_3, \dots) &= \tau_{g^{-1}W}(0), \end{aligned}$$

Then $u = u_W$ will be a solution of the hierarchy (1.1) in the sense that

$$\frac{\partial}{\partial t_k} L_u = [P_u^{(k)}, L_u].$$

Note added July 1984. — We draw the reader's attention to several related papers and preprints [26–33] by Japanese authors, which we have seen since completing this work.

Summary of contents

§2 describes the Grassmannian of Hilbert space and its relationship with loop groups.

§3 describes the determinant line bundle Det on the Grassmannian, and its relationship with a central extension of the loop group. We introduce the τ-function, and calculate it explicitly for the subspaces which correspond to multisolitons.

§4 is an outline of the basic formal theory of the generalized KdV equations.

§5 describes the correspondence between points of the Grassmannian, Baker functions, and differential operators, and works out the simplest examples. We also give a characterization of the class of solutions $\mathcal{C}^{(n)}$.

§6 shows how Krichever's construction fits into the framework of §§2–5. It also contains a discussion of rings of commuting differential operators, and of the "Painlevé property" of the stationary solutions of the KdV equations.

§7 is devoted to the subspaces $\mathrm{Gr}_0^{(n)}$ and $\mathrm{Gr}_1^{(n)}$ of the Grassmannian which were mentioned above.

§8 obtains the expansion of the general τ-function as a sequence of Schur functions.

§9 proves that when W arises from an algebraic curve the τ-function τ_W can be expressed explicitly in terms of the θ-function of the curve.

§10 is an appendix explaining the connections between the theory developed in the paper and the representation theory of loop groups.

2 The Grassmannian and loop groups

In this section we shall describe the Grassmannian of Hilbert space and its relation with loop groups. The material is all fairly well known, and we shall not prove all our assertions. For a much more detailed discussion we refer the reader to [17].

The Grassmannian

Let H be a complex Hilbert space with a given decomposition $H = H_+ \oplus H_-$ as the direct sum of two infinite dimensional orthogonal closed subspaces. We are interested in the Grassmannian of all subspaces of H which are comparable with H_+ in the following sense.

Definition. $\mathrm{Gr}(H)$ is the set of all subspaces W of H such that

(i) the orthogonal projection $\mathrm{pr} : W \to H_+$ is a Fredholm operator (i.e. has finite dimensional kernel and cokernel), and

(ii) the orthogonal projection $\mathrm{pr} : W \to H_-$ is a compact operator.

If W belongs to $\mathrm{Gr}(H)$, then so does the graph of any compact operator from W to $W^\perp$. Thus W lies in a subset of $\mathrm{Gr}(H)$ which is in one to one correspondence with the vector space $\mathcal{K}(W; W^\perp)$ of compact operators $W \to W^\perp$. This makes the Grassmannian into a Banach manifold modeled on $\mathcal{K}(H_+; H_-)$, which is given the operator-norm topology.

If $W \in \mathrm{Gr}(H)$, we shall call the index of the Fredholm operator $\mathrm{pr} : W \to H_+$ the *virtual dimension* of W (recall that the *index* of a Fredholm operator T is $\dim(\ker T) - \dim(\operatorname{coker} T)$). The Grassmannian is not connected: two subspaces belong to the same component if and only if they have the same virtual dimension. In the application to differential equations we shall be interested only in the component consisting of subspaces of virtual dimension zero. These subspaces are precisely the ones which are the images of embeddings $H_+ \to H$ which differ from the standard inclusion by a compact operator.

Because of the restrictions on the subspaces W belonging to $\mathrm{Gr}(H)$, not every invertible operator on H induces a map of $\mathrm{Gr}(H)$. We define the *restricted general linear group* $\mathrm{GL}_{res}(H)$ as follows. Let us write operators $g \in \mathrm{GL}(H)$ in the block form

$$g = \begin{pmatrix} a & b \\ c & d \end{pmatrix} \tag{2.1}$$

with respect to the decomposition $H = H_+ \oplus H_-$. Then $\mathrm{GL}_{res}(H)$ is the closed subgroup of $\mathrm{GL}(H)$ consisting of operators g whose off-diagonal blocks b and c are compact operators. The blocks a and d are then automatically Fredholm. The group of connected components of $\mathrm{GL}_{res}(H)$ is $\mathbb{Z}$, the component being determined by the index of the Fredholm operator a.

Lemma 2.2. *The group* $\mathrm{GL}_{res}(H)$ *acts on* $\mathrm{Gr}(H)$.

Proof. A subspace W belongs to $\mathrm{Gr}(H)$ precisely when it is the image of an embedding $\wedge_+ \oplus w_- : H_+ \to H_+ \oplus H_-$ with w_+ Fredholm and w_- compact. Then its transform by the element g in (2.1) above is the image of $w'_+ \oplus w'_-$, where $w'_+ = aw_+ + bw_-$ and $w'_- = cw_+ dw_-$. But w'_+ is Fredholm and w'_- is compact.

We can read off from the formula for w'_+ that the virtual dimension of gW differs from that of W by the index of a. □

Remark. The action of $\mathrm{GL}_{res}(H)$ on $\mathrm{Gr}(H)$ is easily seen to be transitive.

We now specialize to the case where H is the space of all square-integrable complex valued functions on the unit circle $S^1 = \{z \in \mathbb{C} : |z| = 1\}$. In this space we have a natural orthonormal basis consisting of the functions $\{z^k\}$, $k \in \mathbb{Z}$. We define H_+ and H_- to be the closed subspaces spanned by the basis elements $\{z^k\}$ with $k \geq 0$ and $k \leq 0$, respectively. Then $H = H_+ \oplus H_-$: we shall write simply Gr for the Grassmannian corresponding to this choice of (H, H_+, H_-).

Any continuous non-vanishing function f on S^1 defines an invertible multiplication operator, again written f, on H. This induces an action on Gr in view of the following theorem.

Proposition 2.3. *Let Γ denote the group of continuous maps $S^1 \to \mathbb{C}^\times$, regarded as multiplication operators on H. Then $\Gamma \subset \mathrm{GL_{res}}(H)$.*

If $f : S^1 \to \mathbb{C}^\times$ is twice differentiable, then the off-diagonal blocks of the corresponding operator are of trace class (i.e. nuclear).

Proof. The first assertion follows from the second, as the usual topology on Γ corresponds to the norm topology on the multiplication operators, and for this a limit of operators of trace class is compact.

Now let $f = \sum f_k z^k$ be the Fourier expansion of f. The matrix of the corresponding operator, with respect to the basis $\{z^k\}_{k\in\mathbb{Z}}$ of H, has (i,j)-th entry f_{i-j}. We must show that the blocks $\{i \geq 0, j < 0\}$ and $\{i < 0, j \geq 0\}$ are of trace class. But a matrix (α_{ij}) is certainly of trace class if $\sum |\alpha_{ij}| < \infty$. So what we need is that

$$\sum_{i\geq 0, j<0} |f_{i-j}| < \infty \text{ and } \sum_{i<0, j\geq 0} |f_{i-j}| < \infty,$$

that is,

$$\sum_{k>0} k|f_k| < \infty \text{ and } \sum_{k>0} k|f_{-k}| < \infty.$$

These conditions are satisfied if f is twice differentiable, because the Fourier series of a C^1 function is absolutely convergent. □

In this paper we shall be interested mainly in the action of the subgroup Γ_+ of Γ consisting of all real-analytic functions $f : S^1 \to \mathbb{C}^\times$ which extend to holomorphic functions $f : D_0 \to \mathbb{C}^\times$ in the disc $D_0 = \{z \in \mathbb{C} : |z| \leq 1\}$ satisfying $f(0) = 1$. (Here and elsewhere, when we say that a function defined on a closed set in $\mathbb{C}$ is holomorphic, we mean that it extends to a holomorphic function in a neighbourhood of the set.) We shall also consider the subgroup Γ_- of Γ consisting of functions f which extend to nonvanishing holomorphic functions in $D_\infty = \{z \in \mathbb{C} \cup \infty : |z| \geq 1\}$ satisfying $f(\infty) = 1$. Concerning this subgroup we can assert

Proposition 2.4. *Γ_- acts freely on* Gr.

We shall postpone the proof of this for a moment.

The stratification of Gr

We shall make much use of some special spaces $H_S \in \mathrm{Gr}$ indexed by certain subsets S of the integers: for any $S \subset \mathbb{Z}$ we define H_S to be the closed subspace of H spanned by $\{z^s\}_{s\in S}$. The kernel and cokernel of the orthogonal projection $H_S \to H_+$ are spanned by the functions $\{z^i\}$ with i belonging to $S \setminus \mathbb{N}$ and

$\mathbb{N}\setminus S$, respectively; thus $H_S \in \mathrm{Gr}$ precisely when both $S\setminus\mathbb{N}$ and $\mathbb{N}\setminus S$ are finite. We denote by $\mathcal{S}$ the set of all subsets $S \subset \mathbb{Z}$ of this kind. If $S \in \mathcal{S}$, we call the number

$$\mathrm{card}(S \setminus \mathbb{N}) - \mathrm{card}(\mathbb{N} \setminus S)$$

the *virtual cardinal* of S: it is equal to the virtual dimension of H_S. A set of virtual cardinal d is simply an increasing sequence $S = \{s_0, s_1, s_2, \ldots\}$ of integers such that $s_i = i - d$ for all sufficiently large i. Let us order the set $\mathcal{S}$ by defining

$$S \leq S' \Leftrightarrow s_k \geq s'_k \quad \text{for all } k.$$

Lemma 2.5. *For every $W \in \mathrm{Gr}$, there exist sets $S \in \mathcal{S}$ such that W is the graph of a compact operator $H_S \to H_S^\perp$, or, equivalently, such that the orthogonal projection $W \to H_S$ is an isomorphism. Furthermore there is a unique minimal S with this property.*

We shall omit the straightforward proof of this lemma: it can be found in [17]. Let us only point out that the unique minimal S associated to W consists precisely of those integers s such that W contains an element of order s, i.e. an element of the form $\sum_{k\leq s} a_k z^k$ with $a_s \neq 0$.

A very useful corollary of the lemma is

Proposition 2.6. *In any $W \in \mathrm{Gr}$, the elements of finite order form a dense subspace, which we shall denote by W^{alg}.*

This holds because a projection $W \to H_S$ which is an isomorphism induces an isomorphism between W^{alg} and H_S^{alg}; and the elements of finite order are obviously dense in H_S.

Let us at this point return to give the proof of Proposition 2.4. Suppose that $W \in \mathrm{Gr}$, and that $g \in \Gamma_-$ is such that $gW = W$. Now g is of the form $1 + \sum_{k<0} a_k z^k$. Let $w \in W$ be an element of minimal order s_0. Then $gw - w$ is an element of W or order less than s_0. So $gw = w$, and hence $g = 1$.

The spaces $W \in \mathrm{Gr}$ which are the graphs of operators $H_S \to H_S^\perp$ will be called *transverse* to $H_S^\perp$. They form an open set U_S of Gr; lemma 2.5 asserts that the U_S cover Gr. The set of W such that S is the *minimal* indexing set with $W \in U_S$ form a closed submanifold of U_S denoted by Σ_S. Notice that W belongs to Σ_S precisely when W^{alg} has a basis $\{w_i\}_{i\geq 0}$ (in the algebraic sense) with w_i of order s_i.

The Σ_S constitute a stratification of Gr by manifolds of finite codimension. The codimension of Σ_S is the *length* of S, defined as

$$\ell(S) = \sum_{k\geq 0}(k - s_k - d),$$

where d is the virtual cardinal of S.

Scaling

For each $\lambda \in S^1$, we can consider the operator R_λ on H induced by rotating the circle S^1, that is, the operator defined by

$$R_\lambda f(z) = f(\lambda^{-1} z), \quad (f \in H).$$

If λ is a complex number with $|\lambda| \neq 1$, the operator R_λ defined by this formula is unbounded. Nevertheless, using (2.5), we can see that if $|\lambda| \leq 1$, then the operator R_λ still induces a transformation of Gr. For then the domain of R_λ includes the dense subspace H^{alg} of H consisting of functions of finite order, i.e. those whose Fourier series involve only a finite number of positive powers of z. We can therefore define $R_\lambda W$ to be the closure of the space $R_\lambda W^{\text{alg}}$. To see that $R_\lambda W$ belongs to Gr, we use (2.5): if W is transverse to $H_S^\perp$, then clearly $R_\lambda W$ is too, and is the graph of a compact operator. We shall refer to the operators $\{R_\lambda : |\lambda| \leq 1\}$ as the semigroup of *scaling transformations* of Gr. It should be noticed that $R_\lambda W$ depends continuously both on λ and on W.

The scaling operators R_λ preserve the stratification of Gr by the Σ_S. In the sense of Morse theory, Σ_S is the stable manifold of the point H_S for the scaling flow, i.e. the set of all W such that $R_\lambda W \to H_S$ as $\lambda \to 0$.

Loop groups

We now come to the connection of the Grassmannian with loop groups. Although this will not play a very prominent role in the paper, we regard it as fundamental.

Let $H^{(n)}$ be the Hilbert space of all square integrable functions on S^1 with values in $\mathbb{C}^n$. We break up $H^{(n)}$ as $H_+^{(n)} \oplus H_-^{(n)}$, using Fourier series just as in the case $n = 1$. The group $\mathrm{LGL}_n(\mathbb{C})$ of all continuous maps $\gamma : S^1 \to \mathrm{GL}_n(\mathbb{C})$ acts on $H^{(n)}$ in an obvious way. Generalizing Proposition 2.3 we have

Proposition 2.7. $\mathrm{LGL}_n(\mathbb{C}) \subset \mathrm{GL}_{\text{res}}(H^{(n)})$.

The proof is exactly the same as in the case $n = 1$.

Thus $\mathrm{LGL}_n(\mathbb{C})$ acts on $\mathrm{Gr}(H^{(n)})$. For each $\gamma \in \mathrm{LGL}_n(\mathbb{C})$ we set $W_\gamma = \gamma \cdot H_+^{(n)}$. Then $zW_\gamma \subset W_\gamma$, where z denotes the operation of multiplication by the scalar-valued function z on S^1; for multiplication by z commutes with the action of γ on $H^{(n)}$, and $zH_+^{(n)} \subset H_+^{(n)}$. This leads us to introduce the

Definition. $\mathrm{Gr}^{(n)} = \{W \in \mathrm{Gr}(H^{(n)}) : zW \subset W\}$.

$\mathrm{Gr}^{(n)}$ is a closed subspace of Gr, and $\mathrm{LGL}_n(\mathbb{C})$ acts on it. One reason for its importance is that it is essentially the loop space ΩU_n of the unitary group U_n, i.e. the space of continuous maps $\gamma : S^1 \to U_n$ such that $\gamma(1) = 1$. To be precise, $\gamma \mapsto W_\gamma$ maps ΩU_n injectively onto a dense subspace of $\mathrm{Gr}^{(n)}$; and indeed $\mathrm{Gr}^{(n)}$ can be identified, if one wants, with a certain class of measurable loops in U_n.

The construction by which one associates a loop to a subspace W in $\mathrm{Gr}^{(n)}$ is as follows. One first observes that the quotient space W/zW is n-dimensional. Let $w_1, \ldots, w_n$ by the elements of W which span W/zW. Think of them as functions on the circle whose values are n-component column vectors. Then $(w_1, w_2, \ldots, w_n)$ is a function on the circle with values in $\mathrm{GL}_n(\mathbb{C})$: call it γ. It is obvious that $\gamma \cdot H_+^{(n)} = W$. Unfortunately the matrix entries of γ are a priori only L^2 functions, and it may not be possible to choose them continuous. If the elements $w_1, \ldots, w_n$ are chosen to be an orthonormal basis for the orthogonal complement of zW in W, then it is easy to see that the loop γ takes its values in U_n. Furthermore γ is then unique up to multiplication on the right by a constant unitary matrix.

We should notice that in the correspondence between loops and subspaces the winding number of a loop γ is minus the virtual dimension of W_γ. (This can be seen by deforming γ continuously to a standard loop with the same winding number.)

We shall now identify the Hilbert space $H^{(n)}$ with $H = H^{(1)}$ by letting the standard basis $\{\varepsilon_i z^k : 1 \le i \le n, k \in \mathbb{Z}\}$ for $H^{(n)}$ correspond lexicographically to the basis $\{z^k\}$ for H. (Here $\{\varepsilon_i\}$ denotes the standard basis for $\mathbb{C}^n$.) Thus $\varepsilon_i z^k$ corresponds to z^{nk+i-1}. More invariantly, given a vector valued function with components $(f_0, \ldots, f_{n-1})$, we associate to it the scalar valued function $\tilde{f}$ such that

$$\tilde{f}(z) = f_0(z^n) + z f_1(z^n) + \ldots + z^{n-1} f_{n-1}(z^n).$$

Conversely, given $\tilde{f} \in H$, we have

$$f_k(z) = \frac{1}{n} \sum_{\zeta} \zeta^{-k} \tilde{f}(\zeta)$$

where ζ runs through the n-th roots of z. The isomorphism $H^{(n)} \cong H$ is an isometry. It makes continuous functions correspond to continuous ones, and also preserves most other reasonable classes of functions, for example: smooth, real analytic, rational, polynomial. Multiplication by z on $H^{(n)}$ corresponds to multiplication by z^n on H; and $H_+^{(n)}$ corresponds to H_+. From now on we shall always think of $\mathrm{Gr}^{(n)}$ as the subspace of Gr given by

$$\mathrm{Gr}^{(n)} = \{W \in \mathrm{Gr} : z^n W \subset W\}.$$

Note that $\mathrm{Gr}^{(n)}$ is preserved by the action of the group Γ and also by the semigroup of scaling transformations.

Proposition 2.8. *Let $W \in \mathrm{Gr}^{(n)}$. Then for any complex number λ with $|\lambda| < 1$, the space $R_\lambda W$ corresponds to a real analytic loop.*

The proposition implies that for the purposes of this paper we could perfectly well confine ourselves to the subspace of $\mathrm{Gr}^{(n)}$ consisting of those W that correspond to analytic loops (see (5.10) below). However, most of our arguments apply naturally to the larger space $\mathrm{Gr}^{(n)}$.

Proof of (2.8). We have to see that there is a complementary subspace for $z^n(R_\lambda W)$ in $R_\lambda W$ consisting of analytic functions. Choose a complementary subspace A for $z^n W$ in W such that $A \subset W^{\text{alg}}$ (this is possible because, as we have seen, W^{alg} is dense in W). Then each $f \in A$ has the form

$$f(z) = \sum_{-\infty}^{N} c_i z^i$$

where the series converges for $|z| > 1$; hence the series for $f(\lambda^{-1}z)$ converges for $|z| > |\lambda|$, so that $f(\lambda^{-1}z)$ is an analytic function on S^1. Thus the space $\{f(\lambda^{-1}z) : f \in A\}$ is a complement to $z^n(R_\lambda W)$ in $R_\lambda W$ of the desired kind. □

Rational and polynomial loops

In §7 we shall consider two subspaces $\text{Gr}_1^{(n)}$ and $\text{Gr}_0^{(n)}$ of $\text{Gr}^{(n)}$: they can be defined as the subspaces corresponding to rational and Laurent polynomial loops, respectively. They can also be characterized in another way, which will be more convenient for us.

Proposition 2.9. *The following conditions on a subspace* $W \in \text{Gr}^{(n)}$ *are equivalent.*

(i) $W = W_\gamma$ *for some rational loop* γ *(that is, a loop such that each matrix entry in* γ *is a rational function of* z *with no poles on* S^1*).*

(ii) There exist polynomials p *and* q *in* z *such that*

$$pH_+ \subset W \subset q^{-1}H_+.$$

(iii) W *is* commensurable *with* H_+, *i.e.* $W \cap H_+$ *is of finite codimension in both* W *and* H_+.

We denote by $\text{Gr}_1^{(n)}$ the subspace of $\text{Gr}^{(n)}$ consisting of those W that satisfy the conditions in (2.9). We define Gr_1 to be the subspace of Gr consisting of those $W \in \text{Gr}$ that satisfy condition (ii) in (2.9). Notice that we may assume that the roots of the polynomials p and q all lie in the region $\{|z| < 1\}$; for if $|c| > 1$, then $z - c$ is an invertible operator on H_+.

Example 2.10. For spaces $W \in \text{Gr}$ not belonging to any $\text{Gr}^{(n)}$ the condition of commensurability (2.9) (iii) does not imply condition (2.9) (ii). As an example, consider the subspace W of codimension 1 in H_+ which is the kernel of the linear map $F : H_+ \to \mathbb{C}$ defined by

$$F(f) = \text{residue}_{z=0}(e^{1/z} \cdot f).$$

Obviously there is no polynomial p such that $pH_+ \subset W$.

Proposition 2.11. *The following conditions on a subspace $W \in \mathrm{Gr}^{(n)}$ are equivalent.*

(i) $z^q H_+ \subset W \subset z^{-q} H_+$ *for some positive integer* q.

(ii) $W = W_\gamma$ *for some Laurent polynomial loop* γ *(by this we mean that both* γ *and* γ^{-1} *have finite Laurent expansions).*

We denote by Gr_0 the subspace of Gr consisting of those W that satisfy the condition (2.11) (i), and we set

$$\mathrm{Gr}_0^{(n)} = \mathrm{Gr}_0 \cap \mathrm{Gr}^{(n)}.$$

Then Gr_0 is the union of all the $\mathrm{Gr}_0^{(n)}$.

We note that all the Grassmannians Gr_1, Gr_0, $\mathrm{Gr}_1^{(n)}$ and $\mathrm{Gr}_0^{(n)}$ are invariant under the semigroup of scaling transformations, and also under the action of the group Γ_+ of holomorphic functions in the disc (defined after (2.3)). (Gr_0 and Gr_1 are preserved by Γ_+ because $gH_+ = H_+$ for any $g \in \Gamma_+$.)

It is easy to see that Gr_0 is dense in Gr. As Gr_0 is the union of a sequence of compact finite dimensional algebraic varieties (namely the Grassmannians of $z^{-q}H_+/z^q H_+$), this implies that every holomorphic function on Gr is constant.

Although it will play only a minor role in this paper, we should mention that the space Gr_0 has a cell decomposition into even-dimensional cells indexed by the same set $\mathcal{S}$ as the stratification. For $S \in \mathcal{S}$ the cell C_S consists of all $W \in \mathrm{Gr}_0$ for which W^{alg} has a basis $\{w_s\}_{s \in S}$ with w_s of the form

$$w_s = z^s + \sum_{i>s} \alpha_{si} z^i.$$

The cell C_S is homeomorphic to $\mathbb{C}^{\ell(S)}$. It is a submanifold of Gr transversal to the stratum Σ_S, which it meets in the single point H_S. On Gr_0 the scaling operators R_λ make sense for all $\lambda \in \mathbb{C}^\times$, and C_S is the "unstable manifold" of H_S for the scaling flow, i.e. the set of W such that $R_\lambda W \to H_S$ as $\lambda \to \infty$.

Finally, let us observe that H_S belongs to $\mathrm{Gr}^{(n)}$ if and only if $S + n \subset S$. For such S let us write $C_S^{(n)}$ for $C_S \cap \mathrm{Gr}^{(n)}$. The $C_S^{(n)}$ form a cell decomposition of $\mathrm{Gr}^{(n)}$, and the dimension of $C_S^{(n)}$ is $\sum_i (i - s_i - d)$, where the sum is taken only over the n integers i such that $s_i \notin S + n$, and d is the virtual cardinal of S.

3 The determinant bundle and the τ-function

In this section we are going to construct a holomorphic line bundle Det over *Gr*. For simplicity, we shall confine ourselves to the connected component of the Grassmannian consisting of spaces of virtual dimension zero: the symbol Gr will now denote this component. We think of Det as the "determinant bundle", that is, the bundle whose fibre over $W \in \mathrm{Gr}$ is the "top exterior power" of W. Our first task is to explain how to make sense of this.

On the Grassmannian $\mathrm{Gr}_k(\mathbb{C}^n)$ of k-dimensional subspaces of $\mathbb{C}^n$ the fibre of the determinant line bundle at $W \in \mathrm{Gr}_k(\mathbb{C}^n)$ is $\det(W) = \Lambda^k(W)$. A typical element of $\Lambda^k(W)$ can be written $\lambda w_1 \wedge w_2 \wedge \ldots \wedge w_k$, with $\lambda \in \mathbb{C}$, where $\{w_i\}$ is a basis for W. In analogy with this, an element of $\det(W)$, for $W \in \mathrm{Gr}$, will be an infinite expression $\lambda w_0 \wedge w_1 \wedge w_2 \wedge \ldots$, where $\{w_i\}$ is what we shall call an *admissible basis* for W. The crucial property of the class of admissible bases is that if $\{w_i\}$ and $\{w_i'\}$ are two admissible bases of W then the infinite matrix t relating them is of the kind that has a determinant; for we want to be able to assert that

$$\lambda w_0 \wedge w_1 \wedge w_2 \wedge \ldots = \lambda \det(t) w_0' \wedge w_1' \wedge w_2' \wedge \ldots$$

when $w_i = \sum t_{ij} w_j'$.

Let us recall (see, for example, [19]) that an operator has a determinant if and only if it differs from the identity by an operator of trace class. Now the subspaces W we are considering have the property that the projection $\mathrm{pr} : W \to H_+$ is Fredholm and of index zero. This means that W contains sequences $\{w_i\}$ such that

(i) the linear map $w : H_+ \to H$ which takes z^i to w_i is continuous and injective and has image W, and

(ii) the matrix relating $\{\mathrm{pr}(w_i)\}$ to $\{z^i\}$ differs from the identity by an operator of trace class.

Such a sequence $\{w_i\}$ will be called an *admissible basis.* (A possible choice for $\{w_i\}$ is the inverse image of the sequence $\{z^s\}_{s \in S}$ under a projection $W \to H_S$ which is an isomorphism (see (2.5).)

We shall think of $w : H_+ \to H$ as a $\mathbb{Z} \times \mathbb{N}$ matrix

$$w = \begin{pmatrix} w_+ \\ w_- \end{pmatrix}$$

whose columns are the w_i, and where $w_+ - 1$ is of trace class; the block w_- is automatically a compact operator. Then w is determined by W up to multiplication on the right by an $\mathbb{N} \times \mathbb{N}$ matrix (or operator $H_+ \to H_+$) belonging to the group $\mathcal{G}$ of all invertible matrices t such that $t - 1$ is of trace class. (The topology of $\mathcal{G}$ is defined by the trace norm.) Because operators in $\mathcal{T}$ have determinants we can define an element of $\mathrm{Det}(W)$ as a pair (w, λ), where $\lambda \in \mathbb{C}$ and w is an admissible basis of W, and we identify (w, λ) with (w', λ') when $w' = wt^{-1}$ and $\lambda' = \lambda \det(t)$ for some $t \in \mathcal{T}$, (we could also write (w, λ) as $\lambda w_0 \wedge w_1 \wedge \ldots$)

To be quite precise, the space $\mathcal{P}$ of matrices w should be given the topology defined by the operator norm on w_- and the trace norm on $w_+ - 1$. Then $\mathcal{P}$ is a principal $\mathcal{T}$-bundle on $\mathrm{Gr} = \mathcal{P}/\mathcal{T}$, and the total space of Det is $\mathcal{P} \times_{\mathcal{T}} \mathbb{C}$ where $\mathcal{T}$ acts on $\mathbb{C}$ by $\det : \mathcal{T} \to \mathbb{C}^\times$.

Now we come to the crucial difference between the finite and infinite dimensional cases. The group $\mathrm{GL}_n(\mathbb{C})$ acts on $\mathrm{Gr}_k(\mathbb{C}^n)$, and also on the total space of the line bundle det on it: if $g \in \mathrm{GL}_n(\mathbb{C})$ and $w_1 \wedge \ldots \wedge w_k \in \det(W)$ then $g \cdot (w_1 \wedge \ldots \wedge w_k)$ is defined as $gw_1 \wedge \ldots \wedge gw_k$ in $\det(gW)$. We have seen that the corresponding group which acts on Gr is not the entire general linear group of H but the identity component of the smaller group $\mathrm{GL}_{res}(H)$ of invertible operators in H of the form

$$g = \begin{pmatrix} a & b \\ c & d \end{pmatrix} \tag{3.1}$$

(with respect to the decomposition $H = H_+ \oplus H_-$), where b and c are *compact.* But this action on Gr does not automatically induce an action on Det, for if $\{w_i\}$ is an admissible basis for W then $\{gw_i\}$ is usually not an admissible basis for gW. To deal with this problem we introduce the slightly smaller group $\mathrm{GL}_1(H)$ consisting of invertible operators g of the form (3.1), but where the blocks b and c are of trace class. The topology of $\mathrm{GL}_1(H)$ is defined by the operator norm on a and d, and the trace norm on b and c. We shall see that the action of the identity component $\mathrm{GL}_1(H)^0$ on Gr does lift projectively to Det. In other words there is a central extension $\mathrm{GL}_1^\wedge$ of $\mathrm{GL}_1(H)^0$ by $\mathbb{C}^\times$ which acts on Det, covering the action of $\mathrm{GL}_1(H)^0$ on Gr.

To obtain a transformation of Det we must give not only a transformation g of H but also some information telling us how to replace a non-admissible basis $\{gw_i\}$ of gW by an admissible one. To do this we introduce the subgroup $\mathcal{E}$ of $\mathrm{GL}_1(H)^0 \times \mathrm{GL}(H_+)$ consisting of pairs (g, q) such that $aq^{-1} - 1$ is of trace class, where a is as in (3.1). (We give $\mathcal{E}$ the topology induced by its embedding $(g,q) \mapsto (g, q, aq^{-1} - 1)$ in $\mathrm{GL}_1(H) \times \mathrm{GL}(H_+) \times \{\text{operators of trace class}\}$.) The definition of $\mathcal{E}$ is precisely designed to make it act naturally on the space $\mathcal{P}$ of admissible bases by

$$(g, q) \cdot w = gwq^{-1},$$

and hence act on Det by $(g,q) \cdot (w, \lambda) = (gwq^{-1}, \lambda)$.

The group $\mathcal{E}$ has a homeomorphism $(g, q) \mapsto g$ onto $\mathrm{GL}_1(H)^0$. Its kernel can clearly be identified with $\mathcal{T}$. Thus we have an extension

$$\mathcal{T} \to \mathcal{E} \to \mathrm{GL}_1(H)^0.$$

But the subgroup $\mathcal{T}_0$ of $\mathcal{T}$ consisting of operators of determinant 1 acts trivially on Det, so that in fact the quotient group $\mathrm{GL}_1^\wedge = \mathcal{E}/\mathcal{T}_0$ acts on Det. This last group is a central extension of $\mathrm{GL}_1(H)^0$ by $\mathcal{T}/\mathcal{T}_0 \cong \mathbb{C}^\times$.

The extension

$$\mathbb{C}^\times \to \mathrm{GL}_1^\wedge \to \mathrm{GL}_1(H)^0$$

is a non-trivial fibre bundle: there is no continuous cross-section $\mathrm{GL}_1(H)^0 \to \mathrm{GL}_1^\wedge$, and the extension cannot be described by a continuous cocycle. But on the dense open set GL_1^{reg} of $\mathrm{GL}_1(H)^0$ where a is invertible, there is a cross-section s of $\mathcal{E} \to \mathrm{GL}_1(H)^0$ given by $s(g) = (g, a)$; the corresponding cocycle is

$$(g_1, g_2) \mapsto \det(a_1 a_2 a_3^{-1}),$$

where $g_i = \begin{pmatrix} a_i & b_i \\ c_i & d_i \end{pmatrix}$, and $g_3 = g_1 g_2$. We shall always make the elements of GL_1^{reg} act on Det by means of the section s. Of course, GL_1^{reg} is not a group, and the map s is not multiplicative. But let GL_1^+ by the subgroup of GL_1^{reg} consisting of elements whose block decomposition has the form $\begin{pmatrix} a & b \\ 0 & d \end{pmatrix}$. Then the restriction of s to GL_1^+ is an inclusion of groups $\mathrm{GL}_1^+ \to \mathcal{E}$ and we can regard GL_1^+ as a group of automorphisms of the bundle Det. Similar remarks apply to the subgroup GL_1^-, consisting of elements of GL_1^{reg} whose block decomposition has the form $\begin{pmatrix} a & 0 \\ c & d \end{pmatrix}$. In particular the subgroups Γ_+ and Γ_- of the group of maps $S^1 \to \mathbb{C}^\times$ act on Det, for $\Gamma_\pm \subset \mathrm{GL}_1^\pm$. (Cf. remarks following (2.3).)

The τ-function

We have now reached our main goal in this section, the definition of the τ-function.

Alongside the determinant bundle Det just constructed there is its dual Det*, whose fibres are the duals of the fibres of Det. A point of Det* over $W \in \mathrm{Gr}$ can be taken to be a pair (w, λ), where w is an admissible basis for $W, \lambda \in \mathbb{C}$, and (w, λ) is identified with (w', λ') if $w' = wt$ and $\lambda' = \lambda \det(t)$ for some $t \in \mathcal{T}$. The action of $\mathrm{GL}_1^\wedge$ on Det induces an action on Det*. The line bundle Det* has a canonical global holomorphic section σ, defined by

$$\sigma(W) = (w, \det w_+),$$

where $W \in \mathrm{Gr}$, and w is an admissible basis for W. We can think of $\sigma(W)$ as the determinant of the orthogonal projection $W \to H_+$; note that $\sigma(W) = 0$ if and only if W is not transverse to H_-. The section σ is not equivariant with respect to the action of Γ_+ on Det*. For each $W \in \mathrm{Gr}$, the τ-function of W is the holomorphic function $\tau_W : \Gamma_+ \to \mathbb{C}$ defined by

$$\tau_W(g) = \frac{\sigma(g^{-1}W)}{g^{-1}\delta_W}$$

where δ_W is some non-zero element of the fibre of Det* over W. In general there is no canonical choice of δ_W, so that τ_W is defined only up to a constant factor. However, if W is transverse to H_-, it is natural to choose $\delta_W = \sigma(W)$, so that the τ-function is given by

$$\tau_W(g) \cdot g^{-1}\sigma(W) = \sigma(g^{-1}W) \ \text{(for } W \text{ transverse to } H_-). \qquad (3.2)$$

It is easy to give an explicit formula for τ_W as an infinite determinant.

Proposition 3.3. *Let $g^{-1} \in \Gamma_+$ have the block form*

$$g^{-1} = \begin{pmatrix} a & b \\ 0 & d \end{pmatrix}$$

with respect to the splitting $H = H_+ \oplus H_-$. Then for $W \in \mathrm{Gr}$, we have

$$\tau_W(g) = \det(w_+ + a^{-1}bw_-), \tag{3.4}$$

where w is an admissible basis of W. In particular, if W is transverse to H_- and τ_W is normalized as in (3.2), *then we have*

$$\tau_W(g) = \det(1 + a^{-1}bA), \tag{3.5}$$

where $A : H_+ \to H_-$ is the map whose graph is W.

The proposition follows at once from the definitions.

Example

An interesting example of a space W belonging to $\mathrm{Gr}_1^{(2)}$ is the following one, which, as we shall see, is related to the m-soliton solution of the KdV equation.

Let $p_1, \dots, p_m$ be non-zero complex numbers such that $|p_i| < 1$ and all p_i^2 are distinct; and let $\lambda_1, \dots, \lambda_m$ be also non-zero. Then $W = W_{p,\lambda}$ denotes the closure of the space of function f which are holomorphic in the unit disc except for a pole of order $\leq m$ at the origin, and which satisfy $f(-p_i) = \lambda_i f(p_i)$ for $i = 1, \dots, m$. To calculate τ_W we first determine the map $A : H_+ \to H_-$ whose graph is $W_{p,\lambda}$. This assigns to $f \in H_+$ the polynomial

$$A(f) = \alpha_1(f)z^{-1} + \dots + \alpha_m(f)z^{-m}$$

such that $f + A(f)$ belongs to $W_{p,\lambda}$. Clearly each $\alpha_i(f)$ is a linear combination of $\beta_1(f), \dots, \beta_m(f)$, where

$$\beta_i(f) = \frac{1}{2}\left(\lambda_i^{1/2} f(p_i) - \lambda_i^{-1/2} f(-p_i)\right),$$

for $A(f)$ is zero when the $\beta_i(f)$ vanish. In fact $\beta_i = \sum M_{ij}\alpha_j$, where

$$M_{ij} = \frac{1}{2}\left(\lambda_i^{1/2} - (-1)^j \lambda_i^{-1/2}\right) p_i^{-j};$$

and $W_{p,\lambda}$ is transverse to H_- precisely when $\det(M_{ij}) \neq 0$.

To apply (3.5) we must also calculate the map $a^{-1}b : H_- \to H_+$ corresponding to the element g^{-1} of Γ_+. We write g in the form $\exp \sum_{k>0} t_k z^k$.

Suppose that $a^{-1}b$ takes z^{-k} to $f_k \in H_+$. Before determining f_k let us observe that an infinite determinant of the form

$$\det\left(1 + \sum_{i=1}^{m} f_i \oplus \alpha_i\right)$$

reduces to the determinant of the $m \times m$ matrix whose (i, j)-th entry is

$$\delta_{ij} + \alpha_i(f_j).$$

Thus $\tau_W(\mathbf{t}) = \det(M_{ij})^{-1} \det(M_{ij} + \beta_i(f_j))$.

If $\mathrm{pr} : H \to H_+$ is the projection, we find

$$\begin{aligned} f_k &= g \cdot \mathrm{pr}(g^{-1} z^{-k}) \\ &= z^{-k}\{1 - e^{\sum t_i z^i}(1 + c_1 z + c_2 z^2 + \ldots + c_{k-1} z^{k-1})\}, \end{aligned}$$

where $\sum c_i z^i$ is the expansion of $e^{-\sum t_j z^j}$; and so

$$M_{ij} + \beta_i(f_j) = -\beta_i\{z^{-j} e^{\sum t_k z^k}(1 + c_1 z + \ldots + c_{j-1} z^{j-1})\}.$$

The determinant of this matrix, after the obvious column operations have been performed on it, reduces to

$$(-1)^m \exp\left(\sum i, k t_{2k} p_i^{2k}\right)$$
$$\times \det \begin{pmatrix} p_1^{-1}\varphi_1(\theta_1 + \delta_1) & p_1^{-2}\varphi_2(\theta_1 + \delta_1) & \cdots & p_1^{-m}\varphi_m(\theta_1 + \delta_1) \\ p_2^{-1}\varphi_1(\theta_2 + \delta_2) & \cdots & \cdots & \cdots \\ \cdots & \cdots & \cdots & \cdots \\ p_m^{-1}\varphi_1(\theta_m + \delta_m) & \cdots & \cdots & p_m^{-m}\varphi_m(\theta_m + \delta_m) \end{pmatrix}$$

where $\varphi_i = \cosh$ for i odd and $= \sinh$ for i even,

$$\theta_i = \sum_{k \text{odd}} p_i^k t_k, \quad \text{and}$$
$$\delta_i = \frac{1}{2} \log \lambda_i.$$

The constant factor $(-1)^m \det(M_{ij})^{-1}$ in τ_W can be ignored.

In §5 we shall see that $2\left(\frac{\partial}{\partial t_1}\right)^2 \log \tau_W$ is a solution to the KdV equations. It is usually called the "m-soliton" solution.

The projective multiplier on Γ_+ and Γ_-

The results of this subsection will be used only in §9.

The actions of the groups Γ_+ and Γ_- on Gr obviously commute with each other. However, their actions on Det* do not commute, and we shall need to know the relationship between them. Note that since the discs D_0 and D_∞ are simply connected, the elements $g \in \Gamma_+$ and $\tilde{g} \in \Gamma_-$ can be written uniquely in the form $g = e^f$, $\tilde{g} = e^{\tilde{f}}$, where $f : D_0 \to \mathbb{C}$ and $\tilde{f} : D_\infty \to \mathbb{C}$ are holomorphic maps with $f(0) = \tilde{f}(\infty) = 0$. If γ is an element of either Γ_+ or Γ_-, we shall write $\mathcal{D}(\gamma)$ for the corresponding automorphism of the bundle Det*.

Proposition 3.6. *If $g \in \Gamma_+$ and $\tilde{g} \in \Gamma_-$, then*

$$\mathcal{D}(\tilde{g})\mathcal{D}(g) = c(\tilde{g}, g)\mathcal{D}(g)\mathcal{D}(\tilde{g}),$$

where, if as above $\tilde{g} = e^{\tilde{f}}$ and $g = e^f$, we have

$$c(\tilde{g}, g) = e^{S(\tilde{f}, f)}$$

and

$$S(\tilde{f}, f) = \frac{1}{2\pi i}\int_{S^1} \tilde{f}'(z) f(z)\, dz.$$

Proof. It is immediate from the definition of the actions of $\Gamma_\pm$ on Det* that we have a formula of the kind stated, with

$$c(\tilde{g}, g) = \det(a\tilde{a}a^{-1}\tilde{a}^{-1})$$

where a and $\tilde{a}$ are the $H_+ \to H_+$ blocks of g and $\tilde{g}$. (The commutator has a determinant because, from the fact that g and $\tilde{g}$ commute, it is equal to $1 - b\tilde{c}a^{-1}\tilde{a}^{-1}$, where b and $\tilde{c}$ are the off-diagonal blocks of g and $\tilde{g}$, which are of trace class by (2.3).) The map c is a homeomorphism from $\Gamma_- \times \Gamma_+$ to $\mathbb{C}$; it follows easily that it is of the desired form, with

$$S(\tilde{f}, f) = \operatorname{trace}[\alpha, \tilde{\alpha}],$$

where α and $\tilde{\alpha}$ are the $H_+ \to H_+$ blocks of f and $\tilde{f}$. Now, if $f = \sum a_i z^i$ and $\tilde{f} = \sum b_i z^{-i}$, the (k,k) matrix element of the commutator $[\alpha, \tilde{\alpha}]$ is

$$\sum_{m=1}^{k} a_m b_m - \sum_{m=1}^{\infty} a_m b_m.$$

The trace is therefore

$$-\sum_{m=1}^{\infty} m a_m b_m = \frac{1}{2\pi i}\int_{S_i} \tilde{f}'(z) f(z)\, dz.$$

as stated. □

Lemma 3.7. *The section σ of* Det* *is equivariant with respect to the action of Γ_-, that is, we have*

$$\sigma(\tilde{g}W) = \tilde{g}\sigma(W) \quad \textit{for } \tilde{g} \in \Gamma_-$$

Lemma 3.8. *For $\tilde{g} \in \Gamma_-$, we have*

$$\tau_{\tilde{g}W}(g) = e^{S(\tilde{f}, f)}\tau_W(g),$$

where as before $g = e^f$ and $\tilde{g} = e^{\tilde{f}}$.

Both lemmas follow at once from the definitions.

General remarks

In the theory of loop groups like the group L of smooth maps $S^1 \to \mathrm{GL}_n(\mathbb{C})$ the existence of a certain central extension

$$\mathbb{C}^\times \to \hat{L} \to L$$

plays an important role. This extension (at least over the identity component of L) is the restriction of the central extension $\mathrm{GL}_1^\wedge$ constructed in this section, when L is embedded in the usual way in $\mathrm{GL}_1(H)$.

On the level of Lie algebras the extension can be described very simply for the loop group LG of any reductive group G. The Lie algebra of LG is the vector space $\mathfrak{L}\mathfrak{g}$ of loops in the Lie algebra $\mathfrak{g}$ of G, and the extension is defined by the cocycle

$$\beta : \mathfrak{L}\mathfrak{g} \times \mathfrak{L}\mathfrak{g} \to \mathbb{C}$$

given by

$$\beta(f_1, f_2) = \frac{1}{2\pi} \int_0^{2\pi} \langle f_1'(\theta), f_2'(\theta) \rangle d\theta,$$

where $\langle \ , \ \rangle$ is a suitably normalized invariant bilinear form on $\mathfrak{g}$.

The existence of the corresponding extension of groups is less obvious (cf. [18]), partly because it is topologically non-trivial as a fibre bundle. The discussion in this section provides a concrete realization of $\hat{L}$ as a group of holomorphic automorphisms of the line bundle Det, in the case $\mathrm{G} = \mathrm{GL}_n(\mathbb{C})$. For the elements of $\hat{L}$ above $\gamma \in L$ are precisely the holomorphic bundle maps $\hat{\gamma} : \mathrm{Det} \to \mathrm{Det}$ which cover the action of γ on Gr. (For given γ the possible choices of $\hat{\gamma}$ differ only by multiplication by constants, as any holomorphic function on Gr is constant. (Cf. remark following (2.11)).)

The corresponding central extension of the loop group of any complex reductive group (characterized by its Lie algebra cocycle) can be constructed in a similar way as a group of holomorphic automorphisms of a complex line bundle, and conversely the holomorphic line bundle is determined by the group extension. This is explained in [17]. But in the general case the line bundle does not have such a simple description.

4 Generalized KdV equations and the formal Baker function

The *n-th generalized KdV hierarchy* consists of all evolution equations for $n-1$ unknown functions $u_0(x,t), \dots, u_{n-2}(x,t)$ that can be written in the form $\partial L/\partial t = [P, L]$, where L is the n-th order ordinary differential operator

$$L = D^n + u_{n-2}D^{n-2} + \ldots + u_1 D + u_0$$

and P is another differential operator. (As usual, D denotes $\partial/\partial x$.) The possible operators P are essentially determined by the requirement that $[P, L]$

should have order (at most) $n-2$. A very simple description of them is available if we work in the algebra of *formal pseudo-differential operators*, which we denote by Psd.

A formal pseudo-differential operator is, by definition, a formal series of the form

$$R = \sum_{-\infty}^{N} r_i(x) D^i$$

for some $N \in \mathbb{Z}$. The coefficients $r_i(x)$ are supposed to lie in some algebra of smooth functions of x. To multiply two such operators, we need to know how to move D^{-1} across a function $r(x)$: the rule for this,

$$D^{-1} r = \sum_{j=0}^{\infty} (-1)^j r^{(j)} D^{-1-j},$$

follows easily from the basic rule

$$Dr = r\,D + \partial r/\partial x \tag{4.1}$$

determining the composition of *differential* operators. It is easy to check that this makes Pds into an associative algebra.

Proposition 4.2. *In the algebra* Psd, *the operator L has a unique n-th root of the form*

$$L^{1/n} = Q = D + \sum_{1}^{\infty} q_i D^{-1}.$$

The coefficients q_i are certain universal differential polynomials in the u_i; if we assign to $u_i^{(j)}$ the weight $n-i+j$, then q_i is homogeneous of weight $i+1$.

Proof. Equating coefficients of powers of D in the equality $Q^n = L$, we find that

$$u_{n-i-1} = nq_i + \alpha_i,$$

where α_i is some differential polynomial in $q_i, \ldots, q_{i-1}$ (here we have set $u_j = 0$ if $j < 0$). We claim that if we give $q_i^{(j)}$ weight $i+j+1$ then α_i is homogeneous of weight $i+1$. Granting that, it is clear that the above equations can be solved uniquely for the q_i, and that these have the form stated.

The homogeneity of the α_i is most easily seen as follows. Consider the algebra of formal pseudo-differential operators whose coefficients are differential polynomials in the q_i (which we think of for the moment as abstract symbols, rather than as fixed functions of x). Call such an operator homogeneous of weight r if the coefficient of D^i is homogeneous of weight $r-i$ (thus D has weight 1). From the homogeneity of the basic rule (4.1) it follows at once that the product of two operators that are homogeneous of weights r and s is homogeneous of weight $r+s$. Since Q is homogeneous of weight 1, Q^n must be homogeneous of weight n. □

If $R = \sum r_i D^i$ is a formal pseudo-differential operator, we shall write R_+ for the "differential operator part" $R_+ = \sum i \geq 0 r_i D^i$, and $R_- = \sum_{i<0} r_i D^i$. Thus $R = R_+ + R_-$.

Proposition 4.3. *The equation*

$$\partial L / \partial t = [L_+^{r/n}, L] \tag{4.4}$$

is equivalent to a system of evolution equations

$$\frac{\partial u_i}{\partial t} = f_i$$

for the coefficients $u_0, \ldots, u_{n-2}$ *of* L. *The* f_i *are differential polynomials in the* u_j, *homogeneous of weight* $n + r - i$.

Proof. Note first that $L_+^{r/n}$ denotes $(L^{r/n})_+$; $L^{r/n}$ is defined as Q^r. The only part of the proposition that is not obvious from what precedes is that the commutator in (4.4) is an operator of order at most $n - 2$. But that follows at once from the equality

$$[L_+^{r/n}, L] = [-L_-^{r/n}, L].$$

(Of course $L^{r/n}$ and L commute, because they are both powers of $Q = L^{1/n}$.) □

The equation (4.4) is called the *r-th equation of the n-th KdV hierarchy.* It is trivial if r is a multiple of n, because then $L_+^{r/n} = L^{r/n}$ is just an integral power of L.

It is usual to think of the equations (4.4) as defining flows on some space of functions $\{u_0(x), \ldots, u_{n-2}(x)\}$: it is then a basic fact that the flows corresponding to different values of r commute. For this assertion to make sense, we need to identify some class of functions on which the flows can be proved to exist, that is, we need to prove existence and uniqueness theorems for solutions of the equations (4.4). However, the analytic problems involved here are in a sense irrelevant: the basic "infinitesimal" fact underlying the commutativity can be formulated in a purely algebraic way. We refer to [22] for a very simple proof of this algebraic version of the commutativity. In the present paper none of these questions need concern us, because for the special class of solutions that we are interested in, both the existence of the flows and their commutativity will be clear from the construction.

The formal Baker function

The main idea in all studies of solutions of the equations (4.4) is this: as L changes in time, we try to follow the evolution of the eigenfunctions of L by comparing them with the eigenfunctions of the constant operator D^n. To do that, we find an operator K such that $K^{-1}LK = D^n$; then if ψ_0 is an eigenfunction of D^n, $\psi = K\psi_0$ will be an eigenfunction of L. The algebra Psd enables us to give one realization of this idea.

Proposition 4.5. *There is an operator* $K \in \mathrm{Psd}$ *of the form*

$$K = 1 + \sum_1^\infty a_i(x) D^{-i} \tag{4.6}$$

such that $K^{-1}LK = D^n$. *Such a* K *is unique up to right multiplication by a constant coefficient operator of the form* $1 + c_1 D^{-1} + \ldots$.

Proof. Only constant coefficient operators commute with D^n, so the statement about uniqueness is trivial. To prove existence, we simply compare coefficients of powers of D in the equality $LK = KD^n$; this gives equations $\partial a_i / \partial x = \ldots$, where the right hand side involves only a_j with $j < i$; we can therefore solve these equations successively to get suitable a_i. □

Proposition 4.5 can be reformulated as follows.

Proposition 4.7. *The equation* $L\psi = z^n \psi$ *has a solution in the space of formal series of the form*

$$\psi = e^{xz} \left(1 + \sum_1^\infty a_i(x) z^{-i} \right). \tag{4.8}$$

The solution ψ *is unique up to multiplication by a series with constant coefficients of the form* $1 + c_1 z^{-1} + \ldots$.

The series ψ in (4.8) is called the *formal Baker function* of L. The solutions of the KdV equations that we are going to construct are characterized by the property that this formal series actually converges (for $|z|$ sufficiently large). As we mentioned in the introduction, among these solutions are the rank 1 algebro-geometric solutions of Krichever: it was essentially in that context that the function ψ was originally introduced by Baker [3].

The intuitive reason for the equivalence of (4.5) and (4.7) was explained above: since $K^{-1}LK = D^n$, we expect the solutions of the equation $L\psi = z^n\psi$ to be of the form $\psi = K\psi_0$, where ψ_0 is a solution of $D^n\psi_0 = z^n\psi_0$. If we take $\psi_0 = e^{xz}$, then formally it is clear that $\psi = K\psi_0$ should be given by (4.8). We can make this argument rigorous as follows. Let M be the space of all formal expressions $f = e^{xz}\tilde{f}$, where $\tilde{f}$ is a formal series

$$\tilde{f} = \sum_{-\infty}^{N} f_i(x) z^i \quad \text{(for some } N\text{)}.$$

Differential operators act on M in an obvious way: the action of D on M is given by

$$De^{xz}\tilde{f} = e^{xz}(D + z)\tilde{f}.$$

If we let D^{-1} act on M by

$$D^{-1}e^{xz}\tilde{f} = e^{xz}(D + z)^{-1}\tilde{f},$$

where $(D+z)^{-1}$ is interpreted as the formal series $z^{-1} - Dz^{-2} + \ldots$, it is easy to check that this makes M into a module over the algebra Psd. If $R = \sum r_i(x)D^i \in$ Psd, then

$$Re^{xz} = e^{xz}\left(\sum r_i(x)z^{-i}\right),$$

so that M is in fact a free Psd-module of rank 1, with generator $e^{xz} \in M$.

The KP equations

It will often be convenient for us to regard the n-th KdV hierarchy (for any n) as embedded in a certain "universal KdV hierarchy" of evolution equations in infinitely many variables; for brevity we shall follow [5] and call these equations the KP (for Kadomtsev-Petviashvili) hierarchy. The KP equations are defined as follows. Let Q be a general first order formal pseudo-differential operator of the form

$$Q = D + \sum_1^\infty q_i(x)D^{-i}$$

(in general, such a Q will not be the n-th root of a differential operator for any n).

Proposition 4.9. *The equation*

$$\frac{\partial Q}{\partial t} = [Q_+^r, Q] \tag{4.10}$$

is equivalent to a system of evolution equations

$$\frac{\partial q_i}{\partial t} = f_i$$

for the (infinitely many) functions $q_i(x,t), i \geq 1$. The f_i are certain universal differential polynomials in the q_i, homogeneous of weight $r+i+1$ if we give $q_i^{(j)}$ weight $i+j+1$.

The proof is the same as that of (4.3). We call (4.10) the r-th equation of the KP hierarchy.

Proposition 4.11. *The assignment $L \mapsto L^{1/n} = Q$ sets up a 1-1 correspondence between solutions of the n-th KdV hierarchy and solutions Q of the KP hierarchy such that Q^n is a differential operator.*

Proof. It is trivial that if Q satisfies (4.10) then $L = Q^n$ satisfies (4.4). We refer to [22] for the proof of the converse, which is only slightly harder. □

The scaling transformation

Proposition 4.12. *Let $Q = D + \sum q_i D^{-i}$ be any solution of the r-th equation* (4.10). *For any non-zero complex number λ, let $R_\lambda Q = D + \sum q_i^{(\lambda)} D^{-i}$, where the coefficients $q_i^{(\lambda)}$ are defined by*

$$q_i^{(\lambda)}(x,t) = \lambda^{i+1} q_i(\lambda x, \lambda^r t).$$

Then $R_\lambda Q$ is another solution of (4.10).

Proof. This follows trivially from the assertion in (4.9) about the homogeneity of the f_i.

We call the operation R_λ the *scaling transformation* of the solutions to the KP equations. Notice that each variable gets rescaled by the power of λ corresponding to its weight. The scaling transformations clearly act on the solutions to the n-th KdV hierarchy (for any n). □

Note on the literature — Our construction of the KdV equations follows closely the exposition in [14]. The basic idea of using fractional powers of L first appeared in the 1976 paper of Gel'fand and Dikii [9], and has been used extensively in the literature since then. In [5] this idea is attributed to Sato (1981).

5 The Baker function

In this section Gr and $\mathrm{Gr}^{(n)}$ will denote the component of the Grassmannians consisting of spaces of virtual dimension zero. We are going to associate to each $W \in \mathrm{Gr}$ a "Baker function" ψ_W, and also a sequence of differential operators defined in terms of ψ_W.

We recall from §2 that the group Γ_+ of holomorphic maps $g : D_0 \to \mathbb{C}^\times$ with $g(0) = 1$ acts on Gr. Given a space $W \in \mathrm{Gr}$, we set

$$\Gamma_+^W = \{g \in \Gamma_+ : g^{-1}W \text{ is transverse to } H_-\}.$$

From now on we shall refer to spaces transverse to H_- simply as *transverse.* From §3 it follows that Γ_+^W is the complement of the zero set of the τ-function $\tau_W : \Gamma_+ \to \mathbb{C}$; in particular it is a dense open subset of Γ_+. (We admit for the moment the fact that Γ_+^W is not empty, that is, that the holomorphic function τ_W is not identically zero: this will be proved in §8.)

Proposition 5.1. *For each $W \in \mathrm{Gr}$ there is a unique function $\psi_W(g,z)$, defined for $g \in \Gamma_+^W$ and $z \in S^1$, such that*

(i) $\psi_W(g,\cdot) \in W$ for each fixed $g \in \Gamma_+^W$

(ii) ψ_W has the form

$$\psi_W = g(z)(1 + \sum_1^\infty a_i(g) z^{-i}). \tag{5.2}$$

The coefficients a_i are analytic functions on Γ_+^W; they extend to meromorphic functions on the whole of Γ_+.

The proof of the last sentence depends on the properties of the τ-function, and will be given later in this section. The rest of the proposition is trivial: the infinite series in (5.2) is simply the unique function of that form that lies in the transverse space $g^{-1}W$, that is, it is the inverse image of 1 under the orthogonal projection $g^{-1}W \to H_+$.

We call ψ_W the *Baker function* of W.

Now, each $g \in \Gamma_+$ can be written uniquely in the form

$$g(z) = \exp(xz + t_2 z^2 + t_3 z^3 + \dots) \tag{5.3}$$

with $x, t_i \in \mathbb{C}$. When g is written in this way, we shall write $\psi_W(x, \mathbf{t}, z)$ instead of $\psi_W(g, z)$. Here $\mathbf{t}$ stands for $(t_2, t_3, \dots)$. In this notation, ψ_W is a "function of infinitely many variables" of the form

$$\psi_W(x, \mathbf{t}, z) = \exp(xz + t_2 z^2 + \dots) \left(1 + \sum_1^\infty a_i(x, \mathbf{t}) z^{-i}\right). \tag{5.4}$$

Proposition 5.5. *For each integer $r \geq 2$, there is a unique differential operator P_r of the form*

$$P_r = D^r + p_{r2} D^{r-2} + \dots + p_{r,r-1} D + p_{rr}$$

such that

$$\frac{\partial \psi_W}{\partial t_r} = P^r \psi_W. \tag{5.6}$$

(Here as usual $D = \partial/\partial x$.) The coefficients p_{ri} are certain universal differential polynomials in the functions a_i in (5.4).

Proof. From (5.4), we have

$$\frac{\partial \psi_W}{\partial t_r} = g(z)(z^r + a_1 z^{r-1} + O(z^{r-2})).$$

On the other hand, $D^r \psi_W$ also has this form, and in general we have

$$D^q \psi_W = g(z)(z^q + O(z^{q-1})).$$

Comparing coefficients, we see at once that there is a unique operator P_r of the form stated such that

$$\frac{\partial \psi_W}{\partial t_r} - P_r \psi_W = g(z)(O(z^{-1})). \tag{5.7}$$

Now, since ψ_W lies in W for each fixed $(x, \mathbf{t})$, the same is true of the derivatives $\partial\psi_W/\partial t_r$ and $D^q\psi_W$. Hence the left hand side of (5.7) lies in W for each fixed value of $(x, \mathbf{t})$ for which it is defined, that is, for which the corresponding g belongs to Γ_+^W. But the right hand side of (5.7) belongs to gH_-. As $g^{-1}W$ is transverse, both sides must vanish. □

Now let

$$K = 1 + \sum_1^\infty a_i(x, \mathbf{t}) D^{-i}$$

be the formal integral operator corresponding to ψ_W (see §4). Equation (5.6) can be written in the form

$$\frac{\partial K}{\partial t_r} + KD^r = P_r K, \tag{5.8}$$

so that in particular we have

$$P_r = (KD^r K^{-1})_+ = Q_+^r,$$

where we have set $Q = KDK^{-1}$. Thus Q is a formal pseudo-differential operator of the form

$$Q = D + \sum_1^\infty q_i(x, \mathbf{t}) D^{-i}.$$

Proposition 5.9. *The coefficients q_i of Q satisfy the equations of the KP hierarchy; that is, we have*

$$\frac{\partial Q}{\partial t_r} = [Q_+^r, Q].$$

Each q_i is a meromorphic function of all the variables $(x, \mathbf{t})$.

Proof. Differentiating the relation defining Q and rewriting, we find

$$\frac{\partial Q}{\partial t_r} = [(\partial K/\partial t_r)K^{-1}, Q].$$

On the other hand, from (5.8) we have

$$\frac{\partial K}{\partial t_r} K^{-1} = P_r - KD^r K^{-1} = Q_+^r - Q^r,$$

so the proposition follows at once. □

Recall from §2 that rescaling z induces an action $W \mapsto R_\lambda W$ of the semigroup of complex numbers λ with $|\lambda| \leq 1$ on Gr.

Proposition 5.10. *The Baker function corresponding to the space $R_\lambda W$ is given by*

$$\psi_{R_\lambda W}(x, t_2, t_3, \ldots; z) = \psi_W(\lambda x, \lambda^2 t_2, \lambda^2 t_3, \ldots; \lambda^{-1} z).$$

If Q is the solution of the KP equations corresponding to W, then the solution corresponding to $R_\lambda W$ is $R_\lambda Q$ (see (4.12)).

The proof is trivial.

We now specialize to the case $W \in \mathrm{Gr}^{(n)}$.

Proposition 5.11. *If $W \in \mathrm{Gr}^{(n)}$, then*

$$P_n \psi_W = z^n \psi_W.$$

Moreover, the functions a_i, and hence also all the operators P_r, are independent of $t_n, t_{2n}, t_{3n}, \ldots$.

Proof. From (5.4) and (5.6) we see that

$$P_n \psi_W - z^n \psi_W = g(z) \sum_1^\infty \frac{\partial a_i}{\partial t_n} z^{-i}.$$

For $W \in \mathrm{Gr}^{(n)}$, the left hand side of this expression lies in W for each fixed $(x, \mathbf{t})$; it therefore vanishes by the same argument as in the proof of (5.5). That proves the first statement in the proposition, and also that the a_i are independent of t_n. Since obviously $\mathrm{Gr}^{(n)} \subset \mathrm{Gr}^{(rn)}$ for all $r \geq 1$, the a_i are independent of t_{rn} too. □

Since the a_i are independent of t_n, the operator K is also, so (5.8) gives

$$P_n = K D^n K_{-1} = Q^n.$$

Thus if $W \in \mathrm{Gr}^{(n)}$ then Q^n is a *differential* operator. Write L for $P_n = Q^n$; then L has the form $D^n + u_{n-2} D^{n-2} + \ldots + u_0$. Combining (5.9) and (4.11), we get the main result of this section.

Corollary 5.12. *If $W \in \mathrm{Gr}^{(n)}$, the coefficients of the operator L satisfy the equations of the n-th KdV hierarchy, that is, we have*

$$\frac{\partial L}{\partial t_r} = [L_+^{r/n}, L].$$

Let us reformulate this slightly. For each $W \in \mathrm{Gr}^{(n)}$, let L_W denote the operator L evaluated for $t_2 = t_3 = \ldots = 0$. The coefficients $u_0, \ldots, u_{n-2}$ of L_W are functions of one variable x: they are the "initial values" of the KdV flows. Let $\mathcal{C}^{(n)}$ be the space of all L_W for $W \in \mathrm{Gr}^{(n)}$. The map $\mathrm{Gr}^{(n)} \to \mathcal{C}^{(n)}$ is not one to one: however, from (4.7) we see that $L_W = L_{W'}$ precisely when $W = \gamma W'$, where γ is a function of the form $1 + c_1 z^{-1} + \ldots$. Since multiplication by γ commutes with the action of Γ_+, we can restate (5.12) as follows.

Proposition 5.13. *The action of* Γ_+ *on* $\mathrm{Gr}^{(n)}$ *induces an action on the space* $\mathcal{C}^{(n)}$. *The flow* $W \mapsto \exp(t_r z^r)W$ *on* $\mathrm{Gr}^{(n)}$ *induces the r-th KdV flow on* $\mathcal{C}^{(n)}$.

Since Γ_+ is commutative, it is obvious that the different KdV flows on $\mathcal{C}^{(n)}$ commute.

Examples

To obtain the simplest interesting example of a space in $\mathrm{Gr}^{(n)}$ we choose $p \in \mathbb{C}$ so that $0 < |p| < 1$, and $\lambda \in \mathbb{C}^\times$, and define $W_{p,\lambda}$ as the L^2 closure of the space of functions f which are holomorphic in $|z| \leq 1$ except for a possible simple pole at the origin, and which satisfy $f(-p) = \lambda f(p)$.

The Baker function of $W_{p,\lambda}$ must be of the form

$$\psi(\mathbf{t}, z) = e^{\sum t_k z^k}(1 + a(\mathbf{t})/z).$$

(We write here $\mathbf{t} = (t_1, t_2, \dots)$, where t_1 is identified with x.) From the condition $\psi(\mathbf{t}, -p) = \lambda\psi(\mathbf{t}, p)$ we find

$$a(\mathbf{t}) = -p \tanh(\theta + \alpha),$$

where $\theta = \sum_{k \text{ odd}} p^k t_k$, and $e^{2\alpha} = \lambda$.

The second-order differential operator L such that $L\psi = z^2\psi$ is $D^2 - 2a'$, i.e.

$$D^2 + 2p^2 \operatorname{sech}^2(\theta + \alpha).$$

This is the well known one-soliton solution of the KdV equation.

More generally, we have the subspace $W_{p,\lambda}$ introduced in §3 which depends on m points $p_1, \dots, p_m$ of the disc $|z| < 1$ and m parameters $\lambda_1, \dots, \lambda_m \in \mathbb{C}^\times$. The corresponding solution of the KdV equation is called the *m-soliton* solution. We shall give an expression for it below in terms of the τ-function which was calculated in §3; but let us at present notice the obvious fact that it depends on t only through $e^{\theta_i + \alpha_i}$, where $\theta_i = \sum_{k \text{ odd}} p_i^k t_k$ and $e^{2\alpha_i} = \lambda_i$. This is because the orbit of $W_{p,\lambda}$ under Γ_+ is isomorphic to $(\mathbb{C}^\times)^m$: in fact if $\gamma : D_0 \to \mathbb{C}^\times$ is an element of Γ_+ then $\gamma \cdot W_{p,\lambda} = W_{p,\mu}$, where $\mu_i = \gamma(p_i)\gamma(-p_i)^{-1}\lambda_i$.

The Baker function and the τ-function

We now turn to the proof of the last part of (5.1), concerning the properties of the functions a_i. It depends on the formula (5.14) below, relating the Baker function to the τ-function, which we mentioned in the Introduction as central to the theory. We return to the case of an arbitrary space $W \in \mathrm{Gr}$ (not necessarily in any $\mathrm{Gr}^{(n)}$). Let us write

$$\tilde{\psi}_W(g, z) = 1 + \sum_1^\infty a_i(g) z^{-1}$$

for the infinite series in (5.2). Clearly $\tilde{\psi}_W$ extends to an analytic function of z in the region $|z| > 1$ (for each fixed $g \in \Gamma_+^W$). For $\zeta \in \mathbb{C}$ we write q_ζ for the map

$$q_\zeta(z) = 1 - z/\zeta.$$

Obviously

$$q_\zeta \in \Gamma_+ \quad \text{for } |\zeta| > 1.$$

Proposition 5.14. *For $g \in \Gamma_+^W$ and $|\zeta| > 1$ we have*

$$\tilde{\psi}_W(g,\zeta) = \tau_W(g \cdot q_\zeta)/\tau_W(g).$$

Proof. It follows easily from (3.2) that the right hand side is equal to $\tau_{g^{-1}W}(q_\zeta)$. The left hand side is characterized as the unique function of the form $1+\sum a_i\zeta^{-i}$ whose boundary value as $|\zeta| \to 1$ lies in the transverse space $g^{-1}W$. Hence the proposition follows at once if we apply the next lemma to $g^{-1}W$. □

Proposition 5.15. *Let $W \in \mathrm{Gr}$ be transverse, and let f_0 be the unique element of H_- such that $1 + f_0 \in W$. Then for $|\zeta| > 1$, we have*

$$\tau_W(q_\zeta) = 1 + f_0(\zeta).$$

Proof. We use the formula (3.5). When q_ζ^{-1} is written in the form $\begin{pmatrix} a & b \\ 0 & d \end{pmatrix}$, the map $b : H_- \to H_+$ takes $z^{-k} \to \zeta^{-k} q_\zeta^{-1}$; thus $a^{-1}b$ is the map of rank one that takes $f \in H_-$ to the constant function $f(\zeta)$. The map $a^{-1}bA$ is thus also of rank one, and the infinite determinant

$$\tau_W(q_\zeta) = \det(1 + a^{-1}bA)$$

is equal to

$$1 + \operatorname{trace}(a^{-1}bA).$$

Since A maps 1 to $f_0(z)$, the lemma follows. □

If we write g in the form (5.3), and correspondingly write q_ζ in the form

$$q_\zeta(z) = \exp\log(1 - z/\zeta) = \exp\left(-\sum_1^\infty z^k/k\zeta^k\right),$$

then (5.14) takes the form

$$\tilde{\psi}_W(x,\mathbf{t},\zeta) = \tau_W(x - 1/\zeta, t_2 - 1/2\zeta^2, \ldots)/\tau_W(x,\mathbf{t}). \tag{5.16}$$

The fact that the functions $a_i(x,\mathbf{t})$ are meromorphic follows at once from this formula: indeed, if we expand the numerator in a Taylor series, we see that each a_i has the form

$$a_i = P_i\tau/\tau$$

where P_i is a polynomial differential operator in $\partial/\partial x, \partial/\partial t_2, \ldots, \partial/\partial t_i$. For example, we have

$$a_1 = -(\partial\tau/\partial x)/\tau$$
$$a_2 = \frac{1}{2}(\partial^2\tau/\partial x^2 - \partial\tau/\partial t_2)/\tau.$$

The proof of (5.1) is now complete.

We can be more precise about the orders of the poles of the functions a_i. Let us fix the values of the variables $t_2, t_3, \ldots$, say $t_k = t_k^0$, and regard a_i as a meromorphic function of one variable x.

Proposition 5.17. *The poles of the function $a_i(x, \mathbf{t}^0)$ have order at most i.*

In the case of a_1, this follows at once from the formula above and the fact that τ is analytic. For $i > 1$, however, that is not so; for example, if we had $\tau = x^n + t_2$, then the corresponding function $a_2(x, \mathbf{o})$ would have a pole or order n at the origin. Our proof of (5.17) uses the expansion of the τ-function in terms of Schur functions: it will be given in §8.

Corollary 5.18. *For $W \in \mathrm{Gr}^{(n)}$, the differential operators $L_W \in \mathcal{C}^{(n)}$ have only regular singular points (except for the point at infinity); that is, the coefficient u_i of D^i has poles of order at most $n - i$.*

Proof. Recall that $L_W = KD^nK^{-1}$, where $K = 1 + \sum a_i(x)D^{-i}$. Thus if we give $a_k^{(j)}$ weight $k + j$, then u_i is a homogeneous differential polynomial in the a_k of weight $n - i$ (cf. proof of (4.2)). Thus the corollary follows at once from (5.17). □

Finally, we note that the coefficients u_i of L can be expressed directly in terms of the τ-function. In the case $n = 2$, L has the form $D^2 + u_W$, where

$$u_W = -2\frac{\partial a_1}{\partial x} = 2\frac{\partial^2}{\partial x^2}\log\tau_W. \tag{5.19}$$

However for $n > 2$ the explicit formulae become very complicated.

The class $\mathcal{C}^{(n)}$

We have shown how to associate an n-th order differential operator

$$L_W = D^n + u_{n-2}(x)D^{n-2} + \ldots + u_0(x), \tag{5.20}$$

with meromorphic coefficients and only regular singular points, to a space $W \in \mathrm{Gr}^{(n)}$. We shall now describe the inverse process of associating a space W to a differential operator L. This cannot be done for an arbitrary operator, even one which is meromorphic with regular singular points. We do not know an altogether satisfying description of the desired class $\mathcal{C}^{(n)}$; roughly speaking, it consists of the operators whose formal Baker functions converge for large z.

Suppose that L is of the form (5.20), with coefficients defined and smooth in an open interval I containing the origin. The formal Baker function

$$\psi(x,z) = e^{xz}\{1 + a_1(x)z^{-1} + a_2(x)z^{-2} + \dots\}$$

of L was introduced in §4. It is a formal series whose coefficients a_i are smooth functions defined in the interval I, and it is uniquely determined by L if we normalize it so that $\psi(0,z) = 1$. If the n formal series

$$\psi(0,z), D\psi(0,z), \dots, D^{n-1}\psi(0,z) \tag{5.21}$$

(which belong to the field $C((z^{-1})))$ converge for large z, then by a scaling transformation we can make them converge for $|z| > 1-\varepsilon$, so that they define n elements $\psi_0, \psi_1, \dots, \psi_{n-1}$ of our Hilbert space H. We should like to define the corresponding $W \in \mathrm{Gr}^{(n)}$ as the closed z^n-invariant subspace of H generated by $\psi_0, \dots, \psi_{n-1}$, i.e. as γH_+, where γ is the $(n \times n)$-matrix-valued function $(\psi_0, \psi_1, \dots, \psi_{n-1})$ on the circle. (In regarding γ as a matrix-valued function we are using the identification $H \cong H^{(n)}$ described in §2.) For this to be possible we need to know that γ is a loop of winding number zero in $\mathrm{GL}_n(\mathbb{C})$—otherwise W^{alg} would turn out to be bigger than the space spanned algebraically by $\{z^{nk}\psi_i\}_{k\geq 0, 0\leq i<n}$. Making γ explicit according to the formulae of §2, we find

$$\gamma(z) = \begin{pmatrix} 1 & \zeta_1 & \cdots & \zeta_1^{n-1} \\ 1 & \zeta_2 & \cdots & \zeta_2^{n-1} \\ \vdots & & & \\ 1 & \zeta_n & \cdots & \zeta_n^{n-1} \end{pmatrix}^{-1} \begin{pmatrix} \psi_0(\zeta_1) & \psi_1(\zeta_1) & \cdots & \psi_{n-1}(\zeta_1) \\ \psi_0(\zeta_2) & \psi_1(\zeta_2) & \cdots & \psi_{n-1}(\zeta_2) \\ \vdots & & & \\ \psi_0(\zeta_n) & \psi_1(\zeta_n) & \cdots & \psi_{n-1}(\zeta_n) \end{pmatrix}$$

where $\zeta_1, \dots, \zeta_n$ are the n-th roots of z. But $\psi_k(z) \sim z^k$ as $z \to \infty$. So γ is holomorphic in $|z| > 1-\varepsilon$, and $\gamma(z) \to 1$ as $z \to \infty$. By a further rescaling, if necessary, we can therefore ensure that $\gamma(z)$ is invertible for $|z| > 1-\varepsilon$, as we want.

Let us notice that the series $D^k\psi(0,z)$ depend only on the jets (i.e. Taylor series) at the origin of the coefficients u_i of L, and that conversely the series $D^k\psi(0,z)$ determine the jets of the u_i at 0. The space W which we have just constructed has its own Baker function ψ_W, which in turn defines a differential operator L_W with coefficients meromorphic in the entire complex plane. (For brevity, we shall write $\psi_W(x,z)$ for the Baker function evaluated at $t_2 = t_3 = \dots = 0$.) Because both $D^k\psi(0,z)$ and $D^k\psi_W(0,z)$ belong to W and are of the form z^k + (lower terms), it follows by induction on k that they coincide. The jets of the coefficients of L and L_W at 0 must therefore coincide too. This gives us the first half of the following result.

Proposition 5.22.

(i) If the series (5.21) converge in a neighbourhood of $z = \infty$, then there are meromorphic functions $\tilde{u}_0, \ldots, \tilde{u}_{n-2}$ defined in the entire complex plane such that u_i and $\tilde{u}_i$ have the same Taylor series at $x = 0$.

(ii) If the series $\psi(x, z)$ converges for $|z| > R$ for each x in I then the u_i coincide with the $\tilde{u}_i$ in I.

To prove the second statement, let $\varphi_i(x, z^n)$ be the solution of $L_\varphi = z^n\varphi$ characterized by the initial conditions $D^j\varphi_i(0, z^n) = \delta_{ij}$. Each φ_i is an entire function of z^n for x in I. If z is fixed and $|z| > R$, then $\psi(x, z)$ and

$$\sum_{i=0}^{n-1} \varphi_i(x, z^n) D^i\psi(0, z)$$

are both solutions of $L\varphi = z^n\varphi$ with the same initial conditions. They must therefore coincide, and it follows that $\psi(x, \cdot)$ belongs to W for all $x \in I$. As $\psi(x, \cdot)$ also belongs to $e^{xz}(1 + H_-)$, we can conclude that $\psi(x, z) = \psi_W(x, z)$ for all $x \in I$, and so L and L_W coincide in I.

6 Algebraic curves: the construction of Krichever

In Krichever's construction of solutions to the KdV equations the starting point is a collection of data whose most important constituents are a compact Riemann surface X and a holomorphic line bundle $\mathfrak{L}$ over it. Mumford [16] pointed out that the construction still applies more or less unchanged if we allow X to be any complete irreducible complex algebraic curve (possibly singular), and that in that case it is natural to allow $\mathfrak{L}$ to be, more generally, a rank 1 torsion free coherent sheaf over X. (If X is non-singular, any such sheaf is a line bundle.) One reason for including singular curves is that the n-soliton solutions correspond to rational curves with n double points; and even the solutions coming from torsion free sheaves that are not line bundles seem to have nothing very exotic about them (we shall se examples in §7). The inclusion of torsion free sheaves will not cause us any extra difficulty, and will be essential for the proof of theorem 6.10 below.

As well as X and $\mathfrak{L}$, the construction requires three more pieces of data (x_∞, z, φ). Here x_∞ is a non-singular point of X and z^{-1} is a local parameter on X near x_∞. We shall suppose that z is an isomorphism from some closed neighbourhood X_∞ of x_∞ in X to the disc $D_\infty = \{|z| \geq 1\}$ in the Riemann sphere. That can always be achieved by rescaling z (see remark 6.5 below). Finally, φ is a trivialization of $\mathfrak{L}$ over X_∞. We shall use φ to identify sections of $\mathfrak{L}$ over subsets of X_∞ with complex-valued functions. We shall also identify the unit circle S^1 with its inverse image in X under z. We denote by X_0 the complement of the interior of X_∞: thus the closed sets X_∞ and X_0 cover X, and their intersection is S^1.

To all this data we associate the following subspace W of $H = L^2(S^1, \mathbb{C})$: *W is the closure of the space of analytic functions on S^1 that extend to sections of $\mathfrak{L}$ over X_0.*

Proposition 6.1. *The subspace W belongs to the Grassmannian* Gr. *The virtual dimension of W is equal to $\chi(\mathfrak{L})-1$, where as usual $\chi(\mathfrak{L})$ denotes the Euler characteristic* $\dim H^0(X;\mathfrak{L}) - \dim H^1(X;\mathfrak{L})$.

Proof. We observe first that the projection $W \to H_-$ factorizes

$$W \xrightarrow{R_{\lambda^{-1}}} H \xrightarrow{\mathrm{pr}} H_- \xrightarrow{R_\lambda} H_-$$

for suitable λ with $0 < \lambda < 1$ (here R_λ is the scaling transformation discussed in §2). For λ sufficiently close to 1, the map $R_{\lambda^{-1}} : W \to H$ is bounded: for each $f \in W$ is the boundary value of a holomorphic section of $\mathfrak{L}$ over $X \setminus X_\infty$, and (by assumption) the trivialization φ extends over some open set containing X_∞. Thus $R_{\lambda^{-1}}$ simply assigns to $f \in W$ the function $z \mapsto f(\lambda z)$, i.e. f evaluated on a circle slightly inside the boundary of X_0. Since $R_\lambda : H_- \to H_-$ is compact, the projection $W \to H_-$ is too. It follows easily that the projection $W \to H_+$ has closed range.

It remains to show that the projection $W \to H_+$ is a Fredholm operator of the index stated. We shall prove a more precise statement: the kernel and cokernel of the orthogonal projection $W \to zH_+$ are $H^0(x, \mathfrak{L})$ and $H^1(X, \mathfrak{L})$ respectively. Let U_0 and U_∞ be open sets of X containing X_0 and X_∞, and let $U_{0\infty} = U_0 \cap U_\infty$. Because U_0, U_∞, and $U_{0\infty}$ are Stein varieties, we can calculate the cohomology of X with coefficients in any coherent sheaf from the covering $\{U_0, U_\infty\}$; in particular, we have an exact sequence

$$0 \to H^0(X,\mathfrak{L}) \to \mathfrak{L}(U_0) \oplus \mathfrak{L}(U_\infty) \to \mathfrak{L}(U_{0\infty}) \to H^1(X,\mathfrak{L}) \to 0,$$

where $\mathfrak{L}(U)$ denotes the sections of $\mathfrak{L}$ over a subset U of X. Taking the direct limit of this as U_0 and U_∞ shrink to X_0 and X_∞ gives the exact sequence

$$0 \to H^0(X,\mathfrak{L}) \to \mathfrak{L}(X_0) \oplus \mathfrak{L}(X_\infty) \to \mathfrak{L}(S^1) \to H^1(X,\mathfrak{L}) \to 0.$$

Since $\mathfrak{L}$ is torsion free, its sections over X_0 or X_∞ are determined by their restrictions to S^1; thus we can identify $\mathfrak{L}(X_0)$ and $\mathfrak{L}(X_\infty)$ with subspaces of the space $\mathfrak{L}(S^1)$ of real analytic functions on S^1. The two middle terms in the above exact sequence then become

$$W^{\mathrm{an}} \oplus zH_-^{\mathrm{an}} \to H^{\mathrm{an}},$$

the map being the inclusion on the first factor and minus the inclusion on the second factor (we write V^{an} for the analytic functions in a subspace V of H). The kernel and cokernel of this map are the same as those of the projection

$W^{an} \to zH_+^{an}$, so we have only to see that the kernel and cokernel of this do not change when we pass to the completions $W \to zH_+$. But a function in the kernel of this last projection is the common L^2 boundary value of holomorphic functions defined inside and outside S^1, hence it must be analytic: thus the two kernels coincide. That the cokernels coincide too follows easily from the fact that $W \to H_+$ has closed range. □

The same argument shows that the kernel and cokernel of the orthogonal projection $W \to H_+$ can be identified with $H^0(X, \mathcal{L}_\infty)$ and $H^1(X, \mathcal{L}_\infty)$, where $\mathcal{L}_\infty = \mathcal{L} \oplus [-x_\infty]$ is the sheaf whose sections are sections of $\mathcal{L}$ that vanish at x_∞. In particular, W is transverse if and only if we have $H^0(X, \mathcal{L}_\infty) = H^1(X, \mathcal{L}_\infty) = 0$. For readers of [16, 21], we note that it is the sheaf $\mathcal{L}_\infty$, rather than $\mathcal{L}$, that is considered in those papers.

We are mainly interested in spaces of virtual dimension zero; by (6.1), these arise from sheaves with $\chi(\mathcal{L}) = 1$. If $\mathcal{L}$ is a line bundle, the Riemann-Roch theorem shows that its degree is then the arithmetic genus of X.

Combining the construction above with that of §5, we obtain a solution to the KP equations for each set of data $(X, x_\infty, z, \mathcal{L}, \varphi)$ with $\chi(\mathcal{L}) = 1$. This construction is essentially the same as that of Krichever [10, 11]. To be more precise, Krichever considers the case where X is non-singular, and starts off from a positive divisor $\mathcal{D} = \{P_1, \dots, P_g\}$, with $P_i \in X$, of degree g equal to the genus of X. He assumes that no P_i is the point x_∞, and that $\mathcal{D}$ is non-special. "Non-special" means that the line bundle $\mathcal{L}$ corresponding to $\mathcal{D}$ has a unique (up to a constant multiple) holomorphic section, which vanishes precisely at the points P_i; this section therefore defines a trivialization of $\mathcal{L}$ over the complement of $\{P_i\}$, in particular over a neighbourhood of x_∞. If all the points P_i lie outside the disc X_∞, we can use this trivialization; our construction then reduces exactly to Krichever's.

The correspondence that we have described between algebro-geometric data and subspaces of H is obviously not one to one, for the following reason: suppose $\pi : X' \to X$ is a map which is a birational equivalence (that is, intuitively, the curve X is obtained from X' by making it "more singular"). Then we obtain the same space W from a sheaf $\mathcal{L}'$ on X' and from its direct image $\mathcal{L} = \pi_*(\mathcal{L}')$ on X. We shall avoid this ambiguity by agreeing to consider only *maximal* torsion free sheaves on X, that is ones that do not arise as the direct image of a sheaf on a less singular curve. A perhaps more illuminating description of them is as follows. Recall (see [7]) that the rank 1 torsion free sheaves over X (of some fixed Euler characteristic) form a compact moduli space M on which the generalized Jacobian of X (the line bundles of degree zero) acts by tensor product. We claim that the maximal torsion free sheaves form precisely the part of M on which the Jacobian acts freely. Indeed, if $\mathcal{L}$ is any rank 1 torsion free sheaf on X and L is a line bundle of degree zero, then giving an isomorphism $L \oplus \mathcal{L} \cong \mathcal{L}$ is equivalent to giving an isomorphism $L \cong \mathrm{Hom}(\mathcal{L}, \mathcal{L})$; but $\mathrm{Hom}(\mathcal{L}, \mathcal{L})$ is just the structure sheaf of the "least singular" curve X' such that $\mathcal{L}$ is the direct image of a sheaf on X', hence it is $\mathcal{O}_X$ exactly when $\mathcal{L}$ is maximal. Obviously, any line bundle is a maximal torsion free sheaf; and if all the singularities of X are planar, these are the only ones, for in that case (and only in that case) M is an irreducible variety containing the line bundles as a Zariski open subset (see [34]). However, in general there are many maximal torsion free sheaves that are not line bundles: we shall meet simple examples in §7.

Proposition 6.2. *The construction described above sets up a one to one correspondence between isomorphism classes of data $(X, \mathfrak{L}, x_\infty, x, \varphi)$, with $\mathfrak{L}$ maximal, and certain spaces $W \in \mathrm{Gr}$.*

Proof. Let W be the space arising from data $(X, \mathfrak{L}, x_\infty, z, \varphi)$ with $\mathfrak{L}$ maximal. We have to show how to reconstruct all of this data (up to isomorphism) from W alone. Let us recall from eqrefsw:2.6 to definition of the dense subspace W^{alg} of W, consisting of all elements of finite order. Clearly W^{alg} can be identified with the space of algebraic sections of $\mathfrak{L}$ over $X \setminus \{x_\infty\}$. If A is the coordinate ring of the affine curve $X \setminus \{x_\infty\}$, then W^{alg} is the rank one torsion free A-module corresponding to the sheaf $\mathfrak{L}$ restricted to $X \setminus \{x_\infty\}$. On the other hand, let A_W be the ring of analytic functions f on S^1 such that $f \cdot W^{alg} \subset W^{alg}$. Clearly A_W is an algebra containing A (if we identify functions in A with their restrictions to S^1), and W^{alg} is a faithful A_W-module. As W is torsion-free and of rank one as a module over A, it follows that A_W can be identified with an integral subring of the quotient field of A. This means that $\mathrm{Spec}(A_W)$ is a curve of the form $X' \setminus \{x_\infty\}$ (with X' complete) projecting birationally on to $X \setminus \{x_\infty\}$; and so if $\mathfrak{L}$ is maximal we must have $A_W = A$. Thus we have reconstructed from W the curve X, the point x_∞, and the restriction of $\mathfrak{L}$ to $X \setminus \{x_\infty\}$. Finally, the inclusion $W^{alg} \subset \mathbb{C}[z] \oplus H_-$ defines a trivialization of $\mathfrak{L}$ over $X_\infty \setminus \{x_\infty\}$ (and hence the extension of $\mathfrak{L}$ to X); for if $|\zeta| > 1$ then evaluation at ζ defines a map $W^{alg} \to \mathbb{C}$ which induces an isomorphism of the fibre of $\mathfrak{L}$ at ζ with $\mathbb{C}$. (That is clear, because the fibre is canonically $W^{alg}/\mathfrak{m}W^{alg}$, where $\mathfrak{m} \subset A_W$ is the ideal of functions that vanish at ζ.) □

Remark 6.3. The definition of A_W makes sense for any $W \in \mathrm{Gr}$. In general, however, it will be trivial, i.e. $A_W = \mathbb{C}$. (This is clearly the case, for example, when W is the subspace of codimension one in H_+ which was described in (2.10).) The spaces $W \in \mathrm{Gr}$ which arise from algebro-geometrical data are precisely those such that A_W contains an element of each sufficiently large order, or, equivalently, such that the A_W-module W^{alg} has rank 1. That follows at once from the preceding discussion, in view of the fact that the coordinate rings A of irreducible curves of the form $X \setminus \{x_\infty\}$ (where X is complete and x_∞ is a non-singular point) are characterized as integral domains simply by the existence of a filtration

$$\mathbb{C} = A_0 \subset A_1 \subset A_2 \subset \ldots \subset A$$

such that

(i) $A_i \cdot A_j \subset A_{i+j}$,

(ii) $\dim(A_k/A_{k-1}) \leq 1$ for all k, and

(iii) $\dim(A_k/A_{k-1}) = 1$ for all large k.

Remark 6.4. We should point out that for any $W \in \mathrm{Gr}$ the construction of §5 defines a realization of A_W as a commutative ring of differential operators. More precisely, the proof of (5.11) shows that for any $f \in A_W$ there is a unique differential operator $L(f)$ such that

$$L(f)\psi_W = f(z)\psi_W.$$

If $W \in \mathrm{Gr}^{(n)}$, then $z^n \in A_W$, and $L(z^n) = L_W$. In general, the order of the operator $L(f)$ is equal to the order of f.

Remark 6.5. As we saw in §5, a change of local parameter $z \mapsto cz$ (c a non-zero constant) corresponds to acting on the solution to the KP hierarchy by the scaling transformation. Thus the condition that the validity of the parameter z should extend up to $|z| = 1$ is not a serious restriction in our theory.

Remark 6.6. The solution to the KP hierarchy does not depend on the choice of trivialization φ: for a different choice of φ would just multiply W by a function of the form $c_0 + c_1 z^{-1} + \ldots$ (with $c_0 \neq 0$), which, as we know, does not change the solution. Even the space W does not change if we replace φ by $c\varphi$ where c is a non-zero constant; that does not contradict (6.2), because the quintuples of data $(X, x_\infty, z, \mathcal{L}, c\varphi)$ for different $c \neq 0$ are obviously isomorphic.

Remark 6.7. We get a solution to the n-th KdV hierarchy (i.e. $W \in \mathrm{Gr}^{(n)}$) if and only if $z^n \in A_W$; that is, if the local parameter z is such that z^n extends to a meromorphic function on X with no singularities except for the n-th order pole at x_∞. For fixed n, this of course imposes a restriction on which pairs (X, x_∞) can occur: for example if $n = 2$ then X must be hyperelliptic and x_∞ must be a Weierstrass point.

Remark 6.8. An important part of Krichever's theory is the observation that the KdV (or KP) flows correspond to straight line motion on the Jacobian of X. That is easily seen from our point of view as follows. For each $g \in \Gamma_+$, let L_g be the line bundle obtained by taking trivial bundles over X_0 and X_∞ and glueing them by the transition function g on (an open neighbourhood of) S^1. Thus L_g comes equipped with a trivialization φ_g over X_∞. The natural action of Γ_+ on Gr corresponds to the following action on the data $(X, x_\infty, z, \mathcal{L}, \varphi) : g \in \Gamma_+$ acts trivially on the first three components, and on $(\mathcal{L}, \varphi)$ by tensoring with (L_g, φ_g). The action of Γ_+ on solutions to the KP hierarchy thus corresponds simply to $\mathcal{L} \mapsto \mathcal{L} \oplus L_g$. The assertion about straight line motion is now clear in view of the following result.

Proposition 6.9. *The assignment $g \mapsto L_g$ defines a surjective homeomorphism from Γ_+ to the generalized Jacobian of X (which consists of all holomorphic line bundles on X of degree zero).*

Proof. If L is a line bundle on X then $L|X_0$ and $L|X_\infty$ are trivial, for all bundles on affine curves are analytically trivial, and X_0 and X_∞ are contained in affine open sets of X. So $L = L_g$ for some holomorphic function $S^1 \to \mathbb{C}^\times$ whose winding number is the degree of L. We can change g by any element of Γ_- without affecting L_g; and so if g has winding number zero we can choose it in Γ_+. □

Example

Let us return briefly to the subspace $W = W_{p,\lambda} \in \mathrm{Gr}_1^{(2)}$ which was introduced in §3 and discussed further in §5. In this case A_W consists of all polynomials f in z such that $f(-p_i) = f(p_i)$ for each i. This is the coordinate ring of the affine curve whose completion X_p is obtained from the Riemann sphere by identifying the point p_i with $-p_i$ for each i : X_p is a rational curve with m double points. If we take $\xi = z^2$ and $\eta = z(z^2 - p_1^2)\dots(z^2 - p_m^2)$ as generators of A_W then the equation of X_p is

$$\eta^2 = \xi(\xi - p_1^2)^2 \dots (\xi - p_m^2)^2.$$

We remarked in §5 that the orbit of $W_{p,\lambda}$ under Γ_+ consists of all $W_{p,\mu}$, where μ runs through $(\mathbb{C}^\times)^m$. This conforms with (6.9), as $(\mathbb{C}^\times)^m$ is the generalized Jacobian of X_p.

Commuting differential operators

Our last goal in this section is to point out that our results lead directly to a proof of the so-called "Painlevé property" of the stationary KdV equations. Since these have the form $[P, L] = 0$, the result can be formulated as a statement about commuting differential operators.

Theorem 6.10. *Let $L = D^n + u_{n-2}D^{n-2} + \dots + u_0$ be an ordinary differential operator whose coefficients u_i are defined and smooth in a neighbourhood I of the origin in $\mathbb{R}$. Suppose that there exists a differential operator P of order m relatively prime to n that commutes with L. Then the functions u_i extend to meromorphic functions on the whole complex plane, with poles of order at most $n - i$, so that all the finite singular points of L are regular.*

Note that the condition about relatively prime orders is obviously essential: if we omitted it there would be trivial counterexamples to the theorem where $L = P$, or more generally L and P are both polynomials in some operator of lower order.

It is easy to see (for example by conjugating L into D^n by a formal integral operator as in §4) that any operator P that commutes with L is some linear combination

$$P = \sum_0^N c_r L_+^{r/n}$$

of the operators P_r occurring in the definition of the n-th KdV hierarchy. For each fixed sequence of constants $\{c_r\}$, the stationary KdV equation $[P, L] = 0$ is a system of ordinary differential equations for the coefficients $\{u_0, \ldots, u_{n-2}\}$ of L. Let us call such an equation (or the corresponding P) *admissible* if there is some index r relatively prime to n with $c_r \neq 0$. For example, if n is prime, then every non-trivial stationary KdV equation is admissible. If P is admissible, then the algebra generated by L and P contains operators or order relatively prime to n. Thus (6.10) can be formulated as follows: *every solution $\{u_i\}$ of an admissible stationary KdV equation is of the kind stated in the conclusion of* (6.10).

Theorem 6.10 will follow from (5.18) if we show that every operator L satisfying the hypotheses is of the form L_W for some $W \in \mathrm{Gr}^{(n)}$ arising from an algebraic curve. This is well known, and is proved in [16, 21]; however, for completeness we give a self-contained proof, following the approach of Burchnall and Chaundy [4].

Proposition 6.11. *If L and P are commuting differential operators as in* (6.10), *then:*

(i) There is an irreducible polynomial $F \in \mathbb{C}[\xi, \eta]$ of the form

$$F = \xi^m + \ldots \pm \eta^n$$

such that $F(L, P) = 0$.

(ii) For all but a finite number of points (λ, μ) of the affine curve X_F whose equation is $F(\lambda, \mu) = 0$ there is a unique common eigenfunction $\varphi_{\lambda,\mu}$ of L and P such that $\varphi_{\lambda,\mu}(0) = 1$:

$$L\varphi_{\lambda,\mu} = \lambda\varphi_{\lambda,\mu}, \qquad P\varphi_{\lambda,\mu} = \mu\varphi_{\lambda,\mu}.$$

For any $x \in I$, $\varphi_{\lambda,\mu}(x)$ is a meromorphic function on the curve X_F.

(iii) For $x \in I$ the formula Baker functions $\psi_L(x, z)$ and $\psi_P(x, z)$ of L and P both converge for large z, and then

$$\psi_L(x, \lambda^{1/n}) = \psi_P(x, \mu^{1/m}) = \varphi_{\lambda,\mu}(x).$$

(Notice that $\lambda^{1/n}$ and $\mu^{1/m}$ are local parameters at the point at infinity of X_F.)

We begin by proving assertion (i). For any $\lambda \in \mathbb{C}$ let V_λ be the n-dimensional vector space of solutions of $L\varphi = \lambda\varphi$ on I. A basis for V_λ is given by the functions $\varphi_i(x, \lambda)$ for $0 \leq i < n$ such that $\varphi_i^{(j)}(0, \lambda) = \delta_{ij}$. Notice that for any $\varphi \in V_\lambda$ and any k we have

$$\varphi^{(k)}(0) = \sum_{i=0}^{n-1} p_{ki}(\lambda)\varphi^{(i)}(0)$$

where the $p_{ki}(\lambda)$ are polynomials independent of φ.

The operator P maps V_λ into itself. In terms of the basis $\{\varphi_i\}$ the action of P on V_λ is given by an $n \times n$ matrix P_λ of polynomials in λ. Let $F(\lambda, \mu)$ be the characteristic polynomial $\det(\mu - P_\lambda)$. It is not hard to see that $F(\lambda, \mu)$ is a polynomial of degree m in λ: in fact one can show that (up to sign) it is the same as the polynomial obtained by reversing the roles of P and L in the construction. Thus F has the form stated in (i). Consider the differential operator $F(L, P)$. There is at least one solution of $F(L, P)\varphi = 0$ in each V_λ. As a differential operator can have only finitely many linearly independent solutions, this implies that $F(L, P) = 0$. But the same argument shows that if G is any factor of F then $G(L, P) = 0$; so F must be a power of an irreducible polynomial. As $F(\lambda, \mu)$ contains the monomials λ^m and μ^n, the power must divide both n and m. But these are relatively prime, so F must be irreducible.

We next prove assertion (ii). Because the polynomial F is irreducible there are, for all but finitely many values of λ, n distinct solutions μ of $F(\lambda, \mu) = 0$. For each of these values of μ there is (up to a scalar multiple) a unique eigenvector $\varphi_{\lambda,\mu}$ of P_λ in V_λ with eigenvalue μ. We can choose it so that its coordinates with respect to the basis $\{\varphi_i\}$ of V_λ (i.e. its derivatives at 0) are polynomials in λ and μ: for example we can take the coordinates to be the cofactors of any row of the matrix $\mu - P_\lambda$. The value of $\varphi_{\lambda,\mu}$ at 0 cannot vanish identically, for the eigenvectors of P_λ must span V_λ for almost all λ. This permits us to normalize $\varphi_{\lambda,\mu}$ so that $\varphi_{\lambda,\mu}(0) = 1$, except at a finite number of points (λ, μ). The derivatives $\varphi_{\lambda,\mu}^{(i)}(0)$ will then be rational functions of λ and μ.

To see that $\varphi_{\lambda,\mu}(x)$ is meromorphic on X_F we observe that

$$\varphi_{\lambda,\mu}(x) = \sum_{i=0}^{n-1} \varphi_{\lambda,\mu}^{(i)}(0)\varphi_i(x, \lambda).$$

(Note that $\varphi_i(x, \lambda)$ is an entire function of λ.)

To prove (iii) we first observe that not only do we have $L\psi_L = z^n\psi_L$ by definition, but also $P\psi_L = \mu(z)\psi_L$, where $\mu(z)$ is a formal power series belonging to the field $\mathbb{C}((z^{-1}))$ of formal series of the form $\sum_{i=-\infty}^{N} \alpha_i z^i$. To see this, choose K as in §4 so that $K^{-1}LK = D^n$. Then $K^{-1}PK$ commutes with D^n, and so must be a formal pseudodifferential operator $\mu(D)$ with constant coefficients. Thus $PK = K\mu(D)$. Applying both sides to e^{xz} gives $P\psi_L$, as $Ke^{xz} = \psi_L$.

Now we adopt the following point of view. The operators L and P can be thought of as acting on the vector space of jets of functions (of x) at the origin: in other words we replace functions φ by sequences $\{\varphi^{(i)}(0)\}_{i\geq 0}$. Consider the vector space J of formal jets whose components $\varphi^{(i)}(0)$ belong to the field $\mathbb{C}((z^{-1}))$ of formal series. The operator $L - z^n$ acts on J, and has an n-dimensional kernel J_L spanned by the jets of the functions $\varphi_i(x, z^n)$ already mentioned. (Recall that $\varphi_i^{(j)}(0, z^n)$ is a polynomial in z^n.) Now formal series of the form $e^{xz}\sum a_k(x)z^{-k}$ define jets in J, and the jet of ψ_L belongs to J_L. Furthermore P preserves J_L, and $P\psi_L = \mu(z)\psi_L$. On the other hand we already know that the unique eigenvectors of P in J_L, when normalized at 0, are the jets of $\varphi_{\lambda,\mu}$, where $\lambda = z^n$, and μ runs through the n roots of $F(\lambda, \mu) = 0$, which are distinct for large λ. This proves that $\psi_L^{(i)}(0,z) = \varphi_{z^n,\mu(z)}^{(i)}(0)$ for some point $(z^n, \mu(z)) \in X_F$, and hence that the series $\psi_L^{(i)}(0,z)$ and $\mu(z)$ converge for large z.

In the preceding discussion the role of the origin could have been played by any point $x_0 \in I$. Thus we can conclude that if a formal Baker function ψ_{L,x_0} is calculated at x_0 then

$$\psi_{L,x_0}^{(i)}(x_0, z) = \varphi_{\lambda,\mu}^{(i)}(x_0)e^{x_0 z}\varphi_{\lambda,\mu}(x_0)^{-1}.$$

(The factor $e^{x_0 z}\varphi_{\lambda,\mu}(x_0)^{-1}$ on the right occurs because φ_{L,x_0} is normalized by $\varphi_{L,x_0}(x_0,z) = e^{x_0 z}$.) The space $W_{x_0} \in \mathrm{Gr}^{(n)}$ defined by φ_{L,x_0} is therefore related to the space W defined by ψ_L by

$$W_{x_0} = e^{x_0 z}\varphi_{z^n,\mu(z)}(x_0)^{-1}W.$$

But $e^{-x_0 z}\varphi_{z^n,\mu(z)}(x_0)$ does not vanish for large z, and so (after scaling, if necessary) it defines an element γ of the group Γ_-. Thus W_{x_0} and W define the same meromorphic differential operator. The jets of its coefficients coincide with those of L at x_0 and 0 respectively. This proves (iii).

Remark 6.12. Notice that we have proved that L arises by Krichever's construction from the curve X_F and the torsion-free sheaf $\mathcal{L}$ whose space of sections over $X_F \setminus \{\infty\}$ is the space W^{alg} generated by the $\varphi_{\lambda,\mu}^{(i)}(0)$. In particular, this proves (6.10). It is not hard to show that the fibre of $\mathcal{L}$ at any point (λ,μ) of X_F is canonically isomorphic to the joint (λ,μ)-eigenspace of L and P.

Remark 6.13. We believe that theorem 6.10 is "well known" (except possibly for the assertion about the orders of the poles), but our proof seems to be the first complete one available. Krichever [10] noted that "most" of the solutions (that is, the ones coming from non-singular curves X) of the stationary KdV equations are globally meromorphic; our proof is essentially the same as his except that he used the theta function of X where we use the more general τ-function (see §9 below). It might be interesting to give a direct algebro-geometric proof of the theorem, presumably by introducing suitable "theta functions" for singular curves. However, we note that one would have to define a theta function, not merely for each singular curve, but for each orbit of the Jacobian of such a curve acting on the space of maximal torsion free sheaves.

7 Rational Curves

We recall from §2 that Gr_1 is the subspace of Gr consisting of spaces W such that

$$pH_+ \subset W \subset q^{-1}H_+$$

for some polynomials p, q, and that p and q can be chosen so that all their roots lie in the region $|z| < 1$.

Proposition 7.1. *The construction described in the preceding section gives a one to one correspondence between spaces $W \in \mathrm{Gr}_1$ and isomorphism classes of data $(X, \mathcal{L}, x_\infty, z, \varphi)$ as in* (6.2) *such that*

(i) X is a rational curve

(ii) z is a rational parameter on X

(iii) φ extends to an algebraic trivialization of $\mathcal{L}$ over some Zariski open set containing the disc X_∞.

Before giving the proof, we clarify the term "rational parameter". The curve X being rational means that there is an algebraic map f from the Riemann sphere to X which is an isomorphism outside the inverse image of the singular set of X. We can choose f so that $f(\infty) = x_\infty$. By a rational parameter z on X we mean the inverse of such a map f in some neighbourhood of x_∞: note that its domain in fact extends to the whole non-singular part of X. Note also that the rational parameter is uniquely determined up to a linear change $z \mapsto az + b$: for any two of them differ by a birational, hence genuine, automorphism of the Riemann sphere preserving ∞, which must be linear.

Proof of (7.1).
(i) Let $W \in \mathrm{Gr}_1$, and let p and q be polynomials as above. Let W^{alg} and A_W be as in the proof of (6.2). Clearly we have

$$p\mathbb{C}[z] \subset W^{\mathrm{alg}} \subset q^{-1}\mathbb{C}[z], \tag{7.2}$$

from which it follows easily that $pq\mathbb{C}[z] \subset A_W \subset (pq)^{-1}\mathbb{C}[z]$. Since A_W is a ring, we have in fact

$$pq\mathbb{C}[z] \subset A_W \subset \mathbb{C}[z].$$

Thus the inclusion of A_W in $\mathbb{C}[z]$ induces an isomorphism of quotient fields; that shows that the curve $X \setminus \{x_\infty\} = \operatorname{Spec} A_W$ is rational, and that z is a rational parameter on X. From (7.2) it is clear that the A_W-module W^{alg}, and hence also the corresponding sheaf $\mathcal{L}$ on X, has rank 1. It remains to prove (iii). Let $z_0 \in \mathbb{C}$; then evaluation at z_0 gives a map $e(z_0) : W^{\mathrm{alg}} \to \mathbb{C}$ which is defined provided that z_0 is not a root of q, and non-zero provided that z_0 is not a root of p. Let P_0 be the point of X corresponding to z_0, and let $\mathfrak{m} \subset A_W$ be its maximal ideal. Then $e(z_0)$ defines a map from the fibre $W^{\mathrm{alg}}/\mathfrak{m}W^{\mathrm{alg}}$ of $\mathcal{L}$ over P_0 to $\mathbb{C}$, which is an isomorphism provided that z_0 satisfies the two conditions above and that P_0 lies in the open set of X over which $\mathcal{L}$ is a line bundle. That completes the proof that W gives rise to algebro-geometric data of the kind stated in the proposition.

(ii) Conversely, suppose we are given data $(X, \mathcal{L}, x_\infty, z, \varphi)$ of the kind listed in the proposition: we have to see that the corresponding W belongs to Gr_1. Let $B \subset X$ be the finite set of points over which φ is not defined. If necessary we enlarge B to include all the singular points of X. Let $\{z_1, \ldots, z_r\} \subset \mathbb{C}$ be the inverse image of B under the map $f : s^2 \to X$ whose inverse is the parameter z. Then we can identify the sections of $\mathcal{L}$ over $X \setminus B$ with the sections of a trivialized line bundle over $S^2 \setminus \{z_1, \dots, z_r\}$. Thus W^{alg}, which is the space of sections of $\mathcal{L}$ over $X \setminus \{x_\infty\}$, is identified with a subspace of the space $F(z_1, \dots, z_r; -\nu_1, \dots, -\nu_r)$ of rational functions of z that are holomorphic except for poles of prescribed orders ν_i at the points z_i. More precisely, W^{alg} is the subspace of $F(z_i, -\nu_i)$ consisting of all functions whose Laurent series at the points z_i satisfy some finite set of linear conditions. These conditions are certainly satisfied by all polynomials that vanish to suitably high orders μ_i at the points z_i. It follows that if we set $p = \prod(z - z_i)^{\mu_i}$, $q = \prod(z - z_i)^{\nu_i}$, then we have

$$p\mathbb{C}[z] \subset W^{alg} \subset q^{-1}\mathbb{C}[z].$$

Passing to the L^2 closures, we find $pH_+ \subset W \subset q^{-1}H_+$, as required. □

Next recall that Gr_0 is the subspace of Gr_1 consisting of spaces W for which the polynomials p and q can be taken to be powers of z. If we follow through the above proof in that case, we obtain the following.

Proposition 7.3. *The construction described in §6 gives a one to one correspondence between spaces $W \in \mathrm{Gr}_0$ and isomorphism classes of data $(X, x_\infty, z, \mathcal{L}, \varphi)$ as in (6.2) such that*

(i) X is a rational curve with just one irreducible (i.e. cusp-like) singularity

(ii) z is a rational parameter on X such that the singular point x_0 corresponds to $z = 0$

(iii) φ extends to an algebraic trivialization of $\mathcal{L}$ over the whole non-singular part $X \setminus \{x_0\}$ of X.

The term "irreducible" in (i) means that when we resolve the singularity we still get only one point, so that z is in fact a bijection between X and the Riemann sphere. Note that z and φ are now both uniquely determined up to multiplication by non-zero constants. The fact that φ is unique means that the correspondence between spaces in Gr_0 and solutions to the KP hierarchy is one to one. Indeed it is easy to see directly that if $W \in \mathrm{Gr}_0$ and γ is a function of the form $1 + c_1 z^{-1} + \dots$, then γW cannot belong to Gr_0 unless $\gamma = 1$.

The subspaces $W \in \mathrm{Gr}_0$ provide many simple examples of maximal torsion free sheaves that are not line bundles. Indeed let $W = H_S$, where $S \subset \mathbb{Z}$ is a set of virtual cardinal zero. Then $W \in \mathrm{Gr}_0$, and we claim that the corresponding maximal torsion free sheaf is seldom a line bundle. Here W^{alg} is the vector space spanned by $\{z^s\}_{s \in S}$. Let R be the semi-group of strictly positive integers r such that $S + r \subset S$. Then A_W is the algebra spanned by 1 and $\{z^r\}_{r \in R}$, and the maximal ideal $\mathfrak{m}$ of A_W corresponding to the singular point $z = 0$ is spanned by $\{z^r\}_{r \in R}$. The dimension of the fibre $W^{alg}/\mathfrak{m}W^{alg}$ of the sheaf $\mathcal{L}$ over the singular point is thus the number of elements of $S \setminus S'$, where we have set

$$S' = \bigcup_{r \in R} (S + r).$$

Unless this number is 1, the maximal torsion free sheaf $\mathcal{L}$ is not a line bundle. The simplest example is when $S = \{-1, 0, 2, 3, \ldots\}$; then $R = \{3, 4, 5, \ldots\}$ and $S' = \{2, 3, \ldots\}$. In this case the dimension of the singular fibre of $\mathcal{L}$ is 2. Note that since the algebra $A_W = \mathbb{C}[z^3, z^4, z^5]$ needs more than two generators, the singularity here is not planar: this conforms with our observation in §6 that in the planar case every maximal torsion free sheaf is a line bundle.

The case $n = 2$

In general, the isomorphism classes of data listed in (7.3) are hard to classify. However, if we confine ourselves to the case of $\mathrm{Gr}_0^{(2)}$, then many simplifications take place: perhaps the most important is that the orbits of the group Γ_+ in $\mathrm{Gr}_0^{(2)}$ coincide with the cells in the cell decomposition described in §2. Here we give a brief description of the situation, leaving most of the (easy) proofs to the reader. For simplicity, what follows will refer only to the component of $\mathrm{Gr}_0^{(2)}$ consisting of spaces of virtual dimension zero.

We recall from §2 that $\mathrm{Gr}_0^{(2)}$ has a cell decomposition with cells indexed by the sets $S \in \mathcal{S}$ such that $S + 2 \subset S$. It is easy to see that the only such S are the sets S_k given by

$$S_k = \{-k, -k+2, -k+4, \ldots, k, k+1, k+2, \ldots\}.$$

We denote by C_k the corresponding cell in $\mathrm{Gr}_0^{(2)}$; it has complex dimension k, and consists of all W of virtual dimension zero such that $z^k H_+ \subset W \subset z^{-k} H_+$ and k is the smallest number with this property. It is not hard to see directly that these W form a k-dimensional cell: such a space W contains elements w of the form

$$w = z^{-k} + \alpha_1 z^{-k+1} + \ldots + \alpha_{2k-1} z^{k-1},$$

and $\{w, z^2 w, \ldots, z^{2k-2} w\}$ is then a basis for $W/z^k H_+$. Thus w determines W uniquely. The converse is not true; however, the coefficients α_i can be normalized in various ways, of which the most convenient for us is the following.

Lemma 7.4. *Each $W \in C_k$ contains a unique element w of the form*

$$w = z^{-k} \exp(a_1 z + a_2 z^3 + \ldots + a_k z^{2k-1}).$$

The correspondence $W \leftrightarrow (a_1, \ldots, a_k)$ gives us an explicit isomorphism of the cell C_k with $\mathbb{C}^k$; the centre of the cell (corresponding to the origin in $\mathbb{C}^k$) is the space H_{S_k}, which we shall denote simply by H_k. It is clear from (7.4) that the subgroup $\{\exp(tz^{2r-1})\}$ of Γ_+ acts on C_k by translating the r-th coordinate a_r. In particular, we see that C_k is precisely the orbit of H_k under Γ_+.

It is interesting to see how this description of the orbits of Γ_+ fits in with the algebrogeometric one implicit in (7.3). The main points are as follows. First, if $W = H_k$ then $A_W = \mathbb{C}[z^2, z^{2k+1}] = A_k$, say. Let $X_k = (\mathrm{Spec}\, A_k) \cup \{x_\infty\}$ be the corresponding complete curve, and let J_k be the Jacobian of X_k (parametrizing line bundles of degree zero). If we use the point x_∞ to identify the spaces of line bundles of different degrees, then the torsion free sheaf over X_k corresponding to the space H_k is the neutral element in J_k; indeed, it is clear that $H_k^{\text{alg}} = z^{-k}A_k$. Hence the orbit of H_k under Γ_+, that is, the cell C_k, can be identified with the Jacobian J_k. The fact that the cells C_k exhaust $\text{Gr}_0^{(2)}$ implies that the curves X_k are the only ones that arise from a space $W \in \text{Gr}_0^{(2)}$, and also that every maximal torsion free sheaf over one of the curves X_k is a line bundle. Both of these facts can be seen directly: it is easy to show that the A_k are the only subalgebras of $\mathbb{C}[z]$ containing z^2 and also an odd power of z; and, as we have observed before, the assertion about the sheaves is true for any curve with planar singularities (a simple proof that covers our present case (X degenerate hyperelliptic) can be found in [8]; in fact the assertion for singularities of the type $y^n = x^m$ is implicitly contained in [4(c)]). To see directly that J_k is a k-dimensional cell, we can use the exponential sheaf theory exact sequence: since $H^1(X_k, \mathbb{Z}) = 0$, this gives an isomorphism $H^1(X_k, \mathcal{O}) \cong J_k$. The dimension k of the vector space $H^1(X_k, \mathcal{O})$ can be calculated as the number of "gaps" in the ring A_k, that is, the number of positive integers r such that A_k does not contain a polynomial of order r. The algebras A_k are invariant under $z \mapsto cz$, which implies that the pairs (X_k, cz) for different $c \neq 0$ are isomorphic, so that the scaling transformations can be viewed as flows on the Jacobians J_k. Indeed, from (7.4) we see that the scaling flow on the cell C_k is given by

$$R_\lambda(a_1, \dots, a_k) = (\lambda^{-1}a_1, \dots, \lambda^{-2k+1}a_k).$$

Finally, it is interesting to consider the closure $\overline{C}_k$ of the cell C_k: this the union of all the cells C_r with $r \leq k$. Alternatively, $\overline{C}_k$ consists of all $W \in \text{Gr}_0^{(2)}$ such that $A_k W \subset W$. Hence each point of $\overline{C}_k$ determines a torsion free sheaf (in general not maximal) over X_k; in fact we get a bijective map $\overline{C}_k \to M_k$, where M_k is the moduli space of rank one torsion free sheaves of some fixed Euler characteristic over X_k (see [7]). The closed cell $\overline{C}_k$ is an algebraic variety, for it is an algebraic subset (given by the condition $z^2 W \subset W$) of the Grassmannian of k-dimensional subspaces of $z^{-k}H_+/z^k H_+$, and it is fairly clear that the above construction gives us an algebraic family of sheaves over X_k: that implies that the bijection $\overline{C}_k \to M_k$ is an algebraic map. Unfortunately, we cannot assert that it is an isomorphism of algebraic varieties: for example, $\overline{C}_1$ is a one-dimensional projective space (non-singular), whereas M_1 is isomorphic to the curve X_1, which has a cusp. (We do not know a precise reference for this fact, but P. Deligne and T. Ekedahl have kindly pointed out to us that it follows easily from (2.6.1) in [24].) In general, we expect that $\overline{C}_k$ is the normalization of M_k. The inclusion $A_k \subset A_{k-1}$ induces a map $X_{k-1} \to X_k$, and hence (taking the direct image of sheaves) a map $M_{k-1} \to M_k$. This map corresponds to the inclusion $\overline{C}_{k-1} \subset \overline{C}_k$, and identifies M_{k-1} with the boundary of M_k, that is, with the space of torsion free sheaves over X_k that are not line bundles.

The solutions to the KdV equations corresponding to the points of $\text{Gr}_0^{(2)}$ have been much studied (see [1, 2]): the cell C_k corresponds to the solutions to the KdV hierarchy flowing out of the initial value

$$u(x,0,0,\dots) = -k(k+1)/x^2.$$

(This is the initial value defined by the space H_k, as will become clear in §8, when we describe the τ-functions of the spaces H_S.) It is known that these exhaust the rational solutions to the KdV hierarchy that vanish at $x = \infty$.

8 The τ-function and Schur functions

We have already given explicit formulae (3.4) and (3.5) for the τ-function as an infinite determinant. It is useful for some purposes to make the formula even more explicit by expanding the determinants in a certain way: the result is that the τ-function can be written as an infinite linear combination of Schur functions.

We begin by reviewing the basic definitions concerning partitions and Schur functions (for more details see, for example, [13]). By a *partition* we mean an infinite sequence $\nu = (\nu_0, \nu_1, \nu_2, \dots)$ of non-negative integers such that $\nu_0 \geq \nu_1 \geq \nu_2 \geq \dots$ and all except a finite number of the ν_i are zero. The number $|\nu| = \sum \nu_i$ is called the *weight* of ν. To each partition ν there is associated a *Schur function* F_ν. This is a polynomial with integer coefficients in a sequence of indeterminates $(h_1, h_2, h_3, \dots)$; it is homogeneous of weight $|\nu|$ when h_i is given weight i. One way to define it is as the $r \times r$ determinant

$$F_\nu(\mathbf{h}) = \det(h_{\nu_i - i + j}), \quad (0 \leq i, j \leq r-1)$$

where r is any number sufficiently large so that $\nu_i = 0$ for $i \geq r$. Here it is understood that $h_0 = 1$ and $h_i = 0$ for $i < 0$; it is clear that the value of the determinant does not depend on the choice of r. One reason for the importance of Schur functions is that they are characters of the general linear groups $\mathrm{GL}_N(\mathbb{C})$: to each partition ν there corresponds an irreducible representation of $\mathrm{GL}_N(\mathbb{C})$ (for any large N), and its character χ_ν is given by $\chi_\nu(A) = F_\nu(\mathbf{h})$, where

$$1 + \sum_1^\infty h_i z^i = \{\det(1 - Az)\}^{-1},$$

that is, the h_i are the "complete homogeneous symmetric functions" of the eigenvalues of the matrix A. In our context, however, the Schur functions arise in a purely formal manner, and the representations of $\mathrm{GL}_N(\mathbb{C})$ do not seem to be relevant.

Let $\mathcal{S}_0$ denote the set of all subsets $S \subset \mathbb{Z}$ of virtual cardinal zero (see §2); that is, $\mathcal{S}_0$ consists of all strictly increasing sequences $S = \{s_0, s_1, s_2, \dots\}$ of integers such that $s_i = i$ for all except a finite number of indices i.

Lemma 8.1. *There is a one to one correspondence between elements of* $\mathcal{S}_0$ *and partitions, given by* $S \leftrightarrow \nu$ *where* $\nu_i = i - s_i$.

The proof is trivial. Notice that the weight $|\nu|$ of a partition is equal to the length $\ell(S)$ for the corresponding S; that is, it is the codimension of the stratum $\sum_S$ of Gr. In what follows we shall write F_S for the Schur function of the partition corresponding to an element $S \in \mathcal{S}_0$.

Recall from §2 that if $S \in \mathcal{S}_0$, then $H_S \in \mathrm{Gr}$ is the closed subspace of H spanned by $\{z^s\}_{s\in S}$.

Proposition 8.2. *Let $W = H_S$, where $S \in \mathcal{S}_0$. Then the τ-function of W is given by*

$$\tau_W(g) = F_S(\mathbf{h})$$

where we have set

$$g^{-1} = 1 + \sum_1^\infty h_i z^i.$$

Proof. We use the formula (3.4). As an admissible basis for H_S, we choose $w_i = z^{s_i}$ where $S = \{s_0, s_1, s_2, \dots\}$. Also, the map $(a, b) : H \to H_+$ is just $f \mapsto (fg^{-1})_+$, where the subscript $+$ denotes orthogonal projection onto H_+. Thus if g^{-1} is expanded as in the statement of the proposition, it follows at once that the matrix of the map $aw_+ + bw_- : H_+ \to H_+$ is

$$(h_{j-s_i}) = (h_{(i-s_i)-i+j}), \quad (i, j \in \mathbb{N}).$$

Since $s_i = i$ for large i, this matrix is strictly (that is, with 1's on the diagonal) upper triangular apart from a finite block in the top left corner. The matrix of the map $a : H_+ \to H_+$ is strictly upper triangular, so it follows easily that the τ-function

$$\det(w_+ + a^{-1}bw_-) = \det a^{-1}(aw_+ + bw_-)$$

is equal to the determinant of this finite block. That proves the proposition. □

Now let $W \in \mathrm{Gr}$ be any space of virtual dimension zero. Fix an admissible basis $w = (w_0, w_1, \dots)$ for W. As in §3, we think of w as a $\mathbb{Z} \times \mathbb{N}$ matrix, using the natural basis $\{z^k\}$ for H. For each $S \in \mathcal{S}_0$, let w^S be the determinant of the $\mathbb{N} \times \mathbb{N}$ matrix formed by extracting from w the rows indexed by the numbers $s \in S$; that is, if $w_j = \sum w_{ij} z^i$, we set

$$w^S = \det(w_{ij})_{i\in S, j\in\mathbb{N}}.$$

We call the numbers $\{w^S\}$ the *Plücker coordinates* of W: they are homogeneous coordinates (a different choice of admissible basis for W multiplies them all by the same non-zero constant). As in the finite dimensional case, the Plücker coordinates can be regarded as giving a projective embedding of the Grassmannian (see the appendix §10 below). Notice that w^S is non-zero precisely when W is transverse to $H_S^\perp$: indeed, w^S is just the determinant of the orthogonal projection $W \to H_S$ with respect to the bases $\{w_j\}$ for W and $\{z^s : s \in S\}$ for H_S. In particular, by (2.5), there is a unique S of minimal length such that $w^S \neq 0$. If we choose w so that w_+ has the form $1 +$ (finite rank), then the w^S

reduce to finite determinants. For example, if W is transverse, we can choose w so that $w_+ = 1$, and then if we set $S \setminus \mathbb{N} = A$ and $\mathbb{N} \setminus S = B$ we have

$$w^S \det(w_{ij})_{i \in A, j \in B}.$$

Proposition 8.3. *The τ-function of W is given by*

$$\tau_W(g) = \sum_S w^S F_S(\mathbf{h}),$$

where $\{w^S\}$ are the Plücker coordinates of W, the sum is taken over all $S \in \mathcal{S}_0$, and the variables h_i are related to g as in (8.2).

Proof. We first observe that if v and w are $m \times n$ and $n \times m$ matrices respectively, with $n \geq m$, then we have

$$\det vw = \sum v_S w^S,$$

where the sum is taken over all subsets $S \subset \{1, 2, \dots, n\}$ with m elements, v_S is the determinant formed from the columns of v indexed by the elements of S, and w^S is the determinant formed from the corresponding rows of w. (This identity simply expresses the functoriality of the m-th exterior power.) It is not hard to see that the identity extends to a product of infinite matrices, indexed by $\mathbb{N} \times \mathbb{Z}$ and $\mathbb{Z} \times \mathbb{N}$, of the form

$$(v_+, v_-) \begin{pmatrix} w_+ \\ w_- \end{pmatrix}$$

where $v_+ - 1$, $w_+ - 1$, v_- and w_- are all of trace class and S runs through the indexing sets $S \subset \mathbb{Z}$ of virtual cardinal zero.

We apply this to the determinant (3.4) giving the τ-function, with

$$(v_+, v_-) = (1, a^{-1}b).$$

Then w^S is the Plücker coordinate defined above and v_S is the τ-function of H_S, which we calculated in (8.2). That completes the proof. □

As we saw in §5, for the application to differential equations we have to write the elements of Γ_+ in the form

$$g(z) = \exp\left(\sum_1^\infty t_i z^i\right)$$

(we write t_1 where we wrote x in §5). We shall write $\tau_W(\mathbf{t})$ for the τ-function expressed in terms of these "coordinates" on Γ_+: to calculate $\tau_W(\mathbf{t})$ from (8.2) or (8.3), we have only to substitute the variables t_i for the h_i, using the relation

$$\exp\left(-\sum_1^\infty t_i z^i\right) = 1 + \sum_1^\infty h_i z^i. \tag{8.4}$$

Each t_k is a polynomial in the h_i, homogeneous of weight k if we give h_i weight i. If we regard the h_i as symmetric functions of the eigenvalues $\{\lambda_j\}$ of a matrix, as above, then the t_k are given by

$$-kt_k = \sum_j \lambda_j^k$$

(this differs by a sign from the convention adopted in [5]).

Example. Let $S = \{-1, 0, 2, 3, \dots\}$. The corresponding partition is $\nu = (1, 1, 0, \dots)$, so the Schur function is

$$F_S(\mathbf{h}) = \det \begin{pmatrix} h_1 & h_2 \\ 1 & h_1 \end{pmatrix} = h_1^2 - h_2.$$

From (8.4), we have $h_1 = -t_1$, $h_2 = \frac{1}{2}t_1^2 - t_2$, so by (8.2) the τ-function of the space $W = H_S$ is

$$\tau_W(\mathbf{t}) = \frac{1}{2}t_1^2 + t_2.$$

We end this section with some examples of the use of (8.3). First, note that it is clear that W has only finitely many non-zero Plücker coordinates if and only if it belongs to Gr_0; hence we can read off from (8.3) the following.

Proposition 8.5. *The function $\tau_W(\mathbf{t})$ is a polynomial in (a finite number of) the variables $(t_1, t_2, \dots)$ if and only if W belongs to* Gr_0.

As a more substantial application of (8.3), we shall prove the assertion (5.17) about the orders of the poles of the functions $a_i(x, \mathbf{t}^0)$. We shall continue to write t_1 instead of x. The crucial ingredient in the proof is the fact that the restriction of the τ-function to the one parameter subgroup $\exp(t_1 z)$ of Γ_+ cannot be identically zero. More precisely, we have the following.

Proposition 8.6. *For any $W \in \mathrm{Gr}$, we have*

$$\tau_W(t_1, 0, 0, \dots) = ct_1^\ell + \textit{(higher terms)},$$

where $c \neq 0$ and ℓ is the codimension of the stratum of Gr *containing $W^{(*)}$.*

In particular, the proposition shows that the τ-function cannot vanish identically, a fact that we used implicitly throughout §5.

Proof of (8.6). We first consider the behaviour of a Schur function F_S when we set $t_2 = t_3 = \dots = 0$. Since F_S is a homogeneous polynomial of weight $\ell(S)$ in the t_i, it is clear that we have

$$F_S(t_1, 0, 0, \dots) = d_S t_1^{\ell(S)}$$

$^{(*)}$Added in proof. J. Fay has independently proved an equivalent result in the case when W arises from a Riemann surface. (See his paper "On the even-order vanishing of Jacobian theta functions", *Duke Math. J.*, **51** (1984), 109–132, thm 1.2.)

where d_S is some rational number. We claim that this number is non-zero. Indeed, d_S is equal to $(-1)^{\ell(S)}$ times the reciprocal of a certain positive integer, the "product of the hook lengths" of the partition associated to S (see [13], p. 37, ex. 3). Explicitly, we have

$$(-1)^{\ell(S)} d_S = \prod_{i<j\leq n} (s_j - s_i) / \prod_{i\leq n} (n - s_i)!$$

where n is any number large enough so that $s_{n_1} = n + 1$, and as usual $S = \{s_0, s_1, \ldots\}$ (see [13], p. 9, formula (4)). We have already observed that for any $W \in \mathrm{Gr}$, there is a unique S of minimal length ℓ, say, such that the Plücker coordinate w^S is non-zero; this S is the index of the stratum containing W, and ℓ is the codimension of the stratum. That means that in the expansion (8.3) of τ_W, all the terms have weight at least ℓ; and the terms of minimal weight ℓ form a non-zero multiple of a single Schur function F_S. Thus the proposition follows at once from (8.3) and the fact that $d_S \neq 0$. □

Proof of (5.17). Replacing W if necessary by gW for suitable $g \in \Gamma_+$, we see that it is enough to consider the case where the pole is at the origin $\mathbf{t} = 0$. We already observed in §5 that the functions a_i are quotients of the form

$$a_i = P_i \tau / \tau$$

where P_i is a polynomial differential operator in $\{\partial/\partial t_k\}$; indeed, P_i is the coefficient of z^{-i} in the formal expansion of the expression

$$\exp\left[-\sum_1^\infty \frac{1}{r} z^{-r} (\partial/\partial t_r)\right].$$

It follows at once from this that the operator P_i lowers weight by i (where, as always, t_k has weight k). Thus in the power series expansion of the numerator $P_i\tau$ in the expression for a_i, only terms of weight at least $\ell - i$ can occur. (If $\ell - i < 0$, this statement is vacuous.) Hence when we put $t_2 = t_3 = \ldots = 0$ in the numerator, the lowest power of t_1 that can occur is $t_1^{\ell-i}$ (any terms involving a lower power of t_1 must also involve a higher t_k, and hence vanish when we set $t_2 = \ldots = 0$). Proposition (5.17) follows at once from this and (8.6). In fact the argument shows also that the order of the pole of any a_i cannot be more than ℓ[(†)]. □

9 The τ-function and the theta function

Let X be a compact Riemann surface of genus g, and let J be the Jacobian of X: it is the identity component of the group $H^1(X, \mathcal{O}^\times)$, where $\mathcal{O}$ is the sheaf

[(†)] Added in proof. According to G. Laumon (private communication) the order cannot be more than $-s_0$.

of holomorphic functions on X. We set $U = H^1(X, \mathcal{O})$ and $\Lambda = H^1(X, \mathbb{Z})$. The map $f \mapsto e^f$ induces a sheaf homeomorphism $\mathcal{O} \to \mathcal{O}^\times$ with kernel $2\pi i\mathbb{Z}$, from which we get the exact sequence

$$0 \to \Lambda \to U \to J \to 0$$

(the kernel is really $H^1(X, 2\pi i\mathbb{Z})$, but we identify this with $H^1(X, \mathbb{Z})$ in the obvious way). We recall that U is a g-dimensional complex vector space, Λ is a lattice in U, and $J = U/\Lambda$ is a complex torus.

We denote by $B : \overline{U} \times U \to \mathbb{C}$ the unique Hermitian form whose imaginary part is the $\mathbb{R}$-bilinear extension of the intersection pairing $\Lambda \times \Lambda \to \mathbb{Z}$. We fix a quadratic form $q : \Lambda \to \mathbb{Z}/2\mathbb{Z}$ such that

$$q(\lambda + \mu) - q(\lambda) - q(\mu) = \lambda \cdot \mu \pmod 2,$$

where $\lambda \cdot \mu$ is the intersection pairing. Then the *theta function* of X (see, for example, [15]) is the holomorphic function $\theta : U \to \mathbb{C}$ defined by

$$\theta(\mu) = \sum_{\lambda \in \Lambda} (-1)^{q(\lambda)} e^{-\frac{1}{2}\pi B(\lambda, \lambda + 2u)}.$$

It is characterized (up to a constant factor) by the functional equation

$$\theta(u + \lambda) = (-1)^{q(\lambda)} e^{\frac{1}{2}\pi B(\lambda, \lambda + 2u)} \theta(u) \tag{9.1}$$

(for $u \in U$, $\lambda \in \Lambda$). It follows at once that we have

$$\theta(u + \lambda) = C\theta(u)\theta(\lambda)e^{\pi B(\lambda, u)}$$

(where $C = \theta(0)^{-1}$). We shall use the fact that this relation too characterizes the theta function up to certain simple transformations. More precisely, we have the following.

Lemma 9.2. *Let $\tilde{\theta} : U \to \mathbb{C}$ be a holomorphic function such that*

$$\tilde{\theta}(u + \lambda) = \tilde{C}\tilde{\theta}(u)\tilde{\theta}(\lambda)e^{\pi B(\lambda, u)}$$

for all $u \in U$, $\lambda \in \Lambda$, and some (non-zero) constant $\tilde{C}$. Then we can find a constant A, a $\mathbb{C}$-linear map $\alpha : U \to \mathbb{C}$ and a point $\beta \in U$ such that

$$\tilde{\theta}(u) = Ae^{\alpha(u)}\theta(u - \beta).$$

Proof. Set

$$G(u) = \frac{\tilde{C}\tilde{\theta}(u)}{C\theta(u)}.$$

Then $G(u + \lambda) = G(u)G(\lambda)$, and the restriction of G to Λ is a homeomorphism $\Lambda \to \mathbb{C}^\times$. Choose an $\mathbb{R}$-linear map $\gamma : U \to \mathbb{C}$ such that $G(\lambda) = e^{\gamma(\lambda)}$ for

$\lambda \in \Lambda$. Splitting γ into its $\mathbb{C}$-linear and $\mathbb{C}$-antilinear parts and using the non-degeneracy of the form B, we see that there are α and β as in the statement of the lemma such that

$$G(\lambda) = e^{\alpha(\lambda) - \pi B(\lambda, \beta)}$$

for $\lambda \in \Lambda$. If we set $H(u) = e^{-\alpha(u)}\tilde{\theta}(u + \beta)/\theta(u)$, then $H(u + \lambda) = H(u)$ for all $\lambda \in \Lambda$; hence the holomorphic function

$$\theta_1(u) = e^{-\alpha(u)}\tilde{\theta}(u + \beta)$$

satisfies the same functional equation (9.1) as the theta function, and must therefore be a constant multiple of it. The lemma follows. □

Remark 9.3. Obviously, the constant A is uniquely determined by $\tilde{\theta}$. The α and β are not quite uniquely determined, because the map γ occurring in the proof of the lemma is determined only up to addition of a map γ_0 with $\gamma_0(\Lambda) \subset 2\pi i\mathbb{Z}$. However, it is easy to check that this would change the corresponding β only by a lattice point, so the projection of β onto the Jacobian U/Λ is uniquely determined. Also, α is uniquely determined once we have chosen β.

The τ-function is a function on the group Γ_+; our next task is to explain how we can regard the theta function too as defined on Γ_+, so that it makes sense to compare the two functions. We fix a point $x_\infty \in X$ and a local parameter z as in §6. We shall use z to identify $X_\infty \subset X$ with the disc $D_\infty = \{|z| \geq 1\}$ in the Riemann sphere. We denote by V the vector space of all holomorphic maps $f : D_0 \to \mathbb{C}$ with $f(0) = 0$. As in §5, we identify V with Γ_+ via the map $f \mapsto e^f$, and we shall regard the τ-function as a function on V. Now, any $f \in V$ (indeed, any holomorphic function on S^1) can be regarded as a cocycle for the Cech cohomology group $H^1(X, \mathfrak{U})$, where $\mathfrak{U} = \{U_0, U_\infty\}$ is an open covering of X as in the proof of (6.1). Using again the fact that we can calculate the cohomology of X from any such covering, we get a surjective homeomorphism

$$V \to H^1(X, \mathcal{O}) = U.$$

Thus if K_0 denotes the kernel of this map, we can regard the theta function as a K_0-invariant function on V. Now, K_0 is the linear subspace of V consisting of all functions $k \in V$ which can be written in the form $k = k_0 + k_\infty$, where k_0 and k_∞ are holomorphic functions on X_0 and X_∞, respectively; the splitting is unique if we normalize k_∞ so that $k_\infty(\infty) = 0$. We denote by $\tilde{V}$ the vector space of all such maps k_∞. Let K be the kernel of the composite map $V \to U \to J$; it consists of all functions $k \in V$ such that there is a factorization (necessarily unique)

$$e^k = \varphi_k e^{k_\infty} \tag{9.4}$$

where $k_\infty \in \tilde{V}$ and φ_k is a non-vanishing holomorphic function on X_0. Clearly $K/K_0 \cong \Lambda$, so that K_0 is indeed the identity component of K, as the notation suggests. In the proof of (9.10) below we shall give an explicit description of the integral cohomology class corresponding to an element $k \in K$.

We now fix a line bundle $\mathcal{L}$ of degree g over X and a trivialization φ as in §6; let $W \in \mathrm{Gr}$ be the corresponding space. For simplicity we assume that W is transverse and that the function $\tau = \tau_W : V \to \mathbb{C}$ is normalized as usual by $\tau(0) = 1$. The τ-function is not usually K_0-invariant: however, we show next that a simple modification of it is. We define a map $a : K \to \tilde{V}$ by $a(k) = k_\infty$, where k_∞ is as in (9.4). Clearly a is a homeomorphism, and its restriction to K_0 is a $\mathbb{C}$-linear map.

Lemma 9.5. *Let $f \in V$, $k \in K$. Then we have*

$$\tau(f + k) = \tau(f)\tau(k)e^{S(a(k),f)},$$

where S is the multiplier relating the actions of Γ_+ and Γ_- on the bundle Det* *(see (3.6)).*

Proof. By the definition of the τ-function (see (3.2)), we have

$$\tau(f + k)e^{-f-k}\sigma(W) = \sigma(e^{-f-k}W). \tag{9.6}$$

From the definition of W, it is clear that $\varphi_k W = W$, so we have $e^{-k}W = e^{-a(k)}W$ for $k \in K$. Using this and the fact that σ is Γ_--equivariant (see (3.7)), we find

$$\tau(k)e^{-k}\sigma(W) = e^{-a(k)}\sigma(W). \tag{9.7}$$

The right hand side of (9.6) is equal to

$$e^{-a(k)}\sigma(e^{-f}W) = \tau(f)e^{-a(k)}e^{-f}\sigma(W) = \tau(f)e^{S(a(k),f)}e^{-f}e^{-a(k)}\sigma(W).$$

Inserting (9.7) into this and canceling the non-zero vector $e^{-f-k}\sigma(W)$, we get the lemma. □

If we apply (9.5) when both f and k belong to K, we find that

$$S(a(k), \ell) - S(a(\ell), k) \in 2\pi i\mathbb{Z}$$

for all $k, \ell \in K$. Extend a to an $\mathbb{R}$-linear map $V \to \tilde{V}$; since K spans V over $\mathbb{R}$, the extension is unique, and we have

$$S(a(f), g) - S(a(g), f) \in i\mathbb{R} \tag{9.8}$$

for all $f, g \in V$. Write $a = b + c$, where b is $\mathbb{C}$-linear and c is antilinear. Then (9.8) implies that

$$S(b(f), g) = S(b(g), f),$$
$$S(c(f), g) = \overline{S(c(g), f)}$$

for all $f, g \in V$. Since $a|K_0$ is $\mathbb{C}$-linear, we have $c(K_0) = 0$; thus c, and hence also the Hermitian form $(f,g) \mapsto S(c(f), g)$, are well defined on $U = V/K_0$. Set $\tau_1(f) = \tau(f)e^{-\frac{1}{2}S(b(f),f)}$. Then from (9.5) we have

$$\tau_1(f+k) = \tau_1(f)\tau_1(k)e^{S(c(k),f)}.$$

In particular, the restriction of τ_1 to K_0 is a homeomorphism $K_0 \to \mathbb{C}^\times$. Choose a $\mathbb{C}$-linear map $\eta : V \to \mathbb{C}$ such that $\tau_1(k) = e^{\eta(k)}$ when $k \in K_0$; set $\tau_2(f) = \tau_1(f)e^{-\eta(f)}$. Then $\tau_2(f+k) = \tau_2(f)$ for $k \in K_0$. Thus τ_2 is well defined on U, and it satisfies

$$\tau_2(u+\lambda) = \tau_2(u)\tau_2(\lambda)e^{S(c(\lambda),u)} \qquad (9.9)$$

for $\lambda \in \Lambda = K/K_0$. But now we have the following crucial result.

Proposition 9.10. *For all $k, \ell \in K$, we have*

$$S(c(k),\ell) - S(c(\ell),k) = 2\pi i[k]\cdot[\ell],$$

where $[k]$, $[\ell]$ denote the classes of k, ℓ in the group $K/K_0 = \Lambda = H^1(X,\mathbb{Z})$.

The proposition shows that the Hermitian form occurring in the exponent in (9.9) is π times the form B occurring in the definition of the theta function. We can therefore apply (9.2) to obtain the main result of this section.

Theorem 9.11. *The τ-function $\tau_W : V \to \mathbb{C}$ is related to the theta function by*

$$\tau_W(f) = Ae^{\alpha_W(f)+\frac{1}{2}S(b(f),f)}\theta(\overline{f} - \beta_W),$$

where A is a constant, $\alpha_W : V \to \mathbb{C}$ is a linear map, β_W is a point of U, and $\overline{f}$ denotes the projection of f onto $U = V/K_0$.

Remarks

(i) Note that the quadratic term $\frac{1}{2}S(b(f),f)$ depends only on X and z.

(ii) By (9.3), the projection of β_W onto the Jacobian J is uniquely determined by W. If W moves according to one of the KP flows, then β_W moves along the corresponding straight line in J.

(iii) There seems no point in trying to be more explicit about the map α,since it depends on the choice of trivialization φ (see (3.8)).

It remains to give the proof of (9.10). For this we fix a basis $\Delta = \{\alpha_i, \beta_i\}$, $1 \le i \le g$, for $H_1(X,\mathbb{Z})$ of the standard kind, that is, such that $\alpha_i \cdot \beta_i = 1$ and all other intersections are zero. We can then regard the Riemann surface X in the classical way as a quotient of a polygon Y with $4g$ edges arranged in groups of four $(\alpha_i, \beta_i, \alpha_i^{-1}, \beta_i^{-1})$ (we get X from Y by identifying the two edges corresponding to each element of Δ). We suppose Y chosen so that the disc

X_∞ in X corresponds to a small disc Y_∞ in the interior of Y; let Y_0 be the complement of the interior of Y_∞. If $k \in K$, then $k = k_0 + k_\infty$, where k_0 and k_∞ are functions on Y_0 and Y_∞, respectively. Now, $e^{k_0} = \varphi_k$ is a function on X: that means that the values of k_0 at corresponding points of the two edges of Y corresponding to a generator $\gamma \in \Delta$ differ by an integer multiple of $2\pi i$, say by $2\pi i n(k, \gamma)$. The cohomology class defined by k is then given by

$$[k] = \sum_{\gamma \in \Delta} n(k, \gamma)\gamma *$$

where $\{\gamma *\}$ is the basis of $H^1(X, \mathbb{Z}) = \Lambda$ dual to Δ.

Now, we have

$$\begin{aligned} S(c(k), \ell) - S(c(\ell), k) &= S(a(k), \ell) - S(a(\ell), k) \\ &= \frac{1}{2\pi i} \int_{S^1} (\ell k'_\infty - k \ell'_\infty). \end{aligned}$$

After a short calculation we find that this is equal to

$$\frac{1}{2\pi i} \int_{S^1} k_0 \ell'_0.$$

Since k_0 and ℓ_0 are holomorphic functions on Y_0, we can replace S^1 by the boundary of Y in this integral. The contribution to the integral of a typical set of four edges

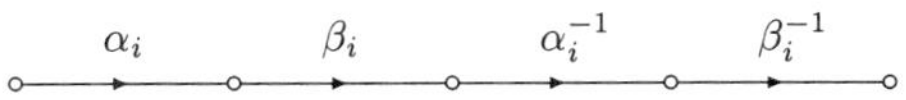

can be reduced to an integral over the middle pair (β_i, α_i^{-1}): we obtain

$$n(k, \beta_i) \int_{\beta_i} \ell'_0 + n(k, \alpha_i) \int_{\alpha_i} \ell'_0 = 2\pi i \{-n(k, \beta_i) n(\ell, \alpha_i) + n(k, \alpha_i) n(\ell, \beta_i)\}.$$

Summing over i and using the fact that the intersection matrix of the basis $\{\alpha_i^*, \beta_i^*\}$ is the same as that of $\{\alpha_i, \beta_i\}$, we see that the integral is indeed equal to $2\pi i [k] \cdot [\ell]$.

The Baker function and the theta function

If we combine (9.11) with (5.14), we obtain a formula expressing the Baker function (of a space W arising from a Riemann surface) in terms of the theta function. As we mentioned in the introduction, such a formula is well known in the Russian literature (see, for example, [10, 11, 36]). However, it is perhaps not immediately obvious that the Japanese formula (5.14) coincides with the Russian one: so at the suggestion of the referee we end this section by offering a fairly detailed comparison of the two formulas.

The Russian formula is expressed in terms of the classical Riemann theta function, whose definition involves a choice of canonical homology basis $\{\alpha_i, \beta_i\}$ as in the proof of (9.10) above: we suppose such a basis fixed from now on. The classical theta function is a function on the dual space R^* of the space R of global holomorphic differentials on X; but R^* is usually identified with $\mathbb{C}^g$ via the basis

$$\left\{\omega \mapsto \int_{\alpha_i} \omega\right\}, \qquad \omega \in R.$$

On the other hand we have the natural pairing

$$H^1(X, \mathcal{O}) \oplus H^0(X, \Omega) \to H^1(X, \Omega) \cong \mathbb{C},$$

where Ω is the sheaf of holomorphic differentials on X, which canonically identifies R^* with the space $U = H^1(X, \mathcal{O})$ on which our theta function was defined. In what follows we shall use without further comment these identifications $U \cong R^* \cong \mathbb{C}^g$. The choice of homology basis $\{\alpha_i, \beta_i\}$ gives a natural choice for the form $q : \Lambda \to \mathbb{Z}/2\mathbb{Z}$ occurring in our version of the theta function, namely, we can choose q to vanish on the basis for $\Lambda \cong H^1(X, \mathbb{Z})$ dual to $\{\alpha_i, \beta_i\}$. It is then easy to check that our theta function differs from the classical one only by a factor $\exp Q(u, u)$, where Q is a symmetric $\mathbb{R}$-bilinear form on U. Thus if we use the classical theta function, theorem 9.11 remains true except that the quadratic form is different. From now on we write θ for the classical theta function.

With these preliminaries, we can now explain the Russian formula relating the Baker function and the theta function. We follow the account given in [36], to which we refer the reader for more details. With Krichever, we fix a non-special positive divisor $\mathcal{D} = \{P_1, \ldots, P_g\}$ on X; without loss of generality (see (6.5)) we suppose the points P_i lie outside the disc $D_\infty \subset X$. We want to write down the Baker function ψ_W, where W is the closure of the space of analytic functions on S^1 which extend to meromorphic functions on X_0 that are regular except for (possible) simple poles at the points P_i. We fix a base point $P_0 \neq x_\infty$ in X, and let $A : X \to R^* \cong \mathbb{C}^g$ be the corresponding Abel map, given by

$$A(P)(\omega) = \int_{P_0}^{P} \omega \qquad (P \in X, \omega \in R).$$

The map A is well defined only modulo the period lattice Λ (because of the choice of path integration). Let $C \in \mathbb{C}^g$ be a constant vector such that the function (on X) $\theta(A(P) - C)$ vanishes precisely when $P = P_1, \ldots, P_g$. For $n = 1, 2, \ldots$, let ω_n be the differential of the second kind which has zero α-periods and is regular except for a singularity at x_∞ with principal part $d(z^n)$. Let $W_n \in \mathbb{C}^g$ be the vector of β-periods of ω_n. Consider the expression

$$\exp\left\{\sum_i t_i \int_{P_0}^{P} \omega_i\right\} \frac{\theta(A(P) + \sum t_i W_i - C)}{\theta(A(P) - C)}. \tag{9.12}$$

It is understood that the path of integration in the first term is the same as that used in the Abel map; it is then easy to check (see [36], ch. 3, §1) that (9.12) is a well defined function of $P \in X$, although the individual terms in it are not. It is obvious that when restricted to $S^1 \subset X$ the function (9.12) belongs to W for each fixed $\mathbf{t}$, and has the form

$$\exp \sum t_i z^i (a_0(\mathbf{t}) + a_1(\mathbf{t}) z^{-1} + \ldots).$$

Thus to get the Baker function ψ_W, we have only to divide by $a_0(\mathbf{t})$. That yields the final formula

$$\psi_W = \exp \left\{ \sum t_i \int_{P_0}^{P} \omega_i \right\} \exp \left\{ -\sum t_i b_{i0} \right\} \frac{\theta(A(P) + \sum t_i W_i - C)\theta(A(x_\infty) - C)}{\theta(A(P) - C)\theta(A(x_\infty) + \sum t_i W_i - C)} \tag{9.13}$$

where the constants b_{i0} are defined from the expansions

$$\int_{P_0}^{z} \omega_n = z^n + \sum_{0}^{\infty} b_{nr} z^{-r}$$

for z near x_∞.

The formula (9.13) is global (P can run over the whole Riemann surface X). We now restrict it to $P \in D_\infty$ and accordingly write z instead of P. We claim that (9.13) can then be identified with the formula obtained by substituting (9.11) into (5.14). Note first that the quotient

$$\theta(A(x_\infty) - C)/\theta(A(z) - C)$$

in (9.13) is nothing but a function of the form $1 + c_1 z^{-1} + \ldots$; it comes from the uninteresting linear term α in (9.11). The exponential terms in (9.13) can be written

$$\exp \left\{ \sum t_i z^i \right\} \exp \left\{ \sum_{i,j=1}^{\infty} t_i b_{ij} z^{-j} \right\};$$

the second factor here is the contribution to (9.13) coming from the quadratic term in (9.11). To complete our check that (5.14) and (9.13) agree, we have still to see two things: (i) that the vectors $W_i \in \mathbb{C}^g$ corresponding to the different t_i agree with those in (9.11) (obtained by regarding the functions z^i as cocycles for $H^1(X, \mathcal{O})$); (ii) that the difference in the arguments of the two remaining theta function terms in (9.13) agrees with the q_ζ in (5.14). For (i), we use the fact that the canonical pairing $U \times R \to \mathbb{C}$ can be derived from the pairing $V \times R \to \mathbb{C}$ given by

$$(f, \omega) \mapsto \frac{1}{2\pi i} \int_{S^1} f\omega;$$

the desired assertion then reduces to something well known (see, for example, [36], (2.1.12)). Concerning (ii), note that the difference in question is

$$A(z) - A(x_\infty) = A_\infty(z),$$

where A_∞ is the Abel map defined using the base point x_∞. Hence the result we need is the following.

Lemma 9.14. *Let* $\Gamma_+ \to U \to J = U/\Lambda$ *be the map used earlier in this section (defined by regarding an element of* Γ_+ *as a transition function for a line bundle on* X*). Then for* $|\zeta| > 1$*, the image of* q_ζ *under this map is* $A_\infty(\zeta)$*.*

Proof. We write q_ζ in the form

$$q_\zeta(z) = (z^{-1} - \zeta^{-1})z.$$

The two factors here are transition functions for the line bundles corresponding to the divisors $[\zeta]$ and $[-x_\infty]$, respectively. Thus the image of q_ζ in the Jacobian is $[\zeta] - [x_\infty]$, which is indeed $A_\infty(\zeta)$. □

Finally, we point out that one can reverse some of the arguments we have just given and *prove* (9.11) by comparing the formulas (5.14) and (9.13). This argument is indicated in [5], and is indeed the only possible one there, because at this point in [5] the τ-function is *defined* in terms of the Baker function by the formula (5.14). In our context, however we have an independent definition of the τ-function, so it seemed to us very desirable to give a direct proof of (9.11), avoiding the use of the Baker function.

10 Appendix: the representation theory of the loop group

In this paper we have not mentioned the representation theory of the loop group $\mathrm{LGL}_n(\mathbb{C})$, whereas the Japanese papers [5] put it in the foreground. The difference, however, is more apparent than real, as we shall now explain. We shall begin by describing the situation without any attempt at justification, and at the end we shall return to give some indications about the proofs.

It will be convenient in this section to let Gr denote the "Hilbert-Schmidt" Grassmannian of H, consisting of closed subspaces W of H such that the projection $W \to H_+$ is Fredholm and the projection $W \to H_-$ is Hilbert-Schmidt. Alternatively, Gr consists of the graphs of all Hilbert-Schmidt operators $H_S \to H_S^\perp$. It is clearly a Hilbert manifold. We shall write $\mathrm{LGL}_n(\mathbb{C})$ for the group of *smooth* loops.

We have seen[‡] that a central extension of $\mathrm{LGL}_n(\mathbb{C})$ by $\mathbb{C}^\times$ acts on the holomorphic line bundle Det* on Gr. This means that $\mathrm{LGL}_n(\mathbb{C})$ acts projectively on the space $\Gamma(\mathrm{Det}^*)$ of all holomorphic sections of Det*. With the topology of uniform convergence on compact sets, $\Gamma(\mathrm{Det}^*)$ is a complete topological vector space. It is the so-called "basic" irreducible projective representation of $\mathrm{LGL}_n(\mathbb{C})$.

[‡]Strictly speaking, in §3 we considered only one component of $\mathrm{LGL}_n(\mathbb{C})$ and Gr; but it is not hard to extend the discussion to include the other components (see [17]).

For any indexing set $S \in \mathcal{S}$ the "Plücker coordinate" $W \mapsto w^S$ (introduced in §8) is an element of $\Gamma(\mathrm{Det}^*)$. We shall denote it by π_S. In fact the π_S span a dense subspace; and there is a natural Hilbert space $\mathcal{H}$ inside $\Gamma(\mathrm{Det}^*)$—it can be thought of as the "square-integrable" holomorphic sections—for which the π_S form an orthonormal basis. The subgroup LU_n of $\mathrm{LGL}_n(\mathbb{C})$ acts by a projective unitary representation on $\mathcal{H}$.

The geometrical significance of $\mathcal{H}$ is that there is a natural antiholomorphic embedding

$$\Omega : \mathrm{Gr} \to P(\mathcal{H})$$

of the infinite dimensional complex manifold Gr in the projective space of $\mathcal{H}$. It assigns to $W \in \mathrm{Gr}$ the ray in $\mathcal{H}$ containing the section Ω_w of Det^* defined by

$$\Omega_w(w') = \det\langle w, w'\rangle,$$

where w is an admissible basis of W. (Here $\langle w, w'\rangle$ denotes the matrix whose (i,j)-th element is $\langle w_i, w'_j\rangle$; and we are thinking of a section of Det^* as a $\mathcal{T}$-equivariant map $\mathcal{P} \to \mathbb{C}$.) The embedding Ω is equivariant with respect to LU_n.

The vector in $\mathcal{H}$ corresponding to H_+ with its standard basis, i.e. the canonical section σ of Det^* (cf. §3), is called the *vacuum vector* Ω_0.

If we think of the representation $\mathcal{H}$ as given, rather than the manifold Gr, then the discussion of §§3 and 5 can be very simply translated. The crucial formula is the definition of the τ-function, which becomes

$$\tau_W(g) = \langle \Omega_0, g\Omega_w\rangle$$

where $g \in \Gamma_+$, and w is an admissible basis for W. This is the definition in parts of [5], except that these authors appear to have in mind only the group of polynomial loops, corresponding to our Grassmannian Gr_0.

Two other realizations of the Hilbert space $\mathcal{H}$ are of importance. To describe the first, notice that the connected components of Gr are indexed by the integers, and that correspondingly

$$\mathcal{H} = \bigoplus_{k \in \mathbb{Z}} \mathcal{H}_k,$$

where $\mathcal{H}_k$ consists of functions on the k-th connected component. We saw in §2 that the group Γ_- of holomorphic functions in the disc $|z| \geq 1$ acts freely on Gr. Let X denote the orbit of H_+ under Γ_-. The restriction of Det^* to X is canonically trivial; so holomorphic sections of Det^* restrict to give complex-valued holomorphic functions on X. Writing a general element of Γ_- in the form

$$1 + h_1 z^{-1} + h_2 z^{-2} + \dots ,$$

we think of functions on X as functions on the infinite sequence of complex variables $h_1, h_2, h_3, \dots$. In fact sections of Det^* over the component of Gr containing H_+ are determined by their restrictions to X, and we have.

Proposition 10.1.

(i) If $S \in \mathcal{S}$ has virtual cardinal zero then the Plücker coordinate $\pi_S \in \Gamma(\mathrm{Det}^*)$ restricts to the Schur function $F_S(h_1, h_2, \ldots)$. (Cf. §8.)

(ii) $\mathcal{H}_0$ can be identified with the completion of the ring of symmetric polynomials $\mathbb{Z}[h_1, h_2, \ldots]$ with respect to its standard inner product [13]; equivalently, it is the space of L^2 holomorphic functions on Γ_- with respect to the Gaussian measure

$$d\mu(g) = e^{-\sum n|a_n|^2} \prod da_n d\bar{a}_n,$$

where $g = \exp \sum a_n z^{-n}$.

The second concrete realization of $\mathcal{H}$ is as the exterior algebra on the Hilbert space $H_+ \oplus \overline{H}_-$. As H_+ and $\overline{H}_-$ have the orthonormal bases $\{z^k\}_{k \geq 0}$ and $\{\bar{z}^k\}_{k<0}$ respectively, the exterior algebra $\Lambda(H_+ \oplus \overline{H}_-)$ has an orthonormal basis of the form

$$\bar{z}^{a_1} \wedge \ldots \wedge \bar{z}^{a_k} \wedge z^{b_1} \wedge \ldots \wedge z^{b_m}, \tag{10.2}$$

where $a_1 < \ldots < a_k < 0 \leq b_1 < \ldots < b_m$. These basis elements correspond exactly to the indexing sets $S \in \mathcal{S}$ with which we are familiar: we write $S \setminus \mathbb{N} = \{a_1, \ldots, a_k\}$ and $\mathbb{N} \setminus S = \{b_1, \ldots, b_m\}$. Thus we can denote the element (10.2) by z^S; the isomorphism $\Lambda(H_+ \oplus \overline{H}_-) \cong \mathcal{H}$ makes z^S correspond to the Plücker coordinate π_S.

A more interesting and also more relevant way of constructing the map $\Lambda(H_+ \oplus \overline{H}_-) \to \mathcal{H}$ is by defining "fermionic field operators" on $\mathcal{H}$. These amount to an operator-valued distribution $\theta \mapsto \varphi(\theta)$ on the circle, satisfying the anticommutation relations

$$[\varphi(\theta_1), \varphi(\theta_2)]_+ = 0.$$
$$[\varphi(\theta_1), \varphi(\theta_2)^*]_+ = \delta(\theta_1 - \theta_2).$$

Then the map $\Lambda(H_+ \oplus \overline{H}_-) \to \mathcal{H}$ is

$$f_1 \wedge \ldots \wedge f_k \wedge \bar{g}_1 \wedge \ldots \wedge \bar{g}_m \mapsto \varphi_{f_1} \cdots \varphi_{f_k} \varphi^*_{g_1} \cdots \varphi^*_{g_m} \Omega_0,$$

where

$$\varphi_f = \frac{1}{2\pi} \int_0^{2\pi} f(\theta) \varphi(\theta)\, d\theta.$$

The highly singular "vertex operator" $\varphi(\theta)$ is constructed from the action of $\Gamma = L\mathbb{C}^\times$ on $\mathcal{H}$ as the limit $\rho \to 1+$ of the action of $p_\zeta q_\zeta$, where $\zeta = \rho e^{i\theta}$, and

$$q_\zeta = 1 - \zeta^{-1} z \in \Gamma_+,$$
$$p_\zeta = (1 - \bar{\zeta}^{-1} z^{-1})^{-1} \in \Gamma_-.$$

The important formula (5.15) for the unique element of $W \cap (1 + H_-)$ can be written

$$\psi_W(0, e^{i\theta}) = \langle \Omega_0, \varphi(\theta)\Omega_w \rangle.$$

this is equivalent to (5.15), because

$$\begin{aligned}\langle \Omega_0, \varphi(\theta)\Omega_w \rangle &= \lim \langle \Omega_0, p_\zeta q_\zeta \Omega_w \rangle \\ &= \langle p_\zeta^* \Omega_0, q_\zeta \Omega_w \rangle \\ &= \langle \Omega_0, q_\zeta \Omega_w \rangle \\ &= \tau_W(q_\zeta).\end{aligned}$$

Remarks about the proofs

Let $H_{m,n} = z^m H_+ / z^n H_+$ when $m \leq n$. Then Gr contains the finite dimensional Grassmannians $Y_n = \mathrm{Gr}(H_{-n,m})$, and the union of the Y_n is dense. The bundle Det* on Gr restricts to the usual Det* on Y_n; so we know that $\Gamma(\mathrm{Det}^*|Y_n)$ can be identified with the exterior algebra $\Lambda(H_{-n,m})$. A section of Det* is determined by its restrictions to the Y_n. Thus we have an inclusion

$$\Gamma(\mathrm{Det}^*) \hookrightarrow \varprojlim_n \Lambda(H_{-n,n}). \tag{10.3}$$

Now $\Lambda(H_{-n,n})$ has a basis indexed by the 2^{2n} sets $S \in \mathcal{S}$ such that

$$[n, \infty) \subset S \subset [-n, \infty).$$

These come from the corresponding Plücker coordinates π_S in $\Gamma(\mathrm{Det}^*)$. This shows that the map (10.3) has a dense image, and also that the π_S span a dense subspace of $\Gamma(\mathrm{Det}^*)$.

To construct the Hilbert space $\mathcal{H}$ we begin by observing that $w \mapsto \Omega_w$ defines an antiholomorphic map

$$\Omega : \mathrm{Det} \to \Gamma(\mathrm{Det}^*)$$

which is antilinear on each fibre of Det. (Notice that if $w = \{z^s\}_{s \in S}$ then Ω_n is the Plücker coordinate π_S.) By transposing Ω we obtain a $\mathbb{C}$-linear map

$$\Omega^* : F \to \Gamma(\mathrm{Det}^*),$$

where F is the antidual of $\Gamma(\mathrm{Det}^*)$, i.e. the space of continuous antilinear maps $\Gamma(\mathrm{Det}^*) \to \mathbb{C}$. This gives us a hermitian form $\overline{F} \times F \to \mathbb{C}$ defined by

$$(\alpha, \beta) \mapsto \alpha(\Omega^* \beta).$$

In fact Ω^* is injective and has dense image, because the Ω_w span $\Gamma(\mathrm{Det}^*)$, and the Hilbert space completion $\mathcal{H}$ of F is sandwiched between F and $\Gamma(\mathrm{Det}^*)$. It is clear that the π_S form an orthonormal basis of $\mathcal{H}$. Because Ω is equivariant with respect to LU_n (or, more accurately, with respect to a central extension of LU_n by the circle), it follows that LU_n acts unitarily on $\mathcal{H}$.

The proof of (10.1) (i) is almost exactly the same as that of (8.2); the second part is then routine.

For a discussion of vertex operators we refer to [17] or [18].

Bibliography

[1] H. Airault, H.P. McKean, J. Moser, *Rational and elliptic solutions of the Korteweg-de Vries equation and a related many-body problem*, Comm. Pure. Appl. Math. **30** (1977), 95–148.

[2] M. Adler and J. Moser, *On a class of polynomials connected with the Korteweg-de Vries equation*, Comm. Math. Phys. **61** (1978), 1–30.

[3] H.F. Baker, *Note on the foregoing paper "Commutative ordinary differential operators", by J.L. Burchnall and T.W. Chaundy*, Proc. Royal Soc. London (A) **118** (1928), 584–593.

[4] J.L. Burchnall, T.W. Chaundy, a) *Commutative ordinary differential operators*, Proc. London Math. Soc. **21** (1923), 420–440; b) *Commutative ordinary differential operators*, Proc. Royal Soc. London (A) **118** (1928), 557–583; c) *Commutative ordinary differential operators II. The identity $P^n = Q^m$*, Proc. Royal Soc. London (A) **134** (1932), 471–485.

[5] E. Date, M. Jimbo, M. Kashiwara, T. Miwa, *Transformation groups for soliton equations*: I. Proc. Japan Acad. **57A** (1981), 342–347; II. Ibid., 387–392; III. J. Phys. Soc. Japan **50** (1981), 3806–3812; IV. Physica **4D** (1982), 343–365; V. Publ. RIMS, Kyoto Univ. **18** (1982), 1111–1119; VI. J. Phys. Soc. Japan **50** (1981), 3813–3818; VII. Publ. RIMS, Kyoto Univ. **18** (1982), 1077–1110.

[6] V.G. Drinfel'd, V.V. Sokolov, *Equations of Korteweg-de Vries type and simple Lie algebras*, Dokl. Akad. Nauk SSSR **258** (1) (1981), 11–16; Soviet Math. Dokl. **23** (1981), 457–462.

[7] C. D'Souza, *Compactification of generalized Jacobians*, Proc. Ind. Acad. Sci. **88A** (1979), 421–457.

[8] F. Ehlers, H. Knörrer, *An algebro-geometric interpretation of the Bäcklund transformation for the Korteweg-de Vries equation*, Comment. Math. Helvetici **57** (1982), 1–10.

[9] I.M. Gel'fand, L.A. Dikii, *Fractional powers of operators and Hamiltonian systems*, Funct. Anal. Appl. **10** (4) (1976), 13–29 (Russian), 259–273 (English).

[10] I.M. Krichever, *Integration of non-linear equations by methods of algebraic geometry*, Funct. Anal. Appl. **11** (1) (1977), 15–31 (Russian), 12–26 (English).

[11] I.M. Krichever, *Methods of algebraic geometry in the theory of non-linear equations*, Uspekhi Mat. Nauk **32** (6) (1977), 183–208; Russian Math. Surveys **32** (6) (1977), 185–213.

[12] B.A. Kuperschmidt, G. Wilson, *Modifying Lax equations and the second Hamiltonian structure*, Inventiones Math. **62** (1981), 403–436.

[13] I.G. Macdonald, *Symmetric functions and Hall polynomials*, Oxford University Press, 1979.

[14] I. Yu. Manin, *Algebraic aspects of non-linear differential equations*, Itogi Nauki i Tekhniki, ser. Sovremennye Problemi Matematiki **11** (1978), 5–152; J. Sov. Math. **11** (1) (1979), 1–122.

[15] D. Mumford, *Abelian varieties*, Oxford University Press, 1974.

[16] D. Mumford, An algebro-geometric construction of commuting operators and of solutions to the Toda lattice equation, Korteweg-de Vries equation and related non-linear equations, *Proceedings of Symposium on Algebraic Geometry* (M. Nagata, ed.), Kinokuniya, Tokyo, 1978.

[17] A. Pressley, G. Segal, *Loop groups and their representations* (Book in preparation; Oxford University Press).

[18] G. Segal, *Unitary representations of some infinite dimensional groups*, Commun, Math. Phys. **80** (1981), 301–342.

[19] B. Simon, *Notes on infinite determinants of Hilbert space operators*, Adv. in Math. **24** (1977), 244-273.

[20] V.V. Sokolov, A.B. Shabat, *(L, A)-pairs and a substitution of Riccati type*, Funct. Anal. Appl. **14** (2) (1980), 79–80 (Russian), 148–150 (English).

[21] J.-L. Verdier, *Equations différentielles algébriques*, Séminaire Bourbaki (1977–1978), *Exposé 512* = Lecture notes in Math. **710**, 101–122.

[22] G. Wilson, *Commuting flows and conservation laws for Lax equations*, Math. Proc. Camb. Phil. Soc. **86** (1979), 131–143.

[23] V.E. Zakharov, A.B. Shabat, *Integration of the non-linear equations of mathematical physics by the inverse scattering method II*, Funct. Anal. Appl. **13** (3) (1979), 13–22 (Russian), 166–174 (English).

[24] P. Deligne, M. Rapoport, *Les schémas de modules de courbes elliptiques*, in *Modular functions of one variable, II* (P. Deligne and W. Kuyk, eds.), Lecture Notes in Math. **349**, Springer, 1973.

[25] H.P. McKean, E. Trubowitz, *Hill's operator and hyperelliptic function theory in the presence of infinitely many branch points*, Comm. Pure Appl. Math. **29** (1976), 143–226.

[26] M. Mulase, *Geometry of soliton equations*, MSRI preprint 035-83, Berkeley (1983).

[27] M. Mulase, *Algebraic geometry of soliton equations I*, MSRI preprint 040-83, Berkeley (1983).

[28] M. Mulase, *Structure of the solution space of soliton equations*, MSRI preprint 041-83, Berkeley (1983).

[29] M. Mulase, *Complete integrability of the Kadomtsev-Petviashvili equation*, MSRI preprint 053-83, Berkeley (1983).

[30] M. Mulase, *Algebraic geometry of soliton equations*, Proc. Japan Acad. **59**, Ser. A (1983), 285–288.

[31] M. Mulase, *Cohomological structure of solutions of soliton equations, isospectral deformation of ordinary differential operators and a characterization of Jacobian varieties*, MSRI preprint 003-84-7, Berkeley (1984).

[32] M. Sato, Y. Sato, *Soliton equations as dynamical systems on infinite dimensional Grassmann manifold*, Preprint, 13 pp. (date unknown).

[33] T. Shiota, *Characterization of Jacobian varieties in terms of soliton equations*, Preprint, 63 pp., Harvard University (1984).

[34] C.J. Rego, *The compactified Jacobian*, Ann. Scient. Ec. Norm. Sup. **13** (1980), 211–223.

[35] G. Wilson, *Habillage et fonctions τ*, C. R. Acad. Sc. Paris, **299**, Sér. I, no. 13 (1984), 587–590.

[36] B.A. Dubrovin, *Theta functions and non-linear equations*, Uspekhi Mat. Nauk **36** (2) (1981), 11–80; Russian Math. Surveys **36** (2) (1981), 11–92.

Scattering and Inverse Scattering for First Order Systems

R. Beals and R. R. Coifman

Introduction

It is well-known that a number of important nonlinear evolution equations are associated to spectral problems for ordinary differential operators (see [1, 4]). The initial value problem for the evolution equation can, in principle, be solved by solving an inverse scattering problem. Schematically, the unknown function $u(\cdot, t)$ (possibly vector-valued) is identified with or transformed into the coefficients $q(\cdot, t)$ of a differential operator L_t. A spectral problem is associated to L_t which carries (at least formally) some asymptotic information called the scattering data $v(\cdot, t)$. The original nonlinear evolution of u, or equivalently of q, corresponds to a trivially solvable linear evolution of the scattering data v.

The analytical theory of scattering and inverse scattering in various cases has been treated, for example, in [1], [6], [10], and other papers of these authors. It should be noted, though, that in much of the literature the expression "solvable by the inverse scattering method" designates evolutions associated to spectral problems for which certain purely formal scattering data would evolve linearly if it existed. The proposed scattering data may exist only for compactly supported or exponentially vanishing q, and the support condition or the vanishing condition may be destroyed by the evolution itself. In short, problems may have been termed "solvable" when neither the scattering map $q \mapsto v$ nor the inverse map $v \mapsto q$ has been seriously investigated. (For such problems one has recipes to produce special solutions, such as soliton or multi-soliton solutions, but the general initial value problem may be untouched.)

This article is reprinted by permission from **Publications Mathématiques de l'Institut des Hautes Etudes Scientifiques 61**

A satisfactory analytical treatment of scattering and inverse scattering for a given spectral problem should aim for the following:

i. to formulate a notion of scattering data v which is meaningful for (essentially) all reasonable coefficients q, such as $q \in L^1$;

ii. to show that $q \to v$ is injective;

iii. to characterize scattering data by determining all the algebraic or topological constraints such data satisfy;

iv. to show that for (essentially) each set of data satisfying the constraints, there is a corresponding q;

v. to discuss the relationship of such analytic properties of q as smoothness or decay at ∞ with corresponding properties of v.

We summarize here some results of this nature on a class of spectral problems sometimes called generalized AKNS-ZS systems (named after [1] and [11]). This class is directly or indirectly related to most of the interesting nonlinear evolution equations which are said to be solvable by the inverse scattering method. The eigenvalue problem has the form

$$\frac{df}{dx} = zJf(x) + q(x)f(x), \quad z \in \mathbb{C}. \tag{1}$$

Here $f : \mathbb{R} \mapsto \mathbb{C}^n$, J is a constant $(n \times n)$ matrix, and q is a matrix-valued function. The (2×2) case was introduced by Zakharov and Shabat [11] in connection with the cubic nonlinear Schrödinger equation and was studied extensively by Ablowitz, Kaup, Newell, and Segur [1]. The formal theory of the $(n \times n)$ case, including the determination of the appropriate nonlinear evolutions of q, has been considered by a number of authors (see [5], [7]). The results described below seem to be new, in some respects, even for the (2×2) case.

Our results on the analytic theory of the scattering and inverse scattering problems for generalized AKNS systems are stated in detail in the first section.

The direct problem is treated in Sections 2–6. The case of compactly supported q is studied in Section 2 and the case of q with small L^1 norm in Section 3. The general case is obtained by limiting or patching methods in Sections 4 and 5. The consequences of smoothness of q or decay of q are studied in Section 6.

Sections 7–11 treat the inverse problem. The problem is reformulated as an integral equation in Section 7. The problem is solved for "small" data in Section 8, with refinements for smooth or decaying data in Section 9. In Sections 10 and 11 a rational approximation is used, together with the result for small data, to reduce the general inverse problem to a purely algebraic problem: a system of linear equations with x-dependent coefficients.

In Section 12 we consider systems with a symmetry and the relations between symmetry conditions on the potential and on the scattering data. We derive a formula of Hirota type (see [4], [9]) for the soliton and multi-soliton potentials for a system with symmetry.

We have benefitted from discussions with B. Dahlberg, P. Deift, C. Tomei, and E. Trubowitz. Several key observations, in particular the relationship of the winding number constraint to asymptotic solvability of the inverse problem, are due to D. Bar-Yaacov [2] in his work on the case when the matrix J is skew adjoint.

1 Summary of Principal Results

We assume throughout that the matrix J in (1) is diagonal, with distinct complex eigenvalues:

$$J = \text{diag}(\lambda_1, \lambda_2, \dots, \lambda_n), \quad \lambda_j \neq \lambda_k \text{ if } j \neq k. \tag{1.1}$$

Let $\mathbf{P}$ denote the Banach space of $(n \times n)$ matrix-valued functions on $\mathbb{R}$ which are integrable and off-diagonal: $\mathbf{P} \ni q = (q_{jk})$, where

$$q_{jj} \equiv 0, \quad q_{jk} \in L^1(\mathbb{R}). \tag{1.2}$$

We refer to $q \in \mathbf{P}$ as a *potential.*

The spectral problem (1) leads to the problem of determining a fundamental matrix $\psi(x, z)$:

$$\frac{d}{dx}\psi(x,z) = zJ\psi(x,z) + q(x)\psi(x,z) \quad \text{a.e. } x, \tag{1.3}$$

$$\det \psi(x,z) \neq 0.$$

The desired solution is normalized to be of the form

$$\psi(x,z) = m(x,z)e^{xzJ}, \tag{1.4}$$

$$m(\cdot, z) \text{ bounded and absolutely continuous,} \tag{1.5}$$

$$m(x,z) \to I \text{ as } x \to -\infty.$$

Equation (1.3) is equivalent to

$$\frac{d}{dx}m = z[J, m] + qm \quad \text{a.e. x.} \tag{1.6}$$

Let Σ be the following union of lines through the origin in $\mathbb{C}$:

$$\Sigma = \{z : \Re(z\lambda_j) = \Re(z\lambda_k), \text{ some } j \neq k\}. \tag{1.7}$$

Theorem A. Suppose q belongs to $\mathbf{P}$.

(a) There is a bounded discrete set $Z \subset \mathbb{C}\backslash\Sigma$ such that for every $z \in \mathbb{C}\backslash(\Sigma \cup Z)$ the problem (1.4)–(1.6) has a unique solution $m(\cdot, z)$ and such that, for every $x \in \mathbb{R}$, $m(x, \cdot)$ is meromorphic in $\mathbb{C}\backslash\Sigma$ with poles precisely at the points of Z. Moreover, on $\mathbb{C}\backslash\Sigma$,

$$\lim_{z\to\infty} m(x, z) = I. \tag{1.8}$$

(b) There is a dense open set $\mathbf{P}_0 \subset \mathbf{P}$ such that if q belongs to $\mathbf{P}_0$, then

Z is finite, (1.9)

the poles of $m(x, \cdot)$ are simple, (1.10)

distinct columns of $m(x, \cdot)$ have distinct poles, (1.11)

in each component Ω of $\mathbb{C}\backslash\Sigma, m(x, \cdot)$ has a continuous extension to $\overline{\Omega}\backslash Z$. (1.12)

The function m is an eigenfunction for the matrix differential equation (1.6); we call it the *eigenfunction associated to q*. The elements of the dense open set $\mathbf{P}_0$ will be called *generic potentials*.

Let $\Omega_1, \Omega_2, \ldots, \Omega_r$ be the sectors which are the components of $\mathbb{C}\backslash\Sigma$, ordered in the positive sense about the origin. Let Σ_ν be the closed ray from the origin which one crosses in passing from Ω_ν to $\Omega_{\nu+1}$ in the positive sense. According to (1.12), if $m(x, \cdot)$ is associated to a generic potential, it gives rise to two continuous functions on Σ_ν:

$$m_\nu^-(x, \cdot) = \text{limit on } \Sigma_\nu \text{ from } \Omega_\nu, \tag{1.13}$$

$$m_\nu^+(x, \cdot) = \text{limit on } \Sigma_\nu \text{ from } \Omega_{\nu+1}, \quad (\Omega_{r+1} = \Omega_1). \tag{1.14}$$

Theorem B. Suppose q is a generic potential with associated eigenfunction m.

(a) For $z \in \Sigma_\nu$ there is a unique matrix $v_\nu(z)$ such that, for all x,

$$m_\nu^+(x, z) = m_\nu^-(x, z) e^{xzJ} v_\nu(z) e^{-xzJ}. \tag{1.15}$$

(b) If $m(x, \cdot)$ has poles at $\{z_1, \cdots, z_N\}$, then for each z_j there is a matrix $v(z_j)$ such that the residue satisfies

$$\text{Res}\ (m(x, \cdot); z_j) = \lim_{z\to z_j} m(x, z) \exp\{xz_jJ\} v(z_j) \exp\{-xz_jJ\}. \tag{1.16}$$

(c) The potential q is uniquely determined by the functions $\{v_\nu\}$, the singularities $\{z_j\}$, and the matrices $\{v(z_j)\}$.

Given q as in Theorem B we denote

$$v = (v_1, \cdots, v_r; z_1, \cdots, z_N; v(z_1), \cdots, v(z_N)) \tag{1.17}$$

and call v the *scattering data* associated to q. Note that

$$v_\nu \in C(\Sigma_\nu), \quad v_\nu(z) \to I \text{ as } z \to \infty. \tag{1.18}$$

Part of the scattering data may be recovered from asymptotic information on the singular set Σ. Let Π_ν be the following projection in the matrix algebra:

$$(\Pi_\nu a)_{jk} = \begin{cases} a_{jk} & \text{if } \Re(z\lambda_j) = \Re(z\lambda_k), \quad z \in \Sigma_\nu, \\ 0 & \text{otherwise.} \end{cases} \tag{1.19}$$

Theorem C. Suppose q is a generic potential with associated eigenfunction m. If z is in Σ_ν, then the limits

$$s_\nu^\pm(z) = \lim_{x \to +\infty} \Pi_\nu(e^{-xzJ} m_\nu^\pm(x,z) e^{+xzJ}) \tag{1.20}$$

exist and uniquely determine $v_\nu(z)$. Moreover, the set of functions $\{s_\nu^\pm\}$ determines the poles $\{z_1, \cdots, z_N\}$ and the columns which have singularities at these points. Conversely, this information determines the $\{s_\nu^\pm\}$.

To describe constraints on the scattering data we introduce additional notation. For any matrix a we let $d_k^+(a)$ and $d_k^-(a)$ denote the upper and lower $(k \times k)$ principal minors:

$$d_k^+(a) = \det((a_{ij})_{i,j \le k}), \tag{1.21}$$
$$d_k^-(a) = \det((a_{ij})_{i,j > n-k}). \tag{1.22}$$

Given $z \in \Omega_\nu$, we introduce an ordering of the eigenvalues $\{\lambda_j\}$ so that $\Re(z\lambda_j)$ is strictly decreasing. Note that the induced ordering of the standard basis gives a new matrix representation of the matrix algebra, denoted

$$a \mapsto a^\nu. \tag{1.23}$$

Thus a^ν is the matrix a after conjugation by a permutation matrix, and J^ν has its diagonal entries occurring in the ν-ordering.

Theorem D. Suppose q is a generic potential with scattering data v. Then

$$\Pi_\nu v_\nu(z) = v_\nu(z), \quad z \in \Sigma_\nu, \tag{1.24}$$
$$v_\nu(0) = a_\nu^{-1} a_{\nu+1}, \tag{1.25}$$

where $(a_\nu)_{jj} = 1$ and $(a_\nu)^\nu$ is upper triangular,

$$d_k^-(v_\nu(z)^\nu) = 1, \quad 1 \le k \le n, z \in \Sigma_\nu, \tag{1.26}$$
$$d_k^+(v_\nu(z)^\nu) \ne 0, \quad 1 \le k \le n, z \in \Sigma_\nu, \tag{1.27}$$

if z_i is in Ω_ν, then $v(z_i)^\nu$ has a single non-zero entry which is in the $(k, k+1)$ position for some $k < n$. (1.28)

Moreover, let $\alpha_{\nu k}$ be the winding number of the k-th upper minor of $(v_\nu)^\nu$:

$$2\pi\alpha_{\nu k} = \int_{\Sigma_\nu} d[\arg d_k^+(v_\nu(z)^\nu)], \tag{1.29}$$

where Σ_ν is oriented from 0 to ∞. Let,

$$\beta_{\nu k}\text{=number of } z_i \in \Sigma_\nu \text{ such that } k\text{-th column of } v(z_i) \text{ is } \neq 0. \tag{1.30}$$

Then the $\{\alpha_{\nu k}, \beta_{\nu k}\}$ satisfy $n-1$ independent homogeneous equations

$$\Sigma(\epsilon_{\nu k,j}\alpha_{\nu k} + \eta_{\nu k,j}\beta_{\nu k}) = 0, \quad 1 \le j < n, \tag{1.31}$$

where the coefficients belong to $\{0, \pm 1\}$.

Some analytic properties of the scattering map are summarized in the next theorem.

Theorem E. Suppose q is a generic potential with scattering data v and suppose k is a non-negative integer.

(a) If the distribution derivatives of q satisfy

$$D^j q \in L^1, \quad 0 \le j \le k, \tag{1.32}$$

then

$$\lim_{z\to\infty} z^k[v_\nu(z) - I] = 0. \tag{1.33}$$

(b) If

$$x^k q \in L^1, \tag{1.34}$$

then

$$v_\nu \in C^k(\Sigma_\nu) \text{ and } D^j(v_\nu - I) \to 0 \text{ as } z \to \infty, \quad 0 \le j \le k. \tag{1.35}$$

Moreover, let $v_{\nu,k}$ be the Taylor polynomial of degree k for v_ν at the origin. Then there are matrix-valued polynomials $a_{\nu k}$ as in (1.25), with

$$v_{\nu,k}(z) = a_{\nu,k}(z)^{-1}a_{\nu+1,k}(z) + O(z^k), \quad |z| \le 1. \tag{1.36}$$

(c) If

$$x^j q \in L^1 \cap L^2, \quad 0 \le j \le k, \tag{1.37}$$

then in addition to (1.35) we have

$$D^j(v_\nu - I) \in L^2(\Sigma_\nu), \quad 1 \le j \le k. \tag{1.38}$$

Let $\mathcal{S}(\mathbb{R})$ denote the usual Schwartz space (of matrix-valued functions) and let $\mathcal{S}(\Sigma_\nu)$ denote the space of functions each of whose derivatives is continuous on the closed ray Σ_ν and is rapidly decreasing at ∞. Theorem E shows that for a generic potential belonging to $\mathcal{S}(\mathbb{R})$, v_ν is rapidly decreasing at ∞ and all its derivatives are bounded. A rapidly decreasing function with bounded second derivative has rapidly decreasing first derivative; thus we have the following:

Theorem E′. *If q is a generic potential belonging to $\mathcal{S}(\mathbb{R})$, then each v_ν belongs to $\mathcal{S}(\Sigma_\nu)$, and* (1.36) *holds for every integer $k \geq 0$.*

For the inverse problem we introduce the space $\mathbf{S}$ of *formal scattering data*, consisting of elements v of the form (1.17), satisfying the following:

$$I - v_\nu, \quad Dv_\nu \in L^2(\Sigma_\nu); \tag{1.39}$$

this implies

$$v_\nu \to I \text{ as } z \to \infty, \tag{1.40}$$

$$\text{conditions (1.24)–(1.28) and (1.31) hold.} \tag{1.41}$$

In (1.39), Dv_ν is the distribution derivative along the ray. Condition (1.39) implies (after correction on a set of measure zero) that v_ν is bounded and belongs to the Hölder space $C^{1/2}(\Sigma_\nu)$, so (1.25) makes sense.

Note that from Theorems D and E it follows that if q is a generic potential and satisfies (1.37) with $k = 2$, then its scattering data belongs to $\mathbf{S}$.

There are two natural ways to topologize $\mathbf{S}$, each of them being, for each fixed N, a product topology. The components v_ν receive the obvious topology, the components $v(z_j)$ the natural matrix topology, and the z_j either the usual $\mathbb{C}$-topology *or* the discrete topology. The map from generic potentials satisfying (1.37) with $k = 2$ to $\mathbf{S}$ is continuous with respect to the first topology on $\mathbf{S}$, while the set $\mathbf{S}_0$ below is open and dense with respect to either topology.

Given $v \in \mathbf{S}$ and $x \in \mathbb{R}$ we look for a matrix-valued function $m(x, \cdot)$ such that

$$m(x, \cdot) \text{ is meromorphic on } \mathbb{C} \backslash \Sigma \text{ with poles at } z_1, \cdots, z_N, \tag{1.42}$$

$$\text{conditions (1.12), (1.15), and (1.16) hold.} \tag{1.43}$$

Theorem F. Suppose v belongs to $\mathbf{S}$.

(a) For any real x there is at most one associated eigenfunction; for $|x|$ large, there is exactly one.

(b) There is a dense open set $\mathbf{S}_0 \subset \mathbf{S}$ such that for every $v \in \mathbf{S}_0$ the associated eigenfunction exists for every real x. Moreover, $m(\cdot, z)$ is absolutely continuous with respect to x for all $z \notin \Sigma \cup \{z_1, \cdots, z_N\}$ and satisfies the differential equation (1.6), where q is an off-diagonal matrix-valued function with

$$(1 + |x|)q \in L^2 + L^\infty. \tag{1.44}$$

The following is a partial converse of Theorem E.

Theorem G. Suppose v belongs to $\mathbf{S}_0$ and suppose q is the corresponding potential.

(a) If $z^k(v_\nu - I) \in L^2(\Sigma_\nu)$, all ν, then the distribution derivatives of q satisfy

$$D^j q \in L^2 + L^\infty, \quad 0 \le j \le k. \tag{1.45}$$

(b) Suppose the distribution derivatives of v satisfy

$$D^j v_\nu \in L^2, \quad 0 < j \le k+1. \tag{1.46}$$

Let $v_{\nu,k}$ be the Taylor polynomial of degree k for v_ν at the origin, and suppose (1.36) holds. Then

$$x^{k+1} q \in L^2 + L^\infty. \tag{1.47}$$

As above, Theorem G has the following consequence for Schwartz class scattering data.

Theorem G′. *Suppose v belongs to $\mathbf{S}_0$. Suppose also that each v_ν belongs to $\mathcal{S}(\Sigma_\nu)$ and that* (1.36) *holds for every integer $k \ge 0$. Then the corresponding potential belongs to $\mathbf{S}(\mathbb{R})$.*

The conditions defining $\mathbf{S}$ are preserved under the type of nonlinear evolution of q which is associated to the spectral problem (1.3). Indeed the scattering data $v(\cdot, t)$ for such an evolution evolves with singularities $\{z_j\}$ fixed with

$$\frac{\partial v_\nu}{\partial t}(z,t) = [\mu(z), v_\nu(z,t)], \tag{1.48}$$

$$\frac{\partial v}{\partial t}(z_j,t) = [\mu(z_j), v(z_j,t)], \tag{1.49}$$

where μ is a diagonal-valued function. Thus

$$v_\nu(z,t) = e^{t\mu(z)} v_\nu(z,0) e^{-t\mu(z)}, \tag{1.50}$$

a similar expression holding for the $v(z_j, t)$. The algebraic conditions on v are clearly invariant under conjugation by a diagonal matrix. Therefore the flow (1.48)–(1.49) maps $\mathbf{S}$ to itself (continuously with respect to either topology) if and only if

$$\mu \text{ is continuous}, \tag{1.51}$$

$$\Re \Pi_\nu \mu(z) = 0, \quad z \in \Sigma_\nu. \tag{1.52}$$

In general, $\mathbf{S}_0$ is not invariant under (1.48), (1.49). For example, if $v_\nu \equiv 0$, all ν, then q is of the form

$$\begin{aligned} q(x,t) &= R(e_1(x,t), \cdots, e_N(x,t)), \\ e_j(x,t) &= \exp\{\tau_j t + \xi_j x\}, \end{aligned} \tag{1.53}$$

where R is a rational function. Depending on the data, q may have algebraic singularities for some real x and t.

2 Compactly Supported Potentials

With the assumptions and notation of Section 1, we investigate the solution of the eigenvalue problem (1.5), (1.6) when the potential has compact support.

Proposition 2.1. *Suppose $q \in \mathbf{P}$ has compact support. For each complex z there is a unique absolutely continuous $m_0(\cdot, z)$ such that*

$$\frac{d}{dx} m_0(x,z) = z[J, m_0(x,z)] + q(x) m_0(x,z), \quad \text{a.e. } x, \tag{2.2}$$

$$m_0(x,z) = I \text{ if } x \ll 0. \tag{2.3}$$

Moreover, $m_0(x, \cdot)$ is an entire function and

$$m_0(x,z) = e^{xzJ} s_0(z) e^{-xzJ}, \quad x \gg 0, \tag{2.4}$$

$$\det m_0 \equiv 1. \tag{2.5}$$

Proof: Clearly m_0 is the (unique) solution of the Volterra integral equation

$$m_0(x,z) = I + \int_{-\infty}^{x} e^{(x-y)zJ} q(y) m_0(y,z) e^{(y-x)zJ}\, dy. \tag{2.6}$$

The dependence on the parameter z is holomorphic. To prove (2.4) we note that

$$\frac{d}{dx} m_0 = z[J, m_0], \quad x \gg 0. \tag{2.7}$$

To prove (2.5) we let $\psi_0(x,z) = m_0(x,z) e^{xzJ}$. Then

$$\frac{d}{dx}\psi_0 = (zJ + q)\psi_0 \tag{2.8}$$

so that

$$\frac{d}{dx}(\det \psi_0) = z \operatorname{tr} J \cdot \det \psi_0 \tag{2.9}$$

and thus $\det m_0$ is constant with respect to x.

Note that (2.6) gives

$$s_0(z) = I + \int_{\mathbb{R}} e^{-yzJ} q(y) m_0(y,z) e^{yzJ}\, dy. \tag{2.10}$$

Proposition 2.11. *Suppose $q \in \mathbf{P}$ has compact support. Then the eigenvalue problem (1.5), (1.6) has a unique solution $m(\cdot, z)$ for every $z \in \mathbb{C} \backslash (\Sigma \cup Z)$, where $Z \subset \mathbb{C} \backslash \Sigma$ is discrete. Moreover, $m(x, \cdot)$ is meromorphic on $\mathbb{C} \backslash \Sigma$.*

Proof: We look for m of the form

$$m(x,z) = m_0(x,z)a_0(x,z). \tag{2.12}$$

Since m_0 is invertible we must have

$$\frac{d}{dx}a_0 = z[J, a_0], \tag{2.13}$$

so a must have the form

$$a_0(x,z) = e^{xzJ}a(z)e^{-xzJ}. \tag{2.14}$$

Conversely, (2.12) and (2.14) imply that m is an eigenfunction. Now

$$m(x,z) = e^{xzJ}a(z)e^{-xzJ}, \quad x \ll 0, \tag{2.15}$$
$$m(x,z) = e^{xzJ}s_0(z)a(z)e^{-xzJ}, \quad x \gg 0. \tag{2.16}$$

In order to have m bounded as $x \to -\infty$ it is necessary and sufficient, in view of (2.15) that

$$a(z)_{jk} = 0 \text{ if } \Re(z\lambda_j) < \Re(z\lambda_k). \tag{2.17}$$

In terms of the ν-representation introduced in Section 1, this condition is

$$a(z)^\nu \text{ is upper triangular if } z \in \Omega_\nu. \tag{2.18}$$

Similarly, (2.16) and boundedness as $x \to +\infty$ become

$$s_0(z)^\nu a(z)^\nu \text{ is lower triangular if } z \in \Omega_\nu. \tag{2.19}$$

Convergence to the identity as $x \to -\infty$ requires

$$a(z)^\nu_{jj} = 1, \quad 1 \le j \le n. \tag{2.20}$$

Thus (2.12), (2.14), (2.18)–(2.20) are necessary and sufficient conditions. The algebraic conditions (2.18)–(2.20) determine a^ν, and thus a, uniquely, provided that the upper minors do not vanish (see [8], Theorem 1.1):

$$d_k^+(s_0(z)^\nu) \neq 0, \quad 1 \le k \le n, z \in \Omega_\nu. \tag{2.21}$$

Thus Z is precisely the set where (2.21) fails.

Remark 2.22. It will be shown in Section 4 that Z is finite. It is clear from this construction that starting from a given sector Ω_ν, the function m has an extension which is meromorphic in all of $\mathbb{C}$. In particular, except for a discrete set Z_ν, in a ray Σ_ν, the limits (1.13) and (1.14)exist. These limits again have determinant 1, and by differentiating the expression $(m_\nu^-)^{-1}m_\nu^+$, we see that it satisfies the equation (2.13). Thus there is $v_\nu(z)$ such that

$$m_\nu^+(x,z) = m_\nu^-(x,z)e^{xzJ}v_\nu(z)e^{-xzJ}, \quad z \in \Sigma_\nu \backslash Z_\nu. \tag{2.23}$$

Definition 2.24. Let q, m, Z be as in Proposition 2.11. A singularity $z \in Z$ is *simple* if it is a simple pole for m and only one column of m is singular at z.

Proposition 2.25. *Suppose $z \in Z \cap \Omega_\nu$ is a simple singularity for m. Then there is a matrix $v(z)$ such that*

$$\text{Res }(m(x,\cdot);z) = \lim_{z' \to z} m(x,z')e^{xzJ}v(z)e^{-xzJ}. \tag{2.26}$$

Moreover, $v(z)^\nu$ is of the form $ce_{k,k+1}$, where c is a constant and $e_{k,k+1}$ a matrix unit.

Proof: For convenience, replace the original matrix representation with the ν-representation. Thus in Ω_ν, m is of the form (2.12), (2.14), where

$$a^\nu = I + u, \quad u \text{ strictly upper triangular.} \tag{2.27}$$

It is easily seen that Res $(m(x,\cdot),z)$ satisfies the differential equation (1.6), and it follows that

$$\text{Res }(m(x,\cdot);z) = m_0(x,z)e^{xzJ}v_0(z)e^{-xzJ}. \tag{2.28}$$

Suppose that it is the column $k+1$ of m which is singular at z. Taking $x \ll 0$ we see that v_0 is the residue of u at z; thus v_0 is strictly upper triangular, with only column $k+1$ non-null. Let

$$p_k = e_{11} + e_{22} + \cdots + e_{kk}.$$

Then up_k has no singularity at z and we may define

$$v(z) = (1 + u(z)p_k)^{-1}v_0(z). \tag{2.29}$$

In view of (2.12), (2.14), and (2.27)–(2.29), it is clear that (2.26) holds. To see that v has the given form we use the fact that, in Ω_ν, m is lower triangular for $x \gg 0$. Therefore,

$$\begin{aligned} 0 &= p_k \text{ Res }(s_0(1+u);z) \\ &= p_k s_0(z)(1+u(z)p_k)v(z) \\ &= p_k b(z) p_k v(z) = p_k b(z) v(z), \end{aligned}$$

where b is lower triangular. By assumption the $(k-1)\times(k-1)$ upper minor of $s_0(z)$ is not zero, so the same is true for b. It follows that the first $k-1$ entries of column $k+1$ of v are zero, and the proof is complete.

Proposition 2.30. *Suppose $q \in \mathbf{P}$ has compact support. In any neighborhood of q there is a potential whose associated eigenfunction m has properties* (1.10)–(1.12).

Proof: Suppose $r \in \mathbf{P}$ also has compact support. Let $M_n(\mathbb{C})$ be the $(n \times n)$ matrix algebra. Given $\zeta \in M_n(\mathbb{C})$, let r_ζ be the potential with entries $\zeta_{jk} r_{jk}$. Let s_0 be the matrix-valued function associated to q as in the proof of Proposition 2.1, and let $s_{0,\zeta}$ be the corresponding function associated to r_ζ. Note that (2.10) implies

$$s_{0,\zeta}(z) = I + \int_{\mathbb{R}} e^{-yzJ} r_\zeta(y) e^{yzJ}\, dy + O(|\zeta|^2). \tag{2.31}$$

Now suppose that the support of r lies to the right of supp (q). Then it is easily seen that the eigenfunction for $q + r_\zeta$ which satisfies (2.3) is

$$m_{0,\zeta}(x,z) e^{xzJ} s_0(z) e^{-xzJ}, \quad x \gg 0, \tag{2.32}$$

where $m_{0,\zeta}$ is the corresponding eigenfunction for r_ζ. Thus the asymptotic matrix s_ζ corresponding to $q + r_\zeta$ is

$$s_\zeta(z) = s_{0,\zeta}(z) s_0(z). \tag{2.33}$$

Consider now the map

$$\varphi : \mathbb{C} \times M_n(\mathbb{C}) \mapsto M_n(\mathbb{C}), \tag{2.34}$$

$$\varphi(z,\zeta) = \text{diag } (1 + \zeta_{11}, \cdots, 1 + \zeta_{nn}) s_\zeta(z).$$

This map is holomorphic. For fixed z, equation (2.31) shows that the differential of $s_{0,\zeta}$ at $\zeta = 0$ is a matrix whose entries are certain dilations of the Fourier-Laplace transforms of the entries of r, evaluated at z. In particular, r may be chosen so that none of these entries vanishes at a given point z, and it follows that $d\varphi$ is surjective at $(z,0)$. Let $\Gamma_j \subset M_n(\mathbb{C})$ consist of all matrices for which at least j distinct minors vanish. This is an algebraic variety of complex codimension j in $M_n(\mathbb{C})$. Thus if $d\varphi$ is surjective at $(z,0)$ it follows that the complex codimension of $\varphi^{-1}(\Gamma_j)$ near $(z,0)$ is j, and the complex codimension of the projection to $M_n(\mathbb{C})$ is at least $j-1$. In particular, this means that if z is a point where two or more distinct minors of s_0 vanish, then there is a neighborhood U of z and a sequence of potentials converging to q such that the minors of the corresponding s_0 have distinct zeros in U. A similar argument based on the real codimension and the restriction of φ to $\mathbb{R} \times M_n(\mathbb{C})$ shows that if a minor of s_0 has a real zero z, then there is a neighborhood U and a sequence of approximating potentials whose s_0 have no minors with real zeros in U.

In the proof of Theorem A (a) below we show that there is a constant C such that the eigenfunction m for q, or for any sufficiently nearby potential, has no singularities in the region $|z| > C$. A regular point remains regular under small perturbations, so the argument just given implies that we may remove singularities on Σ one at a time by arbitrarily small perturbations, and similarly we may split poles of distinct columns which happen to coincide. The last step is to show that any multiple poles can be split into simple poles by small perturbations. If $z \in \Omega_\nu$ is a multiple pole, it corresponds to a multiple zero of an upper minor d_b^+ of s_0^ν. We choose r with support to the right of supp (q), such that the dilated Fourier-Laplace transforms of the r_{jk} have a simple zero at z. Rewriting matrices in the ν representation, and multiplying s_0 on the right by an upper triangular matrix with ones on the diagonal, we may assume that $(s_0)_{ij} = 0$ for $i \geq j$ if $j \leq k$, near z. Since the next minor is not zero at z, $(s_0)_{k+1,k}(z) \neq 0$. If we choose $\zeta_{k,k+1} \neq 0$ but small, and the other $\zeta_{ij} = 0$, then $(s_\zeta)_{kk}$ will have simple zeros near z and so will the corresponding minor. This completes the proof.

3 Small Potentials

It is convenient here to introduce more notation and structure. We let $\mathbb{C}^n$ have the standard hermitian inner product, and let the matrix algebra $M_n(\mathbb{C})$ operate on $\mathbb{C}^n$ in the standard way. Then $M_n(\mathbb{C})$ is a Hilbert space with respect to the trace form

$$(a, b) = \text{tr } b^* a$$

and we denote the norm by $|\ |$. Then the L^1-norm on the space of potentials $\mathbf{P}$ is

$$||q||_1 = \int_{\mathbb{R}} |q(x)| dx. \tag{3.1}$$

Define $\mathfrak{J} : M_n(\mathbb{C}) \mapsto M_n(\mathbb{C})$ by

$$\mathfrak{J}a = adJ(a) = [J, a]. \tag{3.2}$$

Then $\mathfrak{J}$ is a normal operator, and for any complex z the operator

$$\Re(z\mathfrak{J}) = ad(\Re(zJ)) = \tfrac{1}{2}(z\mathfrak{J}) + \tfrac{1}{2}(z\mathfrak{J})^*$$

is selfadjoint. Let

$$\Pi_+^z, \Pi_-^z, \Pi_0^z = M_n(\mathbb{C}) \mapsto M_n(\mathbb{C}) \tag{3.3}$$

denote the orthogonal projections of $M_n(\mathbb{C})$ onto the positive, negative, and null subspaces for $\Re(z\mathfrak{J})$, respectively. Let

$$\Pi_0^z = M_n(\mathbb{C}) \mapsto M_n(\mathbb{C}) \tag{3.4}$$

be the orthogonal projection onto the kernel of $\mathfrak{J}$, the set of diagonal matrices. Then the projections (3.3) are constant on each component of $\mathbb{C}\backslash\Sigma$, while

$$\Pi_0^z = \begin{cases} \Pi_0 & \text{if and only if } z \notin \Sigma, \\ \Pi_\nu & \text{if } z \in \Sigma_\nu \backslash (0). \end{cases} \tag{3.5}$$

Furthermore,

$$\Pi_+^z a = a \Leftrightarrow a^\nu \text{ is upper triangular,} \tag{3.6}$$

where $\Omega_\nu \ni z$, a similar statement holding for Π_-^z and lower triangularity. Note also that

$$\exp\{t\mathfrak{J}\}(a) = e^{tJ} a e^{-tJ}. \tag{3.7}$$

Theorem 3.8. *Suppose $q \in L^1$ has norm $\|q\|_1 < 1$. Then for each $z \in \mathbb{C}\backslash\Sigma$ there is a unique associated eigenfunction $m(\cdot, z)$ satisfying* (1.5) *and* (1.6). *The function m is holomorphic in $\mathbb{C}\backslash\Sigma$ with values $L^\infty \cap C$. On each component of $\mathbb{C}\backslash\Sigma$, m and its inverse extend continuously to the closure. In addition,*

$$m(x, z) \to I \text{ as } z \to \infty \text{ uniformly w.r. to } x, \tag{3.9}$$

$$|m(x, z)| \leq (1 - \|q\|_1)^{-1} \text{ for all } x, z, \tag{3.10}$$

$$|m(x, z)^{-1}| \leq (1 - \|q\|_1)^{-1} \text{ for all } x, z. \tag{3.11}$$

Proof: Given $q \in \mathbf{P}$ and $z \in \mathbb{C}\backslash\Sigma$, let

$$K_{z,q} : L^\infty \mapsto L^\infty \cap C \qquad \text{(matrix-valued)}, \tag{3.12}$$

$$\begin{aligned}[K_{z,q} f](x) = &\int_{-\infty}^{x} e^{(x-y)z\mathfrak{J}} (\Pi_0^z + \Pi_-^z)(q(y) f(y))\, dy \\ &- \int_{x}^{\infty} e^{(x-y)z\mathfrak{J}} \Pi_+^z (q(y) f(y))\, dy.\end{aligned}$$

The exponential operators here have norm at most 1 on the subspaces where they act, and these subspaces are orthogonal and invariant for $\mathfrak{J}$, so the operator norm satisfies

$$\|K_{z,q}\| \leq \|q\|_1. \tag{3.13}$$

Clearly,

$$\frac{d}{dx} K_{z,q} f(x) = z\mathfrak{J} K_{z,q} f(x) + q(x) f(x) \quad \text{a.e.} \tag{3.14}$$

It follows that, for $\|q\|_1 < 1$,

$$m(x, z) = [(Id - K_{z,q})^{-1} I](x) \tag{3.15}$$

is a bounded continuous function of x satisfying the differential equation (1.6) and the estimate (3.10). Moreover, the map $z \to K_{z,q}$ is holomorphic from

$\mathbb{C}\backslash\Sigma$ to the space of bounded operators in $L^\infty \cap C$, so m is holomorphic with values in this space. Similarly, m extends to be continuous on the closure of a component of $\mathbb{C}\backslash\Sigma$, with values in $L^\infty \cap C$.

The dominated convergence theorem implies that, for fixed z,

$$K_{z,q}f(x) \to 0 \text{ as } x \to -\infty, \tag{3.16}$$

uniformly on bounded sets in $L^\infty \cap C$. In particular, since

$$m(x,z) = I + [K_{z,q}m(\cdot,z)](x), \tag{3.17}$$

it follows that m satisfies (1.5).

As in the proof of (2.5) we see that

$$\det m \equiv 1; \tag{3.18}$$

thus m is invertible and $m(\cdot,z)^{-1}$ is bounded. If m_1 were a second solution of (1.5), (1.6), then as in the proof of Proposition 2.11 we would have

$$m_1(x,z) = m(x,z)e^{xz\mathcal{J}}(a(z)), \quad z \notin \Sigma. \tag{3.19}$$

But $\exp\{xz\mathcal{J}\}(a(z))$ would have to be bounded with respect to x, which implies that $a(z)$ is diagonal, and then the asymptotic condition (1.5) for m and m_1 implies $a(z) = I$.

To prove the estimate (3.11) for m^{-1}, let $q_2(x) = -q(x)^*$ and let m_2 be the corresponding eigenfunction. Then

$$\left(\frac{d}{dx} - z\mathcal{J}\right)[m(x,z)m_2(x,-\bar{z})^*] = 0 \tag{3.20}$$

and as above we can conclude that

$$m(x,z)^{-1} = m_2(x,-\bar{z})^* \tag{3.21}$$

which implies (3.11).

Finally, consider the asymptotic behavior in z. Suppose first that dq/dx is also in L^1. Let

$$n(x,z) = I - z^{-1}\mathcal{J}^{-1}q(x), \tag{3.22}$$

where $\mathcal{J}^{-1} : \operatorname{ran}\mathcal{J} \mapsto \operatorname{ran}\mathcal{J}$. (Note that (1.2) implies $q(x) \in \operatorname{ran}\mathcal{J}$ for all x; this is our first essential use of this fact.) Then

$$\left(\frac{d}{dx} - z\mathcal{J}\right)n = qn + z^{-1}f, \tag{3.23}$$

where $f \in L^1$. Then using the asymptotic information and

$$\left(\frac{d}{dx} - z\mathcal{J}\right)(m^{-1}n) = z^{-1}m^{-1}f = g(x,z), \tag{3.24}$$

we obtain

$$(m^{-1}n)(x,z) = I + \int_{-\infty}^{x} e^{(x-y)z\mathfrak{J}}(\Pi_0^z + \Pi_-^z)g(y,z)\,dy \\ - \int_x^{\infty} e^{(x-y)z\mathfrak{J}}\Pi_+^z g(y,z)dy. \tag{3.25}$$

Thus

$$|m(x,z) - n(x,z)| \le C|z|^{-1} \text{ for all } x, z. \tag{3.26}$$

This proves (3.9) when the derivative of q is in L^1. In general we approximate q by such potentials and note that the corresponding eigenfunctions converge uniformly with respect to z and x. In fact, suppose q and q_1 are in L^1 with norm less than 1 and let m, m_1 be the corresponding eigenfunctions. Then

$$\left(\frac{d}{dx} - z\mathfrak{J}\right)(m^{-1}m_1) = m^{-1}(q_1 - q)m_1, \tag{3.27}$$

and we obtain an integral expression analogous to (3.25) which implies

$$|m_1(x,z) - m(x,z)| \le \|q - q_1\|_1 \|m(\cdot,z)^{-1}\|_\infty \|m(\cdot,z)\|_\infty^2. \tag{3.28}$$

4 Proof of Theorem A and Theorem B

Suppose q belongs to **P**. When $\|q\|_1 < 1$, part (a) of Theorem A is just Theorem 3.8, and the set Z of singularities is empty. To complete the proof of part (a) we induce on the least integer $N \ge 0$ such that $\|q\|_1 < 2^N$. Note that the eigenfunction corresponding to a translate of q is the translate (with respect to x) of the eigenfunction. Thus after translation we may assume that

$$\int_{-\infty}^{0} |q(x)|\,dx = \int_0^{\infty} |q(x)|\,dx. \tag{4.1}$$

Let $q_1(x) = q(x)$ for $x < 0$, $q_1(x) = 0$ for $x \ge 0$, and $q_2 = q - q_1$. The induction assumption implies that q_j has an eigenfunction m_j for which Theorem A (a) holds. Any eigenfunction m for q must be of the form

$$m(x,z) = m_1(x,z)e^{xz\mathfrak{J}}a_1(z), \quad x \le 0, \\ = m_2(x,z)e^{xz\mathfrak{J}}a_2(z), \quad x \ge 0. \tag{4.2}$$

For boundedness, continuity, and the asymptotic condition as $x \to -\infty$ it is necessary and sufficient to have

$$m_1(0,z)a_1(z) = m_2(0,z)a_2(z), \tag{4.3}$$

$$\Pi_-^z a_1(z) = 0 = \Pi_+^z a_2(z), \tag{4.4}$$

$$\Pi_0 a_1(z) = I, \quad z \in \mathbb{C}\backslash\Sigma. \tag{4.5}$$

In the matrix representation corresponding to the sector $\Omega_\nu \ni z$, this is a factorization problem:

$$[m_2(0,z)^{-1}m_1(0,z)]^\nu = a_2(z)^\nu[a_1(z)^{-1}]^\nu,$$

$$a_2(z)^\nu \text{ lower triangular}, \quad a_1(z)^\nu \text{ upper triangular}, \tag{4.6}$$

$$a_1(z)_{jj} = 1, \quad 1 \le j \le n.$$

As before this problem has a unique solution as long as the upper minors of $(m_2^{-1}m_1)^\nu$ are non-zero. The latter matrix approaches I as $z \to \infty$, so the factorization problem (4.6) introduces at most a bounded, discrete set of new singularities in the construction of m. Moreover, $a_j(z) \to I$ as $z \to \infty$, so $m(x,z) \to I$ as $z \to \infty$. This completes the proof of Theorem A (a).

To prove that the set $\mathbf{P}_0$ of generic potentials is dense, we note first that the set of compactly supported potentials is dense in $\mathbf{P}$. Second, since the set of singularities is now shown to be bounded, the construction in the proof of Proposition 2.11 shows that any compactly supported potential has only finitely many singularities, including the singularities of the extensions to Σ. Therefore, the potentials obtained in Proposition 2.30 are generic and dense in $\mathbf{P}$.

Finally, we need to show that $\mathbf{P}_0$ is open.

Lemma 4.7. *Suppose $q \in \mathbf{P}$ has associated eigenfunction m. Suppose K is a compact subset of the one-point compactification of the closure of a sector Ω_ν and suppose that m extends to be continuous on $\mathbb{R} \times K$. Given $\epsilon > 0$, there is $\delta > 0$ such that, if $q_1 \in \mathbf{P}$ and $\|q - q_1\|_1 < \delta$, then the associated eigenfunction m_1 extends to $\mathbb{R} \times K$ and $|m - m_1| < \epsilon$ on $\mathbb{R} \times K$.*

Proof: The argument leading to the inequality (3.28) proves this when $\|q\|_1 < 1$, and the inductive construction of this section gives the general result.

Suppose now that q is a generic potential. Lemma 4.7 implies that the eigenfunction for a nearby potential will extend to Σ and will have singularities only near those of the eigenfunction m of q. Moreover, if the potential is sufficiently close, the singularities will be simple poles occurring in one column at a time, by a contour integration argument. Thus $\mathbf{P}_0$ is open, and the proof of Theorem A is complete.

If q is a generic potential with associated eigenfunction m, then by assumption the limits $m_\nu^\pm$ exist on Σ_ν. If z belongs to Σ_ν, then

$$\left(\frac{d}{dx} - z\mathfrak{J}\right)[m_\nu^-(x,z)^{-1}m_\nu^+(x,z)] = 0 \tag{4.8}$$

so (1.15) holds. If $z_j \in \mathbb{C}\backslash\Sigma$ is a singularity for m, to prove (1.16) we approximate q by compactly supported generic potentials. It follows from Lemma 4.7 and the construction in (2.29) that we may pass to the limit in (1.16) for the approximating potentials to obtain (1.16) for m. Note also that the limit $v(z_j)$ is of the form $ce_{k,k+1}$ in the ν-matrix representation.

Finally, to see that a generic potential is uniquely determined by its scattering data, suppose q_1 and q_2 are generic potentials with eigenfunctions m_1, m_2 and having the same scattering data. For any fixed $x \in \mathbb{R}$ the function

$$f(z) = m_1(x,z)m_2(x,z)^{-1} \tag{4.9}$$

is meromorphic on $\mathbb{C} \backslash \Sigma$ and converges to I as $z \to \infty$. It is enough to show that the apparent singularities are removable. On Σ_ν, f has limits $f_\nu^\pm$. A trivial calculation shows that $f_\nu^+ = f_\nu^-$, so f is holomorphic except possibly at the sigularities $\{z_j\}$ of m_1 and m_2. Now

$$m_1(x,z) = (z - z_j)^{-1}a + b + O(|z - z_j|), \tag{4.10}$$

and (1.16) implies

$$av_j = 0, \quad bv_j = a, \tag{4.11}$$

where $v_j = \exp\{xz_j\mathfrak{J}\}[v(z_j)]$. Thus

$$m_1(x,z) = b(I + (z - z_j)^{-1}v_j) + O(|z - z_j|). \tag{4.12}$$

Since $v_j^2 = 0$, (4.12) implies that

$$m_1(x,z)(I - (z - z_j)^{-1}v_j) = m_1(x,z)w_j(z) \tag{4.13}$$

has a removable singularity at z_j. The same is true for m_2, so $f = m_1 m_2^{-1} = (m_1 w_j)(m_2 w_j)^{-1}$ has a removable singularity at z_j. Then $f \equiv I$ and the proof of Theorem B is complete.

5 Proof of Theorem C and Theorem D

Assume first that the generic potential q has compact support. Let m_0 be the eigenfunction of Proposition 2.1 and s_0 the function in (2.4) which gives the asymptotic behavior of m_0. We know that

$$m(x,z) = m_0(x,z)e^{xz\mathfrak{J}}a(z). \tag{5.1}$$

The function a has limits $a_\nu^\pm$ on Σ_ν. Now

$$m(x,z) = e^{xz\mathfrak{J}}s_1(z), \quad x \gg 0, \tag{5.2}$$

where $s_1 = s_0 a$. Again, s_1 has limits on Σ_ν and we set

$$s_\nu^\pm = \Pi_\nu(s_0 a_\nu^\pm) = \Pi_0^z(s_0 a_\nu^\pm), \quad z \in \Sigma_\nu. \tag{5.3}$$

Clearly, (1.20) is satisfied. We may now pass to the limit from compactly supported potentials to obtain the existence of (1.20) in the general case.

To see that the limits $s_\nu^\pm$ determine v_ν, we return to the compactly supported case. Recall that (5.2) and boundedness imply that $(s_1)^\nu$ is lower triangular when z is in Ω_ν. Thus

$$(\Pi_0^z + \Pi_-^z)(s_0 a_\nu^\pm) = s_0 a_\nu^\pm \text{ on } \Omega_\nu. \tag{5.4}$$

For $x \gg 0$ and $z \in \Omega_\nu$,

$$\begin{aligned} v_\nu(z) &= e^{-xzJ}[m_\nu^-(x,z)^{-1} m_\nu^+(x,z)] \\ &= [s_0(z) a_\nu^-(z)]^{-1} s_0(z) a_\nu^+(z). \end{aligned} \tag{5.5}$$

Boundedness implies

$$v_\nu(z) = \Pi_0^z v_\nu(z), \quad z \in \Sigma_\nu. \tag{5.6}$$

From (5.4)–(5.6) we obtain

$$v_\nu = \Pi_0^z (s_0 a_\nu^-)^{-1}(s_0 a_\nu^+) = (s_\nu^-)^{-1} s_\nu^+ \tag{5.7}$$

since Π_0^z is multiplicative on the range of $\Pi_0^z + \Pi_+^z$, which is an algebra. Again passage to the limit gives (5.6) and (5.7) for general generic potentials.

To see that the limits $s_\nu^\pm$ determine the location of the singularities, we return again to the compactly supported case. In the matrix representation corresponding to Ω_ν, a singularity in column $k+1$ occurs at a zero of the k-th upper minor of s_0^ν. Since a^ν is upper triangular with 1's on the diagonal,

$$d_k^+(s_0^\nu) = d_k^+(s_0^\nu a^\nu). \tag{5.8}$$

Now $s_0^\nu a^\nu$ is lower triangular, so its upper minors are the same as those of the corresponding diagonal matrix δ^ν, where

$$\begin{aligned} \delta(z)_{jj} &= (s_0 a)(z)_{jj} \\ &= \lim_{x \to \infty} m(x,z)_{jj}. \end{aligned} \tag{5.9}$$

The elements of δ, being ratios of the minors which are holomorphic in $\mathbb{C}\backslash\Sigma$, are themselves meromorphic in $\mathbb{C}\backslash\Sigma$ with continuous extensions to Σ. They are therefore determined by their restrictions to Σ, which are just the diagonal elements of the $s_\nu^\pm$. Once again we may pass to the limit from compactly supported q.

To complete the proof of Theorem C, we want to show that $\{v_\nu\}$ and the location of the singularities of m determine $s_\nu^\pm$. Again we start with the compactly supported case and pass to the limit. In the ν representation the range of Π_0^z consists of matrices whose non-zero elements occur only in certain diagonal blocks, when $z \in \Omega_\nu$. Moreover, in these blocks $(s_\nu^-)^\nu$ is lower triangular, while

$(s_\nu^+)^\nu$ is upper triangular. Thus, within the range of Π_0^z, (5.7) gives a triangular factorization of v_ν^ν. This factorization is unique up to left multiplication of $s_\nu^\pm$ by a diagonal matrix; hence it is determined once we know the diagonal parts of s_ν^+ or s_ν^-. The factorization (5.7) implies that the upper minors of v_ν^ν are quotients of the upper minors of $(s_\nu^+)^\nu$ and $(s_\nu^-)^\nu$, and this means that v_ν determines the ratios of corresponding diagonal elements of s_ν^- and s_ν^+. Returning to the diagonal matrix δ, we conclude that the ratios of its elements on Σ are determined by the $\{v_\nu\}$, while the zeros and poles are determined by the singularities of m. This information determines δ and thus it determines $s_\nu^\pm$. This completes the proof of Theorem C.

We have already proved (1.24) of Theorem D, as (5.6) above, and also (1.28). When q has compact support and a is as in (5.1), we take $x \ll 0$ to obtain

$$\begin{aligned} v_\nu &= (a_\nu^-)^{-1} a_\nu^+ \\ &= (b_\nu^-)^{-1} b_\nu^+, \end{aligned} \tag{5.10}$$

where

$$b_\nu^\pm = \Pi_0^z a_\nu^\pm, \quad z \in \Omega_\nu. \tag{5.11}$$

In the diagonal blocks where v_ν^ν lives, $(a_\nu^-)^\nu$ is upper triangular and $(a_\nu^+)^\nu$ is lower triangular. Thus $(b_\nu^-)^\nu$ is upper, $(b_\nu^+)^\nu$ is lower, and each has 1's on the diagonal. It follows that (1.25) is true and also that the lower minors of v_ν^ν are $\equiv 1$. Similarly, the factorization (5.7) implies that the upper minors of v_ν^ν are non-zero. Once again, passage to the limit gives (1.25), (1.26), and (1.27) in general.

Finally we come to the constraints (1.31). Let δ be the diagonal matrix (5.9). As we have already observed, the ratios of the limits $(\delta_\nu^\pm)_{jj}$ on Σ_ν are determined by the upper minors of v_ν^ν; in fact, the latter ratios are certain products of the former. On the other hand, the singularities of m and the columns in which they occur determine the zeros and poles of the δ_{jj}. For each j, there is a compatibility condition between the ratios $(\delta_{\nu,jj}^-)^{-1}\delta_{\nu,jj}^+$ and the zeros and poles on δ_{jj}. This condition takes the form (1.31); the n conditions are not independent because of the single constraint

$$\Pi \delta_{jj} = \det \delta \equiv 1, \tag{5.12}$$

which follows from the fact that

$$\delta(z) = \lim_{z \to +\infty} m(x, z). \tag{5.13}$$

In fact, (5.13) is clear from (2.16) when q has compact support and then follows from a limit argument for general generic q. For details on the compatibility condition, see part 1 of the appendix.

6 Proof of Theorem E

Suppose q is a generic potential with eigenfunction m and scattering data v. Part (a) of Theorem E is an immediate consequence of the following.

Theorem 6.1. *Suppose q belongs to* $\mathbf{P}$ *and suppose*

$$D^j q \in L^1, \quad 0 \le j \le k. \tag{6.2}$$

Then there are unique functions

$$m_j : \mathbb{R} \mapsto M_n(\mathbb{C}), \quad 0 \le j \le k, \tag{6.3}$$

such that $m_0 \equiv I$ and

$$\left| m(x,z) - \sum_{j=0}^{k} z^{-j} m_j(x) \right| = O(|z|^{-k}) \tag{6.4}$$

as $z \to \infty, z \in \mathbb{C}\backslash\Sigma$, uniformly with respect to $x \in \mathbb{R}$.

Proof: Uniqueness is clear. Since $m(x,z) \to I$ as $x \to -\infty$ we need

$$m_j(x) \to 0 \text{ as } x \to -\infty \text{ if } j > 0. \tag{6.5}$$

We determine $m_0, \cdots, m_k$ from (6.5) and

$$\frac{d}{dx} m_j - q m_j = \partial m_{j+1} \quad \text{a.e.} \tag{6.6}$$

In fact, given $a \in M_n(\mathbb{C})$, write

$$a = a' + a'', \tag{6.7}$$

where a' is diagonal and a'' is off-diagonal. Then (6.6) determines m''_{j+1}, given m_j. Note also that (6.6) requires

$$\frac{d}{dx} m'_j = (q m_j)' \quad \text{a.e.} \tag{6.8}$$

Now $q = q''$ so that $(qm_j)' = (qm''_j)'$. Thus (6.5) and (6.8) give

$$m_j(x)' = \int_{-\infty}^{x} (q(y) m_j(y)'')' \, dy. \tag{6.9}$$

Hence (6.6) and (6.9) allows us at least formally to determine $m''_1, m'_1, m''_2, m'_2, \cdots, m'_k$. Moreover, $D^j q \in L^\infty$ if $j < k$, and $\partial m_1 = -q$. Inductively we obtain

$$D' m''_j \in L^1 \text{ if } 0 \le r \le k+1-j, \tag{6.10}$$

$$D' m'_j \in L^1 \text{ if } 0 < r < k+2-j. \tag{6.11}$$

Let

$$m^k(x,z) = \sum_{j=0}^{k} z^{-j} m_j(x). \tag{6.12}$$

Then from (6.6) we obtain

$$(D - z\mathfrak{J})m^{-1}m^k = z^{-k} m(x,z)^{-1} f(x) = g(x,z), \tag{6.13}$$

where $f \in L^1$. Therefore for large z we have

$$\begin{aligned}(m^{-1}m^k)(x,z) = I + \int_{-\infty}^{x} e^{(x-y)z\mathfrak{J}} (\Pi_0^z + \Pi_-^z) g(y,z)\, dy \\ - \int_{x}^{\infty} e^{(x-y)z\mathfrak{J}} \Pi_+^z g(y,z)\, dy.\end{aligned} \tag{6.14}$$

This implies that, for all x and all large z,

$$|m(x,z) - m^k(x,z)| \le C|z|^{-k}. \tag{6.15}$$

The map $q \mapsto m_j$ is continuous to L^∞ with respect to the L^1 norms of the derivatives (6.2), and for large z the map $q \mapsto m(\cdot, z)$ is continuous from L^1 to L^∞ uniformly with respect to z. Considering (6.13) for the functions corresponding to two nearby potentials and taking the difference, we conclude that, for large z,

$$|m_1(x,z) - m_2(x,z)| \le C(q_1,q_2)|z|^{-k}, \tag{6.16}$$

where, for fixed q_1, C is small when the derivatives of $q_2 - q_1$ have small L^1 norm. Because of this it is enough to prove (6.4) for a dense set of q. But when, in addition to (6.2), we have $D^{k+1}q \in L^1$, then (6.15) with k replaced by $k+1$ implies (6.4).

We turn now to part (b) of Theorem E.

Lemma 6.17. *Suppose k is a positive integer, and suppose*

$$\|q\|_1 < 1, \quad q_k \in L^1, \tag{6.18}$$

where $q_k(x) = (-x\mathfrak{J})^k q(x)$. Then for x real and $z \in \mathbb{C}\backslash\Sigma$ we have

$$\left| \left(\frac{\partial}{\partial z} - x\mathfrak{J} \right)^k m(x,z) \right| \le C_k (1 - \|q\|_1)^{-k-1} (1 + \|q\|_1 + \|q_k\|_1)^k. \tag{6.19}$$

Proof: Let $K_{z,q}$ be the operator (3.12), and let K_q be the operator on functions of two variables:

$$[K_q f](x,z) = [K_{q,z} f(\cdot, z)](x). \tag{6.20}$$

Let $A = \partial/\partial z - x\mathfrak{J}$. The commutator is

$$[A, K_q] = K_{q_1}, \quad q_1 = -x - \mathfrak{J}q. \tag{6.21}$$

Also,

$$[A, K^N] = \sum_{j=1}^{N} K^{N-j}[A, K]K^{j-1}. \tag{6.22}$$

If I denotes the identity constant function,

$$AK_q^N(I) = [A, K_q^N](I). \tag{6.23}$$

Using (6.21)–(6.23) we obtain

$$|AK_q^N(I)| \le N\|q\|_1^{N-1}\|q_1\|_1 \tag{6.24}$$

on $\mathbb{R} \times (\mathbb{C}\backslash\Sigma)$. Thus we may differentiate the Neumann series term by term to obtain

$$\begin{aligned}\left(\frac{\partial}{\partial z} - x\mathfrak{J}\right)m &= \sum_{N=1}^{\infty}\sum_{j=1}^{N} K_q^{N-j}K_{q_1}K_q^{j-1}(I) \\ &= (1 - K_q)^{-1}K_{q_1}(1 - K_q)^{-1}(I) \\ &= (1 - K_q)^{-1}K_{q_1}m.\end{aligned} \tag{6.25}$$

This gives the estimate

$$\left|\left(\frac{\partial}{\partial z} - x\mathfrak{J}\right)m\right| \le (1 - \|q\|_1)^{-2}\|q_1\|_1. \tag{6.26}$$

The estimate (6.19) for $k > 1$ is obtained by an elaboration of this procedure. The argument just given shows more generally that

$$[A, (1 - K_q)^{-1}] = (1 - K_q)^{-1}A(1 - K_q)^{-1} \tag{6.27}$$

applied to functions f with f and Af bounded. In particular,

$$\begin{aligned}A^2m &= A(1 - K_q)^{-1}K_{q_1}m \\ &= (1 - K_q)^{-1}AK_{q_1}m + (1 - K_q)^{-1}K_{q_1}(1 - K_q)^{-1}K_{q_1}m \\ &= (1 - K_q)^{-1}K_{q_2}m + 2(1 - K_q)^{-1}K_{q_1}(1 - K_q)^{-1}K_{q_1}m.\end{aligned} \tag{6.28}$$

Thus

$$\begin{aligned}\left|\left(\frac{\partial}{\partial z} - x\mathfrak{J}\right)^2 m\right| &\le (1 - \|q\|_1)^{-2}\|q_2\|_1 + 2(1 - \|q\|_1)^{-3}\|q_1\|_1^2 \\ &\le 3(1 - \|q\|_1)^{-3}(1 + \|q\|_1 + \|q_2\|_1)^2.\end{aligned} \tag{6.29}$$

The general case is proved by the obvious elaboration of this argument.

Lemma 6.30. *Suppose* $q \in \mathbf{P}$ *and*

$$\int_{\mathbb{R}} (1+|x|)^k |q(x)|\, dx < \infty. \tag{6.31}$$

Then, for large $z \in \mathbb{C}\backslash\Sigma$ *and all* $x \in \mathbb{R}$,

$$\left| \left(\frac{\partial}{\partial z}\right)^k m(x,z) \right| \le C_q (1+|x|)^k. \tag{6.32}$$

Proof: As in the proof of Theorem A (a), we induce on the least integer $r \ge 0$ such that $\|q\|_1 < 2^r$; the case $r = 0$ is part of Lemma 6.17. If $\|q\|_1 < 2^r$ we choose y so that

$$\int_{\infty}^{y} |q(x)|\, dx = \int_{y}^{\infty} |q(x)|\, dx,$$

and let $q = q_1 + q_2$ with

$$q_1(x) = q(x) \text{ if } x \le y, \quad q_1(x) = 0 \text{ if } x > y.$$

Again the eigenfunction associated to q is of the form

$$m(x,z) = \begin{cases} m_1(x,z) e^{(x-y)z\mathfrak{J}} a_1(z), & x \le y, \\ m_2(x,z) e^{(x-y)z\mathfrak{J}} a_2(z), & x > y, \end{cases}$$

where m_j is associated to q_j. The a_j solve a factorization problem for

$$m_2(y,z)^{-1} m_1(y,z). \tag{6.33}$$

We know that this problem is solvable as $z \to \infty$ since the matrix (6.33) tends to I. The solution has entries which are rational functions of the entries of (6.33), and the denominators are bounded away from 0 for large z. Therefore the induction assumption applied to m_1, m_2 gives the desired estimates for m.

We may now prove Theorem E (b). Given a generic $q \in \mathbf{P}$ satisfying (6.30), we may write $q = q_1 + q_2 + q_3$, where q_1 is supported on $(-\infty, y_1]$, q_2 is supported in $[y_1, y_2]$, and q_3 in $[y_2, \infty)$. Moreover, we may assume

$$\|q_1\|_1 < 1, \quad \|q_3\|_1 < 1, \quad q_2 \in \mathbf{P}_0. \tag{6.34}$$

Let m_j be the associated eigenfunction. Again

$$m(x,z) = m_j(x,z) e^{xz\mathfrak{J}} a_j(z) \tag{6.35}$$

on the support of q_j and the a_j solve a certain algebraic factorization problem. Now since q_2 has compact support, m_2 is meromorphic with respect to z. From Lemma 6.17, m_1 and m_3 are C^k with respect to z, on the closure of any sector

Ω_ν. Therefore, the a_j in (6.35) are C^k with respect to z, and so then is m. Thus

$$v_\nu(z) = m_\nu^-(0,z)^{-1} m_\nu^+(0,z), \quad z \in \Sigma_\nu,$$

is C^k in z on Σ_ν. Lemma 6.30 gives boundedness of the derivative. To prove that $D^j(v_\nu - I)$ converges to zero at ∞, it is enough to show this for a dense set of generic potentials. Part (a) of Theorem E implies that if the generic potential q belongs to the Schwartz class, then $v_\nu - I$ is rapidly decreasing, while we have just shown that the derivatives are bounded. As noted in Section 1, these facts imply that the derivatives also are of rapid decrease. This completes the proof of (1.35).

Finally, let $v_{\nu,k}$ be the Taylor expansion of v_ν at the origin. When q has compact support, let a be the function in (2.18) and let $a_{\nu,k}$ be the Taylor polynomial of a in the sector Ω_ν, at the origin. Then we have (1.25) and (1.36). Note that if we assume that our polynomials are of degree k, then they are uniquely determined by (1.25) and (1.36). In fact, $s = 2p$ is even and the ν and $\nu + p$ orderings are opposite. Thus (letting $a_{\nu+s,k} = a_{\nu,k}$ and $v_{\nu+s,k} = v_{\nu,k}$) we have

$$(a_{\nu+p,k})^\nu = (a_{\nu,k})^\nu \{v_{\nu,k} v_{\nu+1,k} \cdots v_{\nu+p-1,k}\}^\nu + O(z^k), \tag{6.36}$$

an upper- and lower-triangular factorization problem with a unique solution; the condition that it have a solution is that the term in braces has appropriate minors which are $1 + O(z^k)$. Therefore we may pass to the limit from potentials with compact support.

To prove Theorem E (c), we assume first

$$||q||_1 < 1, \quad ||q||_2^2 = \int |q(x)|^2 \, dx < \infty. \tag{6.37}$$

Consider the operator K_q of (6.20); we extend if to $z \in \Sigma_\nu$ from either side of Σ_ν and consider it as mapping the space of matrix-valued functions

$$L^\infty(\mathbb{R}; L^2(\Sigma)) \cap C(\mathbb{R}; L^2(\Sigma)) \tag{6.38}$$

to itself. Then, as a mapping in this space, K is easily seen to have operator norm

$$||K||_q \leq ||q||_1. \tag{6.39}$$

The function

$$g = K_q(I) \tag{6.40}$$

has entries which are Fourier or Fourier-Laplace transforms of products of translates of q with the characteristic function of $\mathbb{R}_\pm$, so g belongs to the space (6.38). Under assumption (6.37),

$$\begin{aligned} m - I &= (Id - K)^{-1}(I) - I \\ &= (Id - K)^{-1} g; \end{aligned}$$

thus $m - I$ belongs to the space (6.38). Then, on Σ_ν,

$$v_\nu(z) - I = m_\nu^-(0,z)^{-1}\{m_\nu^+(0,z) - m_\nu^-(0,z)\}$$

belongs to L^2. We may now induce again on the smallest integer $N \geq 0$ such that $||q||_1 < 2^N$ to establish the following:

Lemma 6.41. *Suppose q is in $\mathbf{P} \cap L^2$. Then there is a bounded set $A_\nu \subset \Sigma_\nu$ such that $m_\nu^\pm$ exists on $\Sigma_\nu \backslash A_\nu$ and*

$$m_\nu^\pm - I \in L^2(\Sigma_\nu \backslash A_\nu). \tag{6.42}$$

Proof: By induction, using the method of proof of Theorem A (a), and noting that since $m_2(0,z)^{-1}m_1(0,z) - I$ is in L^2 near ∞ on any ray, the same is true for $a_1 - I$ and $a_2 - I$.

This lemma and Theorem E (b) give Theorem E (c) when $k = 0$. The extension to positive k is analogous to the argument for Theorem E (b) when $k > 0$, operating again in the space (6.37). We omit the details.

7 A Reformulation of the Inverse Problem

Suppose q is a generic potential with eigenfunction m, and suppose that m has no singularities in $\mathbb{C}\backslash\Sigma$. On Σ_ν, m has an (additive) jump

$$\begin{aligned} g_\nu(x,z) &= m_\nu^+(x,z) - m_\nu^-(x,z) \\ &= m_\nu^-(x,z)[e^{xz\partial}v_\nu(z) - I]. \end{aligned} \tag{7.1}$$

We may expect m to be given by the corresponding Cauchy integral,

$$m(x,z) = I + \Sigma \frac{1}{2\pi i} \int_{\Sigma_\nu} (\zeta - z)^{-1} g_\nu(x,\zeta)\, d\zeta, \tag{7.2}$$

where Σ_ν is oriented from 0 to ∞. In fact, suppose

$$I - v_\nu \in L^2(\Sigma_\nu), \quad Dv_\nu \in L^2(\Sigma_\nu), \tag{7.3}$$

so that g_ν belongs to $L^2(\Sigma_\nu)$ for each x. Then well-known results for $\mathbb{R}$, carried over to the rays Σ_ν, imply that the function defined by (7.2) has the additive jump g_ν on Σ_ν, from which we can deduce that (7.2) is valid.

We want to formulate (7.2) as an integral equation for $m(x,\cdot)$ on Σ, and it is convenient to make the following choice for m on Σ.

Proposition 7.4. *If $q \in \mathbf{P}$ is generic, then for each $z \in \Sigma_\nu\backslash(0)$ there is a unique function $m(\cdot,z)$ satisfying* (1.5) *and* (1.6). *This function has a continuous extension to the closed ray Σ_ν.*

Proof: Suppose first that q has compact support. As in (5.1),

$$m(x,z) = e^{xzJ}a(z), \quad x \ll 0, z \in \mathbb{C}\backslash\Sigma. \tag{7.5}$$

The function a has limits $a_\nu^\pm$ on Σ_ν. As in (5.12), take

$$b_\nu^\pm = \Pi_\nu a_\nu^\pm \tag{7.6}$$

and on Σ_ν set

$$\begin{aligned} m(x,z) &= m_\nu^+(x,z)e^{xzJ}[b_\nu^+(z)^{-1}] \\ &= m_\nu^-(x,z)e^{xzJ}[b_\nu^-(z)^{-1}]; \end{aligned} \tag{7.7}$$

the second equality comes from (5.11). Now

$$\begin{aligned} \Pi_\nu m(x,z) = I \quad &\text{if } x \ll 0, \\ (Id - \Pi_\nu)m \to 0 \quad &\text{as } x \to -\infty. \end{aligned} \tag{7.8}$$

Thus $m \to I$ as $x \to -\infty$.

Recall that $b_\nu^\pm$ are the unique solutions of the factorization problem

$$\begin{aligned} &b_\nu^- v_\nu = b_\nu^+, \quad \Pi_\nu b_\nu^\pm = b_\nu^\pm, \quad (b_\nu^\pm)_{jj} = 1, \\ &(b_\nu^+)^\nu \text{ is lower triangular,} \\ &(b_\nu^-)^\nu \text{ is upper triangular.} \end{aligned} \tag{7.9}$$

In the general case we approximate by compactly supported potentials. The corresponding scattering data converge; thus the solutions to (7.9) converge, and so the eigenfunctions on Σ_ν converge and give the desired eigenfunction. Finally, any other solution would have the form

$$m(x,z)e^{xzJ}c(z). \tag{7.10}$$

But boundedness in x implies $\Pi_\nu c = c$ and the normalization at $-\infty$ implies that $\Pi_\nu c = I$.

Set

$$w_\nu(z) = b_\nu^+(z) - b_\nu^-(z), \quad z \in \Sigma_\nu, \tag{7.11}$$

so that

$$(w_\nu)_{jj} = 1, \quad \Pi_\nu w_\nu = w_\nu. \tag{7.12}$$

Now w_ν is determined from v_ν by (7.9) and (7.11). Conversely, given w_ν satisfying (7.12), let $w_\nu^\pm$ be defined so that $(w_\nu^+)^\nu$ is the lower triangular part of $(w_\nu)^\nu$ and $(w_\nu^-)^\nu$ is the upper triangular part. Then set

$$b_\nu^\pm = I \pm w_\nu^\pm, \quad v_\nu = (b_\nu^-)^{-1}b_\nu^+. \tag{7.13}$$

Then v_ν satisfies the constraints (1.24), (1.26), (1.27), and w_ν is determined from v_ν by (7.9) and (7.11). There is a complete equivalence between scattering data (or formal scattering data)

$$v = (v_1, \cdots, v_r; z_1, \cdots, z_N; v(z_1), \cdots, v(z_N)) \tag{7.14}$$

and *transformed scattering data*

$$w = (w_1, \cdots, w_r; z_1, \cdots, z_N; v(z_1), \cdots, v(z_N)). \tag{7.15}$$

We shall show eventually that when m has no singularities and is extended to Σ as in Proposition 7.4, then it satisfies on Σ an integral equation

$$m(x, \cdot) = I + C_{w,x} m(x, \cdot), \tag{7.16}$$

where w is the transformed scattering data. Here

$$C_{w,x} f = C^+(f w^-(x, \cdot)) + C^-(f w^+(x, \cdot)), \tag{7.17}$$

where

$$w^\pm(x, z) = e^{xz\mathcal{J}} w_\nu^\pm(z), \quad z \in \Sigma_\nu, \tag{7.18}$$

and where $C^\pm$ are suitably normalized Cauchy integrals which we proceed to describe. When $\mu \neq \nu$, we let $C_{\mu,\nu}$ map functions on Σ_ν to functions on Σ_μ (oriented from 0 to ∞) by

$$C_{\mu,\nu} f(z) = \frac{1}{2\pi i} \int_{\Sigma_\nu} (\zeta - z)^{-1} f(\zeta)\, d\zeta, \quad z \in \Sigma_\mu. \tag{7.19}$$

Let $C_\nu^\pm$ map functions on Σ_ν to functions on Σ_ν:

$$C_\nu^\pm f(z) = \lim_{z' \to z} \frac{1}{2\pi i} \int_{\Sigma_\nu} (\zeta - z)^{-1} f(\zeta)\, d\zeta, \tag{7.20}$$

where the limit is taken from $\Omega_{\nu+1}$ for C_ν^+ and from Ω_ν for C_ν^-. These maps will be discussed more thoroughly in later sections. It is classical that

$$C_{\mu,\nu} : L^2(\Sigma_\nu) \mapsto L^2(\Sigma_\mu), \quad \nu \neq \mu, \tag{7.21}$$

$$C_\nu^\pm : L^2(\Sigma_\nu) \mapsto L^2(\Sigma_\nu), \tag{7.22}$$

$$\pm C_\nu^\pm \text{ are complementary orthogonal projections.} \tag{7.23}$$

For a function $f \in L^2(\Sigma)$ write $f = (f_\nu)$, $f_\nu \in L^2(\Sigma_\nu)$, and define $C^\pm f$ by

$$(C^\pm f)_\mu = \sum_{\nu \neq \mu} C_{\mu,\nu} f + C_\mu^\pm f. \tag{7.24}$$

Suppose that $m(x,\cdot)$ is a function on Σ which solves the integral equation (7.16), where $C_{w,x}$ is defined in (7.17), (7.22). We extend m to $\mathbb{C}\backslash\Sigma$ by taking the natural extension of (7.16), which is

$$m(x,z) = I + \Sigma \frac{1}{2\pi i}\int_{\Sigma_\nu} (\zeta - z)^{-1} m(x,z) e^{xz\partial} w(\zeta)\, d\zeta. \tag{7.25}$$

Then (7.23) implies for the limits of m on Σ_ν that

$$\begin{aligned} m_\nu^+ &= I + C^+(m,w) = I + C_{w,x}m + mw_\nu^+(x,\cdot) \\ &= m + mw_\nu^+(x,\cdot) \\ &= m(x,z)(I + e^{xz\partial} w_\nu^+(z)) = m(x,z)e^{xz\partial} b_\nu^+(z). \end{aligned} \tag{7.26}$$

Similarly,

$$m_\nu^-(x,z) = m(x,z)e^{xz\partial} b_\nu^-(z). \tag{7.27}$$

Thus

$$m_\nu^+(x,z) = m_\nu^-(x,z)e^{xz\partial} v_\nu(\zeta) \tag{7.28}$$

as desired.

8 The Inverse Problem with Small Data, I

We begin our study of the integral equation (7.16) with a lemma which is classical; it is convenient for later use to record the proof.

Lemma 8.1. *The operators $C^\pm$ of (7.24) are bounded in $L^2(\Sigma)$. Moreover, if $\alpha_1, \cdots, \alpha_r$ are constants such that*

$$\alpha_\nu z > 0 \text{ if } z \in \Sigma_\nu\backslash(0) \tag{8.2}$$

and if e_λ denotes the function on Σ with

$$e_\lambda(z) = \exp\{i\lambda\alpha_\nu z\}, \quad z \in \Sigma_\nu, \lambda \in \mathbb{R}, \tag{8.3}$$

then for any $f \in L^2(\Sigma)$ we have

$$||C^\pm(e_\lambda f)||_2 \to 0 \text{ as } \pm\lambda \to -\infty. \tag{8.4}$$

Proof: It is enough to consider the operators $C_{\mu,\nu}$ and C_ν^+, and we may rotate and assume $\Sigma_\nu = R_+$. Parametrize Σ_μ by R_+ also. For $\mu \neq \nu$, we see that $C_{\mu,\nu}$ is an integral operator with kernel

$$k(s,t) = (2\pi i)^{-1}(\alpha s + t)^{-1}, \quad \alpha \in \mathbb{C}\backslash\mathbb{R}, s, t \geq 0. \tag{8.5}$$

Then $|k(s,t)| \leq C(s+t)^{-1}$; L^2-boundedness is classical. To see that

$$||C_{\mu,\nu}(e_\lambda f)|| \to 0 \text{ as } |\lambda| \to \infty, \tag{8.6}$$

where $e_\lambda(t) = e^{i\lambda t}$, $\lambda \in \mathbb{R}$, note that it is enough to prove (8.6) when f is smooth and has compact support in $(0,\infty)$; then (8.6) follows from an integration by parts.

When $\Sigma_\nu = R_+$ the operator C_ν^+ can be computed on test functions:

$$\begin{aligned} C_\nu^+ f(s) &= \lim_{\epsilon \searrow 0} \frac{1}{2\pi i} \int (s-t-i\epsilon) f(t)\, dt \\ &= \lim_{\epsilon \searrow 0} \frac{1}{2\pi i} \iint (s-t-i\epsilon)^{-1} e^{it\xi} \hat{f}(\xi)\, d\xi\, dt \\ &= \int_{-\infty}^{0} e^{it\xi} \hat{f}(\xi)\, d\xi = (h_0 \hat{f})\check{}(s), \end{aligned} \tag{8.7}$$

where $h_a(\xi) = 1$ if $\xi \leq a$, $h_a(\xi) = 0$ if $\xi > a$, and $\hat{}, \check{}$ denote the Fourier transform and its inverse. This gives L^2-boundedness. With $e_\lambda(t) = e^{i\lambda t}$, $\lambda \in \mathbb{R}$, the same calculation gives

$$C_\nu^+(e_\lambda f)(s) = (h_\lambda \hat{f})\check{}(s) \tag{8.8}$$

which yields L^2 convergence to 0 as $\lambda \to -\infty$ and to f as $\lambda \to +\infty$.

Theorem 8.9. *Suppose $w = (w_\nu)$ satisfies the conditions (7.12), belongs to $L^2(\Sigma) \cap L^\infty(\Sigma)$, and $w(z) \to 0$ as $z \to \infty$. Let $C_{w,x}$ be the operator defined by (7.17), (7.24), let $C^\pm$ be defined by (7.24), and let*

$$||C^+|| = ||C^-||$$

be the operator norm in $L^2(\Sigma)$. Suppose

$$2||w||_\infty ||C^\pm|| < 1. \tag{8.10}$$

Then for every real x there is a unique function $m(x,\cdot) \in L^2(\Sigma) + L^\infty(\Sigma)$ which satisfies the integral equation (7.16). If m is extended to $\mathbb{C}\backslash\Sigma$ by (7.25), then for each $z \in \mathbb{C}\backslash\Sigma$, $m(\cdot, z)$ is bounded and absolutely continuous with respect to x, and

$$m(x,z) \to I \text{ as } x \to -\infty. \tag{8.11}$$

Let

$$q(x) = \frac{1}{2\pi i} \mathfrak{J} \int_\Sigma m(x,z) e^{xz\mathfrak{J}} w(z)\, dz. \tag{8.12}$$

Then

$$q \in L^\infty + L^2 \tag{8.13}$$

and, for $z \in \mathbb{C}\backslash\Sigma$,

$$\frac{\partial}{\partial x} m(x,z) = z\mathfrak{J} m(x,z) + q(x) m(x,z) \quad \text{a.e. } x. \tag{8.14}$$

Proof: The operator $C_{w,x}$ maps $L^\infty(\Sigma)$ to $L^2(\Sigma)$, since w is assumed to belong to $L^2(\Sigma)$. As an operator in $L^2(\Sigma)$, $C_{w,x}$ has norm dominated by the expression (8.10). Therefore, $Id - C_{w,x}$ is invertible as an operator in $L^2 + L^\infty$, and (7.16) has the unique solution

$$m(x,\cdot) = (Id - C_{w,x})^{-1}(I). \tag{8.15}$$

Then

$$m(x,\cdot) - I = (Id - C_{w,x})^{-1}(g(x,\cdot)), \tag{8.16}$$

where

$$g(x,\cdot) = C_{w,x}(I). \tag{8.17}$$

Now it follows from Lemma 8.1 that the L^2-norm of $g(x,\cdot)$ approaches zero as $x \to -\infty$. From this and from (8.16) we obtain

$$\sup_x \|m(x,\cdot) - I\|_2 < \infty, \tag{8.18}$$

$$\|m(x,\cdot) - I\|_2 \to 0 \text{ as } x \to -\infty. \tag{8.19}$$

An easy consequence of (8.18) is that $m(\cdot, z)$ is bounded as a function of x for every $z \in \mathbb{C}\backslash\Sigma$. Lemma 8.1 and (8.19) imply (8.11) for $z \in \mathbb{C}\backslash\Sigma$.

Let q be defined by (8.12) and write it as a sum of two integrals, involving $m(x,z) - I$ and I, respectively. The integrand in the first integral is a product of L^2 functions with norms bounded as x varies, so the first term is in L^∞. (In fact $x \mapsto C_{w,x}$ is continuous to the strong operator topology, by the dominated convergence theorem, which implies that $x \mapsto m(x,\cdot) - I$ is continuous to L^2, so this term is even continuous.) The entries of the second term are easily seen to be (dilates of) Fourier transforms of the entries of w; thus the second term is in $L^2(\mathbb{R})$.

In this region where (8.10) holds, this construction shows that

$$w \mapsto m \text{ is continuous from } L^2(\Sigma) \cap L^\infty(\Sigma) \text{ to } C(\mathbb{R}; L^\infty(\Sigma) + L^2(\Sigma)), \tag{8.20}$$

$$w \mapsto q \text{ is continuous from } L^2(\Sigma) \cap L^\infty(\Sigma) \text{ to } L^\infty(\Sigma) + L^2(\Sigma), \tag{8.21}$$

Because of (8.20) and (8.21), it is enough to prove (8.14) when w belongs to a dense set. We shall assume, in fact, that w has compact support. In that case it is clear that $x \mapsto C_{w,x}$ is analytic from $\mathbb{R}$ to the bounded operators in $L^\infty + L^2$, and so $x \mapsto m(x,\cdot)$ is analytic to $L^\infty + L^2$. Consider A and C_w as mapping the space

$$C^\infty(\mathbb{R}; L^\infty + L^2) \tag{8.22}$$

to itself,

$$[Af](x,z) = \frac{d}{dx} f(x,z) - z\mathfrak{J} f(x,z), \tag{8.23}$$

$$[C_w, f](x,\cdot) = C_{w,x} f(x,\cdot). \tag{8.24}$$

Then the commutator

$$[A, C_w]f(x,\cdot) = \frac{1}{2\pi i}\eth\int_\Sigma f(x,\zeta)e^{x\zeta\eth}w(\zeta)\,d\zeta \tag{8.25}$$

maps to functions which are constant with respect to z. As in (6.22)–(6.25), since $A(I) = 0$ we have

$$\begin{aligned} Am(x,\cdot) &= (I - C_w)^{-1}[A, C_w](I - C_w)^{-1}(I) \\ &= (I - C_w)^{-1}q. \end{aligned} \tag{8.26}$$

Since $q = qI$ and C_w commutes with left multiplication by functions independent of z, we have

$$\left(\frac{d}{dx} - z\eth\right)(m,\cdot) = q(x)m(x,\cdot) \tag{8.27}$$

as functions in $C^\infty(\mathbb{R}; L^\infty + L^2)$. Now for $z \in \mathbb{C}\backslash\Sigma$ we differentiate (7.25) to obtain

$$\begin{aligned} 2\pi i\left(\frac{\partial}{\partial x} - z\eth\right)m(x,z) &= \int_\Sigma(\zeta - z)^{-1}\left[\frac{\partial}{\partial x} - \zeta\eth + (\zeta - z)\eth\right]m(x,\zeta)e^{x\zeta\eth}w(\zeta)\,d\zeta \\ &= \int_\Sigma(\zeta - z)^{-1}q(x)m(x,\zeta)e^{x\zeta\eth}w(\zeta)\,d\zeta + 2\pi iq(x) \\ &= 2\pi iq(x)\{m(x,z) - I + I\} = 2\pi iq(x)m(x,z). \end{aligned} \tag{8.28}$$

9 The Inverse Problem with Small Data, II

In this section we strengthen the hypotheses on the function w of (8.1), with respect to decay at ∞ and with respect to smoothness, to obtain results corresponding to Theorem G. We consider first the condition

$$(1 + |z|)^k w_\nu \in L^2(\Sigma_\nu). \tag{9.1}$$

Theorem 9.2. *Let w, m, and q be as in Theorem 8.9. If w satisfies (9.1), then*

$$D^j q \in L^\infty + L^2, \quad j \le k. \tag{9.3}$$

Proof: Assume first that w has compact support. As noted above, this implies that m is analytic in x with $L^\infty + L^2$ values, and q is analytic. It is enough to establish bounds on $D^j q$ in $L^\infty + L^2, j \le k$, which (under the assumption (8.10)) depend only on the pair

$$N_k = \{||w||_\infty, ||(1 + |z|)^k w||_2\}. \tag{9.4}$$

We have this result for $k = 0$. Note also that in (8.12) we have

$$2\pi i q(z) = \int (m(x,z) - I)e^{xz\mathfrak{J}} w(z)\, dz + \int e^{xz\mathfrak{J}} w(z)\, dz. \tag{9.5}$$

As pointed out above, the first term is in L^∞. The second term has L^1 norm dominated by

$$\begin{aligned} ||w||_1 &= ||(1+|z|)^{-1}(1+|z|)w||_1 \\ &\le ||(1+|z|)^{-1}||_2 ||(1+|z|)w||_2, \end{aligned}$$

again because it is essentially a Fourier transform. This shows that $||q||_\infty$ is dominated by N_1.

We now induce: suppose we know that $D^j q$ in $L^\infty + L^2$, $j \le k-1$ is controlled by N_{k-1}, and also that $||D^{k-1}q||_\infty$ is controlled by N_k. Repeated differentiation of (8.12) gives an expression for $D^k q$ as a linear combination of integrals with integrands which (apart from occurrences of the operator $\mathfrak{J}$) are of the forms

$$z^j p(x)[m(x,z) - I]e^{xz\mathfrak{J}} w(z), \tag{9.6}$$

$$z^j p(z) e^{xz\mathfrak{J}} w(z), \quad j \le k, \tag{9.7}$$

where p is a product of derivatives of order less than k of q. By the induction assumption, $||p||_\infty$ is controlled by N_k and it can be ignored. The term (9.6) gives a function with L^∞ norm controlled by $||m(x,\cdot) - I||_2$ and N_k, hence by N_k. The term (9.7), as before, has L^2 norm controlled by N_k and L^∞ norm controlled by N_{k+1}. This completes the induction, and the proof.

In order to consider the effect of smoothness of w, we introduce two spaces of functions on Σ and an extension of Lemma 8.1. Recall that

$$D^j f_\nu \in L^2(\Sigma_\nu), \quad 0 \le j \le k+1, \tag{9.8}$$

implies, after correction on a set of measure zero,

$$f_\nu \in C^k(\Sigma_\nu), D^j f_\nu \to 0 \text{ as } z \to \infty, \quad j \le k. \tag{9.9}$$

Definition 9.10. For k an integer greater than or equal to 0, we denote by $H^{k+1}(\Sigma)$ the space of matrix-valued functions $f = \{f_\nu\}$ satisfying (9.8) and such that

$$D^j f_\mu(0) = D^j f_\nu(0) \text{ for all } \mu, \nu, \quad j \le k. \tag{9.11}$$

We denote by $H_0^{k+1}(\Sigma)$ the subspace consisting of f such that

$$D^j f_\nu(0) = 0 \text{ for all } \nu, \quad j \le k. \tag{9.12}$$

The Sobolev norm

$$\|f\|_{2,k+1}^2 = \sum_{j\leq k+1} \|D^j f\|_2^2 \tag{9.13}$$

makes $H^{k+1}(\Sigma)$ a Hilbert space. $H^{k+1}(\Sigma)$ is also an algebra under pointwise multiplication, and $H_0^{k+1}(\Sigma)$ is a closed ideal.

Lemma 9.14. *The operators $C^{\pm}$ of* (7.24) *are bounded from $H_0^k(\Sigma)$ to $H^k(\Sigma)$. Moreover, if $\{\alpha_\nu\}$ and e_λ are as in* (8.1), *then*

$$\|C^{\pm}(e_\lambda f)\|_2 \leq C_k|\lambda|^{-k}\|f\|_{2,k}, \quad f \in H_0^k(\Sigma), \pm\lambda < 0. \tag{9.15}$$

Proof: For a smooth function f with support not containing the origin, it is clear that, along any line through the origin,

$$\frac{d}{dz}C^{\pm}(f) = C^{\pm}\left(\frac{df}{dz}\right). \tag{9.16}$$

Such functions are dense in $H_0^k(\Sigma)$, so the first statement follows from L^2-boundedness of $C^{\pm}$. To prove (9.15) we argue as in the proof of Lemma 8.1. For C_ν^+, note in (8.8) that $f \in H^k(\Sigma)$ implies $(1+|\xi|)^k \hat{f}(\xi) \in L^2(R)$; thus

$$\|h_\lambda \hat{f}\|_2 \leq C_k(1+|\lambda|)^{-k}\|f\|_{2,k}, \quad \lambda < 0.$$

Consider now $C_{\mu,\nu}$ when $\mu \neq \nu$. For ease of notation we suppose $\Sigma_\nu = R_+$, $\Sigma_\mu = R_-$, and then change signs on Σ_μ to consider the map in $L^2(R_+)$ with kernel $(t+s)^{-1}$. We want to estimate

$$h(\lambda, t) = \int_0^\infty (t+s)^{-1} e^{i\lambda s} f(s)\, ds, \quad t \geq 0. \tag{9.17}$$

We have

$$\begin{aligned} i\lambda h(\lambda,t) &= \int_0^\infty \frac{d}{ds}(e^{i\lambda s})(t+s)^{-1} f(s)\, ds \\ &= -\int_0^\infty e^{i\lambda s}(t+s)^{-1} Df(s)\, ds + \int_0^\infty e^{i\lambda s}(t+s)^{-2} f(s)\, ds. \end{aligned} \tag{9.18}$$

The first term on the right has L^2 norm in t dominated by $\|Df\|_2$. The second term is dominated by

$$\int_0^\infty (t+s)^{-2} \int_0^s |Df(u)|\, du = \int_0^\infty (t+u)^{-1}|Df(u)|\, du \tag{9.19}$$

and again the L^2 norm in t is dominated by $\|Df\|_2$. This proves (9.15) for $C_{\mu,\nu}$ when $k=1$, and the argument extends in the obvious way to larger k.

Theorem 9.20. *Let w, m, and q be as in Theorem 8.9. Suppose*

$$w \in H_0^{k+1}(\Sigma), \quad \|w\|_{2,k+1} < \delta_k, \tag{9.21}$$

where $\delta_k > 0$ is sufficiently small. Then, for all x in R,

$$m(x, \cdot) - I \in H^{k+1}(\Sigma), \tag{9.22}$$

$$\|m(x, \cdot) - I\|_2 = O(|x|^{-k-1}), \quad x < 0. \tag{9.23}$$

Moreover, there is a function s on Σ with properties

$$s - I \in H^{k+1}(\Sigma), \tag{9.24}$$

$$\|m(x, \cdot) - s(x, \cdot)\|_2 = O(x^{-k-1}), \quad x > 0, \tag{9.25}$$

where $s(x,z) = e^{xz\partial} s(z)$. Finally,

$$(1 + |x|)^{k+1} q \in L^\infty(R) + L^2(R). \tag{9.26}$$

Proof: Fix x and consider the operator

$$B_x f(z) = \frac{d}{dx} f(z) - x\partial f(z). \tag{9.27}$$

Let

$$\|f\|_{2,k+1,x}^2 = \sum_{j \le k+1} \|B_x f\|^2. \tag{9.28}$$

This is equivalent to the $H^{k+1}(\Sigma)$ norm. Set

$$w_x(z) = e^{xz\partial} w(z). \tag{9.29}$$

Clearly,

$$B_x(w_x f) = w_x B_x f + (Dw)_x f. \tag{9.30}$$

The L^1 norm of the Fourier transform of f can be estimated by the Schwarz inequality to obtain $\|f\|_\infty \le c\|f\|_{2,1,x}$ with c independent of x. Thus, iterating (9.30) and estimating L^2 norms, we get

$$\|w_x f\|_{2,k+1,x} \le c_k \|w_x\|_{2,k+1} \|f\|_{2,k+1,x}. \tag{9.31}$$

Here and below c_k will denote various constants depending only on Σ and k. Recall that $H_0^{k+1}(\Sigma)$ is an ideal in the algebra $H^{k+1}(\Sigma)$. Integration by parts shows that the operators $C_\pm$ of (7.24) map

$$C_\pm : H_0^{k+1}(\Sigma) \mapsto H^{k+1}(\Sigma). \tag{9.32}$$

Thus $C_{w,x}$ maps $H^{k+1}(\Sigma)$ to itself with norm

$$||C_{w,x}f||_{2,k+1,x} \le c_k ||w||_{2,k+1} ||f||_{2,k+1,x}. \tag{9.33}$$

Also, clearly

$$||C_{w,x}(I)||_{2,k+1,x} \le c_k ||w||_{2,k+1}. \tag{9.34}$$

It follows from (9.33) and (9.34) that (9.21) with δ_k small enough gives

$$m(x,\cdot) - I = \sum_{n=1}^{\infty} C^n_{w,x}(I) \in H^{k+1}(\Sigma). \tag{9.35}$$

To get the L^2 estimate (9.23) we note that Lemma 9.14 implies

$$||C_{w,x}(I)||_2 = O(|x|^{-k-1}), \quad x < 0. \tag{9.36}$$

Thus (9.23) follows from the identity (9.35).

To obtain the function s we set

$$\begin{aligned} C_{w,x,0}f(z) &= f(z)e^{xzJ}[w^-(z) - w^+(z)] \\ &= f(z)e^{xzJ}\tilde{w}(z), \end{aligned} \tag{9.37}$$

$$\begin{aligned} C_{w,x,1}f &= C_{w,x}f - C_{w,x,0}f \\ &= C^+(fw^+(x,\cdot)) - C^-(fw^-(x,\cdot)), \end{aligned} \tag{9.38}$$

in the notation of (7.17), (7.18). Dropping the subscripts w and x, we write

$$(C_0 + C_1)^N = C_0^N + \sum_{M=1}^{N} (C_0 + C_1)^{N-M} C_1 C_0^{M-1}. \tag{9.39}$$

It is clear from Lemma 9.14 that the off-diagonal part of $C_1 C_0^{M-1}(I)$ has L^2 norm less than or equal to

$$x^{-k-1} c_k^M ||w||^M_{2,k+1}, \quad x > 0. \tag{9.40}$$

The diagonal part of $C_1 C_0^{M-1}(I)$ is independent of x. We apply $(C_0 + C_1)^{N-M}$ to this diagonal part and use the identity (9.39) with N replaced by $N - M$. At the next occurrence of C_1 we again dominate the L^2 norm of the off-diagonal part by an expression like (9.40), and iterate for the diagonal part. This procedure yields

$$(C_0 + C_1)^N(I) = \sum_{M=0}^{N} C_0^{N-M} \delta_{N,M} + r_N, \tag{9.41}$$

where $\delta_{N,M}$ is diagonal and independent of x, while, for $x > 0$,

$$x^{k+1} ||r_N||_2 \le N c_k^N ||w||^N_{2,k+1}, \tag{9.42}$$

$$||\delta_{N,M}||_{2,k+1} \le c_k^M ||w||^M_{2,k+1}. \tag{9.43}$$

Now we set

$$\begin{aligned} s(z) &= e^{-z\mathfrak{J}}\Big(I + \sum_{N=1}^{\infty}\sum_{M=0}^{N} C_0^{N-M}\delta_{N,M}(z)\Big) \\ &= I + \sum_{N=1}^{\infty} \tilde{w}(z)^{N-M}\delta_{N,M}(z). \end{aligned} \tag{9.44}$$

The estimates (9.43) give (9.24) if δ_k in (9.21) is small enough, and the estimates (9.42) yield (9.25).

Finally, we want to obtain information on the potential q. We use (8.12) again and assume $x \geq 0$; the argument for $x < 0$ is the same but uses I in place of $s(x, z)$. We have

$$\begin{aligned} q(x) &= \mathfrak{J}\int m(x,z)e^{xz\mathfrak{J}}w(x)\,dz \\ &= \mathfrak{J}\int [m(x,z) - s(x,z)]w(x,z)\,dz + \mathfrak{J}\int e^{xz\mathfrak{J}}[s(z)w(z)]\,dz. \end{aligned} \tag{9.45}$$

The L^∞ norm of the first term on the right above is dominated by the L^2 norm of $m(x,\cdot) - s(x,\cdot)$, since $w(x,\cdot)$ is in L^2 uniformly with respect to x. Thus (9.25) gives the desired estimate, $O(x^{-k-1})$. From (9.24) we have $sw \in H_0^{k+1}$. Because of the operator $\mathfrak{J}$, only off-diagonal entries appear, and these are dilates of Fourier transforms of the entries of sw, hence have L^2 norm which is $O(x^{-k-1})$.

10 The Inverse Problem Near $-\infty$

Suppose v belongs to the space $\mathbf{S}$ of formal scattering data,

$$v = (v_1, \cdots, v_r; z_1, \cdots, z_N; v(z_1), \cdots, v(z_N)).$$

We shall see that a rational approximation and the results of Sections 8 and 9 will allow us to reduce the inverse problem for v to a finite set of linear equations, with x a parameter.

Definition 10.1. A matrix-valued function u defined on $\mathbb{C}\backslash\Sigma$ is *piecewise rational* if on each component Ω_ν of $\mathbb{C}\backslash\Sigma$ it coincides with a rational function which has no singularities on the boundary $\Sigma_\nu \cup \Sigma_{\nu+1}$.

As before we denote by u_ν^- and u_ν^+ the limits on Σ_ν from Ω_ν and $\Omega_{\nu+1}$.

Lemma 10.2. *Given $v \in \mathbf{S}$ and $\epsilon > 0$, there is a piecewise rational function u with the properties*

$$u_{jj} \equiv 1, \quad u_j(z)^\nu \text{ is upper triangular in } \Omega_\nu, \tag{10.3}$$

$$u \to I \text{ as } z \to \infty, \tag{10.4}$$

$$\|u_\nu^- v_\nu (u_\nu^+)^{-1} - I\|_\infty < \epsilon, \tag{10.5}$$

$$u_\nu^-(0)v_\nu(0)u_\nu^+(0)^{-1} = I. \tag{10.6}$$

Proof: Let $\{a_\nu\}$ be the (unique) matrices satisfying (1.25). Choose a piecewise rational function a having no singularities, such that a satisfies (10.3) and (10.4), and such that

$$a(z) \to a_\nu \text{ as } z \to 0, \quad z \in \Omega_\nu. \tag{10.7}$$

The matrices

$$[a_\nu^-(z)v_\nu(z)a_\nu^+(z)^{-1}]^\nu, \quad z \in \Sigma_\nu \tag{10.8}$$

have lower minors $\equiv 1$, because of (1.26) for v and (10.3) for a. It follows that there is a unique factorization of (10.8) as

$$[b_\nu^-(z)^{-1}b_\nu^+(z)]^\nu, \tag{10.9}$$

where

$$(b_\nu^\pm)_{jj} = 1, \quad (b_\nu^-)^\nu \text{ is upper triangular;} \tag{10.10}$$

$$\Pi_\nu b_\nu^+ = b_\nu^+, \quad (b_\nu^+)^\nu \text{ is lower triangular.} \tag{10.11}$$

In fact, $(\Pi_\nu b_\nu^\pm)^\nu$ are the triangular factors of the Π_ν projection of (10.9), and b_ν^- is then determined from (10.11) and the equality of (10.8) and (10.9). The uniqueness implies

$$b_\nu^\pm(0) = I. \tag{10.12}$$

From condition (10.11) it follows that $(b_\nu^+)^{\nu+1}$ is upper triangular on Σ_ν. Since $(b_{\nu+1}^-)^{\nu+1}$ is upper triangular on $\Sigma_{\nu+1}$ and both are the identity at the origin, continuous, and approach I at ∞, we may approximate both on the boundary of $\Omega_{\nu+1}$ by a rational function; see part 2 of the Appendix. Thus, given $\delta > 0$, there is a piecewise rational function $c = c_\delta$ which satisfies (10.3) and also

$$c_\nu^\pm(0) = I, \quad \|c_\nu^\pm - b_\nu^\pm\|_\infty < \delta. \tag{10.13}$$

With δ to be chosen later, set

$$u(z) = c(z)a(z). \tag{10.14}$$

Then

$$\begin{aligned} u_\nu^- v_\nu (u_\nu^+)^{-1} &= c_\nu^-[a_\nu^- v_\nu (a_\nu^+)^{-1}](c_\nu^+)^{-1} \\ &= c_\nu^-(b_\nu^-)^{-1}b_\nu^+(c_\nu^+)^{-1}, \end{aligned} \tag{10.15}$$

and (10.13) with δ sufficiently small gives (10.5) and (10.6).

Define

$$v_\nu^\# = u_\nu^- v_\nu (u_\nu^+)^{-1}. \tag{10.16}$$

Because of (10.3), $v_\nu^\#$ satisfies the defining conditions (1.25) and (1.26) for elements of **S**. Because of (10.4), it is also clear that $v_\nu^\# - I$ and its derivative belong to $L^2(\Sigma_\nu)$. It follows that if ϵ in (10.5) is small enough, we may apply Theorem 8.9 and obtain an associated eigenfunction $m^\#$, piecewise holomorphic, with

$$(m^\#)_\nu^+(x,z) = (m_\nu^\#)^-(x,z)e^{xzJ}v_\nu^\#(z), \quad z \in \Sigma_\nu. \tag{10.17}$$

Lemma 10.18. *Suppose $v \in$ **S**, and suppose $v_\nu^\#$ is given by (10.16), where u is as in Lemma 10.2 and ϵ is small enough so that $\{v_\nu^\#\}$ has associated eigenfunction (10.17) for all x. Then, for any $x \le 0$, if v has an associated eigenfunction $m(x,\cdot)$, it is of the form*

$$m(x,z) = r(x,z)m^\#(x,z)e^{xzJ}u(z), \tag{10.19}$$

where $r(x,\cdot)$ is rational.

Proof: First, set

$$m_0(x,z) = m^\#(x,z)e^{xzJ}u(z) = m^\#(x,z)u^x(z) \tag{10.20}$$

and note that

$$\begin{aligned}(m_{0,\nu})^+ &= (m_\nu^\#)^+(u_\nu^x)^+ \\ &= (m_{0,\nu})^- v_\nu^x\end{aligned} \tag{10.21}$$

by (10.17) and (10.16). The differential equation (8.14) and asymptotic condition (8.11) imply once again that

$$\det m^\# \equiv 1 \tag{10.22}$$

and the same is true of m_0. Therefore, if $m(x,\cdot)$ is an eigenfunction associated to v and if we define

$$r(x,z) = m(x,z)m_0(x,z)^{-1}, \tag{10.23}$$

we find that

$$r_\nu^+ = r_\nu^-. \tag{10.24}$$

Clearly, $r(x,\cdot)$ is meromorphic in $\mathbb{C}\backslash\Sigma$ since $m^\#$, m, and u are; hence $r(x,\cdot)$ is rational.

Remark 10.25. The piecewise rational function u has the same singularities as the function c in the proof of Lemma 10.2. The latter function can be chosen to have only simple poles, and the locations can be chosen to be distinct from the $\{z_j\}$ of v and to be distinct for distinct entries. It follows from this that at any singularity in Ω_ν the residue of u is strictly upper triangular in the ν-representation and has only one non-zero row; thus its square is zero. We say that such a function u is *regular*, and we assume that u is chosen to be regular.

We now fix $x \in \mathbb{R}$ and look for a rational function $r(x, \cdot)$ so that when $m(x, \cdot)$ is *defined* by (10.19), it is the associated eigenfunction for v. (We remark at this point that the uniqueness proof in the case $q \mapsto v$ shows that formal scattering data has at most one associated eigenfunction, given x). The function r should have only simple poles and should be I at ∞; thus

$$r(x, z) = I + \sum_{k=1}^{P} (z - z_k)^{-1} a_k, \tag{10.26}$$

where $z_1, \cdots, z_N$ are the singularities of m and $z_{N+1}, \cdots, z_P$ are the singularities of u. Then

$$r(x, z) = (z - z_j)^{-1} a_j + b_j + O(|z - z_j|). \tag{10.27}$$

If $j \le N$, then $m_0 = m^{\#} u^x$ is regular at z_j,

$$m_0(x, z) = c_j + (z - z_j) d_j + O(|z - z_j|^2), \tag{10.28}$$

where $c_j = m_0(x, z_j)$ is invertible. Let

$$v_j = \exp\{x z_j \mathfrak{J}\} v(z_j). \tag{10.29}$$

We would like to have

$$\text{Res}\,(m(x, \cdot), z_j) = \lim_{z \to z_j} m(x, z) v_j, \tag{10.30}$$

which is equivalent to

$$a_j c_j v_j = 0, \quad j \le N, \tag{10.31}$$

$$(a_j d_j + b_j c_j) v_j = a_j c_j, \quad j \le N. \tag{10.32}$$

Note that the condition (1.28) in the definition of formal scattering data implies

$$v_j^2 = 0. \tag{10.33}$$

Therefore, (10.31) is a consequence of (10.32).
If $j > N$, then u is singular at z_j,

$$e^{xz\mathfrak{J}} u(z) = (z - z_j)^{-1} u_j + n_j + O(|z - z_j|). \tag{10.34}$$

Note that n_j^ν is upper triangular if $z_j \in \Omega_\nu$, and the diagonal part is I; thus n_j is invertible. Then, as in the remark above,

$$(u_j n_j^{-1})^2 = 0. \tag{10.35}$$

The function $m^{\#}(x, \cdot)$ is regular at z_j; therefore,

$$m_0(x, z) = (z - z_j)^{-1} a_j u_j + (\beta_j u_j + \alpha_j n_j), \tag{10.36}$$

where $\alpha_j = m^\#(x, z_j)$ is invertible. We want $m(x, \cdot)$ to have a removable singularity at $z_j, j > N$. From (10.27) and (10.36) this is equivalent to

$$a_j \alpha_j u_j = 0, \quad j > N, \tag{10.37}$$

$$b_j \alpha_j u_j + a_j(\beta_j u_j + \alpha_j n_j) = 0, \quad j > N. \tag{10.38}$$

These in turn are equivalent to

$$a_j \alpha_j u_j n_j^{-1} = 0, \quad j > N, \tag{10.39}$$

$$a_j \alpha_j = (b_j a_j - a_j \beta_j) u_j n_j^{-1}, \quad j > N. \tag{10.40}$$

Because of (10.35), equations (10.39) are a consequence of (10.40), or (10.38).

Consider now the necessary and sufficient conditions (10.32), (10.38). The c_j, d_j, α_j, β_j, u_j, and n_j are determined by $m^\#$ and u. We have also

$$b_j = I + \sum_{k \neq j} (z_j - z_k)^{-1} a_k. \tag{10.41}$$

Thus (10.32), (10.38) are Pn^2 equations for the Pn^2 unknown coefficients of the a_k. Since c_j, α_j, and n_j are invertible, these equations would have only the trivial solution $a_k = 0$, all k, if we had

$$v_j = 0, j \leq N, \quad u_j = 0, j > N. \tag{10.42}$$

Thus (10.32), (10.38) have a unique solution for almost all choices of the matrices α_j, β_j, c_j, d_j, u_j, n_j, and the entries are rational functions ρ_i of the entries of these matrices. The functions ρ_i are independent of x.

As $x \to -\infty$, we have, near points $z_j, j \leq N$,

$$m^\#(x, z) \to I, \quad \exp\{x z_j \mathfrak{J}\} u(z) \to I, \quad \exp\{x z_j \mathfrak{J}\} v(z_j) \to 0. \tag{10.43}$$

Thus

$$c_j(x) \to I, \quad d_j(x) \to 0, \quad v_j(x) \to 0. \tag{10.44}$$

Similarly, for $j > N$ as $x \to -\infty$ we have

$$\alpha_j \to I, \quad \beta_j \to 0, \quad u_j \to 0, \quad n_j \to I. \tag{10.45}$$

Remark 10.46. We have proved half of Theorem F (a), namely the fact that there is an associated eigenfunction as $x \to -\infty$. Note that the convergence of v_j and u_j in (10.43) and (10.44) is exponential; examination of (10.32), (10.38) shows that we may conclude that $a_j(x) \to 0$ exponentially at $-\infty$. From this we obtain, for some $\delta > 0$,

$$\|m_\nu^\pm(x, \cdot) - (m^\#)_\nu^\pm(x, \cdot)\|_\infty = O(e^{-\delta|x|}) \text{ as } x \to -\infty. \tag{10.47}$$

We have not used the winding number conditions (1.31). In the next section we show that these conditions allow us to transform the scattering data in a way which corresponds to normalizing the eigenfunctions at $+\infty$ instead of $-\infty$. The renormalized problem may then be handled in an analogous fashion, leading to a linear system with coefficients having limits at $+\infty$. It follows, indeed, that (1.31) implies solvability of (10.32), (10.38) for $x \to +\infty$ as well; however, the coefficients grow exponentially in this direction; hence the renormalized system is easier to study theoretically and to solve in practice.

11 Solvability Near $+\infty$; Theorems F and G

To investigate solvability of the inverse problem at $+\infty$, let us suppose first that m is the eigenfunction for a generic potential q. Let

$$\delta(z) = \lim_{x \to +\infty} m(x,z) \tag{11.1}$$

be the diagonal matrix of (5.9). Then $\tilde{m} = m\delta^{-1}$ is an eigenfunction normalized at $+\infty$. We have, clearly,

$$\tilde{m}_\nu^+(x,z) = \tilde{m}_\nu^-(x,z)e^{xzJ}\tilde{v}_\nu(z), \tag{11.2}$$

$$\tilde{v}_\nu = \delta_\nu^- v_\nu (\delta_\nu^+)^{-1}. \tag{11.3}$$

Thus $\{\tilde{v}_\nu\}$ is the scattering data for the renormalized problem on Σ. The singularities of $\tilde{m}$ are the same as those of m, since δ and δ^{-1} are regular where m is. Consider a singularity $z_j \in \Omega_\nu$. For convenience, we suppose that the ν-ordering of the basis vectors coincides with the original ordering. Suppose m is singular at z_j in column $k+1$. According to the discussion in Section 2, this means that the k-th diagonal entry of δ, δ_k, has a simple zero at z_j; since the $k+1$ upper minor is not zero at z_j, δ_{k+1} must have a simple pole at z_j. Thus

$$\delta_k(z)^{-1} = \alpha(z-z_j)^{-1} + O(1), \tag{11.4}$$

$$\delta_{k+1}(z)^{-1} = \alpha(z-z_j) + O(|z-z_j|^2). \tag{11.5}$$

Also,

$$m(x,z) = (z-z_j)^{-1}a + b + O(|z-z_j|), \tag{11.6}$$

where

$$0 \neq a = bv(z_j) = \beta b e_{k,k+1}, \quad \beta \in \mathbb{C}, \tag{11.7}$$

where e_{ij} is the usual matrix unit. There is a similar expression for $\tilde{m}(x,z)$, with

$$0 \neq \tilde{a} = \tilde{b}\tilde{v}(z_j) = \tilde{\beta}\tilde{b} e_{k,k+1}, \quad \tilde{\beta} \in \mathbb{C}; \tag{11.8}$$

in fact the argument giving the form $\beta e_{k,k+1}$ for $v(z_j)$ gives this corresponding form for $\tilde{v}(z_j)$. Since $\tilde{m} = m\delta^{-1}$, we may infer from (11.4) and (11.5) and inspection of columns k and $k+1$ of $\tilde{m}$ that (11.7), (11.8) imply

$$abe_{kk} = \tilde{a} = \tilde{b}\tilde{v}_j = a\gamma\tilde{v}_j = \gamma b v_j \tilde{v}_j; \tag{11.9}$$

therefore

$$\alpha e_{kk} = \gamma v(z_j)\tilde{v}(z_j) = \gamma\beta\tilde{\beta}e_{kk}. \tag{11.10}$$

This may be written in invariant form, using the trace

$$\text{tr }(v(z_j)\tilde{v}(z_j)) = \text{tr }(\text{Res }(\delta, z_j)) \text{ tr }(\text{Res }(\delta^{-1}, z_j)). \tag{11.11}$$

We have shown that the scattering data for the renormalized eigenfunction $\tilde{m}$ is computable from that for m, once we know the diagonal matrix δ. To determine δ from the scattering data for m we note that the condition corresponding to (1.26) is

$$d_k^+(\tilde{v}_\nu(z)^\nu) = 1, \quad 1 \le k \le n, z \in \Sigma_\nu. \tag{11.12}$$

From (11.3) we see that these conditions determine the ratios

$$d_k^+((\delta_\nu^-)^\nu)/d_k^+((\delta_\nu^+)^\nu) = d_k^+(v_\nu^\nu) \tag{11.13}$$

and therefore the ratios of the $(\delta_k)_\nu^\pm$ on Σ_ν. The zeros and poles of the δ_k are also determined by the scattering data; this information, together with the ratios of the limits of the rays Σ_ν, uniquely determines the δ_k. In fact, the winding number constraints (1.31) are exactly the conditions that all this data be compatible; see part 1 of the Appendix. Thus starting with $v \in \mathbf{S}$ we may determine uniquely the data $\tilde{v}$ which would correspond to a normalization at $+\infty$. Repeating the procedure of Section 10, we reduce to an algebraic problem which is uniquely solvable as $x \to +\infty$. We obtain eigenfunctions $\tilde{m}(x,\cdot)$ associated to $\tilde{v}$; then $m(x,\cdot)$ defined by $m(x,\cdot) = \tilde{m}(x,\cdot)\delta(\cdot)$ is the eigenfunction associated to v.

We have now proved part (a) of Theorem F. To prove part (b) we suppose first that each v_ν has compact support. The same is true of the transformed data $v_\nu^\#$ of Section 10, and it follows that $m^\#$ is analytic with respect to x. Therefore the system of equations (10.32), (10.39) has coefficients depending analytically on x. We know now that the system is solvable for $|x|$ large, and hence for all by finitely many values of x. Consider the map

$$\varphi : \mathbb{R} \times \mathbb{C}^P \mapsto \mathbb{C} \tag{11.14}$$

obtained by taking the determinant of the system (10.32), (10.38) at x, when v_j has been replaced by $\zeta_j v_j, j \le N$, and u_j has been replaced by $\zeta_j u_j, j > N$. For $\zeta_j \cong 1$ this is the system corresponding to a slight perturbation of the original scattering data v. Now $\varphi^{-1}(0)$ has real codimension 2, so its projection on

$\mathbb{C}^P$ has real codimension at most 1, and we conclude that there are arbitrarily small perturbations of v for which the associated eigenfunction exists for every x. Data with compact support are dense, so $\mathbf{S}_0$ is dense in $\mathbf{S}$.

To see that $\mathbf{S}_0$ is open, we note that in the construction in Section 10, the piecewise rational function u can be chosen to vary continuously with v, so $m^{\#}$ will also vary continuously with v. Thus the coefficients of (10.32), (10.38) vary continuously with v and x; the system is solvable for large $|x|$ for all v' near a given v, and it follows that if v is in $\mathbf{S}_0$ and v' is sufficiently close, then v' is in $\mathbf{S}_0$.

Finally, we need to establish the differential equation (1.6) and prove (1.44). The additive jump of $m(x,\cdot)$ across Σ_ν is

$$\begin{aligned} m_\nu^+(x,z) - m_\nu^-(x,z) &= m_\nu^-(x,z)[e^{xz\mathfrak{J}}v_\nu(z) - I] \\ &= m_\nu^-(x,z)w_\nu(x,z), \end{aligned} \tag{11.15}$$

while if we define

$$m(x,z_j) = \frac{1}{2\pi i}\int_{C_j}(\zeta - z_j)^{-1}m(x,\zeta)\,d\zeta, \tag{11.16}$$

where C_j is a small circle with center z_j, then (1.16) is

$$\begin{aligned} \text{Res}\,(m(x,\cdot);z_j) &= m(x,z_j)\exp\{xz_j\mathfrak{J}\}v(z_j) \\ &= m(x,z_j)v_j(x). \end{aligned} \tag{11.17}$$

From (11.15), (11.16), and the asymptotic behavior as $z \to I$ we see that $m(x,\cdot)$ is a solution of

$$\begin{aligned} m(x,z) = I &+ \frac{1}{2\pi i}\int_\Sigma (z-\zeta)^{-1}m^-(x,\zeta)w(x,\zeta)\,d\zeta \\ &+ \Sigma(z-z_j)^{-1}m(x,z_j)v_j(x). \end{aligned} \tag{11.18}$$

Suppose now that w has compact support. It is then obvious that any solution of (11.18) is asymptotically I as $z \to \infty$ and satisfies (11.15), (11.17). The eigenfunction $m(x,\cdot)$ constructed in Section 10 is invertible where it is regular, so we repeat the proof of uniqueness in Theorem B to conclude that $m_1m^{-1} \equiv I$. Now still assuming that w has compact support, once again m is analytic in x and we may differentiate (11.18) to see that $m_2 = (\partial/\partial x - z\mathfrak{J})m$ satisfies

$$\begin{aligned} m_2(x,z) = q(x) &+ \frac{1}{2\pi i}\int_\Sigma (\zeta - z)^{-1}m_2(x,\zeta)w(x,\zeta)\,d\zeta \\ &+ \Sigma(z-z_j)^{-1}m_2(x,z_j)v_j(x), \end{aligned} \tag{11.19}$$

where

$$q(x) = \frac{1}{2\pi i}\mathfrak{J}\int_\Sigma m(x,\zeta)w(x,\zeta)\,d\zeta - \Sigma\mathfrak{J}[m(x,z_j)v_j(x)]. \tag{11.20}$$

Again, this equation implies that m_2 satisfies (11.15) and (11.17), while $m_2 \sim q$ as $z \to \infty$. Consequently, $m_2 m^{-1} \equiv q$, and this is our differential equation.

To complete the proof of Theorem F (b) it is only necessary to estimate the norm of $(1+|x|)q$ in $L^\infty + L^2$ in terms of the norms of v and Dv in $L^2(\Sigma)$, locally, since we may then pass to the limit from compactly supported v. (Observe, in this passage to the limit, that the piecewise rational function u of Lemma 10.2 can be held fixed.) As noted at the end of Section 10, $m_\nu^+(x,\cdot)$ is exponentially close to $(m^\#)_\nu^+(x,\cdot)$ as $x \to -\infty$; the same is true of derivatives with respect to z. Since $v_j(x)$ in (11.20) is also exponentially small at $-\infty$, we may estimate q in the same way here as in Section 8, for $x \le 0$. For $x \ge 0$ we repeat the renormalization at $+\infty$ and have formulas of the same type with exponential convergence at $+\infty$.

Remark 11.21. The arguments here show that if v belongs to $\mathbf{S}$ but not to $\mathbf{S}_0$, the associated eigenfunction $m(x,\cdot)$ exists on an open set and satisfies the differential equation on that open set, again with q given by (11.20).

When the scattering data evolve according to (1.50), we may let the piecewise rational function u of Lemma 10.2 evolve in the same way. It continues to satisfy the algebraic constraints, and in the stable case (1.52) it also satisfies (10.5). In short, the rational approximation only needs to be computed twice (at $-\infty$ and at $+\infty$) for an equation of evolution.

Proof of Theorem G: For part (a) we may argue exactly as in the proof of Theorem 9.2, except for considering separately the cases $x \le 0$ and $x \ge 0$ in order to have exponential decrease in the discrete terms in (11.20).

For part (b) we examine Lemma 10.2. Using the assumption (1.36) we may suppose that the piecewise rational function a is chosen so as to have the correct Taylor expansion to order k at 0 from Ω_ν, so that (1.36) holds also for a. For the factors $b_\nu^\pm$ this will imply that they are $I + O(z^k)$ at the origin. We approximate the b on the boundary of Ω_ν in $C^k(\Omega_\nu)$, and the result is that the new data $\{v_\nu^\#\}$ will have transformed data $\{w_\nu^\#\}$ which satisfies the conditions of Theorem 9.20. Thus $m^\#$ is as in Theorem 9.20. Now once again m is exponentially close to $m^\#$ on Σ or on the circles C_j as $x \to -\infty$, so we may argue as in the proof of Theorem 9.20 to obtain (1.47) for $x \le 0$; again the renormalization at $+\infty$ completes the argument.

12 Systems with Symmetry; Multisolitons

Suppose $a \to a^\sigma$ is an automorphism of the matrix algebra $M_n(\mathbb{C})$, and suppose J is an eigenvector:

$$J^\sigma = \alpha^{-1} J. \tag{12.1}$$

Let $\mathbf{P}_0$ denote the space of generic σ-symmetric potentials:

$$\mathbf{P}_0^\sigma = \{q \in \mathbf{P}_0 : q(x)^\sigma \equiv q(x)\}. \tag{12.2}$$

Theorem 12.3. *Under assumption* (12.1), α *is a root of unity and* Σ *is invariant under multiplication by* α. *If* q *belongs to* $\mathbf{P}_0^\sigma$ *and* $v = \{v_\nu, z_j, v(z_j)\}$ *is the associated scattering data, then*

$$v(\alpha z)^\sigma = v(z), \quad z \in \Sigma, \tag{12.4}$$

$$\{z_j\} \text{ is invariant under multiplication by } \alpha, \tag{12.5}$$

$$\alpha^{-1} v(\alpha z_j)^\sigma = v(z_j). \tag{12.6}$$

Conversely, if q *belongs to* $\mathbf{P}_0$ *and the associated scattering data satisfies* (12.4)–(12.6), *then* q *is in* $\mathbf{P}_0^\sigma$.

Proof: The automorphism is inner:

$$a^\sigma = \pi^{-1} a \pi, \quad \text{some } \pi \in M^n(\mathbb{C}). \tag{12.7}$$

From (12.1) it follows that π maps the eigenspace for J with eigenvalue λ to the eigenspace for eigenvalue $\alpha^{-1}\lambda$, and it follows that α is a root of unity and that Σ is invariant under multiplication by α. For a matrix-valued function defined on a subset of $\mathbb{C}$ invariant under multiplication by α, set

$$f^{\#}(z) = f(\alpha z)^\sigma. \tag{12.8}$$

In particular note that if $f(z) = zJ$, then $f = f^{\#}$. It follows for $q \in \mathbf{P}_0^\sigma$ with associated eigenfunction m that $m(x,\cdot)^{\#}$ satisfies the differential equation also. Therefore, $m = m^{\#}$, and (12.4), (12.5) follow immediately. The residue at a singularity satisfies

$$\text{Res } (m(x,\cdot), \zeta_j) = \alpha^{-1} \text{ Res } (m(x,\cdot), \alpha\zeta_j)^\sigma \tag{12.9}$$

and (12.6) is a consequence.

Conversely, if the scattering data satisfy (12.4)–(12.6), then it is easy to see that $m(x,\cdot)^{\#}$ has the same relationship to the scattering data as $m(x,\cdot)$; since $m(x,\cdot)^{\#}$ also is I at ∞, we have $m \equiv m^{\#}$ and the differential equation implies that $q \equiv q^\sigma$.

We suppose now that α is a primitive n-th root of unity, which is equivalent to assuming that π is a cyclic permutation of the eigenspaces of J. Then π^n is scalar, and we may replace π by a scalar multiple so that $\pi^n = I$. After a change of basis and rescaling of the eigenvalue problem we may assume

$$J = \text{diag } (\alpha, \alpha^2, \cdots, \alpha^{n-1}, 1), \tag{12.10}$$

$$\pi = e_{12} + e_{23} + \cdots + e_{n1}, \tag{12.11}$$

where the e_{jk} are the matrix units in $M^n(\mathbb{C})$.

The key fact is then that the subalgebra fixed by σ,

$$M_n(\mathbb{C})^\sigma = \{a \in M_n(\mathbb{C}) : a^\sigma = a\}, \tag{12.12}$$

is *commutative*: it is the commutator of π and consists of polynomials in π.

Under these assumptions we consider the construction of an eigenfunction for scattering data which vanish on Σ. As above, the problem becomes an algebraic one. In this case the symmetries and the commutativity allow an explicit computation. Let the singularities be

$$\{\alpha^{-k} z_j : 1 \leq j \leq N, 0 \leq k < n\}, \tag{12.13}$$

and let these points be distinct. The symmetry condition implies that if one column of m has a singularity at point z_0, then the last column has a singularity at $\alpha^k z_0$, some k. Therefore we may assume for convenience that it is the last column which is to be singular at $z_1, \cdots, z_N$. The matrix $v(z_j)$ is of the form $c_j e_{d_j n}$ for some constant c_j and some index $d_j < n$. Then

$$\exp\{x z_j \partial\} v(z_j) = \exp\{x(\zeta_j - z_j)\} v(z_j), \tag{12.14}$$

$$\zeta_j = \alpha^{d_j} z_j \neq z_j. \tag{12.15}$$

Given a rational matrix-valued function f, we define as before

$$f(z_j) = \frac{1}{2\pi i} \int_{C_j} (z - z_j)^{-1} f(z)\, dz, \tag{12.16}$$

where C_j is a small circle around z_j. We set

$$C_{v,x} f(x) = \sum_{j=1}^{N} \exp\{x(\zeta_j - z_j)\} [f(z_j) v_j]^{\text{sym}} (zJ - z_j)^{-1}, \tag{12.17}$$

where b^{sym} is the symmetrized version of the matrix b:

$$b^{\text{sym}} = \sum_{k=0}^{n-1} \pi^{-k} b \pi^k. \tag{12.18}$$

Then

$$(C_{v,x} f)^{\#} = C_{v,x} f \tag{12.19}$$

and

$$\text{Res}\ (C_{v,x} f; z_j) = f(z_j) \exp\{x z_j \partial\} v(z_j). \tag{12.20}$$

From the symmetry condition (12.19) we see that (12.20) also holds with z_j replaced by $\alpha^{-k} z_j$. Therefore the desired eigenfunction $m(x, \cdot)$ is precisely the solution of

$$m(x, \cdot) = I + C_{v,x} m(x, \cdot). \tag{12.21}$$

Consider the formal Neumann series solution of (12.21). We have

$$C_{v,x}(I) = \Sigma \exp\{x(\zeta_j - z_j)\} v_j (zJ - z_j)^{-1}, \tag{12.22}$$

where

$$v_j = v(z_j)^{\text{sym}}. \tag{12.23}$$

In general, if f is of the form

$$f(z) = \Sigma a_j (zJ - z_j)^{-1}, \quad a_j = a_j^\sigma, \tag{12.24}$$

then

$$C_{v,x} f(z) = \Sigma b_j (zJ - z_j)^{-1}, \tag{12.25}$$

where

$$b_k = \Sigma a_j A(x)_{jk}, \tag{12.26}$$

$$A(x)_{jk} = \exp\{x(\zeta_k - z_k)\}(\zeta_k - z_j)^{-1} v_k. \tag{12.27}$$

We consider $A(x)$ as an $(N \times N)$ matrix with entries in the commutative algebra $M_n(C)^\sigma$ and write it as a product of such matrices:

$$A(x) = \Delta(x) B(x) V \Delta(x)^{-1}, \tag{12.28}$$

where $\Delta(x)$ and V are diagonal:

$$\Delta(x)^{jj} = \exp\{x z_j\} I, \quad V_{jj} = v_j, \tag{12.29}$$

and

$$B(x)_{jk} = \exp\{x(\zeta_k - z_j)\}(\zeta_k - z_j)^{-1} I. \tag{12.30}$$

Let $\Delta_2(x)$ be the diagonal $M_n(C)^\sigma$-valued matrix with

$$\Delta_2(x)_{jj} = \exp\{x \zeta_j\} I. \tag{12.31}$$

Let $\mathbf{1}$ denote the $M_n(C)^\sigma$-valued row vector with N entries, each of them the identity matrix. Then from the above considerations we see that the formal Neumann series solution of (12.21) is given by

$$m(x, z) = I + \Sigma a_j(x)(zJ - z_j)^{-1}, \tag{12.32}$$

where

$$\begin{aligned} a(x) &= (a_1(x), a_2(x), \cdots, a_N(x)) \\ &= \sum_{s=0}^{\infty} \mathbf{1} V \Delta_2(x) (B(x) V)^s \Delta_1(x)^{-1}. \end{aligned} \tag{12.33}$$

The corresponding potential, as in Section 11, is

$$q(x) = -\mathcal{J} \Sigma \operatorname{Res}\, (m(x, \cdot)) \tag{12.34}$$

and from (12.32) we calculate that the sum of the residues is

$$\Sigma \text{ Res } (m(x,\cdot)) = \Sigma a_j J^{-1}. \tag{12.35}$$

Thus

$$q(x) = \Sigma a_j(x) - J(\Sigma a_j(x))J^{-1}. \tag{12.36}$$

Now we can represent $\Sigma a_j(x)$ as the matrix-valued trace of the matrix-valued matrix

$$\mathbf{1}' \cdot a(x) = \sum_{s=0}^{\infty} \mathbf{1}' \cdot \mathbf{1} V \Delta_2(x)(B(x)V)^s \Delta_1(x)^{-1}. \tag{12.37}$$

Relation (12.30) shows that

$$\frac{d}{dx} B(x) = \Delta_1(x)^{-1} \mathbf{1}' \cdot \mathbf{1} \Delta_2(x). \tag{12.38}$$

Note also that V and $\Delta_2(x)$ commute. Since the trace of (12.37) is unchanged under conjugation by $\Delta_1(x)$, it is the same as the trace of

$$\left(\frac{d}{dx} B(x) V\right)(I - B(x)V)^{-1}. \tag{12.39}$$

The trace of (12.37) is the derivative of the trace of $-\log(I - B(x)V)$, which is the negative of the logarithm of $\det(I - B(x)V)$. Therefore we have the (formal) calculation

$$q = JF^{-1}\frac{dF}{dx}J^{-1} - F^{-1}\frac{dF}{dx}, \tag{12.40}$$

where F is the matrix-valued determinant,

$$F(x) = \det(I - B(x)V). \tag{12.41}$$

When the formal scattering data belongs to $\mathbf{S}$, the exponentials are rapidly vanishing at $-\infty$ and the series (12.33) converges for $x \ll 0$. It follows that (12.40) defines the corresponding potential wherever $m(x,\cdot)$ exists.

Appendix

We sketch the derivation of two facts used above which are extensions of well-known results.

A.1. The Scalar Factorization Problem

As before, let Σ be a union of lines through the origin. Write $\Sigma\backslash(0)$ as a union of open rays $\Sigma_1, \Sigma_2, \cdots, \Sigma_r$, where Σ_ν and $\Sigma_{\nu+1}$ form (with the origin) the boundary of a component Ω_ν of $\mathbb{C}\backslash\Sigma$, and $\Sigma_{r+1} = \Sigma_1$. The Σ_ν are indexed in order of increasing argument. The problem to be considered is the following.

A1. Suppose for $1 \le \nu \le r$ that φ_ν is a continuous nonvanishing complex function on the closure of Σ_ν with $\varphi_\nu - 1$ and $D\varphi_\nu$ in L^2. Find functions δ_ν meromorphic on Ω_ν with simple zeros and simple poles at prescribed points of Ω_ν and no other zeros in Ω_ν such that δ_ν extends continuously to the boundary of Ω_ν, has no zeros on the boundary, and has limit 1 at ∞; moreover $\delta_\nu = \delta_{\nu-1}\varphi_\nu$ on Σ_ν, where $\delta_0 = \delta_r$.

Theorem A2. *Problem* A1 *has at most one solution. A solution exists if and only if*

$$\varphi_1(0)\varphi_2(0)\cdots\varphi_r(0) = 1, \tag{A3}$$

$$\sum_{\nu=1}^{r}\int_{\Sigma_\nu} d(\arg\varphi_\nu) = 2\pi(N-P), \tag{A4}$$

where N is the number of zeros, P the number of poles, and the Σ_ν are oriented from 0 to ∞.

Proof: *Uniqueness.* In a simpler version of the argument at the end of Section 4, the quotient of two solutions has removable singularities at the prescribed zeros and poles and on Σ and is 1 at ∞.

Necessity. Since $\varphi_\nu(0) = \delta_{\nu+1}(0)\delta_\nu(0)^{-1}$, with $\delta_{r+1} = \delta_1$, condition (A3) is immediate. If N_ν and P_ν are the numbers of zeros and poles at Ω_ν, then the argument principle gives

$$-\int_{\Sigma_{\nu+1}} d(\arg\delta_\nu) + \int_{\Sigma_\nu} d(\arg\delta_\nu) = 2\pi(N_\nu - P_\nu). \tag{A5}$$

On Σ_ν, $\arg\delta_\nu = \arg\delta_{\nu-1} + \arg\varphi_\nu$. Inserting this identity into the second term on the left in (A5) and summing, we get (A4).

Sufficiency. It is convenient to consider transformations of the problem. Suppose $f_1, \cdots, f_r$ are rational functions having only simple zeros and poles, having no zeros or poles on Σ, and equal to 1 at ∞. Look for the δ_ν in the form

$$\delta_\nu = f_\nu\delta_\nu^*. \tag{A6}$$

Then the δ_ν^* must solve a similar problem with data φ_ν^*, where

$$\varphi_\nu^* = \varphi_\nu f_\nu^{-1} f_{\nu-1}, \quad f_0 = f_r, \tag{A7}$$

and where the prescribed zeros and poles are altered to take into account those created or destroyed by the f_ν. Condition (A4) will be satisfied for one problem if and only if it is for the other, by (A5) for f_ν. In particular we may choose the

f_ν to have the prescribed zeros and poles, so that the δ_ν^* are to have no zeros and poles. Also, by choosing f_ν with

$$f_\nu(0) = \varphi_1(0)\varphi_2(0)\cdots\varphi_\nu(0), \tag{A8}$$

we may ensure that $\varphi_\nu^*(0) = 1$ for all ν.

Now induce on $s = \frac{1}{2}r$. When $s = 1$ we have a single line which we may assume is the real axis with $\Sigma_1 = \mathbb{R}_+$. Set $\varphi(s) = \varphi_1(s)$, $s \geq 0$, and $\varphi(s) = \varphi(s)^{-1}$, $s < 0$. The problem is a trivial Wiener-Hopf factorization problem: with the zeros and poles removed, we want to find δ_+ and δ_-, holomorphic and non-zero in the upper and lower half-planes, respectively, with $\delta_+ = \varphi\delta_-$ on $\mathbb{R}$. The winding number of φ is zero; thus $\varphi = \exp\psi$ and the solution is obtained by expressing ψ as $\psi = \psi_+ - \psi_-$, where ψ_+ and ψ_- are boundary values of functions holomorphic in the upper and lower half-planes, respectively.

For $s > 1$ we first reduce to the case $\varphi_\nu(0) = 1$, all ν. Having done so we note that

$$\frac{1}{2\pi}\int_{\Sigma_1} d(\arg\varphi_1) + \frac{1}{2\pi}\int_{\Sigma_{s+1}} d(\arg\varphi_{s+1}) \tag{A9}$$

is an integer, since each summand is a winding number. A suitable transformation as above by rational functions will then give us a problem for which the integer (A9) is zero. This means that (A4) will be satisfied for the problem for the configuration Σ' in which the (collinear) rays Σ_1 and Σ_{s+1} have been removed. By the induction assumption this problem has a piecewise meromorphic solution φ' with the prescribed zeros and poles. We look for $\varphi = \varphi' f$ and the problem reduces to the Wiener-Hopf factorization problem for a function on the line $\Sigma_1 \cup (0) \cup \Sigma_{s+1}$.

Remark. If the φ_ν satisfy conditions like

$$(1+|z|)^k[\varphi_\nu(z) - 1] \in L^2(\Sigma_\nu) \text{ or } D^j(\varphi_\nu - 1) \in L^2, \quad j \leq k+1,$$

then the same will be true of δ_ν on Σ_ν and $\Sigma_{\nu+1}$. This follows readily from the construction when $s = 1$, and then inductively. Similarly, if the φ_ν satisfy conditions like

$$D^j(\varphi_\nu - 1) \in L^2(\Sigma_\nu) \text{ or } L^\infty(\Sigma_\nu), \quad 0 \leq j \leq k$$

and (A3) holds to order $k-1$, then the same will be true of δ_ν on Σ_ν and $\Sigma_{\nu+1}$. It follows that the renormalization at $+\infty$ in Section 10 does not destroy these conditions.

A.2. Rational Approximation

Here we consider a single sector Ω bounded by the origin and two rays Σ_1 and Σ_2.

Theorem A10. *Suppose f is a continuous complex function on the boundary of Ω, with limit 0 at ∞. Then f may be approximated uniformly by (restrictions* *to $\partial\Omega$ of) rational functions f_n which vanish at ∞. Moreover, suppose*

$$D^j f \in L^\infty \cap L^2 \text{ on } \Sigma_i, \quad 0 \le j \le k, i = 1,2, \tag{A11}$$

$$\lim_{z\to 0} D^j f(z) = 0 \text{ on } \Sigma_i, \quad 0 \le j < k, i = 1,2. \tag{A12}$$

Then the f_n may be chosen so that

$$D^j f_n \mapsto D^j f \text{ in } L^\infty \cap L^2 \text{ for } j < k,$$

$$D^k f_n \mapsto D^k f \text{ in } L^2, \tag{A13}$$

$$\{D^k f_n\} \text{ is bounded in } L^\infty.$$

Proof: Recall one version of the argument when $\partial\Omega = \mathbb{R}$. Given $\epsilon > 0$, let

$$f_\epsilon(t) = \frac{1}{2\pi i}\int_{\mathbb{R}} [(s-t-i\epsilon)^{-1} - (s-t+i\epsilon)^{-1}] f(s)\,ds. \tag{A14}$$

This is just the convolution of f with the Poisson kernel P_ϵ, which is an approximate identity, so the f_ϵ converge uniformly to f and one has the requisite convergence by derivatives as well. For a fixed ϵ, f_ϵ itself (and derivatives) may be approximated by Riemann sums

$$f_{\epsilon,N}(t) = \sum_{j=-N^2}^{N^2} f(j/N) P_\epsilon(s - j/N), \tag{A15}$$

which are rational functions.

With a little more effort, the same construction works for a general sector. We may assume that the positive imaginary axis bisects Ω and define f_ϵ by (A14) with $\mathbb{R}$ replaced by $\partial\Omega$. We no longer have a convolution kernel, but

$$f_\epsilon(t) = \int_{\partial\Omega} P_\epsilon(t,s) f(s)\,ds, \tag{A16}$$

where P_ϵ has the essential features of an approximate identity:

$$\int |P_\epsilon(t,s)|\,|ds| \le C,$$

$$\int P_\epsilon(t,s)\,ds \to 1 \text{ as } \epsilon \searrow 0, \tag{A17}$$

$$\int_{|t-s|>\delta} |P_\epsilon(t,s)|\,|ds| \to 0 \text{ as } \epsilon \searrow 0 \text{ for all } \delta > 0.$$

Thus $f_\epsilon \to f$ uniformly. Under assumptions (A11) and (A12) we also have appropriate convergence of the derivatives, since (complex) differentiation of f_ϵ can be passed onto f in (A16). Finally, the Riemann sums approximating (A16) are again rational functions which vanish at ∞.

Acknowledgment. This research was supported by NSF Grant MCS-8104234.

Bibliography

[1] Ablowitz, M. J., Kaup, D. J., Newell, A. C., and Segur, H., *The inverse scattering transform - Fourier analysis for nonlinear problems*, Studies in Applied Mathematics 53, 1974, pp. 249–315.

[2] Bar-Yaacov, D., *Analytic properties of scattering and inverse scattering for first order systems*, Dissertation, Yale University.

[3] Beals, R., and Coifman, R., *Scattering, transformations spectrales, et equations d'evolution non linéaires.* Séminaire Goulaouic-Meyer-Schwartz 1980-1981, exp. 22, Ecole Polytechnique, Palaiseau.

[4] Bullough, R. K., and Caudrey, P. J., eds., *Solitons*, Topics in Current Physics no. 17, Springer-Verlag, 1980.

[5] Chudnovsky, D. V., *One and multidimensional completely integrable systems arising from the isospectral deformation*, in *Complex Analysis, Microlocal Analysis, and Relativistic Quantum Theory*, Lecture Notes in Physics no. 126, Springer-Verlag, 1980.

[6] Deift, P., and Trubowitz, E., *Inverse scattering on the line*, Comm. Pure Appl. Math. 32, 1979, pp. 121–251.

[7] Dubrovin, B. A., Matseev, V. B., and Novikov, S. P., *Nonlinear equations of KdV type, finite-zone linear operators, and Abelian varieties.* Uspehi Mat. Nauk 31, 1976, pp. 55–136; Russian Math. Surveys 31, 1976, pp. 59–146.

[8] Faddeev, D. K., and Faddeva, V. N., *Computational Methods of Linear Algebra*, Freeman, 1963.

[9] Hirota, R., *Exact solutions of the modified Korteweg-deVries equation for multiple collisions of solitons.* J. Phys. Soc. Japan 33, 1972, pp. 1456–1458.

[10] Shabat, A. B., *An inverse scattering problem.* Diff. Uravn. 15, 1978, pp. 1824–1834; Diff. Equations 15, 1980, pp. 1299–1307.

[11] Zakharov, V. E., and Shabat, A. B., *A refined theory of two-dimensional self-focussing and one-dimensional self-modulation of waves in nonlinear media.* Zh. Eksp. Teor. Fiz. 61, 1971, pp. 118–134; Soviet Physics JETP 34, 1972, pp. 62–69.

R. Beals
Yale University

R. R. Coifman
Yale University

Textbooks from International Press

Calculus: A Computer Algebra Approach,

by Iris L. Ashel and D. Goldfeld

Uniquely designed for use with computer algebra systems and sophisticated calculators, this course also works well with a computer laboratory. The students are encouraged to use technology for manual computation while they rapidly progress through the concepts of differential and integral calculus, mathematical modeling and optimization, ordinary differential equations, differential calculus for vector valued and multi-variable functions. The students will progress to vector geometry and coordinate systems, two and three dimensional graphical display, multiple integration, vector fields and line integrals, and on to Fourier series and the Fourier expansion theorem.

Basic Partial Differential Equations,

by D. Bleecker and G. Csordas

Using partial differential equations as a tool to predict systems based on underlying physical principles, this text is accessible to students with calculus background without presupposing courses in linear algebra or ordinary differential equations. The text takes an inductive approach, motivating students to see the relevance and applications of the general principles which govern the subject, but rigorous proofs are still available.

The problems included range from routine to very challenging. The problems are constructed to require genuine understanding of principles as well as procedures, and the solutions manual is available only to instructors—not to the general public. Suggestions from the author indicate how the book can be used in one semester, two quarter, or two semester courses.

A First Course in Differential Geometry,

by C.-C. Hsiung

This text treats the traditional topics of curves and surfaces in 3-dimensional Euclidean space with attention equally on local and global properties. Extension to higher dimensional spaces and more general surfaces is available. In addition to analytical approaches, the text gives geometric interpretation to its subjects and encourages the development of geometric intuition. The book uses vector analysis and exterior differential calculus but does not require tensor calculus.

Part 1 reviews point-set topology, advanced calculus, and linear algebra. Part 2 establishes a general local theory for curves in three dimensional Euclidean space including proving the uniqueness theorem for curves in E^3. Part 3 develops a Local theory for surfaces in E^3, and Part 4 covers orientation of surfaces, surfaces of constant Gaussian curvature, and global theorems for surfaces.

Singularities and Complex Geometry, *edited by Qi-Keng Lu, Stephen Yau and Anatoly Libgober* copublished with AMS

Surveys in Differential Geometry, Volume 1, *edited by H. Blaine Lawson and S.T. Yau,* **Volume II, Volume III**, *edited by C.C. Hsiung and S.-T. Yau,* **Volume IV,** *edited by Chuu Lian Terng and Karen Uhlenbeck.*

Topics in Symplectic 4-manifolds, volume 1, First International Press Lecture Series, *edited by Ronald Stern*

Tsing Hua Lectures on Geometry and Analysis, *edited by S.-T. Yau*

Current Developments in Mathematics

Proceedings of the annual conference sponsored by Harvard and MIT

1995 Proceedings: lectures by Henri Darmon, Fred Diamond, and Richard Taylor, Mikhail Lyubich, Ib Madsen, Curtis McMullen, and Gang Tian

1996 Proceedings: lectures by Richard Borcherds, Gerrit Heckman and Eric Opdam, Ehud Hrushovski, and Yves Meyer

1997 Proceedings: lectures by A. Connes, L. C. Evans, P. Sarnak, and W. Soergel and open problems by A. Jaffe, D. Stroock, B. Mazur, C. Taubes, and D. Jerison.

1998 Proceedings: lectures by Brian Greene; Bong H. Lian, Kefeng Liu, and S.-T. Yau; P.B. Kronheimer; Lou Van den Dries; and H.-T. Yau